Intelligent Data Mining in Law Enforcement Analytics

Massimo Buscema • William J. Tastle
Editors

Intelligent Data Mining in Law Enforcement Analytics

New Neural Networks Applied to Real Problems

Springer

Editors
Massimo Buscema
Semeion Research Centre of Sciences of Communication
Rome
Italy

William J. Tastle
Ithaca College
NY
USA

ISBN 978-94-007-4913-9 ISBN 978-94-007-4914-6 (eBook)
DOI 10.1007/978-94-007-4914-6
Springer Dordrecht Heidelberg New York London

Library of Congress Control Number: 2012953015

Printed on acid-free paper

Springer is part of Springer Science+Business Media (www.springer.com)

Assembling the contents of an academic book dealing with some new technology or a sophisticated advancement is the task given over to the academic researcher who typically embraces the challenge with dedication and purpose for it is what makes us unique among our brethren. Libraries are filled with esoteric research that is the product of excellent minds, research that is so arcane and possibly cryptic that it might remain on the shelves for potentially centuries until an application succeeds in being brought forth by some other, equally sophisticated and talented, individuals who have the rare talent of merging new-found knowledge with practical application.

Such is not the case with this academic text for it was immediately observed that this method of data analysis and mining could be brought to bear in helping to solve some very complex problems that have plagued the law enforcement community since the advent of the database and its concomitant assortment of management systems. The easy applications, that is to say, the most trivial but definitely useful, were quickly subsumed by the law enforcement community and began a movement to digitise all past and present case data for easy access and management; for the last few decades, their expectation has usually been met with varying degrees of success.

However, the databases that were built over many years, or decades in some cases, still contained unknown and undiscovered knowledge, but no one knew of its existence until a meeting that occurred with an official of one of the most respected law enforcement agencies in the work, the London Metropolitan Police force (also known as New Scotland Yard), and the principal researcher of one of the most prestigious research institutes in Italy, Semeion Research Center of the Sciences of Communications of Rome.

It is to Sergeant Geoff Monaghan of New Scotland Yard that this book is dedicated for it was he who first taught us about the complex world of crime analysis. Sergeant Monaghan inspired us and motivated Semeion towards the adventure of crime analytics. It was his vision to "see" that knowledge was trapped in huge databases and needed some very sophisticated methods to extract it and make it understandable to the "front line" of police. Over the past 3 years, Semeion has worked closely with Sgt Monaghan, and this book explains, in detail, the successes and methods used to extract this unknown knowledge. From here, extraction of knowledge from other databases can become commonplace, as long as there exist other talented visionaries in other disciplines who are willing to take the risk in creating knowledge.

—Semeion and its staff

Preface

This book was written specifically for the law enforcement community although it is applicable to any organisation/institution possessing a database of activity it seeks to analyse for unknown and undiscovered knowledge. This is typically called data mining and the purpose is to extract useful knowledge. Generally, most organisations typically use structured query language (SQL) to query their database. While this does give information, one must know exactly the questions to ask in order to gather a response, and any question raised by means of a query will have an answer if and only if the answer is already present in the database. This kind of information is called *blatant* for it is conspicuous as opposed to hidden. Unfortunately, the knowledge hidden within databases requires some very sophisticated methods in order to coax it out.

The extraction of only blatant information from a database is too limiting given the demands for useful information in the complex society of the twenty-first century. We need to creatively explore a database to extract its hidden information, that is, the underlying information which produces the structure by which the evident information becomes obvious and available for query. In short, the hidden information is responsible for the blatant information making sense. This special meta-information is hidden and trapped in the blatant information. This hidden information is the condition of existence for the blatant information in the same way that the Kantian "noumenon" is the condition for the perception of the phenomenon. Hidden information is like the sea waves, while the blatant information, explicitly coded in a database, a similar to the foam of the waves. For most forms of analysis, hidden information is considered "noise". But it is within this noise that the genetic code of process, that from which this noise is derived, is encrypted. Our challenge is to successfully decrypt the genetic code; such a decryption is explained in this book.

We name this search for the hidden information trapped in the database intelligent data mining (IDM), and we think that the most advanced artificial adaptive algorithms are able to understand which part of the so-called noise is the treble clef of any database music.

The sophistication of the criminal element is exceptional. Drug cartels and terrorist organisations have the financial strength to purchase, or muscle to coerce, brilliant individuals to work for them, and it is egregious for any law enforcement organisation to underestimate the cleverness of those groups. It is argued that the best we can hope to do is minimise the distance between what they do and how we protect against them. To do so requires us to embrace the maxim *scientia est potentia*. This is Latin for "knowledge is power" and is attributed to Sir Francis Bacon, though it first appeared in the 1658 book, *De Homine*, by his secretary Thomas Hobbes. In order to extract knowledge, one must first have information, and to get information one must have data. There is another word used to describe the extraction of knowledge from data: *semeion*. Its origin is from the Greek, and it means the extraction of a large amount of knowledge from a small amount of data given a prepared mind and the spirit of discovery. Not only can remarkable information be gathered from a database, we show in this book how to harness that information to produce knowledge that can be brought to bear on the criminal element in our efforts to defeat them.

The motivation for this book came out of a cooperative venture with the London Metropolitan Police, well known by its metonym Scotland Yard, and the Semeion Research Center of Rome. In a correspondence from the Assistant Commissioner Tarique Ghaffur of the London Specialist Crime Directorate to the Italian Minister of University Education and Research, the basis for successful cooperation is clearly established:

> From the outset, I [Assistant Commissioner Ghaffur] want to emphasise that the Central Drug Trafficking Database (CDTD) Project is an important element of the Specialist Crime Directorate's (SCD) intelligence strategy and I'm delighted to tell you that the project is going very well. Moreover, the CDTD, which has been designed by Semeion in accordance with specifications laid down by my officers, is working very well. One of the most exciting aspects of this project is the idea of using Artificial Adaptive Systems (AAS) to analyse drug trafficking data. I readily acknowledge that this component is totally dependent on the founder and Director of Semeion, Professor Massimo Buscema, in view of his extensive and pioneering work in the field of artificial intelligence. I know my officers hold Professor Buscema in high regard and I would like to place on record my thanks to him and his colleagues at Semeion, particularly Dr Stefano Terzi, for helping to make our partnership a success.

Operationally, Semeion created a database structure that permitted both the use of traditional SQL queries and analysis using adaptive neural network technology. The outcomes, from the Metropolitan Police perspective, are detailed in the letter:

> By way of background, the CDTD is the first of its kind and has been designed to enable the SCD to produce reliable and objective data to help the MPS and its partners to: (a) assess the extent of the problem in London, and (b) devise appropriate responses to tackle the problem. The information will, in the main, be drawn from 4,500 drug trafficking reports recorded by the MPS in 2004. The reports will be scrutinised and the information validated by specially trained Data Entry Operators (DEOs). Where necessary, additional information will be obtained from the Forensic Science Service, the Police National Computer and a number of other databases. The refined data will then be entered onto the CDTD and new records created. Each record comprises around 500 fields. Subsequent analyses will shed new light on the structure of drug markets in London, how organised criminal networks

> shape and influence these markets, and the effectiveness of police tactics (e.g. stop and search, test purchases and controlled deliveries). Data gleaned from drug seizures – unit prices, purity, and chemical profiling – will also be analysed. The project will also highlight operational successes as well as noting the deficiencies in the recording and investigation of drug trafficking crimes. In sum, the CDTD will produce high-quality intelligence, which will be tailored to the varying needs of decision makers in the MPS from the strategic to the tactical levels.

The last point to address is that of the complementary relationship between traditional statistics and neural network technology. While statistics definitely plays an important role in data analysis, there are other methods that provide an entirely different view of the system under investigation. The London Metropolitan Police recognised the limitations of traditional statistics and sought to apply artificial adaptive systems (AAS) to their analysis.

> During their research, the Project Team will be using conventional statistical programmes. But in order to process the vast volumes of data generated and recognising that comprehensive analyses cannot be done without highly advanced data processing capabilities, the team also wants to use AAS. To this end, SCD has contracted Semeion to design the database management system and to analyse the data using AAS developed by Professor Buscema and his colleagues. Although AAS have been applied to various areas of research, we believe that this is the first time that they have been used to analyse drug trafficking crimes (or indeed any other type of crime) on this scale and in this detail.

The theory, methods and applications described in this book can be utilised by any police agency or modified to fit the needs of any business or organisation seeking to extract knowledge from a database. Non-profit organisations will find that donor/membership databases contain knowledge that could be utilised to enhance fundraising and membership drives. Military databases are typically huge and contain hundreds, if not thousands, of variables. A wealth of unknown knowledge may well be contained within those databases if only these new methods presented in this book were applied to them. Medical databases will benefit from the identification of hidden knowledge and could give scientists valuable insights into novel directions for research. Financial institutions have data on every customer, loan, stock portfolio, etc., and there is new knowledge to be gleaned from an analysis using these very sophisticated methods.

Many individuals, beyond the chapter authors, were involved in the production of this book, and we gratefully acknowledge their contribution:

- Dr Giulia Massini, computer scientist and deputy director of the Semeion Research Center of Sciences of Communication, Rome, Italy
- Dr Guido Maurelli, Semeion Research Center of Sciences of Communication, Rome, Italy
- Marco Intraligi, computer data processing expert, Semeion Research Center of Sciences of Communication, Rome, Italy
- Dr Stefano Terzi, computer scientist, formerly Semeion Research Center, Rome, Italy
- Dr Leslie A King, consultant, formerly head of the Drugs Intelligence Unit, Forensic Science Service, London, UK, and former advisor to the Department

of Health, England, and the European Monitoring Centre for Drugs and Drugs Addiction, Lisbon, Portugal

- Paul Richards, formerly CDTD project manager and inspector, Drugs Directorate, New Scotland Yard, Metropolitan Police Service, London
- Ms Mandeep Kaur Bajwa, Ms Zoe nee Beard and Mr Adam Stevens, formally data entry operators, CDTD project, New Scotland Yard, Metropolitan Police Service, London
- Mr Dean Ames, formerly forensic scientist, Drugs Intelligence Unit, Forensic Science Service, London, UK

A special acknowledgement to:

- Tarique Ghaffur, CBE, QPM, formerly assistant commissioner, Central Operations, New Scotland Yard, Metropolitan Police Service, London
- Andy Baker, deputy director of the Serious Organised Crime Agency, London, UK, and formerly commander, Specialist Crime Directorate, New Scotland Yard, Metropolitan Police Service, London
- Dr Stanley "Sholmo" Einstein, Jerusalem, Israel
- Professor (Emeritus) John G D Grieve CBE, QPM, John Grieve Centre for Policing and Community Safety, Faculty of Social Sciences and Humanities, London Metropolitan University, London, UK, and formerly deputy assistant commissioner, Specialist Crime Directorate, New Scotland Yard, Metropolitan Police Service, London
- Paul Hoare, formerly detective superintendent, Drugs Directorate, New Scotland Yard, Metropolitan Police Service, London

Paolo Massimo Buscema
General Director, Semeion Research Center, Rome, Italy
and
Professor, University of Colorado at Denver
Dept of Mathematical and Statistical Sciences
m.buscema@semeion.it
www.semeion.it

William J. Tastle
Professor, Ithaca College, New York, USA
Research Fellow, Semeion
tastle@ithaca.edu
faculty.ithaca.edu/tastle

Philip Bean
Department of Social Sciences
Midlands Centre for Criminology and Criminal Justice
University of Loughborough
Loughborough, Leicestershire LE11 3TU, UK

Teresa Nemitz
Department of Social Sciences
Midlands Centre for Criminology and Criminal Justice
University of Loughborough
Loughborough, Leicestershire LE11 3TU, UK

All images in this book are depicted in black and white and, consequently, the color detail is lost. The color images can be viewed at http://www.semeion.it or in the electronic publication of the book on www.springerlink.com.

Contents

Chapter 1
Introduction to Artificial Networks and Law Enforcement Analytics

William J. Tastle

The word "semeion" is derived from the Greek and defined as a knowledge extraction process that utilizes a small amount of data to achieve a large quantity of knowledge given a prepared mind and the spirit of discovery.

"Intelligence, in itself, does not make up part of the specific attributes of crime, but when it is present, it increases danger immeasurably and causes it to become organized crime." This phrase appears in the book both as an affirmation of the difficulty of the problem as well as a point of departure for finding solutions because what is intelligent is not causal, and what is not causal is foreseeable. Since intelligence must be fought with intelligence, the forces of order require a good dose of "civil intelligence" to fight the "uncivil intelligence" of criminals.

The book is a thorough description and summary of the means currently available to the law enforcement investigators to utilize artificial intelligence in making criminal behavior (both individual and collective) foreseeable and for assisting their investigative capacities. Concretely, there are five cognitive activities carried out by an investigator: (1) the classification of criminal situations; (2) the spatial visualization of where the events occurred; (3) a prediction of how the events developed; (4) the construction of somatic, sociological, and psychological profiles; and finally (5) hypotheses regarding links between events, persons, and clues. Yet, *all* five cognitive activities can be explained (and often in more than one way) by adaptive artificial systems, furnishing a second opinion regarding the analysis of criminal events.

The artificial adaptive systems are efficacious for two reasons: in the first place, they keep in "mind" all the data; the human mind, in contrast, must make a selection from among the various pieces of data before being able to reason; in the

W.J. Tastle (✉)
School of Business, Ithaca College, New York, USA

Semeion Research Center, Rome, Italy
e-mail: tastle@ithaca.edu

M. Buscema and W.J. Tastle (eds.), *Intelligent Data Mining in Law Enforcement Analytics: New Neural Networks Applied to Real Problems*, DOI 10.1007/978-94-007-4914-6_1,

second place, they successfully confront complex phenomena in which there are no relational links and before which the human mind, in its tendency to simplify, finds itself in difficulty because it tends to reason precisely in terms of such relations.

The work arises from collaboration between the Semeion Research Center and New Scotland Yard in a joint project and contains the multiple articles typical of a group effort. It passes from an analysis of the drug-trafficking situation in London by a Scotland Yard investigator to the description of how a reliable database was crafted, arriving finally at the practical application of various algorithms upon criminal events registered in the database. The specialists in artificial intelligence at Semeion have not only supervised the construction of a database by Scotland Yard, as well offered their capacity in applying it, but have also created original algorithms for the precise purpose of making crime predictable.

The results obtained, in width and depth, can be considered the basis for the construction of an "artificial investigator," an integrated support system that functions on the various levels of the police organization: strategic, directive, and operative. On each level, specific types of software support must be furnished; for the upper levels, what is needed is the capacity for prediction on a wide scale and for making a synthesis of all the facts available and, for lower levels, facility and speed of usage.

The value of the ideas and methods presented in this book goes beyond the area strictly linked to crime, to which reference is made in order to have a concrete field of application. A reading of the book can thus be useful, not only for those concerned with investigation or specialists in the algorithms of artificial intelligence but also for those who work in the vast field of the social sciences.

1.1 Navigating the Book

One can approach the reading of this book in the traditional way, from front to back. However, the theoretical chapters are more mathematically demanding than the application-oriented chapters, so there are some tracts one could use to guide their reading. The first track is the theoretical chapters tract and consists of Chaps. 2, 3, 4, 7, 8, 9, 11, 13, 15, 17, and 19. For those individuals already adept at neural network methodologies, these chapters capture the details of the algorithms and provide the basis for which similar algorithms can be created.

Those individuals who seek only to understand how adaptive neural networks can be applied to law enforcement problems can focus their attention on Chaps. 5, 6, 10, 12, 14, 16, and 18.

The last two tracks involve a merger of both the theoretical and applied chapters. To adequately be able to interface with software engineers/programmers who might be creating specialized programs for use in your facility, the reader can take two approaches, each of which is a combination of theoretical and applied chapters. First involves Chaps. 2, 3, 4, 5, 6, 8, 9, 10, and 15, 16, 17, 18, 19 and is a focus on theory augmented with certain applied chapters that bring life to the theoretical chapters.

The other track involves Chaps. 2, 3, 4, 7, 8, 9, 11, 12, 13, 14, and 18, 19. These chapters combine selected theoretical chapters with applied chapters to give one the tools and understanding needed to customize the algorithms to specific needs.

1.2 The Chapters

To properly prepare you to maximize your understanding of some of the very complex methods presented, the following chapters are brief summaries to give you an opportunity to experience the entire flavor of the book and perhaps direct your attention to some specific areas of interest. The actual usage of these algorithms is rather complex and would require the services of someone who is knowledgeable in both mathematics and a skilled programming to develop an interface for use by a particular police agency. On the other hand, an interested enforcement agency could simply seek advice from the director and staff of Semeion.

Chapter 2 ("Law Enforcement and Artificial Intelligence") is a high-level description of the motivation for the work in enforcement analytics. It establishes a justification for the use of adaptive neural networks, briefly explains how artificial learning occurs, and explains why traditional mathematics and methods of analysis with which we have become comfortable are no longer able to serve our needs as they did in the past. The criminal element is very smart, and they have the funds to do smart and innovative things to thwart the efforts of the "traditional" law enforcement community. When enforcement ceases to be creative, those on the other side profit at our expense. As enforcement becomes increasingly creative in its war on crime, the other side must expend increasing resources on higher levels of creativity, and the cycle will continue until one side surrenders. It is unlikely that those who unlawfully profit will be the first to capitulate.

Chapter 3 ("The General Philosophy of Artificial Adaptive Systems") describes the philosophy of the artificial adaptive system and compares it with our natural language. Some parallels are striking. The artificial sciences create models of reality, but how well they approximate the "real world" determines their effectiveness and usefulness. At the conclusion of this chapter, one will have a clear understanding of expectations from using this technology, an appreciation for the complexities involved, and the need to continue forward with a mind open to unexpected and unknown potential. The word "algorithm" is used almost continuously throughout the book. It is a very common word and can be interpreted as a simple "set of steps" used to attain an answer. Each step is very precise and acted upon by a computer as one line of instructional code. Once the computer has completed the running of the algorithm, an answer is provided to the user either in the form of screen output or sometimes as a hard-copy report. Most programs today use the screen as the mechanism for displaying output.

Chapter 4 ("A Brief Introduction to Evolutionary Algorithms and the Genetic Doping Algorithm") is an introduction to evolutionary algorithms, a commonly used method by which solutions to problems that might otherwise be impossible to solve are solved. One such method is that of the genetic algorithm. One of its strengths is its ability to solve problems in a relatively short time period that would otherwise not be solvable with the fastest computers working since the beginning of time. Such problems might be called NP or NP-hard problems, meaning that the time required for a computer to solve them is very, very long. On the downside, the answer provided by the genetic algorithm may not be optimal, but it is an "adequate" answer. To ensure an optimal solution, a computer would have to complete an examination of every possible solution, then select from the list the single winner. That solution would be optimal, but what if it took a supercomputer working at maximum speed a few years to deliver that answer; is it reasonable to expect one to wait that long a period of time for the exact answer? Or if an answer could be provided in a relatively short time period of a few hours (or minutes) and is "close" to optimal be acceptable? These approximate solutions are found to be quite useful and do provide for confident decision-making.

Sometimes evolutionary algorithms are based on what is called heuristics, or rules of thumb. They are guidelines for solutions that work; there are no mathematical proofs of their effectiveness; they just work well. Consequently, methods incorporating heuristics are deemed to be "weak." The word is unfortunate for it conveys a sense of inaccuracy or approximation, but it is, in fact, responsible for some excellent solutions. These weaker methods use less domain knowledge and are not oriented toward specific targets. In law enforcement analytics, the existence of such methods has been shown to be very advantageous. The chapters up through four are an excellent review of the operations of the genetic algorithm, and these are well known in the AI field. Chapter 4 presents a new genetic algorithm that is much more effective, the genetic doping algorithm (GenD). The word "dope" is unfortunate for it congeries up images of narcotics or a stupid person, but it actually means information gotten from a particularly reliable source. In this case, the reliable source is the data, and the effort is to extract maximal information from it.

GenD analyzes the data as though it were a tribe of individuals in which not everyone engages in crossover. To anecdotally explain, individuals in the tribe who are old or weak do not necessarily marry someone from that tribe (crossover does not occur in all genes); the fitness score (a calculated value that determines the ordering of the individuals in the tribe) is calculated on the basis of vulnerability and connectivity, and instead of dealing with the separate genes as individuals, GenD transforms the dataset into a dynamic structure and attempts to more closely mimic a genotype. A detailed and easy-to-read explanation of the differences between traditional genetic algorithms and GenD is given.

Chapter 5 ("Artificial Adaptive Systems in Data Visualization: Proactive Data") addresses the issue of the visualization of data modeled by artificial adaptive systems and one relatively easy visualization if that of the tree structure. A tree is a graph that contains a root, trunk, and leaves given a suitable imagination. Essentially, it is a diagram in which each point is connected to another point but without any circuits or loops anywhere in the graph. Thus, one can move from one point (called a vertex or node) to another following the lines (called edges or arcs) that connect the nodes and never come back over one's previous track. The structure is very important to understanding some very complex datasets. One of the ways it simplifies visualization is in its "dimensionality."

To see the visual representation of one single variable, we need only to plot a point on the x-axis of a graph, say a variable with a value of 12. At 12 units from the origin, we can place a dot to represent that variable. If we expand to two variables, say variable A has a value of 12 and variable B has a value of 3, then we can visualize this by placing a point in the XY coordinate plan that is located at the intersection of $X = 12$ and $Y = 3$. Similarly, we can add another variable to the mix, say variable $C = 2$, but visualization becomes somewhat more of a challenge for we must create a three-dimensional diagram on a sheet of paper (or computer screen). This can be easily done, and now we can see a point in position $X = 12$, $Y = 3$, and $Z = 2$ where X, Y, and Z are the x-axis, y-axis, and z-axis. So we have gone from one dimension, a line, to two dimensions, a plane, to three dimensions, a cube. Suppose we now add a fourth variable, or a fifth, or a 100th variable to the dataset. Visualization becomes a challenge to "see" the structure of the answer. Tree structures are one way by which many dimensions can be reduced to representation in two or three dimensions. While it takes some practice getting used to correctly reading and interpreting the graphs, the outcome is well worth the effort.

This chapter makes it clear that when one has a mass of data, possibly collected over years and on which SQL queries have been repeatedly made to the point that one might not think there is any more information that can be gleaned from further mining, it is the artificial neural network set of tools that come into play to explain the interactions and relationships existent among the data. The rules that connect the various sets of data within the database will very likely be fuzzy and dynamic. As the data submitted to the ANN are updated, it will adjust its "rules" in accordance, integrating the old data with the new, permitting us to correctly generalize new, dirty, incomplete, or future data.

Chapter 6 ("The Metropolitan Police Service Central Drug-Trafficking Database: Evidence of Need") discloses the problems inherent in large database systems, the errors that are entered into it by nontrained or only partially trained data input operators, the inconsistencies in the data that further thwart efforts to glean useful information using traditional methods, and the absence of a recognition that correct DB input, though time consuming, can be an ardent partner in the identification of relationships and the generation of profiles as a definite source of help and assistance to the enforcement community. It becomes apparent that the police, local, national, and international, have at their disposal access to information that could revolutionize the ways in which their jobs are performed, if only they had the knowledge, foresight, funding, and incentive to utilize it.

Chapter 7 ("Supervised Artificial Neural Networks: Back Propagation Neural Networks") becomes technical with a description of one of the most basic neural networks, that of the back propagation network. To understand it requires first a familiarity with the feedforward backpropagation artificial neural network (FF BP ANN). The first half of this chapter is a relatively low-level introduction to the theory FF BP, but it does get into some more challenging mathematics in the second half. If one has a background in calculus and differential equations, the math will be easy to follow. If not, one can simply accept that the mathematics are correct and read "around" the equations. In this manner, one can learn the theory and get a basic understanding how it works. This is probably the pivotal chapter for all the remaining algorithms; most everything else builds on this content.

Chapter 8 ("Preprocessing Tools for Nonlinear Datasets") addresses the most difficult, and arguably the most important, problem in artificial neural networks, the training and testing of the network to ensure the best possible outcome. ANNs must first be "trained" to understand the data and establish the relationships among the variables, and it is a task that the algorithm must do itself. In the classical sense, the dataset would simply be randomly partitioned into two or more subsets, and one subset would be used to train the network, another to test it, and finally one subset on which to actually run the network. There are problems inherent in this method, especially when the database is extremely large, as is typically the case with enforcement DBs, and when the data is "noisy." Noise is the existence of data that does not have any strong relationships with other variables. If a network is overtrained, the noise is incorporated as if it was strongly tied to other variables, and hence, new evaluated data would consider the noise to be an important part of the system. This would yield an incorrect interpretation of the data. Noise must be eliminated so that the network is properly trained. This chapter discusses how best to perform that action.

One way of eliminating noise, or at least reducing its impact, is addressed by two new algorithms called the training and testing algorithm (T&T) and the training and testing reverse algorithm (T&Tr). These are preprocessing systems that permit procedures to be far more effective in training, testing, and validating ANN models. This chapter presents the concept and mathematics of the algorithm and then illustrates their effectiveness with an example.

Chapter 9 ("Metaclassifiers") describes methods by which data can be classified. There are many methods which purport to classify data, and each one performs the classification in a different manner and typically with differing results. The variation in outcome can be explained by saying that the different mathematics associated with each method views the data from various different perspectives, assigning data to classifications that can be, and usually are, different. A metaclassifier, however, is a method by which the results of these individual classifiers are considered as input to an ANN that forms the classifications based on the differing views and perspectives of the individual ANNs. In short, the different perspectives of the individual ANNs are brought together to produce a single, superior classification taking into account the various algorithms that produce certain views of the data.

Chapter 10 ("Auto-identification of a Drug Seller Utilizing a Specialized Supervised Neural Network") is a comprehensive illustration of the application of pattern recognition on a law enforcement database of drug-related data using the metaclassification algorithm discussed in the previous chapter. This chapter is more accessible to the nontechnician and gives an exciting, and detailed, description of how the metaclassifier can be used to identify unknown relationships.

Chapter 11 ("Visualization and Clustering of Self-organized Maps") describes a type of neural network that has been around for some 30 odd years, the self-organizing map. The main significance of this type of ANN is that it can take high-dimensional data and produce a diagram (map) that displays it in one or two dimensions. In short, humans can visualize interactions when displayed in one, two, or three dimensions, but not four or more dimensions. Data composed of only one variable can "see" a point on an x-axis diagram; data composed of two variables can be displayed on an x-y-axis diagram; data composed of three variables can be displayed on an x-y-z-axis diagram, and the visualization stops here. We simply cannot visualize diagrams in four or more dimensions, and that is where the self-organizing map comes into play. It has the ability of analyzing data in an unsupervised way (without any preconceived indication of the number of patterns present in the data) and placing the resulting analysis in a one- or two-dimensional diagram. While some information might be lost in the translation, it is more than made up with the insights that one can glean from the resulting diagram.

This type of ANN is continued in Chap. 12 ("Self-organized Maps: Identifying Nonlinear Relationships in Massive Drug Enforcement Databases") with its use in the analysis of a massive drug enforcement database collected over time by the Scotland Yard Metropolitan Police. Throughout this chapter, the theory of the self-organizing map, as presented in Chap. 12, is explained in substantial detail ending with many visualizations of the drug industry in London. The results yield a "profile" that can be used by law enforcement agencies to target their investigations, monitoring, etc. Since the profile is the result of a mathematical algorithm, an argument that a particular ethnic group is being targeted can and should be dismissed, for the data speak for themselves.

Chapter 13 ("Theory of Constraint Satisfaction Neural Networks") is a description of the constraint satisfaction neural network (CS ANN). Problems typically have some constraints that limit a decision, and we have this situation regularly occurring. For example, a search of a database for the owner of a particular car whose license begins with ABC is a constraint imposed on the solution. A search for a male whose height is between 5′6″ and 5′8″ (167.6 and 172.7 cm) and weight is 220 lb (100 kg) is a constraint problem. Thus, the constraint satisfaction ANN involves finding a solution given the imposition of a series of conditions on it.

Chapter 14 ("Application of the Constraint Satisfaction Network") is an extension of the previous chapter and describes the application of the CS ANN on a dataset composed of 144 variables on 1,120 cases involving drug trafficking within the boroughs of London. The examples show the level of detail that can be derived from data using this method of analysis, and the results are graphically shown in tree diagrams for which the interpretation of which is also provided. Law enforcement officers will get a very good understanding as to the kinds of information, and the depiction of the results of the analysis, that may be available in databases. There is a richness of information that very likely has not been mined, and the methods described here should excite the reader as to possible results.

Chapter 15 ("Auto-contractive Maps, H Function, and the Maximally Regular Graph: A New Methodology for Data Mining") describes an original artificial neural network called the auto-contractive map (AutoCM). This specialized analytical method produces a graphical image that displays the overall relationships that exist among the variables at the most fundamental level of system construction. That is, a set of variables that constitute some system under investigation possess some degree of connectivity among all the variables. To draw a graph in which each point is connected to every other point does, in fact, represent the overall system, though it also gives absolutely no information as to the structure of the underlying, basic structure. We seek to understand the structure but at the point when the individual variables have enough of a relationship with each other to form an initial linkage. Thus, the AutoCM shows a graph in which each point (called a node or vertex) is connected to another node but without the creation of any loops or circuits in the graph. This is called a minimal spanning tree (MST). The benefit of this kind of graph is that it permits us to see how one variable is related to another from a hierarchical perspective. For example, if we possess a large number of records on individuals and we seek to understand how these people are related (in the mathematical sense) to each other, the MST will give us exactly that information. It is this ANN that can be used, for example, to determine the structure of a drug network: who are the pushers, who supplies the pushers, who supplies the suppliers, etc.

An embellishment of this method is the inclusion of the maximal regular graph, the linking of individual nodes that show the strongest degree of "relationship," which creates circuits in parts of the MST. These circuits typically end up creating a perfect (or almost perfect) subgraph. An example of the AutoCM is given using the characters from West Side Story, the Jets, and the Sharks.

Chapter 16 ("Analysis of a Complex Dataset Using the Combined MST and Auto-contractive Map") uses the auto-contractive map ANN described in Chap. 15 to analyze the drug activity in London. The main value in this chapter is the interpretation of the various graphs, taking the visual representation of the mathematical relationships and putting them into words that can yield some action. Anyone seeking to use the AutoCM ANN should spend the time necessary to master this chapter.

Chapter 17 ("Auto-contractive Maps and Minimal Spanning Tree: Organization of Complex Datasets on Criminal Behavior to Aid in the Deduction of Network Connectivity") is a comprehensive example of using adaptive neural networks to solve interesting and important law enforcement-related problems and should be read very carefully. Essentially, let us assume that we have a very large database composed of data on individuals that include gender, age, nationality, where arrested, number of previous convictions, arrests and offenses, type of drugs seized, behavior of the individual, and much more. Missing from the database is any information about associations with others, and that is the item of interest we seek to discover: Which individuals belong to the same "gang" or are involved in the same drug-trafficking "circle"? Using the auto-contractive map and minimal spanning tree methods, this chapter shows how mathematics can answer this question. For law enforcement personnel seeking to get some handle on gang activity or conspiracy activity, guidance toward a solution can be found here.

The illustration of the usage of this ANN technology to solve interesting problems continues with Chap. 18 ("Data Mining Using Nonlinear Auto-associative Artificial Neural Networks: The Arrestee Dataset"). Using the drug database, this chapter explores how to apply nonlinear auto-associative systems (nonsupervised ANNs) to data analysis. The results of the analysis are presented in various MSTs and the graphs interpreted. While fascinating and informative graphs are created and discussed, these structures are the results of the application of mathematical algorithms to records of data, and it must be emphasized that the analysis depends directly on the quality of the data entry and that the results should be interpreted as a point of departure for anyone using these methods for investigative purposes.

The final Chap. 19 ("Artificial Adaptive System for Parallel Querying of Multiple Databases") addresses the interesting problem of analyzing multiple databases that do not possess a similar data structure. An analogy is made with different wineries in the same community. Each winery produces its own special wines, and each wine has its own set of characteristics, but they all come from the same geographical area. Hence, once could want to understand the complex interactions that occur among the different wineries. Similarly, different police organizations may have their data stored in databases whose data structures, or the kinds of data placed in the fields that make up the records, differ across the organizations.

1.3 Collaborative Opportunities

The Semeion Research Center in Rome, Italy, the research facility responsible for the discovery and development of all the methods described in this book, regularly works with companies and organizations who seek to implement some of the methods described here in their own organization but do not have the funding necessary to create their own systems. Arrangements can be made with Semeion to utilize their software under a licensing arrangement. Also, it is not uncommon for organizations to pass their database directly to Semeion where the analysis occurs and a detailed report is provided. The analysis is completed at Semeion and the results are explained, in whatever detail is necessary to ensure proper understanding of the results, to the organizational officials.

Chapter 2
Law Enforcement and Artificial Intelligence

Paolo Massimo Buscema

2.1 Data and Methods

The original purpose of several of the chapters in this book was to acquaint the Metropolitan Police Service (MPS) of London, England, with the interim findings arising from the analyses of data drawn from drug trafficking crimes recorded by the MPS in 2004. Since then, the purpose has evolved into a book that would provide new and exciting guidance as to the possibilities of numerous nontrivial neural network applications to the many different fields of law enforcement. Law enforcement agencies typically collect huge quantities of data and might remain underutilized for numerous reasons, that is, insufficient time for investigators to mine the data, insufficient resources to engage in creative and exploratory methods of analytically gathering useful information from the data, or simply of not being aware of the various forms by which data can be utilized to create new insights into how mathematics, computer science, and systems analysis can be brought to bear on critical problems. These analyses were undertaken by the Semeion Research Center of Sciences of Communication (hereafter *Semeion*) using an array of artificial adaptive systems. As a prerequisite to the testing, training, and analytical phases, the report also discusses the importance of:

- Collecting, organizing, and validating data
- The necessity of designing and constructing a database "fit for purpose"

The report further points out that having a wealth of well-organized data and a sophisticated database is of little use unless the analytical tools used are underpinned by robust mathematics. Without good mathematics, there is a high risk that any analyses of data will generate contradictory and arbitrary gobbledegook.

P.M. Buscema (✉)
Semeion Research Center of Sciences of Communication, Via Sersale 117, Rome, Italy
e-mail: m.buscema@semeion.it

M. Buscema and W.J. Tastle (eds.), *Intelligent Data Mining in Law Enforcement Analytics: New Neural Networks Applied to Real Problems*, DOI 10.1007/978-94-007-4914-6_2,

Data are read by tools, and suitable tools are provided by mathematics. The basic statistics through which often complex database are read are not sufficient to understand the secrets that data freeze in the form of numbers. Frequency tables, means, variances, two-variable correlations, etc., are excellent tools for "getting an idea" of the data that we have before us. But these basic analyses do not reveal anything about the complex relations concealed in an important database. And the reason is simple: in any database, any individual datum interacts in parallel with all the others, and it is only this global "many-to-many" interaction that generates the meaning of each individual datum (Grossi et al. 2007; Dzeroski and Lavrac 2001).

Even more complex statistics sometimes might miss the subtle associations that constitute the framework of any database. Every linear multivariate analysis, in fact, links data through simplified relations. But most natural, biological, and cultural phenomena do not follow plain "cause-effect" relationships. Any structure that has feedback, from a thermostat to human brain circuits, often violates in some way the dynamics of these statistics. This means that in many databases nonsignificant linear relationships may conceal the keystone of the entire system. To be unaware of them may mean not to understand at all how that system will actually respond to our actions.

A database is in fact like a living system frozen in one or more moments of one's life. To understand it means reactivating the interactions between each of its individual data and every other datum. A mathematics that is suitable for activating this reanimation process must consider each individual datum as an agent whose aim may consist of either of the following:

1. Negotiating the value of its relationship with all the others, in order to maintain its own identity, that is, its original numerical value
2. Defining its new identity, in order to maintain its relationships with the others, if these are binding

To operate on a database in this way means understanding the history that has produced it and the future on which it is focused.

Analyses of this type cannot be done through statistical tools that review all the data just once to establish simplified relationships that may entirely miss the mark. A more complex mathematics is required, able to consider each datum as an agent that develops its linear and nonlinear relationships with all the others over time and in parallel. A mathematics operating in a manner similar to that of operation of the human brain is required, that is, locally, in parallel and through continuous feedback between basic units.

Artificial adaptive systems are the mathematical tools suitable for these types of analysis.

These are systems whose:

1. Basic units interact in parallel and locally with one another
2. Units, therefore, negotiate their interactions over time and according to highly nonlinear logics

3. Units are connected to one other and the value of these connections is adapted and modified over time, until the entire system spontaneously reaches its own stability
4. Interacting units modify the entire system from bottom to top, thus revealing the global complexity of their relationships

Therefore, artificial adaptive systems are the appropriate tools to generate each database from within its own specific model, concealed in its data.

A frequent error in the analysis of data is that of entering into the database a model that is external to it and then wondering how much the model explains the database. Artificial adaptive systems are, on the contrary, models for generating the models that are intrinsic to each database.

Their approach to data, therefore, is in a sense "maieutic"; they help the database to give birth to their hidden relationships.

Artificial neural networks (ANNs) and evolutionary algorithms (EAs) are the two classes of artificial adaptive systems that we shall use in this report to make the data "speak." ANNs are information processing paradigms inspired by the analytical processes of the human brain. Such systems are able to modify their internal operating structure and, therefore, the resulting analysis in relation to a defined goal, question, or function objective. *They learn to recognize the complex patterns existing between the input signals and the corresponding outputs.* ANNs are particularly suited for solving problems of the nonlinear type and to analyze complex datasets.

Nowadays, this analytical style is known in the scientific world as *data mining*, the central activity in the process of knowledge discovery in databases which is concerned with finding patterns in data.

2.2 Drug Trafficking and Artificial Adaptive Systems

The purpose of this chapter, therefore, is summarized in two points:

1. To show how to construct a significant relational database on drug trafficking in a city such as London
2. To explain how the artificial adaptive systems may be the best tools for *revealing* from these data, information that is *hidden from and invisible* to other tools and, at the same time, is important for making strategic and operative decisions concerning the fight against drug trafficking

However, this dual aim encounters a problem: artificial adaptive systems are complex mathematical tools, and a person who deals with crime does not normally have sufficient mathematical training. This difficulty is accentuated by the fact that many of the artificial systems that will be used in this book are unpublished and therefore not even known to the experts on artificial neural networks and evolutive algorithms.

In other words, we have the very difficult problem of successfully explaining at least the philosophy of complex mathematics to crime experts and at the same time being able to explain, in sufficiently adequate mathematical terms, to experts on artificial adaptive systems the new models that we have created. We must also show to both of these types of different domain experts how these complex systems can be used in real situations, deriving a practical and theoretical benefit from them.

In practice, this seems to be an impossible task though we take it on in this book.

However, since the introduction to a book is always written at the conclusion of the chapters in order to adequately explain the problems that need to be addressed, the reader will gain an understanding of the philosophical principles that render a potential solution, the methods used to address it, and the success or lack thereof to guide others in benefitting from this effort. It is new ground that is traversed here, very important and fully transferrable to other disciplines to solve other kinds of problems, though one requires knowledge of the contents of this book and a motivation to change existing systems in order to make large jumps forward.

This book is divided into three parts:

1. The initial contribution of the first part is an attempt to explain why various types of artificial adaptive systems can be used as artificial detectives capable of performing the intelligent analysis of data linked to the world of crime. A real detective often *classifies* the various criminal situations; tries to form a *visual map* of the events; makes *predictions* as to how those events could develop; builds a somatic, sociological, and psychological *profile* of the criminal players whom he is pursuing; and, furthermore, *imagines links* between events, players, and clues, in order to reconstruct the complex network of a criminal activity. These five cognitive activities can also be implemented in various artificial adaptive systems: intelligent classification systems, visual clustering systems, prediction systems, prototype generation systems, and systems creating network connections between objects. In this part of the book, we show how artificial adaptive systems can represent a second look and a second opinion on the analysis of criminal acts. This part is easy to understand and contains examples of simplistic applications to the world of crime. The second contribution of this section describes the scenario within which our research moves, specifically, to address drug trafficking in London. It is important to understand the kind of real problems London's Metropolitan Police Service encounters every day in the fight against drug trafficking, what laws govern their actions, and the procedures they implement. But above all, we hope that this section will help police officers, criminologists, and policy makers to better understand how artificial intelligent systems could be fundamental in confronting a criminal network whose complexity ranges from the street pusher to the plethora of innocuous-appearing bank accounts scattered throughout the world.
2. The second part of the book presents the mathematical models of the artificial adaptive systems used to analyze the data on drug trafficking. The presentation

logic that we have followed is a logic guided by the mathematical model: the first article explains the architecture, mathematics, and the specific algorithms contained in the model. There then follows a series of articles in which the presented model is applied to various datasets extracted from the central database. These articles show how the model in question is able to reveal information from the data that is relevant but hidden; it is, in fact, information that other models would be unable to extract. Many of the models contained in these articles are new to the experts themselves, and therefore, we have taken care to present them in such a way that their conceptual validity and mathematics are clear and understandable. However, the applicative contributions should be comprehensible even to a reader who is not mathematically savvy. Our faith in logic, however, is such that we believe that the theoretical presentation of the models should also be comprehensible to a nonspecialist reader; it is sufficient if one wishes to skip the formulae but carefully reads the examples that are proposed in the theoretical discourse.

3. The third section of this book is devoted to a more in-depth study. Each reader may choose which contributions of this section to read and the order in which to read them. We did, in fact, think it advisable to provide a contribution explaining what we mean by artificial adaptive systems and how they are placed in relation to the other artificial intelligence systems existing today. It is only a conceptual proposal and not a complete theory. A second article in this section is devoted to a brief introduction to artificial neural networks which alone represents the majority of the artificial adaptive systems. It is a brief explanatory introduction that was written with the intent of being intelligible to all and which we deem as useful to anyone who wishes to have a more detailed and formal explanation. The third article is along the same lines as the preceding one, but it addresses the family of evolutive algorithms, headed by an explanation on genetic algorithms. In this case, too, the aim is to be explanatory and the language intentionally nontechnical (or at least as much as possible). The fourth article in this section is devoted to the design and implementation of the relational database on drug trafficking that we constructed with the assistance of the Metropolitan Police in London. This chapter is fundamental to understanding how much an intense collaboration between information technologists and detectives is decisive for producing a DB whose quality is the result of professional experiences that are as diverse as they are complementary. Any intelligent analysis is possible from a good database, even though a good database does not guarantee an intelligent analysis. The final chapter is devoted to a basic description of the thousands of datum items collected in the database that have made it possible to generate all the preceding analyses. This contribution provides a picture of the quantity and quality of the data that the group of Metropolitan Police Service's data entry operators (DEOs) at New Scotland Yard[1] have been able to process.

[1]New Scotland Yard is the headquarters of the MPS.

2.3 Intelligence and Crime

Individual intelligence is optional for someone to possess in the commission of a crime, but when it exists, it greatly increases the dangers. For civil society and, therefore, for the police forces, individual intelligence is, however, a prerequisite for its very existence. A society without intelligence contained in its members cannot be civil either.

The final purpose of this book consists of offering civil society one more intelligence tool against the brutality of uncivilized intelligence, that is, against organized crime.

The police forces of democratic countries already have in their service men and women of considerable intelligence and selflessness. However, when criminal networks exceed a certain level of complexity, their dynamic outstrips the capacity for calculation and imagination of any human brain, including the one of criminal individuals who believe that they are the main players in this dynamic.

"Technological intelligence," the historic precipitate of advanced scientific research, must be considered as a precursor of human intelligence; a second brain less complete than the natural one but more specialized in certain activities; a second brain capable of supplying one hypothesis more than the usual ones; a second brain capable of shading light in areas where our capabilities see just a dark invisible corner; and a second brain that was, in short, born of an intangible thought in our brain.

The mathematics, information technology tools, and their applications presented in this book have their origin in a single certainty: we are condemned forever to possess logic. Even behaviors that may seem random, if performed by human individuals and repeated, will sooner or later assume a form. Whoever understands it first can try to make the future become the effect of the present. It would be desirable that civil society imagined our future.

References

Dzeroski, S., & Lavrac, N. (2001). *Relational data mining*. Berlin/New York: Springer.

Grossi, E., Mancini, A., & Buscema, M. (2007). International experience on the use of artificial neural networks in gastroenterology. *Digestive and Liver Disease, 39*, 278–285.

Chapter 3
The General Philosophy of Artificial Adaptive Systems

Paolo Massimo Buscema

3.1 Artificial Adaptive Systems

Artificial adaptive systems (AAS) form part of the vast world of *natural computation* (NC) which is itself a subset of the *artificial sciences* (AS). Artificial sciences are those sciences for which an understanding of natural and/or cultural processes is achieved by the recreation of those processes through automatic models.

We shall use an analogy to explain the difference between artificial science and natural language; the *computer* is to the artificial sciences as writing is to *natural language*. That is, the AS consists of a *formal algebra* used for the generation of *artificial models* which are composed of structures and processes, and natural languages are composed of semantics, syntax, and pragmatics for the generation of texts. Through each of these very different systems, a level of independence is created; in natural languages, the utterances of sounds are fully dependent on the time in which the utterances are made, but by representing those utterances with writing, they become independent from time, for written documents (in the form of books, manuscripts, typewritten pages, computer-generated output in the form of both digital and hardcopy, etc.) exist outside the dimension of time. They exist in the spatial dimension. Similarly, the computer achieves independence from the physical system through the creation of a model. Such models are automations of the original system and permit one to study the natural/physical system at any time, even if the original system no longer exists. An example of such a system is the active eruption of a volcano or the tremors of an earthquake. Through extensive measurements of variables, a model can be constructed that permits researchers to recreate the original volcanic activity or earthquake in a completely controlled environment by which variables of choice can be controlled. Using

P.M. Buscema (✉)
Semeion Research Center of Sciences of Communication, via Sersale 117, Rome, Italy
e-mail: m.buscema@semeion.it

M. Buscema and W.J. Tastle (eds.), *Intelligent Data Mining in Law Enforcement Analytics: New Neural Networks Applied to Real Problems*, DOI 10.1007/978-94-007-4914-6_3,

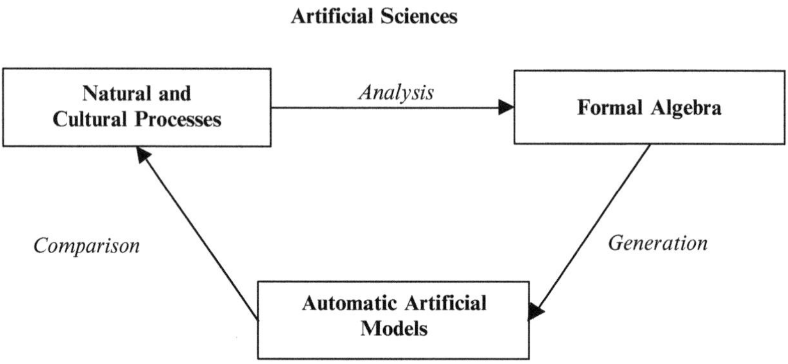

Fig. 3.1 The diagram shows how the analysis of natural and/or cultural processes that need to be understood starts from a theory which, adequately formalized (formal algebra), is able to generate automatic artificial models of those natural and/or cultural processes. Lastly, the generated automatic artificial models must be compared with the natural and/or cultural processes of which they profess to be the model and the explanation

writing as an extension of a natural language permits the creation of cultural objects that, before onset of writing, were unthinkable. Such cultural objects are stories, legal texts, manuals, historical records, etc. In a similar manner, the AS can create models of complexity that, before the construction of computers, were unthinkable.

Natural languages and artificial sciences, in the absence of writing and the computer, are therefore limited. But a written document not based on a natural language, or an automatic model not generated by formal algebra, is little more than a set of scribbles (Fig. 3.1).

In the artificial sciences, the understanding of any natural and/or cultural process occurs in a way that is proportional to the capacity of the automatic artificial model to recreate that process. The more positive the outcome of a comparison between an original process and the generated model, the more likely it is that the artificial model has correctly explained the functioning rules of the original process. However, this comparison cannot be made simple-mindedly. Sophisticated analytical tools are needed to make a reliable and correct comparison between an original process and an artificial model. Most of the analytical tools useful for this comparison consist of comparing the dynamics of the original process and the dynamics of the artificial model when the respective conditions in the surroundings are varied.

In sum, it could be argued that:

1. Varying the conditions in the surroundings yields a greater variety of response dynamics obtained both in the original process and in the resulting artificial model.

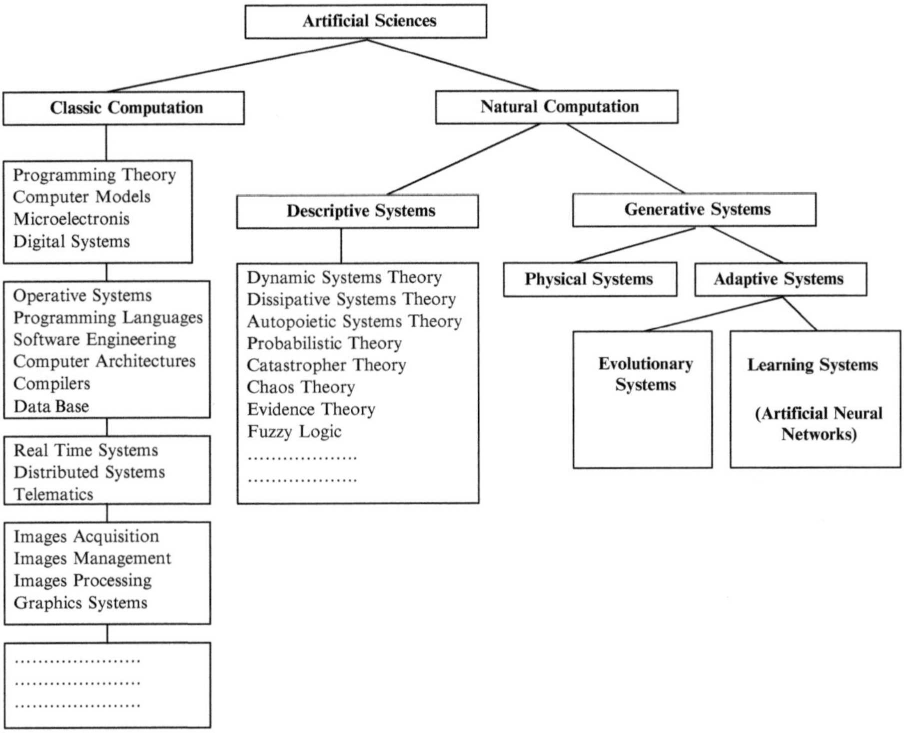

Fig. 3.2 Taxonomic tree of the disciplines that make up the artificial sciences system

2. *The more* these dynamics between the original process and resulting artificial model are homologous, the more probable it is that the artificial model is a good explanation of the original process.
3. *The more* probable it is that the artificial model is a good explanation of the original process.

In Fig. 3.2, we propose a taxonomic tree for characterization of the disciplines that, through natural computation and classic computation, make up the artificial sciences system.

Natural computation (NC) refers to that part of the artificial sciences (AS) responsible for the construction of automatic models of natural and/or cultural processes through the local interaction of nonisomorphic microprocesses. In NC, it is therefore assumed that:

1. Every process is, more or less, contingent on the result of more basic processes that tend to self-organize in time and space.
2. None of the microprocesses are themselves informative concerning the function that they will assume with respect to others, nor the global process of which it will be part.

This computational philosophy, which is very economic for the creation of simple models, can be used effectively to create any type of process or model that is inspired by complex processes. NC in fact deals with the construction of artificial models that do not simulate the complexity of natural and/or cultural processes through rules but, rather, through commitments that, depending on the space and time through which the process takes form, autonomously create a set of contingent and approximate rules. NC does not try to recreate natural and/or cultural processes by analyzing the rules which make them function, and thus formalizing them into an artificial model. On the contrary, NC tries to recreate natural and/or cultural processes by constructing artificial models able to create local rules dynamically and therefore capable of change in accordance with the process itself.

The links that enable NC models to generate rules dynamically are similar to the Kantian transcendental rules: these are rules that establish the conditions of possibility of other rules.

In NC, dynamics such as *learning to learn* are implicit in the artificial models themselves, while in classical computation additional rules are required.

Natural computation can be decomposed into the following:

- *Descriptive systems* (DS) are derived from disciplines that have developed, whether or not intentionally, a *formal algebra* that has proved particularly effective in drawing up appropriate functioning links of artificial models generated within NC (e.g., the theory of the dynamic systems, the theory of autopoietic systems, fuzzy logic).
- *Generative systems* (GS) are theories of NC that have explicitly provided a formal algebra aimed at generating artificial models of natural and/or cultural processes through links that create dynamic rules in space and in time.

In turn, generative systems can be broken down into:

- *Physical systems* (PS): a grouping of those theories of natural computation whose *generative algebra* creates artificial models comparable to physical and/or cultural processes, only when the artificial model reaches given evolutive stages (limit cycles type). While not necessarily the route through which the links generate the model, it is itself a model of the original process; in brief, in these systems the generation time of the model is not necessarily an artificial model of the evolution of the process time (e.g., fractal geometry).
- *Artificial adaptive systems* (AAS) are theories of natural computation whose generative algebra creates artificial models of natural and/or cultural processes, whose birth process is itself an artificial model comparable to the birth of the original process. They are therefore theories assuming the emergence time in the model as a formal model of the process time itself.

In short, for these theories, each phase of artificial generation is a model comparable to a natural and/or cultural process.

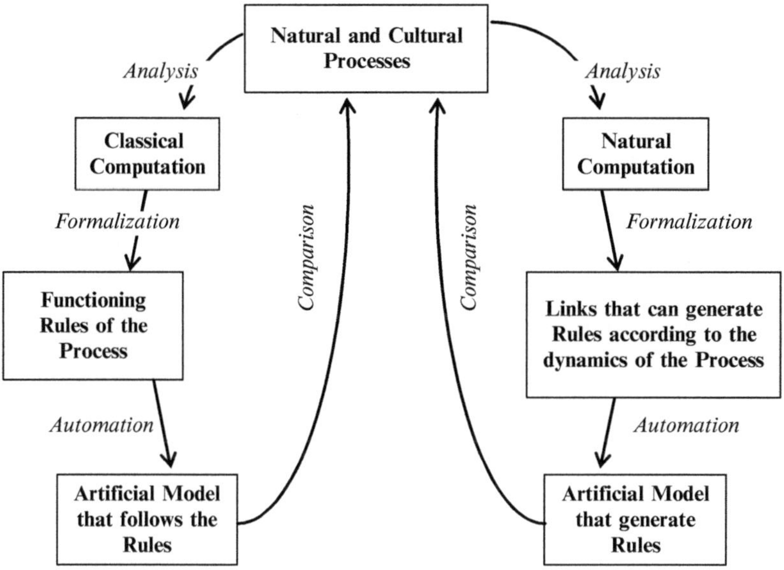

Fig. 3.3 The diagram shows in more detail the formalization, automation, and comparison between natural and/or cultural processes and automatic artificial models seen from two points of view (classical computation and natural computation). Each point of view can be seen as a cycle that can repeat itself several times. This allows one to deduce that the human scientific process characterizing both the cycles resembles more the natural computation than the classical computation one

Artificial adaptive systems in turn comprise:

- *Learning systems (artificial neural networks* – ANNs): these are algorithms for processing information that allow for the reconstruction, in a particularly effective way, the approximate rules relating a set of "explanatory" data concerning the considered problem (the input), with a set of data (the output) for which it is requested to make a correct forecast or reproduction in conditions of incomplete information.
- *Evolutionary systems* (ES): this means the generation of adaptive systems changing their architecture and their functions over time in order to adapt to the environment into which they are integrated or comply with the links and rules that define their environment and, therefore, the problem to be simulated. Basically, these are systems that are developed to find data and/or optimum rules within the statically and dynamically determined links and/or rules.

The development of a genotype from a time t_i to a time $t_{(i+n)}$ is a good example of the development over time of the architecture and functions of an adaptive system (Figs. 3.3 and 3.4).

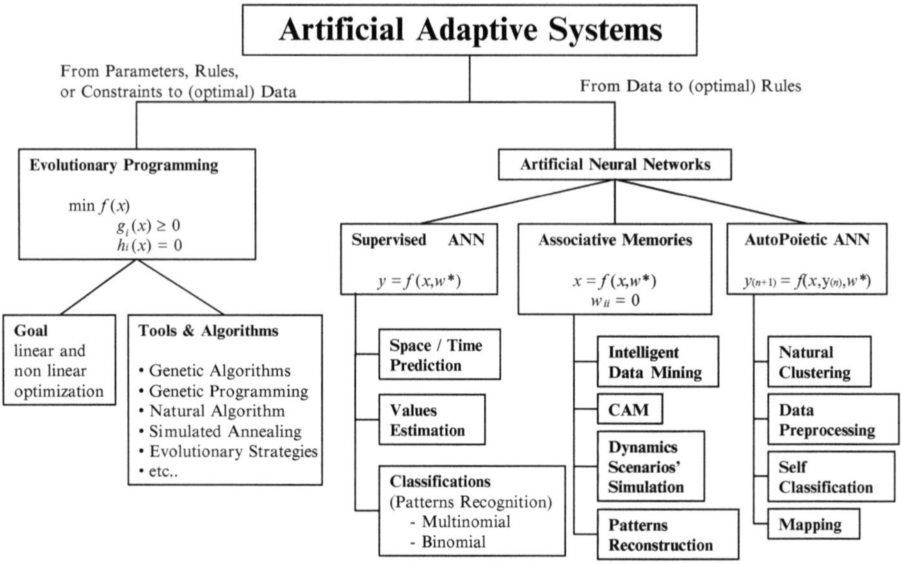

Fig. 3.4 Artificial adaptive systems – general diagram

3.2 A Brief Introduction to Artificial Neural Networks

3.2.1 *Architecture*

ANNs are a family of methods created to simulate the workings of the human brain.

Currently, ANNs comprise a range of very different models, but they all share the following characteristics:

- The fundamental elements of each ANN are the *nodes*, also known as processing elements (pe), and their *connections*.
- Each node in an ANN has its own *input*, through which it receives communications from the other nodes or from the environment, and its own *output*, through which it communicates with other nodes or with the environment. Finally it has a function, $f(\cdot)$, by which it transforms its global input into output.
- Each connection is characterized by the force with which the pair of nodes excite or inhibit each other: positive values indicate excitatory connections, and negative ones indicate inhibitory connections.
- Connections between nodes may change over time. This dynamic triggers a *learning process* throughout the entire ANN. The way (the law by which) the connections change in time is called the "learning equation."
- The overall dynamic of an ANN is linked to *time*: in order for the connections of the ANN to properly change, the environment must act on the ANN several times.

- When ANNs are used to process data, these latter are their environment. Thus, in order to process data, these latter data must be subjected to the ANN several times.
- The overall dynamic of an ANN depends exclusively on the local interaction of its nodes. The final state of the ANN must, therefore, evolve "spontaneously" from the interaction of all of its components (nodes).
- Communications between nodes in every ANN tend to occur in *parallel*. This parallelism may be *synchronous* or *asynchronous*, and each ANN may emphasize it in a different way. However, an ANN *must* present some form of parallelism in the activity of its nodes.
- From a theoretical viewpoint, this parallelism does not depend on the hardware on which the ANNs are implemented.

Every ANN must present the following architectural components:

- Type and number of nodes and their corresponding properties
- Type and number of connections and their corresponding location
- Type of signal flow strategy
- Type of learning strategy

3.2.2 The Nodes

There can be *three types* of ANN nodes, depending on the position they occupy within the ANN:

- *Input nodes*: the nodes that (also) receive signals from the environment outside the ANN.
- *Output nodes*: the nodes whose signal (also) acts on the environment outside the ANN.
- *Hidden nodes*: the nodes that receive signals only from other nodes in the ANN and send their signal only to other nodes in the ANN.

The number of input nodes depends on the way the ANN is intended to *read* the environment. The input nodes are the ANN's *sensors*. When the ANN's environment consists of data the ANN should process, the input node corresponds to a sort of data *variable*. The number of output nodes depends on the way one wants the ANN to *act* on the environment. The output nodes are the *effectors* of the ANN. When the ANN's environment consists of data to process, the output nodes represent the variables sought or the results of the processing that occurs within the ANN. The number of hidden nodes depends on the complexity of the function one intends to map between the input nodes and the output nodes. The nodes of each ANN may be grouped into classes of nodes sharing the same properties. Normally, these classes are called layers.

Fig. 3.5 Types of possible connections

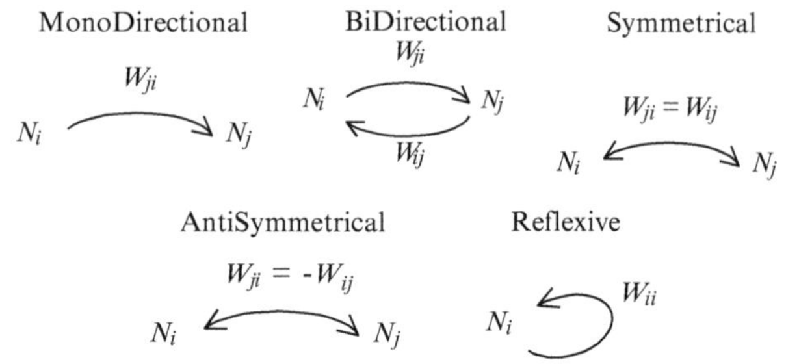

Various types can be distinguished:

- MonoLayer ANNs: all nodes of the ANN have the same properties.
- MultiLayer ANNs: the ANN nodes are grouped in functional classes; for example, nodes that (a) share the same signal transfer functions or (b) receive the signal only from nodes of other layers and send them only to new layers.
- Nodes sensitive ANNs: each node is specific to the position it occupies within the ANN; e.g., the nodes closest together communicate more intensely than they do with those further away.

3.2.3 *The Connections*

There may be various types of connections: MonoDirectional, BiDirectional, Symmetrical, AntiSymmetrical, and Reflexive (Fig. 3.5):

The number of connections is proportional to the memory capabilities of the ANN. Positioning the connections may be useful as methodological preprocessing for the problem to be solved, but it is not necessary. An ANN in which the connections between nodes or between layers are not all enabled is called an ANN with *dedicated connections*; otherwise it is known as a *maximum gradient* ANN.

In each ANN, the connections may be:

- *Adaptive*: they change depending on the learning equation.
- *Fixed*: they remain at fixed values throughout the learning process.
- *Variable*: they change deterministically as other connections change.

3.2.4 *The Signal Flow*

In every ANN, the signal may proceed in a direct fashion (from input to output) or in a complex fashion.

Thus, we have two types of flow strategy:

- *Feed forward ANN*: the signal proceeds from the input to the output of the ANN passing all nodes only once.

- *ANN with feedback*: the signal proceeds with specific feedbacks, determined beforehand, or depending on the presence of particular conditions.

The ANNs with feedback are also known as *recurrent ANNs* and are the most plausible from a biological point of view. They are often used to process timing signals, and they are the most complex to deal with mathematically.

In an industrial context, therefore, they are often used with feedback conditions determined a priori (in order to ensure stability).

3.3 Learning in the Artificial Neural Network

Every ANN can learn, over some period of time, the properties of the environment in which it is immersed or the characteristics of the data which it presents. This is accomplished in basically one of two ways (or mixture of both):

- By reconstructing approximately the probability density function of the data received from the environment, compared with preset constraints
- By reconstructing approximately the parameters which solve the equation relating the input data to the output data, compared with preset constraints

The first method is known in the context of ANNs as *vector quantization*; the second method is *gradient descent*. The vector quantization method articulates the input and output variables in *hyperspheres* of a defined range. The gradient descent method articulates the input and output variables in *hyperplanes*.

The difference between these two methods becomes evident in the case of a feed forward ANN with at least one hidden unit layer. With vector quantization, the hidden units encode *locally* the more relevant traits of the input vector. At the end of the learning process, each hidden unit will be a *prototype* representing one or more relevant traits of the input vector in definitive and exclusive form.

With gradient descent, the hidden units encode *in a distributed manner* the most relevant characteristics of the input vector.

At the end of the learning process, each hidden unit will tend to represent the relevant traits of the input in a fuzzy and nonexclusive fashion.

Summing up, the vector quantization develops a *local* learning; the gradient descent develops a *distributed* or *vectorial* learning.

Considerable differences exist between the two approaches:

- Distributed learning is computationally more efficient than local learning. It may also be more biologically plausible (not always or in every case).
- When the function that connects input to output is nonlinear, distributed learning may "jam" on local minimums due to the use of the gradient descent technique.
- Local learning is often quicker than distributed learning.
- The regionalization of input on output is more sharply defined using vector quantization than when using gradient descent.

- When interrogating an ANN trained with vector quantization, the ANN responses cannot be different from those given during learning; in the case of an ANN trained with gradient descent, the responses may be different from those obtained during the learning phase.
- This feature is so important that families of ANNs treating the signal in two steps have been designed: first with the quantization method and then with the gradient method.
- Local learning helps the researcher to understand how the ANN has interpreted and solved the problem; distributed learning makes this task more complicated (though not impossible).
- Local learning is a competitive type; distributed learning presents aspects of both competitive and cooperative behavior between the nodes.

3.4 Artificial Neural Network Typology

ANNs may, in general, be used to resolve basically three types of complex problems, and, consequently, they can be classified into three subfamilies.

3.4.1 Supervised ANNs

The first type of problem with which an ANN can deal is expressed as follows: *given N variables, about which it is easy to gather data, and M variables, which differ from the first and about which it is difficult and costly to gather data, assess whether it is "possible to predict" the values of the M variables on the basis of the N variables.*

This family of ANNs is named *supervised* ANNs (*SV*) and their prototypical equation is:

$$y = f(x, w^*) \tag{3.1}$$

where y is the vector of the M variables to predict and/or to recognize(target), x is the vector of N variables working as networks inputs, w is the set of parameters to approximate, and $f()$ is a nonlinear and composed function to model.

When the M variables occur in time subsequent to the N variables, the problem is described as a *prediction* problem; when the M variables depend on some sort of typology, the problem is described as one of *recognition* and/or *classification* (this is also sometimes referred to as the *proscription problem*).

Conceptually, it is the same kind of problem: *using values for some "known variables" to predict the values of other "unknown variables".*

To correctly apply an ANN to this type of problem, we need to run a *validation protocol.*

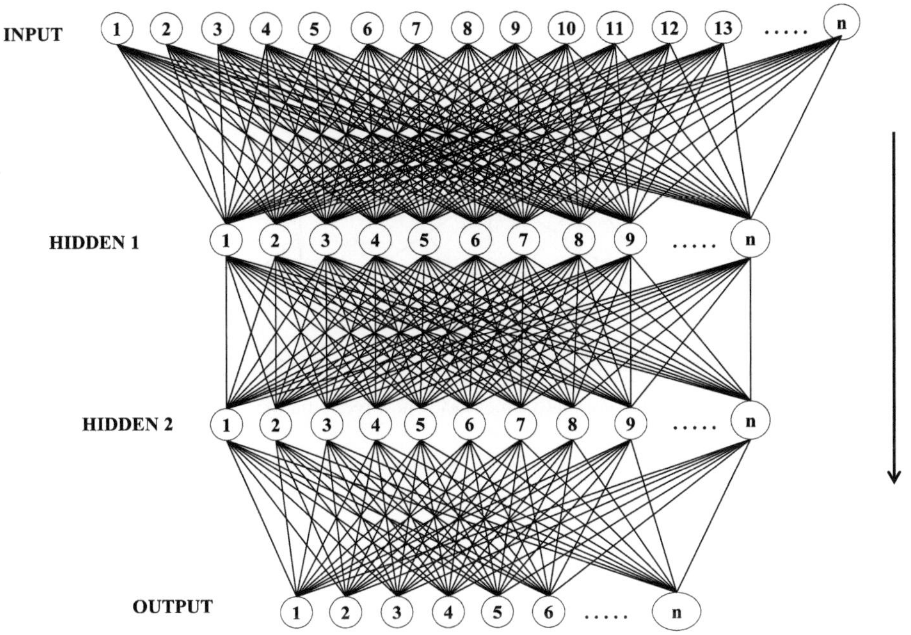

Fig. 3.6 Example of supervised ANN

We must start with a good sample of cases, in each of which the *N* variables (known) and the *M* variables (to be discovered) are both *known* and *reliable*.

The sample of complete data is needed in order to:

- *Train* the ANN
- *Assess* its predictive performance

The validation protocol uses *part* of the sample to *train* the ANN (training set), while the *remaining cases* are used to assess the *predictive capability* of the ANN (testing set or validation set).

In this way, we are able to test the reliability of the ANN in tackling the problem *before* putting it into operation (Fig. 3.6).

3.4.2 Dynamic Associative Memories

The second type of problem that an ANN can be expressed as follows: *given N variables defining a dataset, find out its optimal connections matrix able to define each variable in terms of the others and consequently to approximate the hypersurface on which each data-point is located.*

This second subfamily of ANNs is named *dynamic associative memories (dam).*

The specificity of these ANNs is incomplete pattern reconstruction, dynamic scenario simulation, and possible situations prototyping.

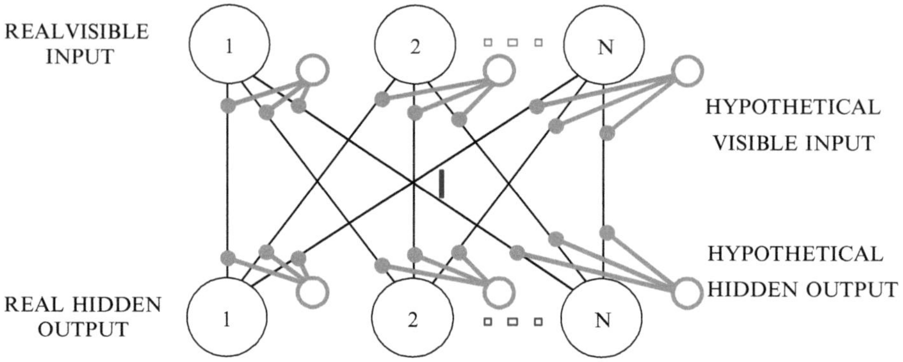

Fig. 3.7 Example of dynamic associative memory – new recirculation ANN

Their representative equation is:

$$x^{[n+1]} = f(x^{[n]}, w^*) \tag{3.2}$$

where $x^{[n]}$ is the N variables evolving in the ANNs internal time, w^* is the connection matrix approximating the parameters of the hypersurface representing the dataset, and $f()$ is some suitable nonlinear and eventually composed function governing the process.

DAM ANNs after the training phase need to be submitted to a validation protocol named "Data Reconstruction Blind Test". In this test, the capability of a DAM ANN to rebuild complete data from uncompleted ones is evaluated from a quantitative point of view (Fig. 3.7).

3.4.3 Autopoietic ANNs

The third type of ANNs can be described as follows: *given N variables defining M records in a dataset, evaluate how these variables are distributed and how these records are naturally clustered in a small projection space K (K < <N) according to their most important relationships.*

These ANNs are named unsupervised or also autopoietic ANNs (US). Their specificity is the nonlinear extraction of the similarities among records in a database, using all the variables at the same time. One important feature of these ANNs is also the possibility that some of them have to visualize in a two- or three-dimensional map the geographical similarities among records and among variables.

The prototypical equation of the US ANNs is

$$y^{[n+1]} = f(y^{[n]}, x, w^*) \tag{3.3}$$

where y is the projection result along the time, x is the input vector (independent variables), and w is the set of parameters (codebooks) to be approximated.

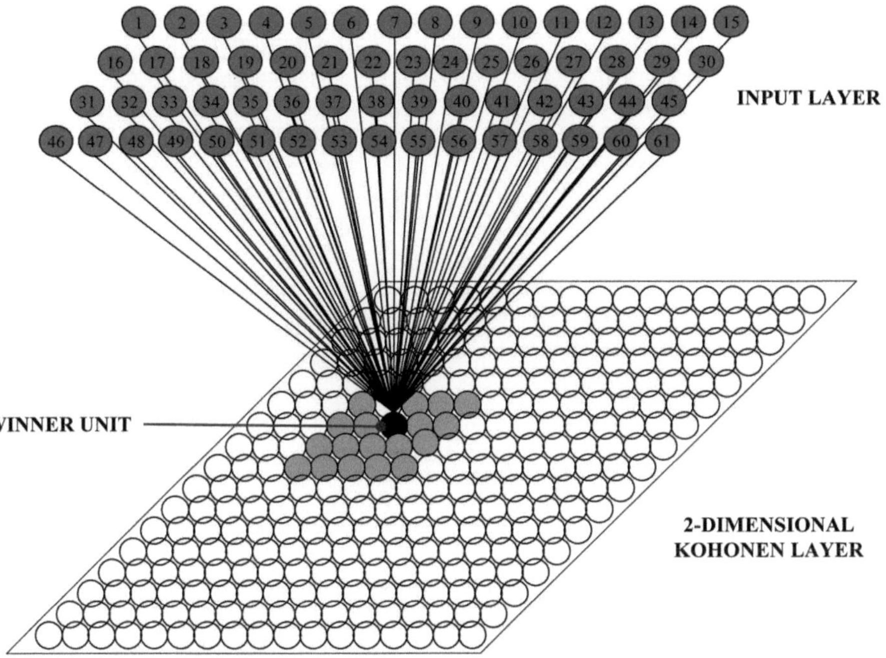

Fig. 3.8 Example of unsupervised ANN for natural clustering – self-organizing map

In US ANNs, the codebooks (w) after the training phase represent an interesting case of cognitive abstraction: in each codebook, the ANN tends to develop its abstract cognitive representation of some of the data which it learned (Fig. 3.8).

3.5 Summary Table

Table 1	*Type*	*Dynamic*	*Properties*
Nodes	Input Output Hidden	Type of equation $I \rightarrow 0$	No Layer (each node is distinct from every other) MultiLayer (each node is the same as those of its own layer) MonoLayer (each node is the same as the others)
Connections	Symmetrical AntiSymmetrical MonoDirectional BiDirectional Reflexive	Adaptive Fixed Variables	Maximum connections Dedicated connections

(continued)

(continued)

Table 2	*Flow strategy*	*Learning strategy*
Type of ANN	Feed forward with parametric or adaptive Feedback	Approximation of the function:
	IntraNode	Gradient descent
	IntraLayer	Vector quantization
	Among layers	Learning conditions of the function:
	Among ANNs	Supervised
		Dynamic associative memories
		Unsupervised or autopoietic

Tables 1 and 2 Semantics and syntax of ANNs

Bibliography

Anderson, J. A., & Rosenfeld, E. (Eds.). (1988). *Neurocomputing foundations of research.* Cambridge, MA: MIT Press.

Arbib, M. A. (Ed.). (1995). *The handbook of brain theory and neural networks, A Bradford book.* Cambridge, MA/London: MIT Press.

Buscema, M. (1998). *Artificial neural networks and complex social systems* (Special issue: I. Theory: *33*(1), II. Models: *33*(2), III. Applications: *33*(3), *Substance Use & Misuse*), New York: Marcel Dekker.

Buscema, M., & Semeion Group. (1999). *Reti Neurali Artificiali e Sistemi Sociali Complessi* (vol. I: Teoria e Modelli, vol. II: Applicazioni). Milan: Franco Angeli.

Carpenter, G. A., & Grossberg, S. (1991). *Pattern recognition by self-organizing neural network.* Cambridge, MA: MIT Press.

Grossberg, S. (1978). How does the brain build a cognitive code? *Psychological Review, 87*, 1–51.

Hopfield, J. J. (1984). Neurons with graded response have collective computational properties like those of two-state neurons. *Proceedings of the National Academy of Sciences USA, Bioscience, 81*, 3088–3092.

Hopfield, J. J. (1988). Neural networks and physical systems with emergent collective computational abilities. *Proceedings of the National Academy of Sciences, 79*, 2554–2558. (Reprinted from *Neurocomputing foundations of research*, by J. A. Anderson & E. Rosenfeld, Eds., 1988, Cambridge, MA: MIT Press)

Hopfield, J. J., & Tank, D. W. (1985). Neural computation of decisions in optimization problems. *Biological Cybernetics, 52*, 141–152.

Hopfield, J. J., & Tank, D. W. (1986). Computing with neural circuits: A model. *Articles Science, 233*, 625–633.

Kohonen, T. (1988). Correlation matrix memories. *IEEE Transactions on Computers, C-21*, 353–359. (Reprinted from *Neurocomputing foundations of research*, by J. A. Anderson & E. Rosenfeld, Eds., 1988, Cambridge, MA: MIT Press.)

Minsky, M. (1954). *Neural nets and the brain-model problem.* Doctoral dissertation, Princeton University.

Minsky, M., & Papert, S. (1988). *Perceptrons*. Cambridge, MA: The MIT Press (Extended edition).

Reetz, B. (1993, May). Greedy solution to the Travelling Sales Person Problem. *ATD, 2.*

Rosenblatt, F. (1962). *Principles of neurodynamics*. New York: Spartan.

Rumelhart, D. E., & Mcclelland, J. L. (Eds.). (1986). *Parallel distributed processing* (vol. 1: Foundations, explorations in the microstructure of cognition; Vol. 2: Psychological and biological models). Cambridge, MA/London: The MIT Press.

Werbos, P. (1974). *Beyond regression: New tools for prediction and analysis in behavioral sciences*. Ph.D. thesis, Harvard University, Cambridge, MA.

Chapter 4
A Brief Introduction to Evolutionary Algorithms and the Genetic Doping Algorithm

Massimo Buscema and Massimiliano Capriotti

4.1 Introduction

Genetic algorithms (GAs) constitute a subset of evolutionary algorithms, a generic term that indicates a range of systems of problems for which resolution is based on the use of the computer and seeks solutions using methods that are similar to evolutionary processes. Apart from genetic algorithms, they include evolutionary programming, evolutionary strategies, classifying systems, genetic programming, and genetic doping algorithms.

In general, the algorithms used in the disciplines of artificial intelligence work on the research of a global minimum or maximum in a finite space on the basis of bounds on the space of the solutions. From a formal point of view, we can state that, given an element X belonging to a Cartesian space D (where n is the cardinality of D and X is a vector), and given a function $f : D \rightarrow R$ called objective function, then the identification of the global optimum is the search for $X*$ that maximizes this function, that is, $X* \in D \text{and} \forall X \in D : f(X) \leq f(X*)$. Factors such as the presence of more points of local maximum bounds on the domain D (i.e., the nonlinearity) can make the research very difficult, and the problem could be not solvable in an acceptable timeframe. Under these computing conditions, we must use algorithms of a heuristic type that, even solving the problem with high degrees of uncertainties and without assuring convergence of the search on a solution, potentially yields a solution in acceptable time. The solution may not be optimal but it usually is acceptable, and it can be argued that a suboptimal solution to a complex problem is better than no solution.

With this in mind, we now consider the distinction between "strong" and "weak" methods. The former are oriented to the solution of a specific problem, on the basis

---

M. Buscema (✉) • M. Capriotti
Semeion Research Center of Sciences of Communication, via Sersale 117, Rome, Italy
e-mail: m.buscema@semeion.it

M. Buscema and W.J. Tastle (eds.), *Intelligent Data Mining in Law Enforcement Analytics: New Neural Networks Applied to Real Problems*, DOI 10.1007/978-94-007-4914-6_4,

of the knowledge of the particular domain and of the inner representation of the system under examination. The good solutions obtained are hardly adaptable to other tasks and provide the researcher with unsatisfying results. The weak methods use less knowledge of the domain; they are not oriented to a specific target and solve a wide range of problems. The evolutionary algorithms are algorithms based on a heuristic research and therefore considered weak methods. However, the new typology of weak evolutionary methods has recently introduced methods having initially little knowledge of the domain but that during their evolution acquire a greater awareness of the problem by implementing some characteristics of the strong methods (sometimes referred to as "emerging intelligence").

4.2 Genetic Algorithms

Between the end of the 1950s and the beginning of the 1960s, the researchers in the field of evolutionary computation began taking interest in the natural systems, convinced that they could build a model for the new algorithms of optimization. It was expected that the mechanisms of evolution could be adapted in order to solve some of the most interesting computational problems, those related to the search for the solution among a large number of alternatives. For instance, to solve the problem of protein design with the help of the computer, it is necessary to build an algorithm that locates a protein with certain characteristics among a very high number of possible sequences of amino acids. Similarly, we can search a group of rules or equations that allow us to foresee the behavior of the financial markets. Algorithms of this type have to be adaptive because they "interact" with a changing environment.

From this point of view, the organisms can be considered as very good problem solvers. As they are able to survive in their own environment, they develop behaviors and skills that are the result of natural evolution. The biologic evolution is similar to a method of research inside a very large number of solutions, constituted by the set of all the genetic sequences, whose results, that is, the desired solutions, are highly adapted organisms with a strong capacity for survival and reproduction in a changeable environment that will transmit their genetic material to the future generations. Essentially, the evolution of a species is thus ruled by two fundamental processes: natural selection and sexual reproduction. The latter determines the recombination of the genetic material of the parents generating an evolution much more rapidly than the one that might be obtained if all the descendants contained simply a copy of genes of one parent, modified randomly by a mutation. It is a process with a high degree of parallelism: it does not work on a species at the time, but it tries and changes millions of species in parallel.

In short, a genetic algorithm (GA) is an iterative algorithm that operates on a population of individuals encoding the possible solutions of a given problem. The individuals are evaluated through a function that measures the capacity to solve the problem and identifies the most suitable individuals that can be used for reproduction. The new population evolves on the basis of random operators motivated by sexual reproduction and mutation. The complete cycle is repeated until, after reaching a given criteria, evolution ceases. The use of these algorithms is essentially linked to the programming of the artificial intelligence in robotics, to biocomputation, to the study of the evolution of parallel cell systems, and to particular problems of management and systems of optimization in engineering.

The GAs thus possess these strong points:

- A distinct possibility of solving complex problems without possessing advanced knowledge of the precise method of solution
- A capacity of automodification on the basis of the mutation of the problem
- A capacity of simulating some phenomena given a structure and operative modalities similar to those of the biological evolution

The first attempts of designing instruments of optimization, the evolutionary strategies of Rechemberg (Rechemberg 1973; Voight et al. 1996) and the evolutionary planning of Fogel et al. (1966), did not produce interesting results such as the tests of biology of the early 1960s because they only highlighted the operator on the mutation rather than the reproductive process necessary for the generation of new genes. In the mid-1960s, John Holland's proposal marked meaningful progress with genetic algorithms that underlined, for the first time, the importance of sexual reproduction.

In some applications, the GAs found acceptable solutions in reasonable time. In others, they might take days, months, or even years to find an acceptable solution. But since they worked with populations of independent solutions, it was possible to distribute the computational load on more computers, producing simultaneously different hypothesis with the consequent reduction of the calculation time.

4.3 Natural Evolution and Artificial Evolution

4.3.1 Natural Evolution

The modalities of action of the Darwinian principle of the natural selection can be summarized as follows:

> Natural evolution acts on genetic material (the genotype) of an individual and not on his physical characteristics, the phenotype. Each variation that promotes the adaptation of an individual emerges from the genetic property, not from what the parents have learned in their life.

The natural selection favors the reproduction of the individuals that improve the adaptability to the changing environment and eliminates the individuals with less reproductive potentials. From the genetic point of view, the natural selection promotes those particular genetic combinations that give birth to a more efficient organism, selecting the genotype, not the phenotype.

The reproduction is the central core of the evolutionary process: the generational variability of a species is determined by the genetic recombination and by the little random mutations of the genetic code. The differences between individuals and parents are established. The variability is an essential condition of evolution. Natural evolution operates over entire populations through cyclic and generational processes and is determined exclusively by the environment and by the interactions among various organisms.

The terminology used draws inspiration directly from the studies on the natural biological evolution.

The combination of the Darwinian hypothesis with genetics gave birth to principles that constitute the basis of population genetics, that is, the explanation of the evolution of populations at a genetic level.

A population is defined as a group of individuals of the same species operating and propagating in the same geographical region. In biology, the chromosomes are the filaments of DNA acting on the organism. Each chromosome is composed of genes, each gene having been encoded with a particular protein that determines a specific characteristic of the organism, such as the color of the eyes or the length of a foot. The position of the genes inside the chromosome is called a locus, and the different configurations of the proteins are defined as *alleles*. Most of the organisms have more than a single chromosome; the set of chromosomes is called the *genome*. The term *genotype* is used to mean the set of the genes contained in the genome. The final result of evolution in the individual is defined as the *phenotype*.

The sexual reproduction consists in the recombination (crossover) of the genetic material of the parents that produces a new complete genetic material for the progeny mutations on single parts of DNA may occur. The *fitness* is the suitability of the individual, the probability that he lives enough to reproduce. The natural selection promotes the individuals having the most suitable phenotypes as parents – encoded from particular genotypes – for the next generation. It can be directional, if it helps the increase of frequency of an extreme form of the character; stabilizing, if it helps the individuals carrying an intermediate form of a certain character; and diverging, if the extreme forms of a character are favored at the expense of the intermediate ones.

The evolution is based on the following mechanisms:

- Mutation of allele: primary source of genetic variability.
- Genetic flux: variation of the frequencies of the alleles due to the migratory movements of some individuals, with a consequent introduction or removal of certain genotypes.
- Genetic drift: unpredictable variations in the frequency of the alleles if a population has a small number of components. Actually, from a probabilistic point of view, it is common that less probable events can occur in a small population resulting in bigger effects.

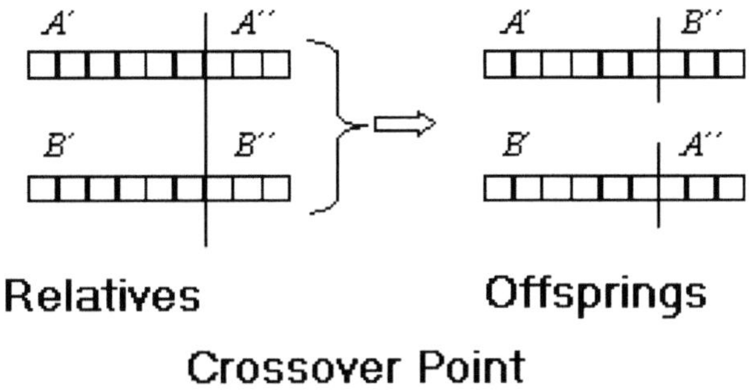

Fig. 4.1 Illustration of the identification of a crossover point and the resulting recombination of chromosomes

4.3.2 *Artificial Evolution*

In the terminology of the genetic algorithms, the chromosome encodes a candidate solution for a given research problem. In Holland's (1992) model with a binary encoding, the chromosome identifies a string of bits, the genes are the bits of the string, and the alleles, as property of the genes, can be 1 or 0. The *crossover* is the recombination of the genetic material of two parents composed by a single chromosome and the mutation in the random variation of the value of the alleles in each *locus* of the chromosome. The phenotype is the meaning of the chromosome that is the decoding of the candidate solution of the problem. In the common applications, both the individuals at a single chromosome are used; thus, in terms of genotype, chromosome and individual are equivalent. The GA randomly calculates a point in a selected chromosome and takes a portion of it up to that calculated point and appends it to another section of a different chromosome beginning at the previously calculated random point. When the coding of the chromosome represents directly a candidate solution, as in some applications where the chromosome is a string of real numbers rather than the bits, the terms genotype and phenotype can coincide, too.

4.3.3 *The Holland Model*

Stating that the genetic algorithms are adaptively complex procedures created for a specific purpose in resolving some research activity requiring optimization is equivalent to saying that they are procedures searching for the maximum point of a certain function where the field is excessively complex to be rapidly maximized with analytical techniques and when a procedure randomly exploring the space of the solutions is inconceivable. The GA selects the best solutions and recombines them with different modalities so that they evolve toward a maximum point (see Fig. 4.1). The function to be maximized is called the *fitness*. The term has many variations: it may mean "adaptation," "adaptability," "biologic success," "competition," etc.

The original model by Holland is based on a population of n strings of bits with a fixed length l (n, $l \in N$), generated randomly. The set of the binary strings with a length l has 2^l elements and represents the space of the solutions of the problems. Each string (genotype) is the binary coding of a candidate solution (phenotype). In general, the function of *fitness* is given in the following form:

$$F = f(x_1, x_2, ..., x_n).$$

Through this function, at each genotype, g_i of the initial population $P(t=0)$ is associated with a value $F_i = F(g_i)$ which represents the capacity of the individual to solve the given problem. In order to determine the value of adaptability, the function of *fitness* receives a genotype in input; it decodes it into the correspondent phenotype and tests it on the given problem. Once the phase of evaluation of the individuals belonging to the initial population is concluded, a new population $P(t+1)$ of new n candidate solutions, obtained applying the operators of selection, crossover, mutation, and inversion, is generated.

4.3.3.1 Selection

Within a population, a probability of selection linked to the fitness is associated to each individual. The selection of an operator generates a random number $c \in [1,0]$ to determine which individual will be chosen. The chosen individual is copied in the so-called *mating pool*. The mating pool is filled with n copies of the selected individuals, at the time $P(t=0)$. The new population $P(t+1)$ is obtained through the operator of crossover, mutation, and inversion. The operator of selection, choosing the individuals that have the possibility of generating descendants with a higher degree of fitness, plays the role of natural selection for the living organisms in the context of the genetic algorithm.

4.3.3.2 Crossover

Within the mating pool, two individuals, called parents, and a point of cut called point of crossover, are randomly chosen. The portions of genotype on the right of the crossover point are exchanged generating two descendants, as in the picture:

The so-called operator of *crossover point* is applied $n/2$ times to obtain n descendants on the basis of a given probability p. If the crossover is not applied, the descendants coincide with the parents.

Another technique used is the *two points' crossover*: the individuals are not represented by linear strings, but by circles. A portion of circle of an individual is substituted with the one of another, selecting two points of crossover.

The *uniform crossover* states that for each couple of parents, a binary string of the same length, called a *mask*, is generated. The descendant is generated by copying the bit of the father or of the mother if the corresponding position of the mask is respectively 0 or 1.

The crossover is a metaphor of the sexual reproduction where the genetic material of the descendants is a combination of one of the parents.

4.3.3.3 Mutation

This operator is created to satisfy the rare case of variation in the elements of the living creatures' genome during the evolution. On the basis of a little probability p, the value of the bits of each individual is changed (from 0 to 1 and vice versa).

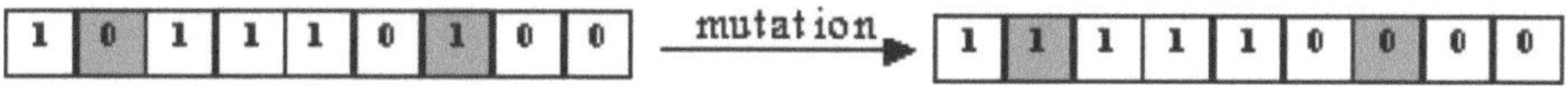

As it happens in nature, the mutation adds a "noise" or randomness to the whole procedure in order to assure that starting from a population generated randomly there are no points within the solution space that is not explored.

4.3.3.4 Inversion

On the basis of a fixed probability p, two points in the string are chosen. The string encodes the individual and inverts the bits between two positions.

In a large initial population, it is difficult to estimate which probability values of the crossover and mutation will give the best performances. Experience shows that there is a strong dependence on the type of problem. Generally, the probability of crossover is between 60 and 80 %, while in the case of mutation it fluctuates between 0.1 and 1.0 %. If the probabilities for selection of an individual are proportional to its fitness (if f is the value of fitness of a solution and F is the sum of the values of fitness of all the population, the probability might be f/F), it is probable that, after the crossover, the best individuals are recombined with the consequent loss of the best chromosome. In order to avoid this and to speed up the convergence times, the best individual of a generation may be cloned. Through this technique, called *elitism*, which retains a high number of populations, it is possible to clone more individuals in the next generation, continuing with the others in the classical way.

4.3.4 Types of Encoding of Artificial Genome

In general, it is possible to encode the solutions of a problem with binary strings. In other cases, representations of a higher level are used, and operators of crossover and mutation are defined to be able to operate on such representations. Besides the binary encoding, it is possible to use another type that is based on real numbers.

The binary encoding is important not only from the historical point of view but also because the more relevant theoretical results have been obtained with models based on it. The structure of the data is a vector of bits with a length equal to l, which possesses a space of 2^l possible solutions. We need to define a function that encodes the genotype, or parts of it, with one or more of the real values. The most used operator of crossover is, in this case, the crossover n points (n points of cut).

The encoding based on floating point numbers is the most natural for problems of optimization of real parameters. The structure of the data is a vector of length l in which each element is a real number. Each candidate solution is a point in the research space, and it is not necessary to foresee decoding functions of the genotype. The operator in crossover can be the classic one point of cut, but for the mutation operator, the elements of another vector are added in order to alter the genes of the individuals.

4.4 Other Evolutionary Algorithms

4.4.1 Evolutionary Programming

Evolutionary programming (EP) is a stochastic strategy of optimization similar to the GAs, based on the definition of population, fitness, and the selection of the "best." But while these try mainly to simulate the operators of crossover and mutation as it happens in nature, evolutionary programming concentrates on the connection existing between parents and offspring, a modification of the parent. The basic method consists of three steps: to randomly choose an initial population (the greater the number of individuals, the faster the convergence to a solution); each individual in the previous population is copied to a new population and also permitted to undergo some probability generated mutation, at which time it is referred to as a son; the degree of fitness of each individual is calculated, and through a tournament with stochastic selection, a set of N possible solutions are chosen to be the next population.

In particular, each individual in the population is an FSM (finite state machine) formed by a series of inner states belonging to a finite alphabet. The FSM receives as input a series of symbols and returns as output a series of states on the basis of the current states and input. The objective is the forecast of the next configuration of the system, not through the operator of crossover (as in the GAs) but entrusting exclusively to the mutation: it alters the initial state, modifies the transition, or changes an inner state. The basic characteristic of this type of algorithm is that the offspring have behavior similar to that of their parents.

4.4.2 Evolutionary Strategies

Evolutionary strategies are techniques similar to the previous one but developed originally for problems concerning civil and structural engineering. The principal difference consists of the selection of individuals in evolutionary programming: they are selected for mutation with a certain probability that is proportional to the fitness, as in the case of the GAs, while in this case, the worst individuals are deterministically rejected. The method of optimization is based on the choice of a strategy that is then applied to a population. The two main strategies are known as plus strategy $(m+l)$ and comma strategy (m,l). In the first case, the parents can participate in the selection of the next generation, while in the second case, it is only possible to select offspring when the parents die. m represents the number of individuals in the population, while l is the number of offspring conceived for each generation. An individual in the population consists of a genotype that represents a point in the solution space (i.e., the space of all possible solutions). To each point, it is possible to associate:

- Object variables, x_i, on which the operators of crossover and mutation will be applied until an optimal solution of the problem is reached.
- Strategy variables, S_i, that determine the "mutability" of x_i. They represent the standard deviation of a Gaussian distribution $(0,S_i)$. With an expectation value equal to zero, the parents will produce, on average, offspring that are similar to them. This strategy works because sooner or later the individual, possessing a good value as determined by the objective function, will be favored, and the recombination that occurs among them should yield better offspring. The value of the objective function $f(x)$ represents the phenotype (fitness) that we will take into consideration in the selection. In the Plus strategy, the better m individuals over $(m+1)$ will survive and will become parents in the next generation, while in the *Comma* strategy, the selection happens only among the offspring.

4.4.3 Classifying Systems

Classifying systems are operators working in an environment from which they receive input that classify according to rules which permit them to generate instructional output. The instructions belong to the type *if…then*. For example, the problem might be the optimization of a productive process done by machinery controlled by a computer; this machinery receives a series of inputs from its sensors, like the temperature of the machinery, the external pressure, or the type of material, and it acts according to a set of starting rules. Such instructions are not fixed, however, but if they are encoded in a binary form, they evolve like populations of GAs and their fitness is the performance of the machinery.

4.4.4 Genetic Programming

The technique of genetic programming is similar to that of genetic algorithms, but in this case, the population is not constituted by bit strings but rather by programs that evolve, combine, reproduce, or change when they are executed in order to create other programs that constitute better solutions of a specific problem. These programs are encoded with a tree structure where the inner nodes are functions and the leaves are the terminal symbols of the program. For example, the expression *IF (TIME>10) THEN return 1+2+3 ELSE return 1+2+4 can be rewritten as+(1 2 (IF>TIME 10)3 4)* (language LISP) and transferred in a tree as in the figure:

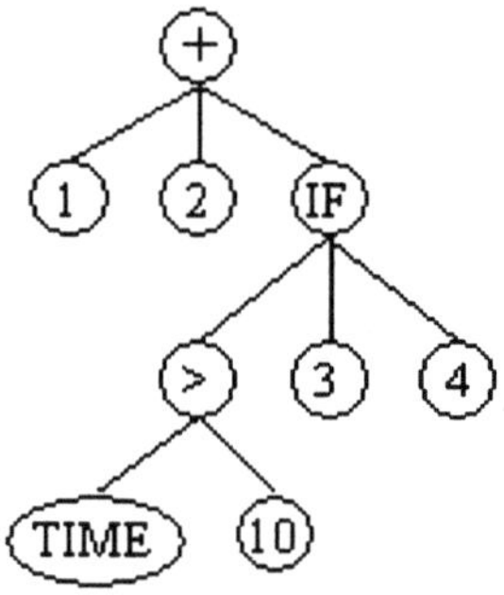

The space of research is constituted by all the programs composed by the terminals and by the functions defined for a specific problem.

Genetic programming has a degree of complexity greater than that of genetic algorithms because programming requires the selection of many more parameters, such as the generation of the initial population, the set of the basic functions and terminals, the type of selection, the dimension of the population and the maximum number of generations, and the criteria of termination.

The crossover operator is the dragging force of the algorithm; we randomly take two subtrees from selected individuals on the basis of their fitness, and we recombine them giving birth to two offspring trees, with parameters establishing limits on the maximum dimension of the population's tree. Other operators, such as the mutation, permutation, editing, encapsulating, and decimation, are used in particular cases. In general, the operators determine the syntactic accuracy of the generated trees, but not the semantic correctness.

To compare the genetic algorithm with genetic programming, the genetic algorithms optimize a solution defined and parameterized by the user, and thus the optimization works on the representation of the parameters of a function whose structure is known, but genetic programming works at a higher level, as the user defines the elements of a grammar (operators and terminal symbols) used to generate functions that are required to evolve. The optimization of the fitness consists not only in the manipulation of the code of the parameters but, above all, in the manipulation of the functions.

The extension of the search space (the space of all the functions that satisfy the grammar defined by the user) and the reduction in the efficiency of the mapping in the memory of the representation used permit the genetic programming to require a higher computational load and a high occupation of memory. It results in a lower usage of genetic algorithms and the necessity to realize a parallel implementation of these algorithms on environments of a distributed type.

4.4.5 Comparison with Other Techniques

As in the GAs, there are other techniques of general approach that operate from an initial start by a fitness function that has to be maximized. Some of them are applicable only to limited domains, such as the dynamic programming, where the fitness function is the sum of the fitness functions calculated in each phase of the problem and there is no interaction among the various phases.

In the method of gradient, the information on the gradient of the function is used to lead the direction of the research. However, the function must be continuous; otherwise, the derivative cannot be calculated. In general, these methods are called *hill climbing,* and in case of functions with only one peak, or less for multimodal functions, it is not sure that the first hill climbed is the one with the highest peak. In the following figure, there is an example of the problem; starting from a random point *X*, with *uphill* movements, the maximum B point is a local maximum, but A and C are not identified.

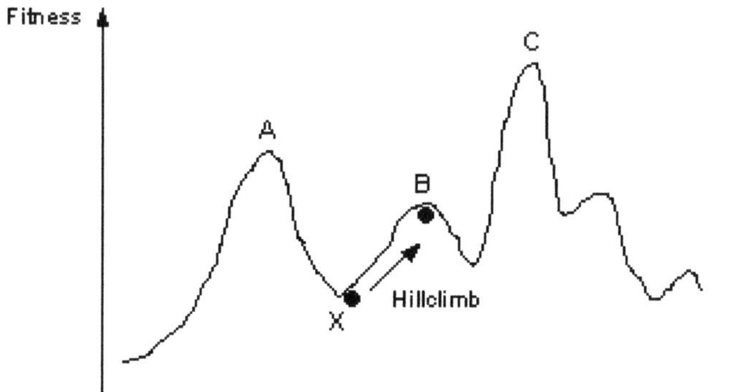

In the iterative research, the method of the gradient is combined with random research, where the points of the research space are randomly chosen. Once the peak is found, the climb starts again from another point randomly chosen. The technique has the advantage of being simple and gives good results with functions that obviously do not have many local maxima. However, since each test is isolated, a total traverse of the form of the domain is not obtained, and while the random search continues, the same tests led both in regions where high values of fitness are not found and in regions with a low value of fitness continue to be allocated. A genetic algorithm, instead, operates starting from an initial random population and carries out attempts in regions with a higher fitness. It might be a disadvantage if the maximum is found in a little region surrounded by regions with a low fitness. But this type of function is hardly optimized with any method.

The technique of the simulated annealing is a modified version of hill climbing. From a random point, a random movement is initiated; if this yields a higher point, it is accepted with a probability $p(t)$, where t is the time. At the beginning, the value of $p(t)$ is close to 1 but it gradually tends to zero. If at the beginning each movement is accepted, the probability of accepting a negative movement decreases with the passage of time. Sometimes, the negative movements are needed to avoid local maxima, but if there are too many, they may deviate from the maximum. In general, this technique works only with one candidate solution at a time; it does not build a total figure of the research space and the information of the previous movements, used to lead us toward the solution, and is not saved.

4.5 The Genetic Doping Evolutionary Algorithm

In the genetic algorithms (GAs), the principle of reproduction assumes an evolutionary criterion external to the system, that each generation identifies the best and the worst individuals.

Such criteria are reminiscent of the technique of exams in school, rather than an evolutionary principle intrinsic to the natural evolution. Similarly, the percentages of crossover and mutations foreseen in every generation cannot be two fixed parameters, established externally by the experimentalist; it is preferable to figure them out as adaptive parameters linked to the health of the whole "population" in each generation.

Besides, in the traditional GA, the crossovers have fixed rules, typical of the overpopulated and advanced societies (monogamies, prohibition of incest, etc.). It would be proper that a population in a phase of "cultural starting" had more flexible sexual rules and proportioned to its rapid expansion and genetic differentiation.

Just like the traditional genetic algorithms, GenD is able to find solutions to problems of optimization, conceiving the possible solutions as individuals of restricted populations, letting the population evolve generation after generation on the basis of specific genetic operators and a selection based on fitness, the more suitable individuals, and to the consequent disappearance of the worst ones.

After a certain number of generations, the resulting population includes a good approximation of the solution to the problem of optimization. Unlike the traditional genetic algorithms, the evolution of GenD is due to its inner instability which generates a continuous evolution and a natural increase of the biodiversity, thanks to specific characteristics such as the conceptualization of the population as a structured organization of individuals (tribes).

4.6 Theory

In order to increase the speed and the quality of the solutions that have to be optimized, the algorithm GenD makes the evolutionary process of the artificial populations more natural and less centered on the culture of the individual liberalism.

In the first phase, the algorithm calculates the fitness score of each individual on the basis of the function we want to optimize. The value of the average health of the whole population is not a simple index. In GenD, the average health constitutes firstly the criteria of *vulnerability* and lastly those of *connectivity* of all the individuals of the population. It results in the basic unit of the algorithm not being the individual, unlike the classical GAs and many other evolutionary algorithms.

The reference unit is the species, acting on the evolution of the individuals in terms of the average health for each generation.

The feedback loop among the individuals and the average health of the population (species) allows the GenD to transform, in evolutionary terms, the whole population from a sparse set of individuals (like the traditional models) into a dynamic structure of subjects.

4.7 The Criteria of Vulnerability

All the individuals whose health is less than or equal to the average health of the population are registered onto a list of vulnerability. They are not eliminated and may continue participating in the process but are flagged.

The number of vulnerable individuals automatically establishes the maximum number of crossovers allowed in that generation. Thus, the number of possible crossovers in each generation is variable and is a function of the average health of the population. The whole population has this possibility. The maximum number of random calls to the coupling is calculated so that all the individuals flagged as vulnerable might, theoretically, be substituted by new individuals. When the type of crossover operator used generates only a new individual, this number corresponds to the number of vulnerable individuals; otherwise, it is equal to its half when the type of the crossover operator used generates two individuals.

4.8 Criteria of Connectivity

In order to let a coupling generate offspring, at least one of the proposed individuals must have a fitness whose value is close to the average fitness of the whole population (average $\pm k$, where $2k$ defines the width of the coupling band).

Alternatively, GenD may adopt other criteria of connectivity. Each couple of individuals may generate offspring if the fitness of at least one of the two is greater than the average fitness. The individual that satisfies the condition is named "candidate qualified for the crossover."

The effects of these two criteria are similar; the normal distribution of the fitness of each individual with respect of the average fitness operates a band of dynamic crossover in each generation.

The algorithm *GenD*, thus, assumes that the individuals "sui generis," that is to say, too weak or too healthy, tend not to marry among each other. In practice, the crossovers are not suitable for the best or the worst. It is the "most normal" subjects that tend to crossover. Furthermore, there are no restrictions on crossovers. Each individual may marry whomever he or she chooses.

The offspring of each crossover occupies the places of the subjects marked in the list of vulnerability. It may happen, therefore, that a weak individual may continue existing through his own offspring.

4.9 The Criteria of the Last Chance

The number of possible crossovers is a function of the number of subjects marked as vulnerable; the latter is a function of the average health of the population. The criteria of coupling, however, urge the system to retain a number of crossovers not necessarily giving birth to offspring. The difference between possible crossovers and realized crossovers defines the number of mutations. That is, the subjects present in the list of vulnerability that were not substituted by the offspring generated by the realized crossovers are changed. A last chance to reenter the evolutionary game is granted to this variable number of weak subjects through a mutation.

4.10 The New Crossover

The criteria of coupling allows crossovers only if at least one of the individuals of the couple is a qualified candidate for the crossover, that is, if the parent enjoys of level of health close to the average health of the population (first type) or greater (second type). If we indicate with F_i the health of the ith individual, $\overline{F}$ the average health of the population, and σ^2 the variance of the health of the population, then F_i is a candidate qualified for marriage of the first type if:

$$\left(\overline{F} - k\right) \leq F_i \leq \left(\overline{F} + k\right).$$

$$k = 1 - \sigma^2$$

He is a candidate of the second type if $F_i \geqslant \overline{F}$.

However, in *GenD*, the coupling does not consist in the simple interchange of genes between husband and wife around a crossover point; that is,

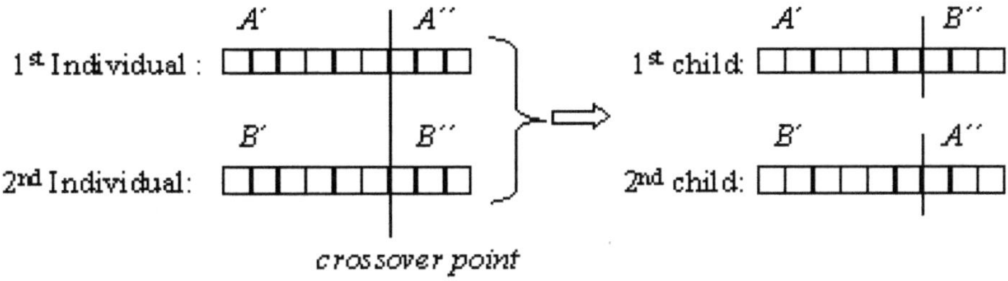

The coupling of genes between parents is carried out in a selective way, in two manners:

1. A logic crossover: when the repetitions are allowed
2. An opportunistic crossover: when the repetitions are not allowed

4.10.1 The Logic Crossover

The logic crossover comprises four cases:

1. The health of the father and mother is greater than the average health of the whole population:

$$\left(F_{\mathrm{F}} > \overline{F}\right) \text{ and } \left(F_{\mathrm{M}} > \overline{F}\right).$$

2. The health of both parents is lower than the average health of the whole population:

$$\left(F_{\mathrm{F}} < \overline{F}\right) \text{ and } \left(F_{\mathrm{M}} < \overline{F}\right).$$

3. and 4. The health of one of the parents is lower than the average, and the health of the other is greater than the average health of the whole population:

$$\left(F_{\mathrm{F}} > \overline{F}\right) \text{ and } \left(F_{\mathrm{M}} < \overline{F}\right) \quad \text{or} \quad \left(F_{\mathrm{F}} < \overline{F}\right) \text{ and } \left(F_{\mathrm{M}} > \overline{F}\right).$$

In the first case, the generation of two offspring (suppose for simplicity the case with two offspring and only one crossover point) happens in the traditional way:

$$\left(F_{\mathrm{F}} > \overline{F}\right) \text{and} \left(F_{\mathrm{M}} > \overline{F}\right): \quad \underbrace{\begin{matrix} A' + A'' \\ B' + B'' \end{matrix}}_{\text{parents}} \Rightarrow \underbrace{\begin{matrix} A' + B'' \\ B' + A'' \end{matrix}}_{\text{children}}.$$

In the *second case*, the generation of two offspring happens through the *negation* ($\sim$) of the parents' genes:

$$\left(F_{\mathrm{F}} < \overline{F}\right) \text{ and } \left(F_{\mathrm{M}} < \overline{F}\right): \quad \underbrace{\begin{matrix} A' + A'' \\ B' + B'' \end{matrix}}_{\text{parents}} \Rightarrow \underbrace{\begin{matrix} \sim A' + \sim B'' \\ \sim B' + \sim A'' \end{matrix}}_{\text{children}}.$$

In the third and fourth case, only the parent whose health is greater than the average health transmits his own genes, while the genes of the other are denied:

$$\left(F_{\mathrm{F}} > \overline{F}\right) \text{ and } \left(F_{\mathrm{M}} < \overline{F}\right): \quad \underbrace{\begin{matrix} A' + A'' \\ B' + B'' \end{matrix}}_{\text{parents}} \Rightarrow \underbrace{\begin{matrix} A' + \sim B'' \\ \sim B' + A'' \end{matrix}}_{\text{children}}$$

or

$$\left(F_{\mathrm{F}} < \overline{F}\right) \text{ and } \left(F_{\mathrm{M}} > \overline{F}\right): \quad \underbrace{\begin{matrix} A' + A'' \\ B' + B'' \end{matrix}}_{\text{parents}} \Rightarrow \underbrace{\begin{matrix} \sim A' + B'' \\ B' + \sim A'' \end{matrix}}_{\text{children}}$$

The concept of genetic negation in *GenD* does not correspond to the cancellation of the weaker parent's genes and therefore to their random replacement. Rather, it deals with a genetic substitution carried out using the criteria of a "moving window" running from right or left.

If, for example, a certain gene shows only two alternatives $g[0,1]$, then the *moving window* has the following form:

↓	1	0
T	0	1

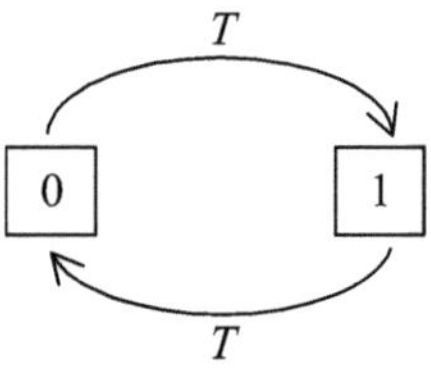

In the same way, we will proceed with a gene that presents four possible states; g[A,B,C,D]:

↓	A	B	C	D
T	B	C	D	A

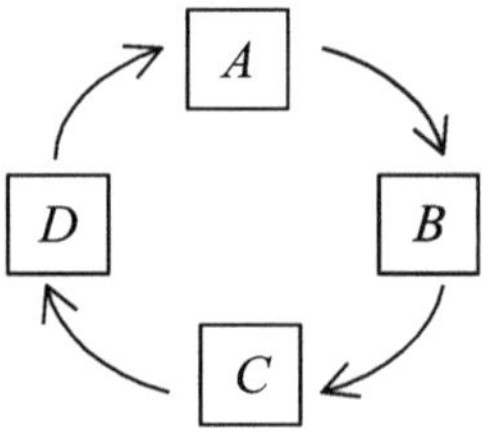

The negation criterion through the moving window is also applicable when the various possible states of a gene are not ordered among them. This method, in fact, is based on the systematic exploration of the phase space and on keeping the same systematic criteria.

4.10.2 The Opportunistic Crossover

This operator acts when the repetitions are not allowed. The parents are overlapped with respect to a random crossover point. After selecting the most effective gene of the parent, the offspring are generated *s*. This operation is repeated until all the genes of the offspring are completed.

This rule may be considered as a variant of the Greedy algorithm. In short, the number of marriages and mutations in GenD are not externally fixed parameters, but adaptive variables defining themselves internally, starting from the global trends of the population system.

Differences between traditional GA and GenD

Traditional GAs	GenD
Assesses individual health	Assesses individual health
Creates a wheel of probabilities	Calculates the average
	Calculates the variance
	Creates a band around the average
Creates a new population based on the wheel of probabilities criterion	Creates a vulnerability list based on the average criterion
Undertakes a fixed % of marriages according to the wheel of probabilities – (the best marry)	Undertakes a variable % of marriages according to the vulnerability list (number) and the average (quality) – ("normal" individuals marry)
Each marriage has *N* crossover and hybridizes parents' genes (produces possible improvements)	Each marriage has *N* crossover and consists of a search for possible states between parents and the population (produces possible improvements and increases biodiversity)
The percentage of mutations is fixed and serves to produce biodiversity	The percentage of mutations is variable, depending on marriages not undertaken, and serves to offer a final opportunity only to some of the most vulnerable individuals
Biodiversity is provided on the basis of errors (mutations) and decreases as the average rises	Biodiversity is generated by marriages and rises as the average goes up
The system tends toward stability	The system becomes more unstable as it approaches stability
There is no evolution of evolution	There is evolution of evolution
Worst and average individuals tend not to reproduce	Worst and best individuals tend not to reproduce

References

Fogel, L. J., Owens, A. J., & Walsh, M. J. (1966). *Artificial intelligence through simulated evolution*. New York: Wiley.
Holland, J. H. (1992). *Adaptation in natural and artificial systems* (2nd ed., MIT Press ed.). Ann Arbor: University of Michigan Press.
Rechemberg, I. (1973). *Evolutionstrategie: Optimierung TechnisherSysteme nach prinzipien des Biologishen Evolution*. Stuttgart: Fromman-Hozlboog Verlag.
Voight, H. M., Ebeling, W., Rechemberg, E., & Schwefel, H. P. (Eds.). (1996). *Proceedings of the 4th international conference on parallel problem solving from nature (PPSN IV), Berlin, Germany* (Lecture notes in computer science, Vol. 1141, pp. 720–729). Berlin: Springer.

Bibliography

Amstrup, B., Tóth, G.J., Szabo, G., Rabitz, H., & Lõerincz, A. (1995). Genetic algorithm with migration on topology conserving maps for optimal control of quantum systems. *J. Phys. Chem., 99*(14), 5206–5213.
Baker, J. E. (1999). Adaptive selection methods for genetic algorithms. In J. J. Grefenstette (Ed.), *Proceedings of the first conference on genetic algorithms and their application*. Hillsdale: Erlbaum.
Buscema, M. (2004). Genetic doping algorithm (GenD): Theory and applications. *Expert Systems, 21*(2), 63–79.
Buscema, M., & Semeion Group. (1999a). *Reti Neurali Artificiali e Sistemi Sociali Complessi, Vol. I: Teoria e Modelli* [Artificial neural networks and social complex systems, Vol. I: Theory and models]. Milan: Franco Angeli.
Buscema, M., & Semeion Group. (1999b). *Reti Neurali Artificiali e Sistemi Sociali Complessi, Vol. II: Applicazioni* [Artificial neural networks and social complex systems, vol. II: Applications], Milan: Franco Angeli.
Cantu-Paz, E. (2000). *Efficient and accurate parallel genetic algorithms*. Boston: Kluwerr Academic.
Davis, L. (1993). Parametrization with genetic algorithms. *Advanced Technology for Developers, 2*, 1–5.
Dawid, H. (1999). *Adaptive learning by genetic algorithms, analytical results and applications to economic models*. Heidelberg/New York: Springer.
Eldredge, N. (1995). *The great debate at the high table of evolutionary theory*. New York: Wiley.
Friederich, C. M., & Moraga, C. (1997). Using genetic engineering to find modular structures and activation functions for architectures of artificial neural networks. In *Computation intelligence, theory and applications, Proceedings of the fifth fuzzy days (LCNS 1226)*. Dortmund: University of Dortmund.
Gen, M., & Cheng, R. (1997). *Genetic algorithms and engineering, design*. New York: Wiley.
Goldberg, D. E. (1989). *Genetic algorithms in search, optimization and machine learning*. Reading: Addison-Wesley.
Harp, S. A., Samad, T., & Guha, A. (1990). Designing application-specific neural networks using the genetic algorithm. In D. S. Touretzky (Ed.), *Connectionist models, Proceedings of the 1990 Summer School*. San Mateo: Morgan Kaufmann.
Klimasauskas, C. C. (1993a). Genetic algorithm optimizes 100-city route in 21 minutes on a PC! *Advanced Technology for Developers, 2*, 9–17.
Klimasauskas, C. C. (1993b). Simulated annealing and the traveling salesperson problem. *Advanced Technology for Developers, 2*, 6–16.

Koza, J. R. (1992). *Genetic programming: On the programming of computers by means of natural selection*. Cambridge, MA: MIT Press.
Mitchell, M. (1996). *An introduction to genetic algorithms*. Cambridge, MA: MIT Press.
Nolfi, S. (1995). Genotypes for neural networks. In M. A. Arbib (Ed.), *The handbook of brain theory and neural networks*. Cambridge, MA: MIT Press, Bradford.
Quagliarella, D., Periaux, J., Polani, C., & Winter, G. (Eds.). (1998). *Genetic algorithms and evolution strategies in engineering and computer science*. New York: Wiley.
Quinlan, J. R. (1987). Simplifying decision trees. *International Journal of Man–Machine Studies, 27*(3), 221–234.
Quinlan, P. T. (1992). *C4.5: Programs for machine learning*. San Mateo: Morgan Kaufmann.
Rao, S. S. (1996). *Engineering optimization: Theory and practice*. New York: Wiley.
Rawling, G. (1991). *Foundations of genetic algorithms*. San Mateo: Morgan Kaufmann.
Reetz, B. (1993). Greedy solutions to the travelling sales persons problem. *Advanced Technology for Developers, 2*, 8–14.
Rogers, D. (1990). Predicting weather using a genetic memory: A combination of Kanerva's sparse distributed memory with Holland's genetic algorithms. In D. S. Touretzky (Ed.), *Connectionist models, Proceedings of the 1990 Summer School*. San Mateo: Morgan Kaufmann.
Todd, P., & Miller, G. F. (1997). Biodiversity through sexual selection. In C. G. Langton & K. Shimohara (Eds.), *Artificial life V. Proceedings of the fifth international workshop on the synthesis and simulation of living systems*. Cambridge, MA: MIT Press, Bradford, pp. 289–299.
Touretzky, D. S. (Ed.). (1999). *Advances in neural information processing systems* (Vol. 2). San Mateo: Morgan Kaufmann.
Whitley, L. D. (Ed.). (1993). *Foundation of genetic algorithms 2*. San Mateo: Morgan Kaufmann.
Whitley, L. D., & Vose, M. D. (Eds.). (1995). *Foundation of genetic algorithms 3*. San Mateo: Morgan Kaufmann.

Chapter 5
Artificial Adaptive Systems in Data Visualization: Proactive Data

Massimo Buscema

5.1 Introduction

An artificial neural network (ANN) is a modeling mechanism particularly suited to *solving nonlinear problems* and to *discovering the approximate rules*, which govern the *optimal* solution to these problems (Hopfield 1982; Rumelhart and McClelland 1986a, b; McClelland and Rumelhart 1988; Anderson and Rosenfeld 1988; Werbos 1994; Arbib 1995; Buscema 1997, 1998a, 2002).

In technical terms, we can say that a system is not complex when the function representing it is linear, that is, when these two equations apply:

$$f(cx) = cf(x) \quad \text{and} \quad f(x_1 + x_2) = f(x_1) + f(x_2).$$

A nonlinear system violates one or both of these conditions.

ANNs are data processing mechanisms which *do not follow specific rules* in order to process data but which use the data they receive to *discover the rules* governing them. This makes ANNs particularly useful in solving a problem in which we possess a dataset but have no information or insight as to how *those data are related* to one another (Fig. 5.1) (Chauvin and Rumelhart 1995).

Compared with other analysis techniques, ANNs are useful when one has a problem with a mass of data but no good theory to explain the interactions and relationships among the data.

ANNs dynamically discover fuzzy rules which connect various sets of data. This means that if they receive certain data in one phase, they focus on certain rules; but if they later receive new and different data, ANNs will adjust their rules in accordance, integrating the old data with the new, and they do this without any external instruction.

M. Buscema (✉)
Semeion Research Center of Sciences of Communication, via Sersale 117, Rome, Italy
e-mail: m.buscema@semeion.it

M. Buscema and W.J. Tastle (eds.), *Intelligent Data Mining in Law Enforcement Analytics: New Neural Networks Applied to Real Problems*, DOI 10.1007/978-94-007-4914-6_5,

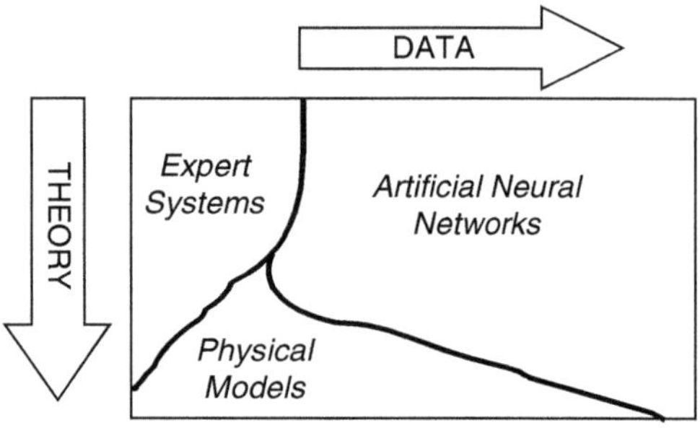

Fig. 5.1 Schematically comparing ANN with other analysis techniques

The continuous updating of data under their management creates a *dynamic bank* whose rules are automatically refined by the ANNs as the problem under investigation evolves through time.

This passage from an early categorization to a later, finer, and more complex one is managed by the ANN *alone*, using *new cases* as data in order to learn about the new category.

Once an ANN has been trained with suitable data to find the hidden rules governing a certain phenomenon, it is then able to *correctly generalize* new, dirty, incomplete, or future data (Churchland and Churchland 1990; Churchland and Sejnowski 1992).

5.2 Classification: Discovering the Hidden Information in the Search Space

Classifying information into categories and subcategories permits police investigators and analysts to identify the characteristics of criminals, victims, and crime trends. Classification systems thus help the investigator and analyst to understand how criminals operate, where they operate, and the steps they take to avoid detection. Intelligent classification is a fundamental task for police investigations. The best decisions and the more appropriate actions depend on the quality of some kind of previous classification: suspect persons are a class of people presenting specific features (aspect, behavior, etc.) according to precedent professional experiences.

Police investigators involved in the examination of crime scenes or carrying out risk assessments rely heavily on classification systems for some details of the crime scene. Using the benefit of previous experiences, anticrime units classify places and persons with different grades of vulnerability and risk with respect to terrorist attacks. In short, police effectiveness is a reflection of the dynamic classification ability of police itself. One might say that investigators investigate in a manner similar to the way in which they classify. But to correctly classify is a nontrivial task.

A good classifier is a person with a great deal of experience and expertise in a specific field. Experience is what one has learned from the past; expertise is the way one has learned to learn from his experience. Using experience and expertise, a good classifier is able to understand how a new fact is connected with precedent facts.

But the implicit knowledge and meta-knowledge of a good classifier are limited by human brain constraints: storage limitations, cognitive expectations, cultural habits, age, personal problems, computational capability (no more than seven variables at the time), ideology, emotions, cognitive dissonance (I see only what I believe), formal education, and the like.

For these reasons alone, it is argued that every police department should construct an artificial intelligent classifier (AIC), which is a computational device able to learn from various experiences and capable of properly classifying new facts using a meta-learning capability that is characteristic of an AIC (Kohonen et al. 1988; Schalkoff 1992; Pao 1990; Bishop 1995; Ripley 1996; Kuncheva 2004; Duda et al. 2001; Witten and Frank 2005; Buscema et al. 2010). An AIC would work in coordination with police experts as a second, impartial examiner of events and provide a second opinion on the interpretation of the facts.

5.2.1 An Example of Intelligent Classification

The following example is used to illustrate that it is possible to build an AIC using a small dataset. The Glass Identification Database (GIU) comprises 163 records.[1]

Each record is defined by nine independent variables:

1. RI: refractive index
2. Na: sodium (unit measurement: weight percent in corresponding oxide, as are attributes 3–9)
3. Mg: magnesium
4. Al: aluminum
5. Si: silicon
6. K: potassium
7. Ca: calcium
8. Ba: barium
9. Fe: iron

[1]Title: *Glass identification database.*

Sources:

(a) Creator: B. German, Central Research Establishment, Home Office Forensic Science Service, Aldermaston, Reading, Berkshire RG7 4PN.

(b) Donor: Vina Spiehler, Ph.D., DABFT, Diagnostic Products Corporation, (213) 776–0180 (ext 3014).

(c) Date: September, 1987. Past Usage: Rule Induction in Forensic Science, Ian W. Evett and Ernest J. Spiehler, Central Research Establishment, Home Office Forensic Science Service, Aldermaston, Reading, Berkshire RG7 4PN. Unknown technical note number. General results: nearest neighbor held its own with respect to the rule-based system. Relevant information: Vina conducted a comparison test of her rule-based system, Beagle, the nearest-neighbor algorithm, and discriminant analysis. Beagle is a product available through VRS Consulting, Inc.; 4676 Admiralty Way, Suite 206; Marina Del Ray, CA 90292 (213) 827–7890 and FAX: 3189.

Each record is a window glass (building windows and vehicle windows) and can belong to one or two possible classes:

- Class A: float-processed (87 records)
- Class B: non-float-processed (76 records)

The study of the classification of types of glass was motivated by criminological investigation. At the scene of the crime, the glass left can be used as evidence, if it is correctly identified. Below is shown the complete dataset (Table 5.1):

The purpose of the application is to construct an intelligent classifier able to recognize in a blind way if a glass shard belongs to the float-processed class or to the non-float-processed class, using only the nine independent variables listed before.

In this test, we have compared four different classifiers:

(a) A linear discriminant analysis (LDA)
(b) A nearest-neighbor algorithm (NN)
(c) An evolutionary algorithm (Beagle)
(d) An artificial adaptive system (supervised ANNs)[2]

In order to validate the four classifiers in blind way, two relatively easy classifier algorithms, the Live One Out protocol and a harder K-fold cross-validation test (with $K = 2$), are used.

Here are the results in terms of average accuracy, sensitivity, and specificity (Table 5.2):

Upon studying the outputs, the usual comment on them is the recognition that not every linear and nonlinear classifier works in the same way. Performances depend on the kind of problem, the kind of data in possession, the preprocessing tools available through which we transform the dataset, and on the quality of the algorithm we apply to the classification task.

Data classification is fundamental, but intelligent classification is better. These two qualities are not mutually exclusive.

5.3 Prediction: Theory and Need

Predictive ability is the necessary prerequisite for crime prevention. No crime prevention is possible without a robust capability to predict what, where, and when a specific kind of event might happen.

---

[2]The results from LDA, NN, and Beagle are taken from literature (see note 1). The Supervised ANNs is an artificial organism created at Semeion Research Center for Sciences of Communication, via Sersale 117–119, 00128, Rome, Italy, (see www.semeion.it).

Table 5.1 The glass identification database

Independent variables									Dependent variable	
1. RI	2. Na	3. Mg	4. Al	5. Si	6. K	7. Ca	8. Ba	9. Fe	Float	Nonfloat
1.5174	13.27	3.62	1.24	73.08	0.55	8.07	0	0	1	0
1.5174	13.3	3.6	1.14	73.09	0.58	8.17	0	0	1	0
1.5176	13.15	3.61	1.05	73.24	0.57	8.24	0	0	1	0
1.5157	12.72	3.46	1.56	73.2	0.67	8.09	0	0.24	1	0
1.5176	12.8	3.66	1.27	73.01	0.6	8.56	0	0	1	0
1.5159	12.88	3.43	1.4	73.28	0.69	8.05	0	0.24	1	0
1.5175	12.86	3.56	1.27	73.21	0.54	8.38	0	0.17	1	0
1.5176	12.61	3.59	1.31	73.29	0.58	8.5	0	0	1	0
1.5176	12.81	3.54	1.23	73.24	0.58	8.39	0	0	1	0
1.5173	13.02	3.54	1.69	72.73	0.54	8.44	0	0.07	1	0
1.5197	14.77	3.75	0.29	72.02	0.03	9	0	0	1	0
1.5172	13.38	3.5	1.15	72.85	0.5	8.43	0	0	1	0
1.5179	13.21	3.48	1.41	72.64	0.59	8.43	0	0	1	0
1.5178	12.85	3.48	1.23	72.97	0.61	8.56	0.09	0.22	1	0
1.5175	12.57	3.47	1.38	73.39	0.6	8.55	0	0.06	1	0
1.5178	12.69	3.54	1.34	72.95	0.57	8.75	0	0	1	0
1.5191	13.89	3.53	1.32	71.81	0.51	8.78	0.11	0	1	0
1.5221	14.21	3.82	0.47	71.77	0.11	9.57	0	0	1	0
1.5179	12.79	3.5	1.12	73.03	0.64	8.77	0	0	1	0
1.5179	12.73	3.43	1.19	72.95	0.62	8.76	0	0.3	1	0
1.519	13.49	3.48	1.35	71.95	0.55	9	0	0	1	0
1.5267	13.99	3.7	0.71	71.57	0.02	9.82	0	0.1	1	0
1.5193	13.2	3.33	1.28	72.36	0.6	9.14	0	0.11	1	0
1.5184	13.14	2.84	1.28	72.85	0.55	9.07	0	0	1	0
1.5178	13.21	2.81	1.29	72.98	0.51	9.02	0	0.09	1	0
1.5182	12.87	3.48	1.29	72.95	0.6	8.43	0	0	1	0
1.5191	13.6	3.62	1.11	72.64	0.14	8.76	0	0	1	0
1.5223	14.17	3.81	0.78	71.35	0	9.69	0	0	1	0
1.521	13.69	3.59	1.12	71.96	0.09	9.4	0	0	1	0
1.5215	13.05	3.65	0.87	72.22	0.19	9.85	0	0.17	1	0
1.5215	13.05	3.65	0.87	72.32	0.19	9.85	0	0.17	1	0
1.523	13.31	3.58	0.82	71.99	0.12	10.17	0	0.03	1	0
1.516	13.02	3.56	1.54	73.11	0.72	7.9	0	0	0	1
1.5165	13.44	3.61	1.54	72.39	0.66	8.03	0	0	0	1
1.5163	13	3.58	1.54	72.83	0.61	8.04	0	0	0	1
1.5159	12.86	3.52	2.12	72.66	0.69	7.97	0	0	0	1
1.5159	13.25	3.45	1.43	73.17	0.61	7.86	0	0	0	1
1.5165	13.41	3.55	1.25	72.81	0.68	8.1	0	0	0	1
1.5159	13.09	3.52	1.55	72.87	0.68	8.05	0	0.09	0	1
1.5141	14.25	3.09	2.08	72.28	1.1	7.08	0	0	0	1
1.5163	13.36	3.58	1.49	72.72	0.45	8.21	0	0	0	1
1.5164	12.55	3.48	1.87	73.23	0.63	8.08	0	0.09	0	1
1.516	12.9	3.44	1.45	73.06	0.44	8.27	0	0	0	1

(continued)

Table 5.1 (continued)

Independent variables									Dependent variable	
1. RI	2. Na	3. Mg	4. Al	5. Si	6. K	7. Ca	8. Ba	9. Fe	Float	Nonfloat
1.5163	12.71	3.33	1.49	73.28	0.67	8.24	0	0	0	1
1.5184	13.02	3.62	1.06	72.34	0.64	9.13	0	0.15	0	1
1.5182	12.62	2.76	0.83	73.81	0.35	9.42	0	0.2	0	1
1.5273	13.8	3.15	0.66	70.57	0.08	11.64	0	0	0	1
1.5241	13.83	2.9	1.17	71.15	0.08	10.79	0	0	0	1
1.5313	10.73	0	2.1	69.81	0.58	13.3	3.15	0.28	0	1
1.5222	14.43	0	1	72.67	0.1	11.52	0	0.08	0	1
1.5266	11.23	0	0.77	73.21	0	14.68	0	0	0	1
1.5185	13.1	3.97	1.19	72.44	0.6	8.43	0	0	0	1
1.5185	13.41	3.89	1.33	72.38	0.51	8.28	0	0	0	1
1.5171	13.72	3.68	1.81	72.06	0.64	7.88	0	0	0	1
1.5166	12.93	3.54	1.62	72.96	0.64	8.03	0	0.21	0	1
1.5171	13.48	3.48	1.71	72.52	0.62	7.99	0	0	0	1
1.5218	13.2	3.68	1.15	72.75	0.54	8.52	0	0	0	1
1.5208	13.78	2.28	1.43	71.99	0.49	9.85	0	0.17	0	1
1.5218	13.75	1.01	1.36	72.19	0.33	11.14	0	0	0	1
1.518	13.71	3.93	1.54	71.81	0.54	8.21	0	0.15	0	1
1.5179	13.19	3.9	1.3	72.33	0.55	8.44	0	0.28	0	1
1.5181	13	3.8	1.08	73.07	0.56	8.38	0	0.12	0	1
1.5171	12.89	3.62	1.57	72.96	0.61	8.11	0	0	0	1
1.5167	12.87	3.56	1.64	73.14	0.65	7.99	0	0	0	1
1.5169	13.33	3.54	1.61	72.54	0.68	8.11	0	0	0	1
1.5185	13.2	3.63	1.07	72.83	0.57	8.41	0.09	0.17	0	1
1.5166	12.85	3.51	1.44	73.01	0.68	8.23	0.06	0.25	0	1
1.5171	13	3.47	1.79	72.72	0.66	8.18	0	0	0	1
1.5166	12.99	3.18	1.23	72.97	0.58	8.81	0	0.24	0	4
1.5184	12.85	3.67	1.24	72.57	0.62	8.68	0	0.35	0	1
1.5161	13.33	3.53	1.34	72.67	0.56	8.33	0	0	1	0
1.5166	13.14	3.45	1.76	72.48	0.6	8.38	0	0.17	1	0
1.5178	13.64	3.65	0.65	73	0.06	8.93	0	0	1	0
1.5169	12.86	3.58	1.31	72.61	0.61	8.79	0	0	1	0
1.5165	13.04	3.4	1.26	73.01	0.52	8.58	0	0	1	0
1.5165	13.41	3.39	1.28	72.64	0.52	8.65	0	0	1	0
1.5212	14.03	3.76	0.58	71.79	0.11	9.65	0	0	1	0
1.5178	13.53	3.41	1.52	72.04	0.58	8.79	0	0	1	0
1.518	13.5	3.36	1.63	71.94	0.57	8.81	0	0.09	1	0
1.5183	13.33	3.34	1.54	72.14	0.56	8.99	0	0	1	0
1.5193	13.64	3.54	0.75	72.65	0.16	8.89	0.15	0.24	1	0
1.5221	14.19	3.78	0.91	71.36	0.23	9.14	0	0.37	1	0
1.521	13.64	4.49	1.1	71.78	0.06	8.75	0	0	1	0
1.5176	13.89	3.6	1.36	72.73	0.48	7.83	0	0	1	0
1.5162	13.53	3.55	1.54	72.99	0.39	7.78	0	0	1	0
1.5177	13.21	3.69	1.29	72.61	0.57	8.22	0	0	1	0

(continued)

Table 5.1 (continued)

Independent variables									Dependent variable	
1. RI	2. Na	3. Mg	4. Al	5. Si	6. K	7. Ca	8. Ba	9. Fe	Float	Nonfloat
1.516	12.79	3.61	1.62	72.97	0.64	8.07	0	0.26	1	0
1.5192	14.04	3.58	1.37	72.08	0.56	8.3	0	0	1	0
1.5175	13	3.6	1.36	72.99	0.57	8.4	0	0.11	1	0
1.5178	12.68	3.67	1.16	73.11	0.61	8.7	0	0	1	0
1.522	14.36	3.85	0.89	71.36	0.15	9.15	0	0	1	0
1.5191	13.9	3.73	1.18	72.12	0.06	8.89	0	0	1	0
1.5175	12.82	3.55	1.49	72.75	0.54	8.52	0	0.19	1	0
1.5174	12.78	3.62	1.29	72.79	0.59	8.7	0	0	1	0
1.5175	12.81	3.57	1.35	73.02	0.62	8.59	0	0	1	0
1.5176	12.98	3.54	1.21	73	0.65	8.53	0	0	1	0
1.5172	12.87	3.48	1.33	73.01	0.56	8.43	0	0	1	0
1.5177	12.56	3.52	1.43	73.15	0.57	8.54	0	0	1	0
1.5178	13.08	3.49	1.28	72.86	0.6	8.49	0	0	1	0
1.5177	12.65	3.56	1.3	73.08	0.61	8.69	0	0.14	1	0
1.5175	12.84	3.5	1.14	73.27	0.56	8.55	0	0	1	0
1.5157	13.29	3.45	1.21	72.74	0.56	8.57	0	0	1	0
1.518	12.74	3.48	1.35	72.96	0.64	8.68	0	0	1	0
1.5221	14.21	3.82	0.47	71.77	0.11	9.57	0	0	1	0
1.5175	12.71	3.42	1.2	73.2	0.59	8.64	0	0	1	0
1.5178	13.21	3.39	1.33	72.76	0.59	8.59	0	0	1	0
1.5221	13.73	3.84	0.72	71.76	0.17	9.74	0	0	1	0
1.5187	13.19	3.37	1.18	72.72	0.57	8.83	0	0.16	1	0
1.5222	13.21	3.77	0.79	71.99	0.13	10.02	0	0	1	0
1.519	13.58	3.35	1.23	72.08	0.59	8.91	0	0	1	0
1.5232	13.72	3.72	0.51	71.75	0.09	10.06	0	0.16	1	0
1.5181	13.43	2.87	1.19	72.84	0.55	9.03	0	0	1	0
1.5177	12.45	2.71	1.29	73.7	0.56	9.06	0	0.24	1	0
1.5122	12.99	3.47	1.12	72.98	0.62	8.35	0	0.31	1	0
1.5175	13.48	3.74	1.17	72.99	0.59	8.03	0	0	1	0
1.5175	13.39	3.66	1.19	72.79	0.57	8.27	0	0.11	1	0
1.5198	13.81	3.58	1.32	71.72	0.12	8.67	0.69	0	1	0
1.5217	13.51	3.86	0.88	71.79	0.23	9.54	0	0.11	1	0
1.5217	13.48	3.74	0.9	72.01	0.18	9.61	0	0.07	1	0
1.5215	13.12	3.58	0.9	72.2	0.23	9.82	0	0.16	1	0
1.5157	14.86	3.67	1.74	71.87	0.16	7.36	0	0.12	0	1
1.5185	13.64	3.87	1.27	71.96	0.54	8.32	0	0.32	0	1
1.5159	13.09	3.59	1.52	73.1	0.67	7.83	0	0	0	1
1.5163	13.34	3.57	1.57	72.87	0.61	7.89	0	0	0	1
1.5159	13.02	3.58	1.51	73.12	0.69	7.96	0	0	0	1
1.5161	13.92	3.52	1.25	72.88	0.37	7.94	0	0.14	0	1
1.5159	12.82	3.52	1.9	72.86	0.69	7.97	0	0	0	1
1.5157	13.24	3.49	1.47	73.25	0.38	8.03	0	0	0	1
1.5165	13.4	3.49	1.52	72.65	0.67	8.08	0	0.1	0	1

(continued)

Table 5.1 (continued)

Independent variables									Dependent variable	
1. RI	2. Na	3. Mg	4. Al	5. Si	6. K	7. Ca	8. Ba	9. Fe	Float	Nonfloat
1.5162	13.01	3.5	1.48	72.89	0.6	8.12	0	0	0	1
1.5184	12.93	3.74	1.11	72.28	0.64	8.96	0	0.22	0	1
1.5159	13.12	3.41	1.58	73.26	0.07	8.39	0	0.19	0	1
1.5159	13.24	3.34	1.47	73.1	0.39	8.22	0	0	0	1
1.5186	13.36	3.43	1.43	72.26	0.51	8.6	0	0	0	1
1.5174	12.2	3.25	1.16	73.55	0.62	8.9	0	0.24	0	1
1.5169	12.67	2.88	1.71	73.21	0.73	8.54	0	0	0	1
1.5181	12.96	2.96	1.43	72.92	0.6	8.79	0.14	0	0	1
1.5165	12.75	2.85	1.44	73.27	0.57	8.79	0.11	0.22	0	1
1.5173	12.35	2.72	1.63	72.87	0.7	9.23	0	0	0	1
1.5247	11.45	0	1.88	72.19	0.81	13.24	0	0.34	0	1
1.5339	12.3	0	1	70.16	0.12	16.19	0	0.24	0	1
1.5182	13.72	0	0.56	74.45	0	10.99	0	0	0	1
1.5274	11.02	0	0.75	73.08	0	14.96	0	0	0	1
1.5278	12.64	0	0.67	72.02	0.06	14.4	0	0	0	1
1.5189	13.46	3.83	1.26	72.55	0.57	8.21	0	0.14	0	1
1.5183	13.24	3.9	1.41	72.33	0.55	8.31	0	0.1	0	1
1.5167	13.3	3.64	1.53	72.53	0.65	8.03	0	0.29	0	1
1.5165	13.56	3.57	1.47	72.45	0.64	7.96	0	0	0	1
1.5184	13.25	3.76	1.32	72.4	0.58	8.42	0	0	0	1
1.5169	13.23	3.54	1.48	72.84	0.56	8.1	0	0	0	1
1.5187	12.93	3.66	1.56	72.51	0.58	8.55	0	0.12	0	1
1.5167	12.94	3.61	1.26	72.75	0.56	8.6	0	0	0	1
1.5207	13.55	2.09	1.67	72.18	0.53	9.57	0.27	0.17	0	1
1.5202	13.98	1.35	1.63	71.76	0.39	10.56	0	0.18	0	1
1.5261	13.7	0	1.36	71.24	0.19	13.44	0	0.1	0	1
1.5181	13.43	3.98	1.18	72.49	0.58	8.15	0	0	0	1
1.5181	13.33	3.85	1.25	72.78	0.52	8.12	0	0	0	1
1.5167	12.79	3.52	1.54	73.36	0.66	7.9	0	0	0	1
1.5177	13.65	3.66	1.11	72.77	0.11	8.6	0	0	1	0
1.5167	13.24	3.57	1.38	72.7	0.56	8.44	0	0.1	1	0
1.5164	12.16	3.52	1.35	72.89	0.57	8.53	0	0	1	0
1.5213	14.32	3.9	0.83	71.5	0	9.49	0	0	1	0
1.5161	13.42	3.4	1.22	72.69	0.59	8.32	0	0	1	0

RI = refractive index Na = Sodium Mg = Magnesium Al = Aluminum Si = Silicon K = Potassium Ca = Calcium Ba = Barium Fe = Iron

Prediction is the capability to use the past to anticipate the future. Every effective police team uses some kind of prediction in an effort to attempt to prevent crime events from occurring.

Prophecies exist to creatively articulate possible futures based on personal beliefs and/or personal reputation (the prophet). Expectations could be considered merely

Table 5.2 Results in terms of average accuracy

	Float processed (%)	Non float processed (%)	Aritm. mean accuracy (%)	Pond. mean accuracy (%)
Beagle	88.51	75.00	81.75	82.21
KNN	86.21	78.95	82.58	82.82
LDA	75.86	71.05	73.46	73.62
SV ANN	*92.87*	*88.32*	*90.60*	*90.87*

common sense prophecies. Predictions are the result of a scientific processing of past data in an effort to identify the most probable immediate future for a defined and specific control setting.

Prediction in crime analytics is possible if criminal activity is not a random occurrence or, to use the language of mathematics, if criminal activity is not a random walk. As crime activity produces an economic and/or symbolic and/or imaginary power to someone, it is not a random walk.

So, predictive activity is a duty of the police forces. In a world where the time dimension exists, if one does not make *good* predictions, it means that one is making *wrong* predictions.

Consequently, the discipline of law enforcement needs to construct a set of artificial intelligent predictors (AIP) to be able to apply against new criminal activity in order to see relationships that might/would go unnoticed. These predictors need to be continuously trained and retrained with new data and validated in blind (unbiased) way, each time, according to the prerequisite of probability theory (Mena 2003; Buscema and Benzi 2011).

Prediction and prevention are strictly linked: in a predictive activity, the past is used to give sense to future events, and in a preventive activity, one acts in the future to modify actions that have occurred in the past.

5.4 Prediction: Discover the Hidden Information in the Time

In 1996, an Italian public office asked Semeion to apply ANNs to predict the real income of selected trader categories using the statement of income that traders annually declare as a description of all their financial activities that have occurred during the year.

To test the possibility of the prediction of real income, Semeion was given a dataset of a selected sample of 1,001 bakeries. Each record was defined by 65 variables, considering the different kind of the expenses that a particularly bakery had occurred along the year and various quantitative representations of each bakery shop. Obviously, it is not possible to show, with any degree of detail, the particular set of variables used.

The dependent variable was the total annual income for each bakery.

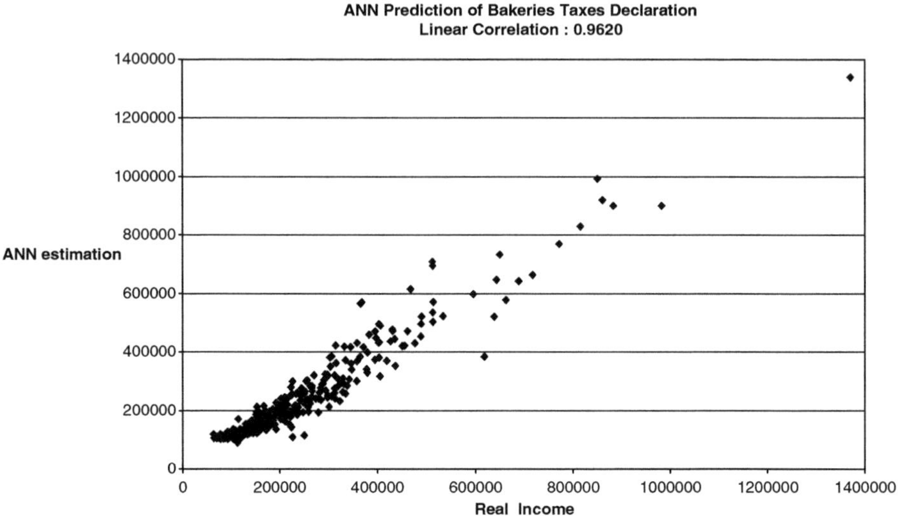

Fig. 5.2 Correlation between ANN estimations and the "real" incomes declared

The Italian office was confident about the reality and the consistency of this variable because the sample set of bakeries was a selected sample from a joint commission of bakery representatives and public officers.

The purpose of this research was to evaluate the capability of ANNs to select among the massive quantity of statements of income a sample of those considered to be most probable false. Consequently, if an ANN had been able to predict the real income in a sample of "true" tax declarations, the same ANN would had been also used as a detector for identifying the "suspect" tax declarations of the whole population of bakeries.

The Italian population of private workers had been divided by the Italian experts into categories of different working activities (specifically the various sectors to be studied), and bakeries were the first category that Semeion was asked to test.

The sample was divided randomly into two subsets: the first subset (703 records) was used to train the ANN; the second one (301 records) was used as an unbiased validation set.

The results were very good: the correlation between the ANN estimations and the "real" incomes declared from the bakeries was around the 96.20 % (Fig. 5.2):

The next graph (Fig. 5.3) shows a comparison between the real incomes and the predicted ones in 50 of the 301 records used as testing set.

All the different cost functions applied to the testing set to determinate the dimension of the estimation error were very good (Table 5.3):

The results were astoundingly accurate. In fact, perhaps this exceptionally high level of accuracy is the reason why this research project was prematurely stopped. No one in authority was prepared to believe that results this good could possibly be achieved.

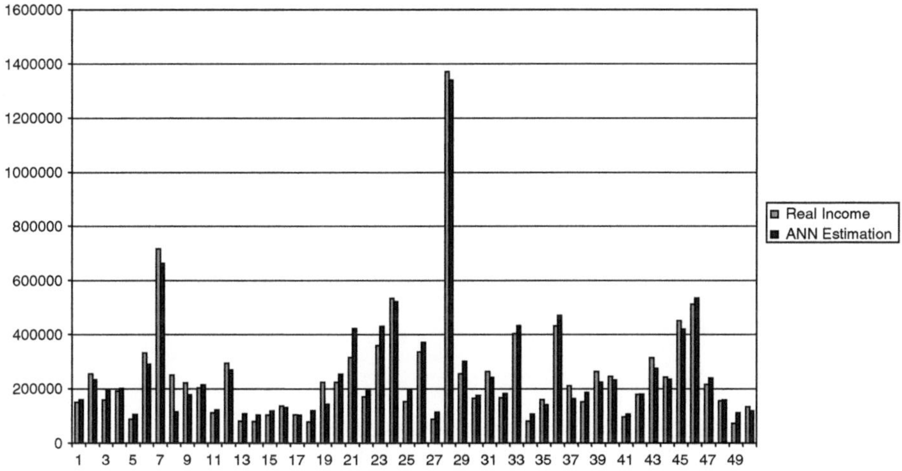

Fig. 5.3 Comparison between the real and predicted incomes

Table 5.3 Different cost functions

Root mean square error	0.0107
Real error	−0.0004
Absolute error	0.0106
Error variance	0.0188
Normalized mean squared error	0.0790

In any case, this is a good example of how the power of predictive tools can create serious difficulties for delinquent activities.

5.5 Simulation and Profiling: Discover the Abstract Prototype in the Hidden Data

Simulation is one way of testing the consequences of possible actions without having to pay for them. Simulations provide law enforcement agencies with strategic information about the map of possible choices. It means that the ability to simulate possible actions is a strategic step in decision making.

Obviously, this book proposes a specific concept for simulation:

(a) First, we consider the simulation of a dynamic system, that is, a system whose dynamic is a sequence of temporal (time-related) and ordered (not randomized) states.
(b) Second, we consider the simulation to be a nonstationary process, characterized by nonlinear feedback loops.
(c) Third, of the kind of simulation described here involves a process that evolves with parallel interaction and modification of the whole set of involved variables.

In short, simulation in this application is a mathematical engine able to learn from the past and generate possible future worlds beginning with present input or possibilities (Buscema 1998b, 2007; Buscema and Diappi 2004).

This kind of artificial intelligent simulator (AIS) is a fundamental device to test the effectiveness of tactics and the corresponding side effect that occur in law enforcement anticrime activity and strategy. Note that "anticrime" as used here also subsumes terrorism, gang activity, rouge state activities, and the like.

From this point of view, simulation for law enforcement essentially means the placement of one's head, as it were, into the future, with their feet fixed and saved in the present, with the ability to report back to the present what was "seen" occurring in the future.

5.5.1 An Example of Prototype Generation

To illustrate some of the abilities of a dynamic scenario simulation technology with respect to gang activity, we use a database of individuals inspired by the musical "West Side Story." The dataset is composed of 27 records, five variables, and 27 identification names, one for each record (Table 5.4).

Some preprocessing operations are useful to transform this dataset in a more suitable way in order to be processed by ANNs. Note the presence of states rather than continuous data; that is, the age of the gang members is represented by the categories or states of 20s, 30s, or 40s.

The dataset presents five variables, each one composed of a class of options:

- Gang = {Jets, Sharks}
- Age = {20s, 30s, 40s}
- Education = {Junior School, High School, College}
- Status = {Married, Single, Divorced}
- Profession = {Pusher, Bookie, Burglar}

The raw data in Table 5.4 is transformed such that the option associated with each variable is shown as a single new variable. This new variable will be binary (1 = presence of the feature, 0 = absence of the feature), and it will be considered as completely independent from the others.

The auto-associative ANN used to process this database will transform every option into a neural network node, able to be activated in a continuous fuzzy[3] way from 0 to 1.

---

[3]Fuzziness is a new branch of mathematics by which degrees of involvement or belief can be determined. An example of fuzziness can perhaps best be described by the activities that occur during jury deliberation. A defendant must be judged guilty or not guilty. There is no ambiguity in this decision; everyone must come to the same conclusion. Even if someone is only 51 % sure of the degree of guilt or innocence, that juror must side entirely with either the guilty or not guilty position. It is the middle ground between guilty and not guilty that fuzzy theory plays its role. One can make a determination from 0 to 100 % as to their degree of belief.

Table 5.4 The gang database

Name	Gang	Age	Education	Status	Profession
ART	Jets	40	Junior School	Single	Pusher
AL	Jets	30	Junior School	Married	Burglar
SAM	Jets	20	College	Single	Bookie
CLYDE	Jets	40	Junior School	Single	Bookie
MIKE	Jets	30	Junior School	Single	Bookie
JIM	Jets	20	Junior School	Divorced	Burglar
GREG	Jets	20	High School	Married	Pusher
JOHN	Jets	20	Junior School	Married	Burglar
DOUG	Jets	30	High School	Single	Bookie
LANCE	Jets	20	Junior School	Married	Burglar
GEORGE	Jets	20	Junior School	Divorced	Burglar
PETE	Jets	20	High School	Single	Bookie
FRED	Jets	20	High School	Single	Pusher
GENE	Jets	20	College	Single	Pusher
RALPH	Jets	30	Junior School	Single	Pusher
PHIL	Sharks	30	College	Married	Pusher
IKE	Sharks	30	Junior School	Single	Bookie
NICK	Sharks	30	High School	Single	Pusher
DON	Sharks	30	College	Married	Burglar
NED	Sharks	30	College	Married	Bookie
KARL	Sharks	40	High School	Married	Bookie
KEN	Sharks	20	High School	Single	Burglar
EARL	Sharks	40	High School	Married	Burglar
RICK	Sharks	30	High School	Divorced	Burglar
OL	Sharks	30	College	Married	Pusher
NEAL	Sharks	30	High School	Single	Bookie
DAVE	Sharks	30	High School	Divorced	Pusher

Consequently, the auto-associative ANN will have 14 nodes [Gang(2) + Age(3) + Education(3) + Status(3) + Profession(3)].

In Table 5.5, the dataset after the binary transformation is shown:

Note that each state that describes a gang member is shown as column headers. Thus, a member belongs to one and only one grouping. After the training phase, officials will be able to ask the ANN three kinds of questions that could not be asked in any manageable way using the standard structured query language of relational databases:

5.5.1.1 Prototypical Questions

One can ask the ANN to generate an abstract profile of the people belonging to a specific class of variables. In this situation, law enforcement wishes to know the

Table 5.5 Gang database after the binary transformation

Gang 14 × 27	Jet	Sharks	20	30	40	JH	COL	HS	Single	Married	Divorced	Pusher	Bookie	Burglar
ART	1	0	0	0	1	1	0	0	1	0	0	1	0	0
AL	1	0	0	1	0	1	0	0	0	1	0	0	0	1
SAM	1	0	1	0	0	0	1	0	1	0	0	0	1	0
CLYDE	1	0	0	0	1	1	0	0	1	0	0	0	1	0
MIKE	1	0	0	1	0	1	0	0	1	0	0	0	1	0
JIM	1	0	1	0	0	1	0	0	0	0	1	0	0	1
GREG	1	0	1	0	0	0	0	1	0	1	0	1	0	0
JOHN	1	0	1	0	0	1	0	0	0	1	0	0	0	1
DOUG	1	0	0	1	0	0	0	1	1	0	0	0	1	0
LANCE	1	0	1	0	0	1	0	0	0	1	0	0	0	1
GEORGE	1	0	1	0	0	1	0	0	0	0	1	0	0	1
PETE	1	0	1	0	0	0	0	1	1	0	0	0	1	0
FRED	1	0	1	0	0	0	0	1	1	0	0	1	0	0
GENE	1	0	1	0	0	0	1	0	1	0	0	1	0	0
RALPH	1	0	0	1	0	1	0	0	1	0	0	1	0	0
PHIL	0	1	0	1	0	0	1	0	0	1	0	1	0	0
IKE	0	1	0	1	0	1	0	0	1	0	0	0	1	0
NICK	0	1	0	1	0	0	0	1	1	0	0	1	0	0
DON	0	1	0	1	0	0	1	0	0	1	0	0	0	1
NED	0	1	0	1	0	0	1	0	0	1	0	0	1	0
KARL	0	1	0	0	1	0	0	1	0	1	0	0	1	0
KEN	0	1	1	0	0	0	0	1	1	0	0	0	0	1
EARL	0	1	0	0	1	0	0	1	0	1	0	0	0	1
RICK	0	1	0	1	0	0	0	1	0	0	1	0	0	1
OL	0	1	0	1	0	0	1	0	0	1	0	1	0	0
NEAL	0	1	0	1	0	0	0	1	1	0	0	0	1	0
DAVE	0	1	0	1	0	0	0	1	0	0	1	1	0	0

prototype of a Jets gang member and/or a prototype of a Sharks gang member. It would be most useful to know the variables that identify a "better" Jet or a "better" Shark and which of the actual Jets are more or less Jet-like than the others, considering the prototype of Jets guy (the same for the Sharks). In essence, an official seeks to understand those qualities of Jet-ness and Shark-ness. These kinds of problems are quite typical for the human brain to attempt to solve but very fundamental for law enforcement activity: What kind of people should I focus my attention on in the street to maximize the probability of stopping the right person? These kinds of important problems are impossible to solve using the classic theory of computation.

When we ask a trained ANN (Buscema 1998a) to generate the cognitive prototype of a Jets member, the ANN in less than a second answers in this way (Table 5.6):

While this answer sounds good, there is only one Jet gang member that exactly fits this profile in the dataset (see Table 5.6). This means that the ANN made a data abstraction, individualizing the "center of gravity" to be a Jet; in other words, Table 5.6 describes the state of Jet-ness. Furthermore, this profile of a typical Jet is what the ANN represents at the end of a very complex process that simulates the interaction among all the variables of the dataset as graphically shown in Fig. 5.4:

Watching this process, one can begin to gather some understand as to the complex calculations the ANN develops in order to answer the question. It is in this manner that the ANN determines a solution to the questions asked.

The ANN is also able to define which individuals (records) are more representative of the Jets gang in this dataset. These are the ANN final answers (Table 5.7):

According to the ANN, not every Jets member is a Jet in the same way. For example, Greg and Al are weak Jets. At the same time, some Sharks, such as Ike, seem more Jet than Shark. Checking these records in the original dataset, we can see that there is an evidence for this suggestion. It is also important to note that in this case, the decision process of the ANN is not linear (Fig. 5.5):

If we ask the ANN to define the prototypical features of the Sharks member, we have this type of profile (Table 5.8):

Table 5.6 Variables profiling the Jets gang

Jets variables prototype		
Field	Variable	Plausibility
Age	20s	0.99
Education	Junior School	1.00
Status	Single	1.00
Profession	Bookie	1.00

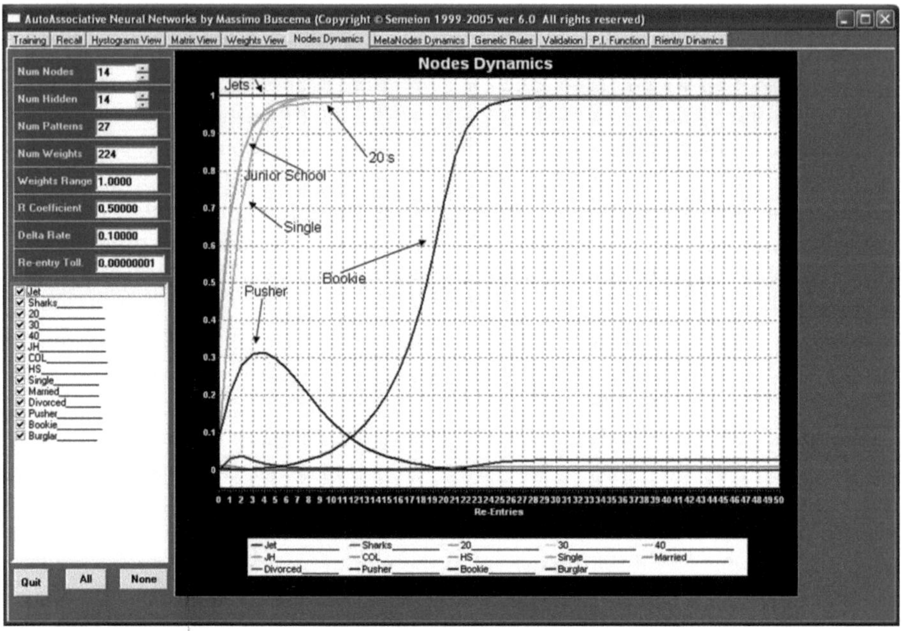

Fig. 5.4 Interaction among all the variables during the recall phase

Table 5.7 Records close to the Jets prototype

Jets records prototype			
Name	Jets membership of Jets	Name	Jets membership of Sharks
ART	*0.91*	PHIL	*0.00*
AL	*0.09*	IKE	*0.91*
SAM	*1.00*	NICK	*0.00*
CLYDE	*1.00*	DON	*0.00*
MIKE	*1.00*	NED	*0.00*
JIM	*0.91*	KARL	*0.00*
GREG	*0.08*	KEN	*0.09*
JOHN	*0.91*	EARL	*0.00*
DOUG	*0.91*	RICK	*0.00*
LANCE	*0.91*	OL	*0.00*
GEORGE	*0.91*	NEAL	*0.08*
PETE	*1.00*	DAVE	*0.00*
FRED	*0.90*		
GENE	*0.91*		
RALPH	*0.91*		

Similarly, the dynamics of the process are nonlinear (Fig. 5.6):

It is interesting to observe the dynamics of the system decision about the "Education": at the beginning, the ANN believes that "High School" could be a good descriptor of the Sharks type, but in a second step, it changes its mind focusing on the "College" variable as the best feature.

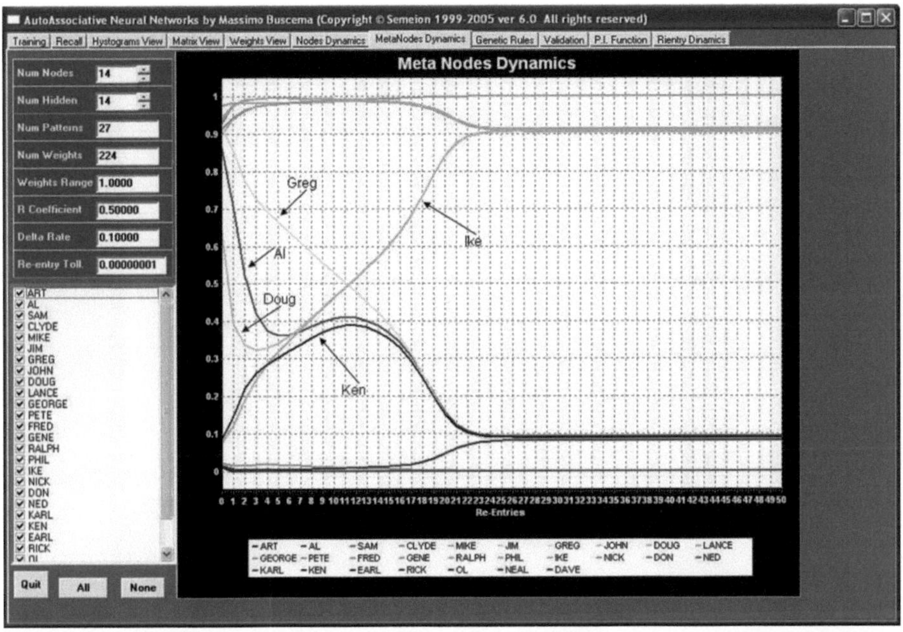

Fig. 5.5 Interaction among all the records during the recall phase

Table 5.8 Variables profiling the Sharks gang

Sharks variables prototype		
Field	Variable	Plausibility
Age	30s	1.00
Education	College	1.00
Status	Married	1.00
Profession	Burglar	1.00

It is also interesting to see the list provided by the ANN concerning which Sharks are more or less representative of the Sharks prototype (Table 5.9):

The Sharks gang is less compact than the Jets gang, and only one member of the Jets gang could be confused as being a Shark (Al). This situation appears more informative when we look at the whole process adopted by the ANN to arrive at this decision (Fig. 5.7):

5.5.1.2 Impossible Questions

Some important kinds of questions are impossible to answer when it involves associations of variables not anticipated by the structure of the database. In the structure of the gang database, for instance, a new member is not allowed to belong simultaneously to the Jets and to the Sharks. This kind of question is impossible for classic computation, but they are fundamental for human beings. If one seeks

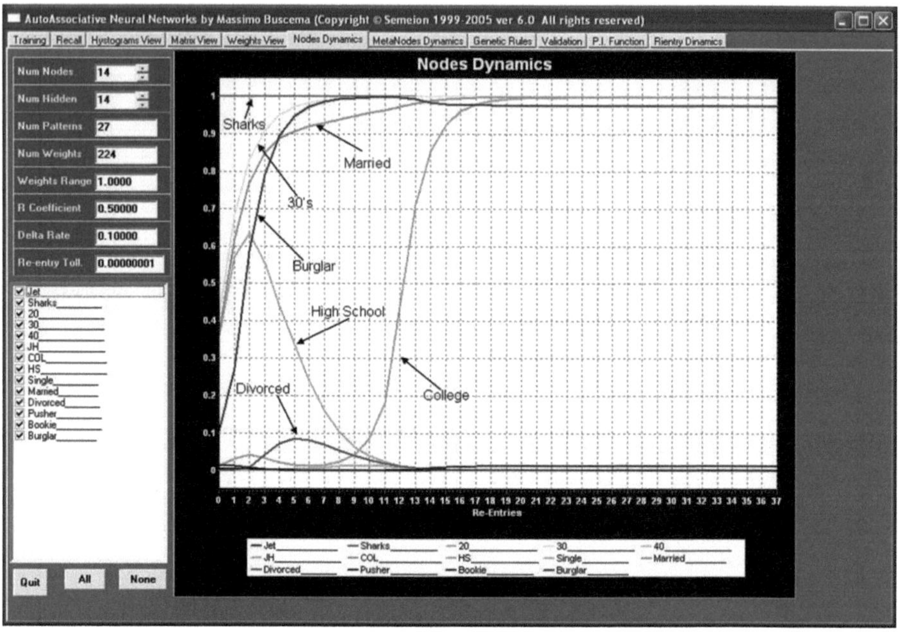

Fig. 5.6 Interaction among all variables during the recall phase

to create a pool of Jets and Sharks able to understand each other, one must know (a) to which variables are the members of a particular gang more similar and (b) which gang members are more similar among the dataset to be associated with a hybrid combination of Jet-ness and Shark-ness. An ANN is able to manage this kind of question. If we ask the ANN to select the hyper-prototype of a person who is both a Jet and a Shark at the same time, the response given is in terms of variables (Table 5.10):

And this is the answer in terms of individual membership of each record of the dataset (Table 5.11):

The ANN establishes a new profile to select new members whose features fuse the gangs Jets and Sharks together, and it considers two of the existing Jets and one of the Sharks as the best candidates for this new class.

5.5.1.3 Virtual Questions

There are occasions when one would like to pose a theoretical question for which it is known that no record in the database actually satisfies the question. Such a question is called a *virtual question*.

In the case of the gang database, for example, there is no record presenting simultaneously the combination of the educational feature "college" and the age of 40s. This association is obviously allowed and possible, but since there is no actual record of such a combination, it is a virtual query.

Table 5.9 Records close to the Sharks prototype

Sharks records prototype			
Name	Sharks membership of Jets	Name	Sharks membership of Sharks
ART	*0.00*	PHIL	*1.00*
AL	*0.90*	IKE	*0.09*
SAM	*0.00*	NICK	*0.09*
CLYDE	*0.00*	DON	*1.00*
MIKE	*0.00*	NED	*1.00*
JIM	*0.00*	KARL	*0.09*
GREG	*0.00*	KEN	*0.08*
JOHN	*0.08*	EARL	*0.90*
DOUG	*0.00*	RICK	*0.91*
LANCE	*0.08*	OL	*1.00*
GEORGE	*0.00*	NEAL	*0.09*
PETE	*0.00*	DAVE	*0.09*
FRED	*0.00*		
GENE	*0.00*		
RALPH	*0.00*		

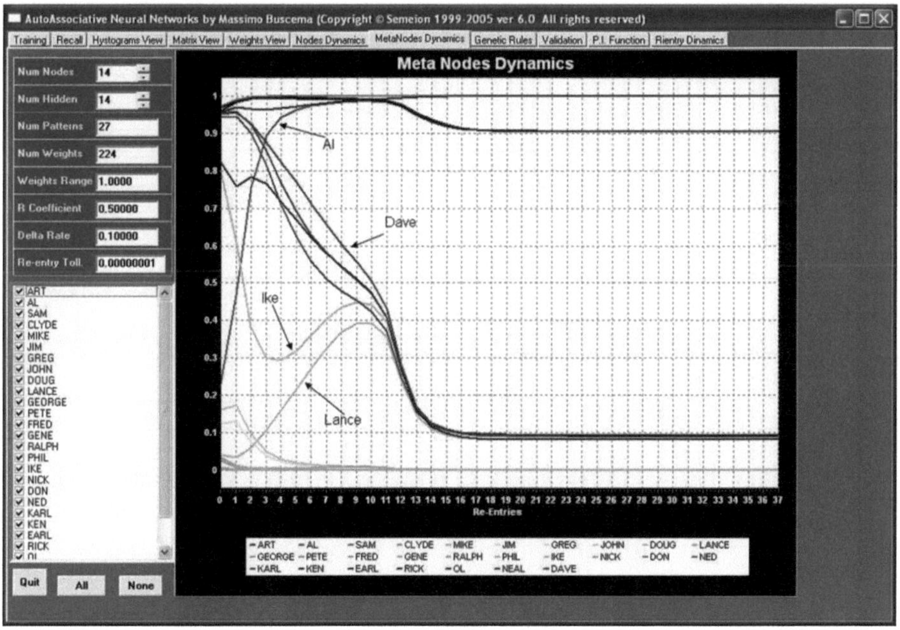

Fig. 5.7 Interaction among all records during the recall phase

Table 5.10 Variables shared by the Jets and the Sharks

Jets and Sharks prototype		
Field	Variable	Plausibility
Age	20s	0.99
Education	College	1.00
Status	Married	1.00
Profession	Burglar	1.00

Table 5.11 Records close to the intersection between Jets and Sharks

Jets and Sharks Records prototype			
Name	Jets belonging to the new Jets and Sharks Class	Name	Sharks belonging to the new Jets and Sharks Class
ART	*0.00*	PHIL	*0.44*
AL	*0.51*	IKE	*0.00*
SAM	*0.49*	NICK	*0.00*
CLYDE	*0.00*	DON	*0.99*
MIKE	*0.00*	NED	*0.39*
JIM	*0.50*	KARL	*0.01*
GREG	*0.61*	KEN	*0.45*
JOHN	*0.99*	EARL	*0.45*
DOUG	*0.00*	RICK	*0.01*
LANCE	*0.99*	OL	*0.44*
GEORGE	*0.50*	NEAL	*0.00*
PETE	*0.01*	DAVE	*0.00*
FRED	*0.01*		
GENE	*0.54*		
RALPH	*0.00*		

Table 5.12 A virtual question – the variables

College and 40s		
Field	*Variable*	*Plausibility*
Gang	Sharks	0.99
Age (1)	30s	0.97
Age (2)	40s	0.51
Status	Married	1.00
Profession	Pusher	0.97

The question that would very likely be raised in the mind of the reader might be, "What is the point of identifying something that does not exist?" It could be very useful to know *which other variables have to be associated* with these two variables (*college* and age in the *40s* – see Table 5.12), while preserving the fuzzy logic of the database, and which of the existing records are closer to this virtual combination of features.

If we ask the ANN the above question, the answer in terms of variables will be the following:

The ANN is not able to completely satisfy our constraints, and it shows how the age of 30s is the closest feature to the prototype which we seek. This behavior is

Table 5.13 A virtual question – the records

College and 40s Records prototype			
Name	Jets belonging to the new class	Name	Sharks belonging to the new class
ART	*0.00*	PHIL	*1.00*
AL	*0.03*	IKE	*0.03*
SAM	*0.00*	NICK	*0.72*
CLYDE	*0.00*	DON	*1.00*
MIKE	*0.00*	NED	*1.00*
JIM	*0.00*	KARL	*0.26*
GREG	*0.03*	KEN	*0.00*
JOHN	*0.00*	EARL	*0.26*
DOUG	*0.00*	RICK	*0.03*
LANCE	*0.00*	OL	*1.00*
GEORGE	*0.00*	NEAL	*0.03*
PETE	*0.00*	DAVE	*0.72*
FRED	*0.00*		
GENE	*0.03*		
RALPH	*0.03*		

important because it clearly shows the capability of the ANN to provide an answer to our question according to its learning experience while fully taking into account the original features of the dataset.

The selection of the more suitable individuals contained in the dataset for this prototype is thus (Table 5.13):

The database has been constructed such that each membership in any particular category is either 0 or 1. For example, a member belongs to set of ages 20, 30, or 40. Two of these three categories will contain a 0 and only one will contain a 1. Since there are only two choices per category (0 or 1), questions raised are referred to as "binary" questions. The number of binary questions that is possible to ask of this ANN is in the order of 2^N, where N is the number of considered variables in the dataset. However, as the ANN is also able to process "fuzzy questions," the number of questions is theoretical infinitive.

Therefore, through the ANN simulation, law enforcement forces can completely explore any complex dataset.

5.6 Visualization and Clustering: How to Cluster Analyzed Data for Understandability

The human eye is a powerful tool and hand-eye coordination is a strategic advantage in human technological evolution. In law enforcement activities, sight may be considered the railroad to intelligence, for visualization is often the first step in understanding.

Unfortunately, not every situation is easy to visualize. It is common for an investigation to involve a considerable number of aspects, each aspect having at least one variable to define it, and collectively the possible interactions can easily result in the visualization task being a hard task to accomplish. While our sight works well with two or three dimensions, it is rare, if not impossible, to view higher dimensions. In these cases, one needs assistance to help bring the pieces together into some coherent while, and artificial systems are able to very nicely accomplish this task. An ANN can reduce a dataset of many variables into an intelligent projection of the same dataset into a space of two or three dimensions (Buscema and Terzi 2006). This projection has to be "intelligent" because of the following:

(a) During the projection process, the system has to select the more informative features of the dataset while simultaneously eliminating the noise contained in the data.
(b) The system has to organize the data into the new smaller space (i.e., from 4, 5, 6, or more dimension space to 2 or 3 dimensions) in order to preserve the original metric distances and relationships that the data-points had into the original space.
(c) The system has to rearrange the data-points onto the projection space so that a human expert can easily cluster them according tǒ his/her own insights, knowledge, subjective feelings, and overall intelligence.

Consequently, law enforcement forces should possess a powerful artificial intelligence visualizer (AIV) able to represent complex data relationships in self-evident bidimensional or three-dimensional maps. These tools can support the law enforcement task of investigation by transforming the verbal language representation of complex situations into a more complete geographical understanding of all the concepts involved in the situations. To use analogy, a grouping of trees can be either a forest or a set of individuals, but the only way to get an overall understanding of these trees is to see them as a forest.

5.7 Visualization and Clustering: An Example

The Glass Identification Database that was mentioned earlier is composed of 167 records, 9 continuous independent variables, and 1 dependent variable (recall the float windows versus non-float windows) for a total of ten variables.

If we scale all the variables between 0 and 1 and then maximize all them with this equation:

$$\mathrm{NewVar}_i = 1 - \mathrm{OldVar}_i$$

we will have 20 variables (the original ten and their complements).

At this point, the new dataset represents 167 records in a tenth dimensional space (note that the addition of ten new variables does not yield a 20-dimensional space because the new ten variables are the complement of the original ones and are thus not independent). An interesting question could be raised: How can tenth dimensional space be projected onto a more comfortable two-dimensional space, preserving the most important relationships among all variables? By having a two-dimensional diagram, one could view it on a computer screen or as hard copy, share it with others, and engage in the synergy that comes from the cross fertilization of ideas and knowledge. This reduction in dimension can be done using a special class of artificial adaptive systems used for multidimensional scaling and also known as topographic mapping (see below).

These techniques are powerful tools for data visualization and have the advantage of improving the capability of humans to understand the relationships between variables and data in a very complex dataset. To illustrate the effectiveness of these tools, the PST (Pick and Squash Tracking in Buscema and Terzi 2006) algorithm is used because it has continuously proved itself to be a more powerful algorithm than the other algorithms presented in the literature. The objective of PST is to preserve the characteristics of the "geometrical structure" of data in a representation with reduced dimensionality.

The concept of "geometrical structure" is connected to the concept of distance. Therefore, to preserve the geometric structure of the original data, it is necessary that, after applying the algorithm of mapping, the elements that used to be "close" in the original space also find themselves to be similarly close in the subspace. It also means that those distant elements in the original space remain separated in the final destination space. The important distinction here is that the new space is typically easily understood by anyone.

PST algorithms obtained a mapping fitness in excess of 90 % to make the projection of the glass identification database into two dimensions (Fig. 5.8). The conceptual power of this visualization should be evident:

Now we can see how the quantity of potassium (the minimum and maximum values of potassium are shown in the ovals) could be critical to distinguishing between float and non-float windows (shown in the squares). One might also be able to use sodium, iron, and aluminum to distinguish between float and non-float windows. At the same time, there are many other variables that seem to be irrelevant with respect to this distinction (refractive index, barium, magnesium, calcium, etc.).

Another example to illustrate the power of data visualization can come from the gang database. We already explored this dataset using dynamic associative memory, but we have no idea about the positions the 14 variables could take when mapped to a two-dimensional surface (which variable is closest to another particular variable – see Fig. 5.9). The PST algorithm can help us in visualizing the dataset from the point of view of its many variables:

We can use the PST algorithm to project and visualize the gang database from the records point of view. In this way, one can analyze the similarity of each gang member with respect to the others (see Fig. 5.10). Through this technique, we can also force the PST algorithm to group all gang members into two clusters. This is

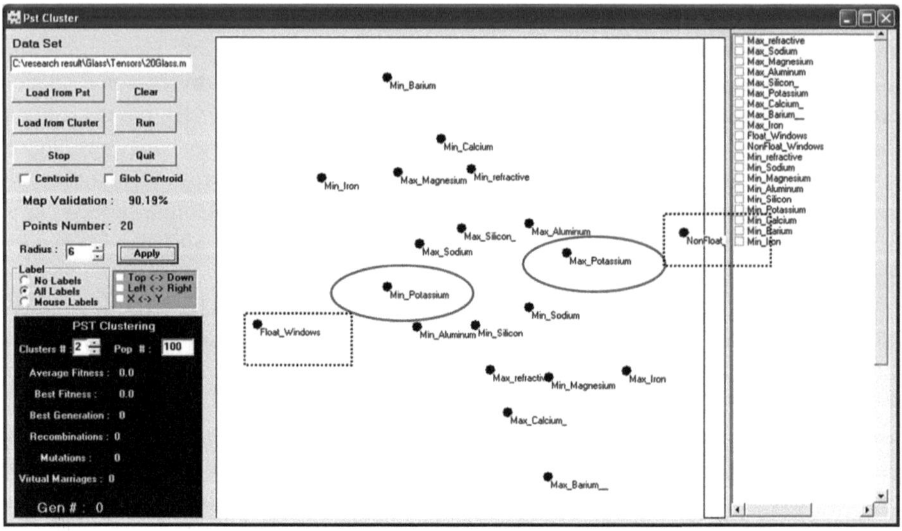

Fig. 5.8 The PST clustering

useful to check if the algorithm is really able to understand the real membership of each member with its original gang.[4]

From this map, we can extract many pieces of relevant information:

1. The PST algorithm unbiasedly creates two clusters whose composition is the same for the two gangs representative of the whole dataset (black color for Jets and red color for Sharks), except for one.
2. The only error of classification is Doug, who is formally a Jet but was associated by the algorithm as a Shark. However, in the preceding example, Doug was shown to be a "strange" Jet and therefore is probably an extreme outlier.
3. We can also observe from this (Fig. 5.10) picture something we have previously noted with the ANN. From the visual point of view, the Jets gang is more compact than Sharks gang. This may suggest that it might be easier to infiltrate the Sharks, or perhaps if a strong bond can be established between an undercover operative and only well placed Jet, then entry into the "inner circle" might be eased.
4. From this perspective, it is quite evident who is similar to whom, who is more representative of his gang and who might lie on the hidden boundary between the two gangs.

Finally, one can conclude that the best place to begin understanding the complexities of a convoluted, ambiguous, and dynamic situation is to view the data in ways in which it is easy to understand.

[4]In Fig. 5.9, Lance and George are overlapped; the same happens for Ol and Phil.

Fig. 5.9 Gang Database: PST projection of 14 dimensions onto 2 dimensions

5.8 Hypothesizing Links: Rebuilding the Net from Scattered and Single Information

Even a rich database often contains data that are not expressly linked with each other. For example, it is possible to check a database of persons arrested in the last four years in London for offenses directly or not directly linked to the drug trafficking and to know details about each of them (sex, age, ethnicity, nationality, committed offenses, served sentences, kind of seized drugs, neighborhood where the arrest took place, residence of the subject, modality of the arrest, etc.). Regardless of the amount of data collected, it is still possible to be unable to find some information aimed at how to understand whether an arrested person is connected with the others in the database; that is to say, how many and which individuals contained in the database are also members of the criminal nets of London and if they are part of that net, which of these arrested individuals can be considered a representative sample of an organized group.

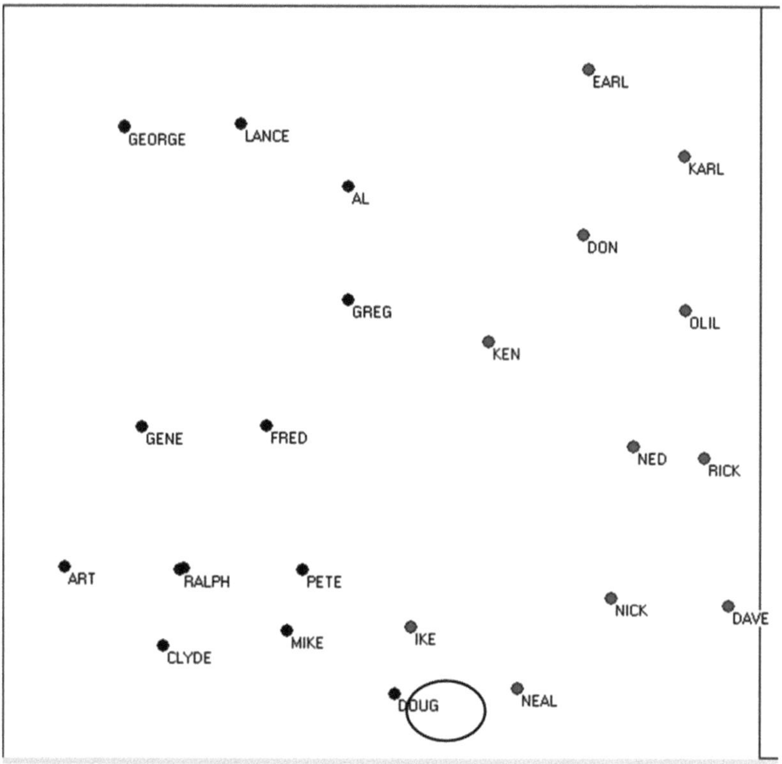

Fig. 5.10 Gang database: PST projection of 27 dimensions into two dimensions (The Jets are *black* and the Sharks are *red*)

Typically, these links among individuals result from data collected from other databases, for it is impossible to otherwise identify other connectivity links among the subjects. However, it can be done from a probabilistic viewpoint, using a different theoretical framework.

First, we assume that in a dataset, all the attributes and all the relations defining any record have to be considered in the same way: the record "Mary," defined as a "female" (attribute) of "24 years old" (attribute), "friend of Clyde" (relation) becomes Mary: {"Female", "24 years old", "Friend of Clyde"}. Consequently, "attributes," "relations," and others features defining a record are all translated in a list of variables of the same rank (they are all blatant data).

Second, once all the records are coded as a list of variables (dummy, numerical, etc.), we assume that the greater the global similarity among the records (i.e., among their own variables), the more plausible it is that those records are linked or will be linked (i.e., hidden connections exist). This assumption comes from the biological sciences: if two very similar molecules meet, it is highly probable they will join together.

Table 5.14 Gang dataset

Name	Gang	Age	Education	Status	Profession
ART	Jets	40	Junior School	Single	Pusher
AL	Jets	30	Junior School	Married	Burglar
SAM	Jets	20	College	Single	Bookie
CLYDE	Jets	40	Junior School	Single	Bookie
MIKE	Jets	30	Junior School	Single	Bookie
JIM	Jets	20	Junior School	Divorced	Burglar
GREG	Jets	20	High School	Married	Pusher
JOHN	Jets	20	Junior School	Married	Burglar
DOUG	Jets	30	High School	Single	Bookie
LANCE	Jets	20	Junior School	Married	Burglar
GEORGE	Jets	20	Junior School	Divorced	Burglar
PETE	Jets	20	High School	Single	Bookie
FRED	Jets	20	High School	Single	Pusher
GENE	Jets	20	College	Single	Pusher
RALPH	Jets	30	Junior School	Single	Pusher
PHIL	Sharks	30	College	Married	Pusher
IKE	Sharks	30	Junior School	Single	Bookie
NICK	Sharks	30	High School	Single	Pusher
DON	Sharks	30	College	Married	Burglar
NED	Sharks	30	College	Married	Bookie
KARL	Sharks	40	High School	Married	Bookie
KEN	Sharks	20	High School	Single	Burglar
EARL	Sharks	40	High School	Married	Burglar
RICK	Sharks	30	High School	Divorced	Burglar
OL	Sharks	30	College	Married	Pusher
NEAL	Sharks	30	High School	Single	Bookie
DAVE	Sharks	30	High School	Divorced	Pusher

In the social world, we can formulate a similar principle: if two persons are defined in very similar way (richness, style of life, age, habit, preferences, desires, targets, house address, etc.), their probability of linkage is very high.

A similar analogy can be made in the criminal world. Two or more criminals group together because in their world, cooperation and aggregation is fundamental to survival. This tendency is named the "criminal elective affinity."

5.8.1 An Example of Rebuilding the Net Link

The gang dataset is a small dataset composed of 27 records and 5 variables (Table 5.14):

Recall the structure of the dataset is:

- Gang = {Jets, Sharks}
- Age = {20s, 30s, 40s}

- Education = {Junior School, High School, College}
- Status = {Married, Single, Divorced}
- Profession = {Pusher, Bookie, Burglar}

The first step is to transform each variable string into a Boolean value through an enumeration of all possible states and the assignment of a 0 to denote exclusion and 1 for inclusion (see Table 5.15):

The new dataset is now composed of 14 binary variables (represented by the columns), most of which are orthogonal.

As we want to use an AutoCM ANN (Buscema 2007; Buscema and Grossi 2008, 2009; Buscema et al. 2008; Buscema and Sacco 2010) to process the records, we must transpose this matrix so that those labels of interest, that is, the names of the gang members, are at the top (Table 5.16):

AutoCM ANN will analyze this dataset using the variables as hyperpoints and the records as coordinates of the hyperpoints.

After about 30 epochs, the AutoCM is completely trained (root mean square error = 0.00000000), and the weights matrix is ready (Table 5.17):

Then we transform the weights matrix into a distance matrix (done automatically by the AutoCM program) (Table 5.18):

At this point, we calculate the minimum spanning tree (MST; Kruskal 1956) of the matrix distance, and the whole dataset is drawn as a tree graph in which every node is a single person (record) (Fig. 5.11):

Through the Delta H Function Hubness, we know that removing Rick, Mike, Neal, or Don from the graph makes the complexity of the graph, and consequently its entropy, increase; while if we remove Al from the global graph, the complexity of the graph, and consequently its entropy, decreases. These actions are not absolutely evident if we analyze the same graph comparing the local indexes. From a naïve point of view, it is possible to think exactly the opposite; since Mike is a big Hub (5 links), if he is removed, one might expect the global network must become simpler. But from a global perspective, the rearrangement of the networks without some of its nodes actually works in a completely opposite way to make the graph far more complex. Hence, one might conclude that by breaking up a known drug ring might result in the appearance of another ring composed of individuals about whom the law enforcement community has no knowledge. That is a serious trade-off that should be discussed by law enforcement officials.

If we remove the vertices within the red circles from Fig. 5.12, the new MST will show a more complex structure, while if we remove the vertex within the blue circle, the new MST will be simpler (see Fig. 5.13 and Fig. 5.14).

From this, we may conclude that once the architecture or structure of a network is known, law enforcement initiatives can choose from being reactive to proactive.

Table 5.15 Binary gang dataset

Gang 14 × 27	Jet	Sharks	20'	30'	40'	JS	COL	HS	Single	Married	Divorced	Pusher	Bookie	Burglar
ART	1	0	0	0	1	1	0	0	1	0	0	1	0	0
AL	1	0	0	1	0	1	0	0	0	1	0	0	0	1
SAM	1	0	1	0	0	0	1	0	1	0	0	0	1	0
CLYDE	1	0	0	0	1	1	0	0	1	0	0	0	1	0
MIKE	1	0	0	1	0	1	0	0	1	0	0	0	1	0
JIM	1	0	1	0	0	1	0	0	0	0	1	0	0	1
GREG	1	0	1	0	0	0	0	1	0	1	0	1	0	0
JOHN	1	0	1	0	0	1	0	0	0	1	0	0	0	1
DOUG	1	0	0	1	0	0	0	1	1	0	0	0	1	0
LANCE	1	0	1	0	0	1	0	0	0	1	0	0	0	1
GEORGE	1	0	1	0	0	1	0	0	0	0	1	0	0	1
PETE	1	0	1	0	0	0	0	1	1	0	0	0	1	0
FRED	1	0	1	0	0	0	0	1	1	0	0	1	0	0
GENE	1	0	1	0	0	0	1	0	1	0	0	1	0	0
RALPH	1	0	0	1	0	1	0	0	1	0	0	1	0	0
PHIL	0	1	0	1	0	0	1	0	0	1	0	1	0	0
IKE	0	1	0	1	0	1	0	0	1	0	0	0	1	0
NICK	0	1	0	1	0	0	0	1	1	0	0	1	0	0
DON	0	1	0	1	0	0	1	0	0	1	0	0	0	1
NED	0	1	0	1	0	0	1	0	0	1	0	0	1	0
KARL	0	1	0	0	1	0	0	1	0	1	0	0	1	0
KEN	0	1	1	0	0	0	0	1	1	0	0	0	0	1
EARL	0	1	0	0	1	0	0	1	0	1	0	0	0	1
RICK	0	1	0	1	0	0	0	1	0	0	1	0	0	1
OL	0	1	0	1	0	0	1	0	0	1	0	1	0	0
NEAL	0	1	0	1	0	0	0	1	1	0	0	0	1	0
DAVE	0	1	0	1	0	0	0	1	0	0	1	1	0	0

Table 5.16 Gang dataset transposed

Gang 27x14	ART	AL	SAM	CLYDE	MIKE	JIM	GREG	JOHN	DOUG	LANCE	GEORGE	PETE	FRED	GENE	RALPH	PHIL	IKE	NICK	DON	NED	KARL	KEN	EARL	RICK	OL	NEAL	DAVE
Jet	1	1	1	1	1	1	1	1	1	1	1	1	1	1	1	0	0	0	0	0	0	0	0	0	0	0	0
Sharks	0	0	0	0	0	0	0	0	0	0	0	0	0	0	0	1	1	1	1	1	1	1	1	1	1	1	1
20'	0	0	1	0	0	1	1	1	0	1	1	1	1	1	0	0	0	0	0	0	0	1	0	0	0	0	0
30'	0	1	0	0	1	0	0	0	1	0	0	0	0	0	1	1	1	1	1	1	0	0	0	1	1	1	1
40'	1	0	0	1	0	0	0	0	0	0	0	0	0	0	0	0	0	0	0	0	1	0	1	0	0	0	0
JS	1	1	0	1	1	1	0	1	0	1	1	0	0	0	1	0	1	0	0	0	0	0	0	0	0	0	0
COL	0	0	1	0	0	0	0	0	0	0	0	0	0	1	0	1	0	0	1	1	0	0	0	0	1	0	0
HS	0	0	0	0	0	0	1	0	1	0	0	1	1	0	0	0	0	1	0	0	1	1	1	1	0	1	1
Single	1	0	1	1	1	0	0	0	1	0	0	1	1	1	1	0	1	1	0	0	0	1	0	0	0	1	0
Married	0	1	0	0	0	0	1	1	0	1	0	0	0	0	0	1	0	0	1	1	1	0	1	0	1	0	0
Divorced	0	0	0	0	0	1	0	0	0	0	1	0	0	0	0	0	0	0	0	0	0	0	0	1	0	0	1
Pusher	1	0	0	0	0	0	1	0	0	0	0	0	1	1	1	1	0	1	0	0	0	0	0	0	1	0	1
Bookie	0	0	1	1	1	0	0	0	1	0	0	1	0	0	0	0	1	0	0	1	1	0	0	0	0	1	0
Burglar	0	1	0	0	0	1	0	1	0	1	1	0	0	0	0	0	0	0	1	0	0	1	1	1	0	0	0

Table 5.17 AutoCM weights matrix

Weights	ART	AL	SAM	CLYDE	MIKE	JIM	GREG	JOHN	DOUG	LANCE	GEORGE	PETE	FRED	GENE	RALPH	PHIL
ART	6.1962	5.2850	5.4981	5.9948	5.8317	5.3123	5.3767	5.2930	5.4659	5.2930	5.3123	5.4435	5.8808	5.9064	6.0280	3.6752
AL	5.2850	6.1962	4.3788	5.2894	5.9166	5.8041	6.5038	6.1319	5.6655	6.0319	5.8041	4.3625	4.3556	4.3721	5.9061	5.3856
SAM	5.4981	4.3788	6.1962	5.8730	5.8248	5.3103	5.3025	5.2865	5.8171	5.2865	5.3103	6.0214	5.8339	6.0422	5.4381	2.9256
CLYDE	5.9948	5.2894	5.8730	6.1962	6.0135	5.3167	4.3493	5.2974	5.8203	5.2974	5.3167	5.8386	5.4637	5.5178	5.8291	0.0001
MIKE	5.8317	5.9166	5.8248	6.0135	6.1962	5.2798	4.3265	5.2605	6.0657	5.2605	5.2798	5.7903	5.4067	5.4608	6.0735	4.2319
JIM	5.3123	5.8041	5.3103	5.3167	5.2798	6.1962	5.3253	6.0110	4.4193	6.0110	6.0871	5.2763	5.2654	5.2946	5.2715	0.0001
GREG	5.3767	5.5038	5.3025	4.3493	4.3265	5.3253	6.1962	5.8737	5.5393	5.8737	5.3253	5.8674	6.0418	5.7770	5.2577	5.1466
JOHN	5.2930	6.0319	5.2865	5.2974	5.2605	6.0110	5.8737	6.1962	4.3875	6.1257	6.0110	5.2525	5.2416	5.2708	5.2522	3.8849
DOUG	5.4659	5.6655	5.8171	5.8203	6.0657	4.4193	5.5393	4.3875	6.1962	4.3875	4.4193	6.0384	5.9002	5.4753	5.9471	4.1552
LANCE	5.2930	6.0319	5.2865	5.2974	5.2605	6.0110	5.8737	6.1257	4.3875	6.1962	6.0110	5.2525	5.2416	5.2708	5.2522	3.8849
GEORGE	5.3123	5.8041	5.3103	5.3167	5.2798	6.0871	5.3253	6.0110	4.4193	6.0110	6.1962	5.2763	5.2654	5.2946	5.2715	0.0001
PETE	5.4435	4.3625	6.0214	5.8386	5.7903	5.2763	5.8674	5.2525	6.0384	5.2525	5.2763	6.1962	6.0521	5.8318	5.3835	0.0001
FRED	5.8808	4.3556	5.8339	5.4637	5.4067	5.2654	6.0418	5.2416	5.9002	5.2416	5.2654	6.0521	6.1962	6.0342	5.8126	3.5168
GENE	5.9064	4.3721	6.0422	5.5178	5.4608	5.2946	5.7770	5.2708	5.4753	5.2708	5.2946	5.8318	6.0342	6.1962	5.8382	4.7576
RALPH	6.0280	5.9061	5.4381	5.8291	6.0735	5.2715	5.2577	5.2522	5.9471	5.2522	5.2715	5.3835	5.8126	5.8382	6.1962	5.2719
PHIL	3.6752	5.3856	2.9256	0.0001	4.2319	0.0001	5.1466	3.8849	4.1552	3.8849	0.0001	0.0001	3.5168	4.7576	5.2719	6.1962
IKE	5.3281	5.2891	5.1172	5.7535	6.0217	3.6919	0.0001	3.6763	5.7314	3.6763	3.6919	5.0173	3.9287	4.0800	5.8611	5.3551
NICK	5.2430	4.0979	4.0358	4.0623	5.4345	0.0001	4.9881	0.0001	5.8464	0.0001	0.0001	5.2777	5.7640	5.1931	5.8022	5.7766
DON	0.0001	5.8303	3.0786	0.0001	4.2896	3.6271	3.6827	5.1309	4.2129	5.1309	3.6271	0.0001	0.0001	2.9942	4.2407	6.0007
NED	0.0001	5.3848	4.8009	3.5016	5.2802	0.0001	3.8553	3.9285	5.2011	3.9285	0.0001	3.4447	0.0001	2.8590	4.1425	6.0006
KARL	2.3591	3.7673	3.5629	4.4991	3.3814	0.0001	5.3836	4.0077	5.0541	4.0077	0.0001	5.2104	3.9247	0.0001	0.0001	5.3013
KEN	4.2137	3.4048	5.2858	4.1760	4.0223	5.0342	5.0804	4.9435	5.2297	4.9435	5.0342	5.7954	5.8026	5.3038	4.0523	3.9815
EARL	2.5185	5.0313	0.0001	2.3812	0.0001	3.6180	5.3490	5.1529	3.8881	5.1529	3.6180	4.0174	4.0084	0.0001	0.0001	5.2474
RICK	0.0001	5.3627	0.0001	0.0001	4.2792	4.4973	3.9957	3.5176	5.4355	3.5176	4.4973	3.9606	3.9516	0.0001	4.2303	5.3638
OL	3.6752	5.3856	2.9256	0.0001	4.2319	0.0001	5.1466	3.8849	4.1552	3.8849	0.0001	0.0001	3.5168	4.7576	5.2719	6.1162
NEAL	4.0836	4.1174	5.1090	5.1164	5.7922	0.0001	3.7810	0.0001	6.0064	0.0001	0.0001	5.7288	5.3049	4.0871	5.4243	5.3000
DAVE	3.6474	4.2389	0.0001	0.0001	4.2475	2.3525	5.1893	0.0001	5.4023	0.0001	2.3525	3.8954	5.1880	3.4981	5.2690	5.8502

(continued)

Table 5.17 (continued)

Weights	IKE	NICK	DON	NED	KARL	KEN	EARL	RICK	OL	NEAL	DAVE
ART	5.3281	5.2430	0.0001	0.0001	2.3591	4.2137	2.5185	0.0001	3.6752	4.0836	3.6474
AL	5.2891	4.0979	5.8303	5.3848	3.7673	3.4048	5.0313	5.3627	5.3856	4.1174	4.2389
SAM	5.1172	4.0358	3.0786	4.8009	3.5629	5.2858	0.0001	0.0001	2.9256	5.1090	0.0001
CLYDE	5.7535	4.0623	0.0001	3.5016	4.4991	4.1760	2.3812	0.0001	0.0001	5.1164	0.0001
MIKE	6.0217	5.4345	4.2896	5.2802	3.3814	4.0223	0.0001	4.2792	4.2319	5.7922	4.2475
JIM	3.6919	0.0001	3.6271	0.0001	0.0001	5.0342	3.6180	4.4973	0.0001	0.0001	2.3525
GREG	0.0001	4.9881	3.6827	3.8553	5.3836	5.0804	5.3490	3.9957	5.1466	3.7810	5.1893
JOHN	3.6763	0.001	5.1309	3.9285	4.0077	4.9435	5.1529	3.5176	3.8849	0.0001	0.0001
DOUG	5.7314	5.8464	4.2129	5.2011	5.0541	5.2297	3.8881	5.4355	4.1552	6.0064	5.4023
LANCE	3.6763	0.0001	5.1309	3.9285	4.0077	4.9435	5.1529	3.5176	3.8849	0.0001	0.0001
GEORGE	3.6919	0.0001	3.6271	0.0001	0.0001	5.0342	3.6180	4.4973	0.0001	0.0001	2.3525
PETE	5.0173	5.2777	0.0001	3.4447	5.2104	5.7954	4.0174	3.9606	0.0001	5.7288	3.8954
FRED	3.9287	5.7640	0.0001	0.0001	3.9247	5.8026	4.0084	3.9516	3.5168	5.3049	5.1880
GENE	4.0800	5.1931	2.9942	2.8590	0.0001	5.3038	0.0001	0.0001	4.7576	4.0871	3.4981
RALPH	5.8611	5.8022	4.2407	4.1425	0.0001	4.0523	0.0001	4.2303	5.2719	5.4243	5.2690
PHIL	5.3551	5.7766	6.0007	6.0006	5.3013	3.9815	5.2474	5.3638	6.1162	5.3000	5.8502
IKE	6.1962	5.8856	5.3577	5.7929	5.1612	5.5253	3.9717	5.3406	5.3551	6.0298	5.3400
NICK	5.8856	6.1962	5.2953	5.2697	5.1870	5.8699	5.2320	5.8235	5.7766	6.0526	6.0164
DON	5.3577	5.2953	6.1962	5.9998	5.2239	5.2634	5.7808	5.8687	6.0007	5.3026	5.3659
NED	5.7929	5.2697	5.9998	6.1962	5.8447	3.9611	5.2565	5.3408	6.0006	5.7581	5.3403
KARL	5.1612	5.1870	5.2239	5.8447	6.1962	5.3089	6.0012	5.2920	5.3013	5.7242	5.2792
KEN	5.5253	5.8699	5.2634	3.9611	5.3089	6.1962	5.8487	5.8227	3.9815	5.8720	5.2537
EARL	3.9717	5.2320	5.7808	5.2565	6.0012	5.8487	6.1962	5.8565	5.2474	5.2474	5.3243
RICK	5.3406	5.8235	5.8687	5.3408	5.2920	5.8227	5.8565	6.1962	5.3638	5.8328	5.9915
OL	5.3551	5.7766	6.0007	6.0006	5.3013	3.9815	5.2474	5.3638	6.1962	5.3000	5.8502
NEAL	6.0298	6.0526	5.3026	5.7581	5.7242	5.8720	5.2474	5.8328	5.3000	6.1962	5.8237
DAVE	5.3400	6.0164	5.3659	5.3403	5.2792	5.2537	5.3243	5.9915	5.8502	5.8237	6.1962

Table 5.18 AutoCM distance matrix

Distances	ART	AL	SAM	CLYDE	MIKE	JIM	GREG	JOHN	DOUG	LANCE	GEORGE	PETE	FRED	GENE	RALPH	PHIL
ART	0.0000	0.9112	0.6981	0.2014	0.3645	0.8839	0.8195	0.9032	0.7303	0.9032	0.8839	0.7527	0.3154	0.2898	0.1682	2.5210
AL	0.9112	0.0000	1.8174	0.9068	0.2796	0.3921	0.6924	0.1643	0.5307	0.1643	0.3921	1.8337	1.8406	1.8241	0.2901	0.8106
SAM	0.6981	1.8174	0.0000	0.3232	0.3714	0.8859	0.8937	0.9097	0.3791	0.9097	0.8859	0.1748	0.3623	0.1540	0.7581	3.2706
CLYDE	0.2014	0.9068	0.3232	0.0000	0.1827	0.8795	1.8469	0.8988	0.3759	0.8988	0.8795	0.3576	0.7325	0.6784	0.3671	6.1961
MIKE	0.3645	0.2796	0.3714	0.1827	0.0000	0.9164	1.8697	0.9357	0.1305	0.9357	0.9164	0.4059	0.7895	0.7354	0.1227	1.9643
JIM	0.8839	0.3921	0.8859	0.8795	0.9164	0.0000	0.8709	0.1852	1.7769	0.1852	0.1091	0.9199	0.9308	0.9016	0.9247	6.1961
GREG	0.8195	0.6924	0.8937	1.8469	1.8697	0.8709	0.0000	0.3225	0.6569	0.3225	0.8709	0.3288	0.1544	0.4192	0.9385	1.0496
JOHN	0.9032	0.1643	0.9097	0.8988	0.9357	0.1852	0.3225	0.0000	1.8087	0.0705	0.1852	0.9437	0.9546	0.9254	0.9440	2.3113
DOUG	0.7303	0.5307	0.3791	0.3759	0.1305	1.7769	0.6569	1.8087	0.0000	1.8087	1.7769	0.1578	0.2960	0.7209	0.2491	2.0410
LANCE	0.9032	0.1643	0.9097	0.8988	0.9357	0.1852	0.3225	0.0705	1.8087	0.0000	0.1852	0.9437	0.9546	0.9254	0.9440	2.3113
GEORGE	0.8839	0.3921	0.8859	0.8795	0.9164	0.1091	0.8709	0.1852	1.7769	0.1852	0.0000	0.9199	0.9308	0.9016	0.9247	6.1961
PETE	0.7527	1.8337	0.1748	0.3576	0.4059	0.9199	0.3288	0.9437	0.1578	0.9437	0.9199	0.0000	0.1441	0.3644	0.8127	6.1961
FRED	0.3154	1.8406	0.3623	0.7325	0.7895	0.9308	0.1544	0.9546	0.2960	0.9546	0.9308	0.1441	0.0000	0.1620	0.3836	2.6794
GENE	0.2898	1.8241	0.1540	0.6784	0.7354	0.9016	0.4192	0.9254	0.7209	0.9254	0.9016	0.3644	0.1620	0.0000	0.3580	1.4386
RALPH	0.1682	0.2901	0.7581	0.3671	0.1227	0.9247	0.9385	0.9440	0.2491	0.9440	0.9247	0.8127	0.3836	0.3580	0.0000	0.9343
PHIL	2.5210	0.8106	3.2706	6.1961	1.9643	6.1961	1.0496	2.3113	2.0410	2.3113	6.1961	6.1961	2.6794	1.4386	0.9243	0.0000
IKE	0.8681	0.9071	1.0790	0.4427	0.1745	2.5043	6.1961	2.5199	0.4648	2.5199	2.5043	1.1789	2.2675	2.1162	0.3351	0.8411
NICK	0.9532	2.0983	2.1604	2.1339	0.7617	6.1961	1.2081	6.1961	0.3498	6.1961	6.1961	0.9185	0.4322	1.0031	0.3940	0.4196
DON	6.1961	0.3659	3.1176	6.1961	1.9066	2.5691	2.5135	1.0653	1.9833	1.0653	2.5691	6.1961	6.1961	3.2020	1.9555	0.1955
NED	6.1961	0.8114	1.3953	2.6946	0.9160	6.1961	2.6409	2.2677	0.9951	2.2677	6.1961	2.7515	6.1961	3.3372	2.0537	0.1956
KARL	3.8371	2.4289	2.6333	1.6971	2.8148	6.1961	0.8126	2.1885	1.1421	2.1885	6.1961	0.9858	2.2715	6.1961	6.1961	0.8949
KEN	1.9825	2.7914	0.9104	2.0202	2.1739	1.1620	1.1158	1.2527	0.9665	1.2527	1.1620	0.4008	0.3936	0.8924	02.1439	2.2147
EARL	3.6777	1.1649	6.1961	3.8150	6.1961	2.5782	0.8472	1.0433	2.3081	1.0433	2.5782	2.1788	2.1878	6.1961	6.1961	0.9488
RICK	6.1961	0.8335	6.1961	6.1961	1.9170	1.6989	2.2005	2.6786	0.7607	2.6786	1.6989	2.2356	2.2446	6.1961	1.9659	0.8324
OL	2.5210	0.8106	3.2706	6.1961	1.9643	6.1961	1.0496	2.3113	2.0410	2.3113	6.1961	6.1961	2.6794	1.4386	0.9243	0.0800
NEAL	2.1126	2.0788	1.0872	1.0798	0.4040	6.1961	2.4252	6.1961	0.1898	6.1961	6.1961	0.4674	0.8913	2.1091	0.7719	0.8962
DAVE	2.5488	1.9573	6.1961	6.1961	1.9487	3.8437	1.0069	6.1961	0.7939	6.1961	3.8437	2.3008	1.0082	2.6981	0.9272	0.3460

(continued)

Table 5.18 (continued)

Distances	IKE	NICK	DON	NED	KARL	KEN	EARL	RICK	OL	NEAL	DAVE
ART	0.8681	0.9532	6.1961	6.1961	3.8371	1.9825	3.6777	6.1961	2.5210	2.1126	2.5488
AL	0.9071	2.0983	0.3659	0.8114	2.4289	2.7914	1.1649	0.8335	0.8106	2.0788	1.9573
SAM	1.0790	2.1604	3.1176	1.3953	2.6333	0.9104	6.1961	6.1961	3.2706	1.0872	6.1961
CLYDE	0.4427	2.1339	6.1961	2.6946	1.6971	2.0202	3.8150	6.1961	6.1961	1.0798	6.1961
MIKE	0.1745	0.7617	1.9066	0.9160	2.8148	2.1739	6.1961	1.9170	1.9643	0.4040	1.9487
JIM	2.5043	6.1961	2.5691	6.1961	6.1961	1.1620	2.5782	1.6989	6.1961	6.1961	3.8437
GREG	6.1961	1.2081	2.5135	2.3409	0.8126	1.1158	0.8472	2.2005	1.0496	2.4152	1.0069
JOHN	2.5199	6.1961	1.0653	2.2677	2.1885	1.2527	1.0433	2.6786	2.3113	6.1961	6.1961
DOUG	0.4648	0.3498	1.9833	0.9951	1.1421	0.9665	2.3081	0.7607	2.0410	0.1898	0.7939
LANCE	2.5199	6.1961	1.0653	2.2677	2.1885	1.2527	1.0433	2.6786	2.3113	6.1961	6.1961
GEORGE	2.5043	6.1961	2.5691	6.1961	6.1961	1.1620	2.5782	1.6989	6.1961	6.1961	3.8437
PETE	1.1789	0.9185	6.1961	2.7515	0.9858	0.4008	2.1788	2.2356	6.1961	0.4674	2.3008
FRED	2.2675	0.4322	6.1961	6.1961	2.2715	0.3936	2.1878	2.2446	2.6794	0.8913	1.0082
GENE	2.1162	1.0031	3.2020	3.3372	6.1961	0.8924	6.1961	6.1961	1.4386	2.1091	2.6981
RALPH	0.3351	0.3940	1.9555	2.0537	6.1961	2.1439	6.1961	1.9659	0.9243	0.7719	0.9272
PHIL	0.8411	0.4196	0.1955	0.1956	0.8949	2.2147	0.9488	0.8324	0.0800	0.8962	0.3460
IKE	0.0000	0.3106	0.8385	0.4033	1.0350	0.6709	2.2245	0.8556	0.8411	0.1664	0.8562
NICK	0.3106	0.0000	0.9009	0.9265	1.0092	0.3263	0.9642	0.3727	0.4196	0.1436	0.1798
DON	0.8385	0.9009	0.0000	0.1964	0.9723	0.9328	0.4154	0.3275	0.1955	0.8936	0.8303
NED	0.4033	0.9265	0.1964	0.0000	0.3515	2.2351	0.9397	0.8554	0.1956	0.4381	0.8559
KARL	1.0350	1.0092	0.9723	0.3515	0.0000	0.8873	0.1950	0.9042	0.8949	0.4720	0.9170
KEN	0.6709	0.3263	0.9328	2.2351	0.8873	0.0000	0.3475	0.3735	2.2147	0.3242	0.9425
EARL	2.2245	0.9642	0.4154	0.9397	0.1950	0.3475	0.0000	0.3397	0.9488	0.9488	0.8719
RICK	0.8556	0.3727	0.3275	0.8554	0.9042	0.3735	0.3397	0.0000	0.8324	0.3634	0.2047
OL	0.8411	0.4196	0.1955	0.1956	0.8949	2.2147	0.9488	0.8324	0.0000	0.8962	0.3460
NEAL	0.1664	0.1436	0.8936	0.4381	0.4720	0.3242	0.9488	0.3634	0.8962	0.0000	0.3725
DAVE	0.8562	0.1798	0.8303	0.8559	0.9170	0.9425	0.8719	0.2047	0.3460	0.3725	0.0000

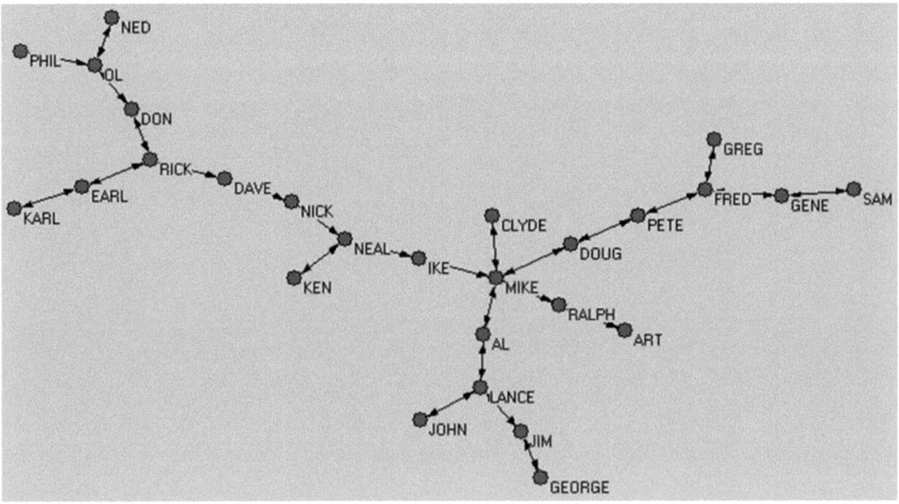

Fig. 5.11 The MST of the global networks

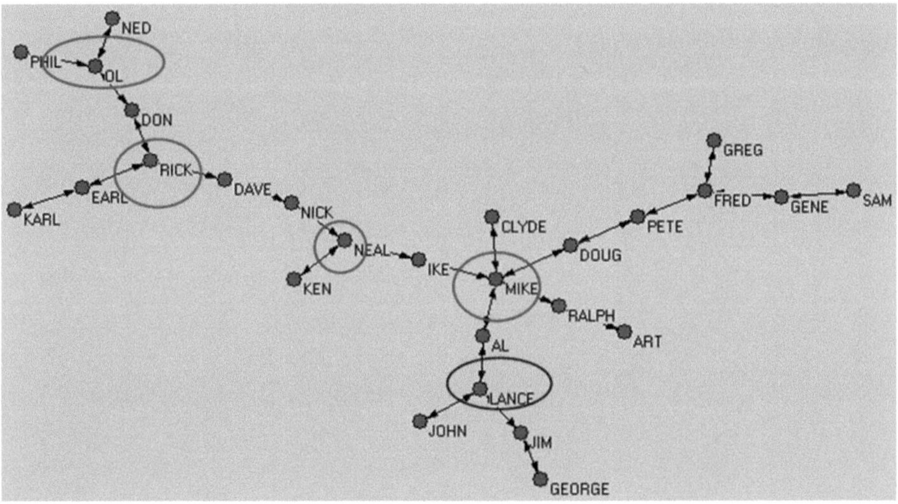

Fig. 5.12 The MST of the global networks marked

5.8.2 Conclusion

This chapter discusses a novel way of taking currently existing data that is residing in a law enforcement database, and, through some straightforward restructuring of the data and the application of the AutoCM neural network, views of the linkages between elements can be inspected. This process could direct law enforcement initiatives into certain, heretofore unknown or unexpected, investigative directions.

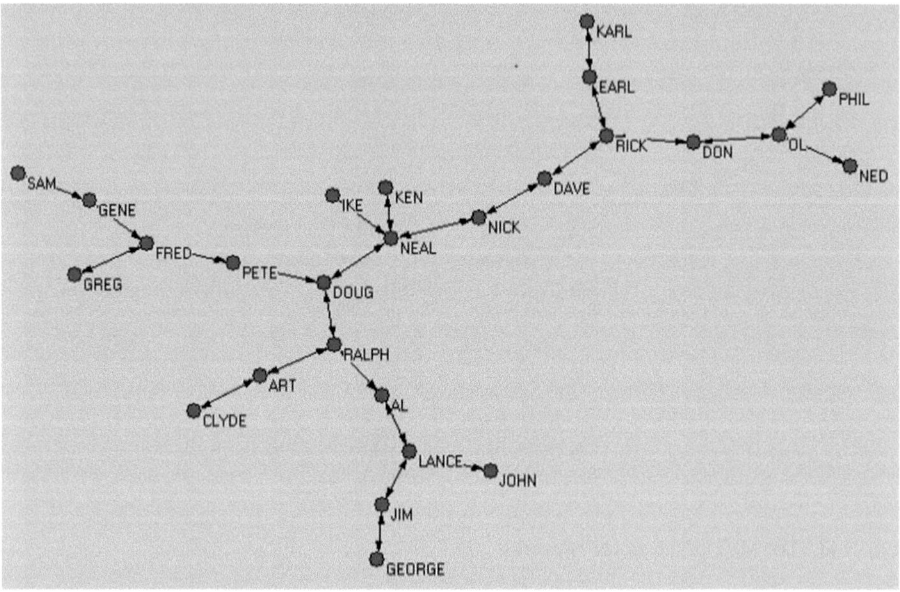

Fig. 5.13 New MST without Mike

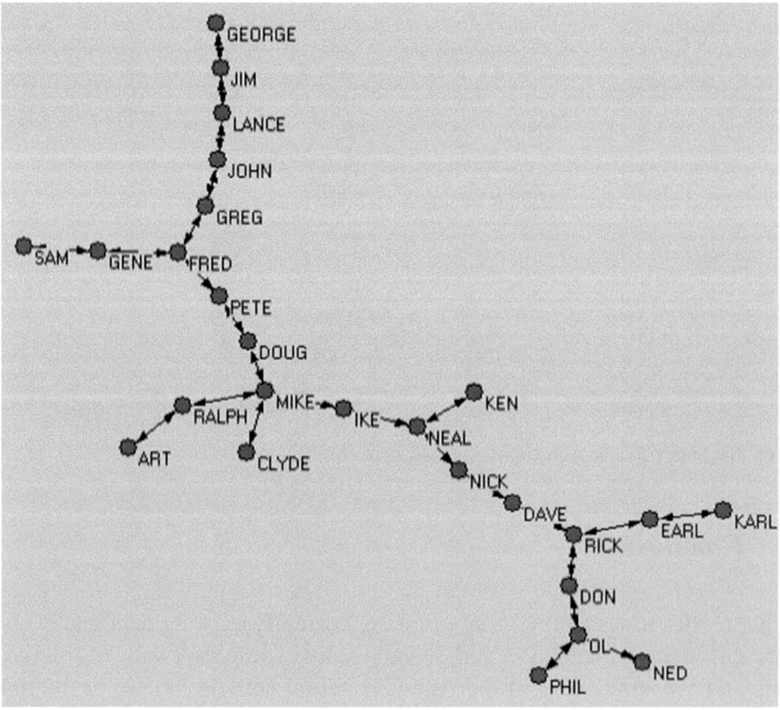

Fig. 5.14 New MST without AL

References

Anderson, J. A., & Rosenfeld, E. (Eds.). (1988). *Neurocomputing foundations of research*. Cambridge, MA: MIT Press.

Arbib, M. A. (Ed.). (1995). *The handbook of brain theory and neural networks*. Cambridge, MA/London: MIT Press, A Bradford Book.

Bishop, C. M. (1995). *Neural networks for pattern recognition*. New York: Oxford University Press.

Buscema, M. (1997). *A general presentation of artificial neural networks*. New York: Marcel Dekker. In *Substance Use & Misuse, The International Journal of the Addictions, 32*(1), 97–112.

Buscema, M. (1998a). *Artificial neural networks and complex social systems*. New York: Marcel Dekker. Special issue of *Substance Use & Misuse, 33*(1), 2, 3.

Buscema, M. (1998b). *Recirculation neural networks*. New York: Marcel Dekker. In *Substance Use & Misuse, (Models), 33*(2),383–388.

Buscema, M. (2002). *A brief overview and introduction to artificial neural networks*. New York: Marcel Dekker. In *Substance Use & Misuse*, Special issue on "The Middle Eastern summer institute on drug use – Proceedings: 1997/1999, June–August 2002", *37*(8–10), 1093–1148.

Buscema, M. (2007). *Squashing theory and contractive map network* (Semeion Technical Paper #32), Rome.

Buscema, M., & Benzi, R. (2011). Quakes prediction using a highly non linear system and a minimal dataset. In M. Buscema & M. Ruggieri (Eds.), *Advanced networks, algorithms and modeling for earthquake prediction* (River Publisher series in information science and technology). Aalborg: River Publisher.

Buscema, M., & Diappi, L. (2004). Improved understanding of urban sprawl using neural networks. In J. P. Van Leeuwen & H. J. P. Timmermans (Eds.), *Recent advances in design and decision support systems in architecture and urban planning*. Dordrecht: Kluwer Academic.

Buscema, M., & Grossi, E. (2008). The semantic connectivity Map: An adapting self-organizing knowledge discovery method in data bases. Experience in gastro-oesophageal reflux disease. *International Journal of Data Mining and Bioinformatics, 2*(4), 362–404.

Buscema, M., & Grossi, E. (Eds.). (2009). *Artificial adaptive systems in medicine* (pp. 25–47). Sharjah: Bentham e-books.

Buscema, M., & Sacco, P. L. (2010). Auto-contractive maps, the H function, and the maximally regular graph (MRG): A new methodology for data mining (Chapter 11). In V. Capecchi et al. (Eds.), *Applications of mathematics in models, artificial neural networks and arts*. Dordrecht/London: Springer. doi:10.1007/978-90-481-8581-8_11.

Buscema, M., & Terzi, S. (2006). Pst: A new evolutionary approach to topographic mapping. *WSEAS Transactions on Information Science and Applications, 3*(9), 1704–1710.

Buscema, M., Terzi, S., Maurelli, G., Capriotti, M., & Carlei, V. (2006). The smart library architecture of an orientation portal. *Quality & Quantity, 40*, 911–933.

Buscema, M., Grossi, E., Snowdon, D., & Antuono, P. (2008). Auto-contractive maps: An artificial adaptive system for data mining. An application to Alzheimer disease. *Current Alzheimer Research, 5*, 481–498.

Buscema, M., Terzi, S., & Tastle, W. (2010, Juy 12–14). *A new meta-classifier*. NAFIPS 2010, Toronto.

Chauvin, Y., & Rumelhart, D. E. (Eds.). (1995). *Backpropagation: Theory, architectures, and applications*. Hillsdale: Lawrence Erlbaum.

Churchland, P. M., & Churchland, P. S. (1990). Could a machine think? *Scientific American, 262*(1).

Churchland, P. S., & Sejnowski, T. J. (1992). *The computational brain*. Cambridge, MA: MIT Press.

Duda, R. O., Hart, P. E., & Stork, D. G. (2001). *Pattern classification*. New York: Wiley.

Hopfield, J. J. (1982). Neural networks and physical systems with emergent collective computational abilities. *Proceedings of the National Academy of Sciences, 79*, 2554–2558.
Kohonen, T., Barna, G., & Chrisley, R. (1988). Statistical pattern recognition with neural networks: Benchmarking studies. *Proceedings of the international conference on neural networks* (Vol. I, pp. 61–68). New York: IEEE Press.
Kruskal, J. (1956). On the shortest spanning sub tree and the traveling salesman problem. *Proceedings of the American Mathematical Society, 7*, 48–50.
Kuncheva, L. I. (2004). *Combining pattern classifiers: Methods and algorithms*. Hoboken: Wiley.
McClelland, J. L., & Rumelhart, D. E. (1988). *Explorations in parallel distributed processing*. Cambridge, MA: MIT Press.
Mena, J. (2003). *Investigative data mining for security and criminal detection*. Boston: Elsevier.
Pao, Y. H. (1990). *Pattern recognition and neural network*. Reading: Addison Wesley.
Ripley, B. D. (1996). *Pattern recognition and neural networks*. Cambridge/New York: Cambridge University Press.
Rumelhart, D. E., & McClelland, J. L. (1986a). *Parallel distributed processing: Foundations, explorations in the microstructure of cognition* (Vol. I). Cambridge, MA/London: MIT Press.
Rumelhart, D. E., & McClelland, J. L. (Eds.). (1986b). *Parallel distributed processing: Psychological and biological models* (Vol. II). Cambridge, MA/London: MIT Press.
Schalkoff, R. J. (1992). *Pattern recognition: Statistical, structural and neural approaches*. New York: Wiley.
Werbos, P. J. (1994). *The roots of backpropagation*. New York: Wiley.
Witten, I. H., & Frank, E. (2005). *Data mining*. Amsterdam/Boston: Morgan Kaufmann.

Software

Buscema, M. (1999–2007). AutoAssociative Neural Networks, software #13, v. 8.
Buscema, M. (2006–2008). Minimum Spanning Tree, software #38, v. 5.
Buscema, M. (2006–2008). PST Cluster, software #34, v. 5.2.
Buscema, M. (2007). PST, software #11, v. 7. System protected under international patent. PCT/EP2004/051190, receipt 26.06.04, applicant Semeion Centro Ricerche.
Buscema, M. (2007). Supervised ANNs and Organisms, software #12, v. 12.5.
Massini, G. (2007). Trees Visualizer, software #40, v.3.

Chapter 6
The Metropolitan Police Service Central Drug-Trafficking Database: Evidence of Need

Geoffrey Monaghan and Stefano Terzi

6.1 Introduction

Law enforcement agencies are often drowning in data and require their officers to enter details of their actions into computers that seem to transmit the data into a black hole. An increasing number are fortunate enough to have systems that have search capabilities so that some value can be gleaned from the input effort; however many others use fairly rudimentary systems that produce little more than generalized statistical reports. This chapter describes the work of the Metropolitan Police Service (MPS) in London to improve the quality of its data on drug trafficking offences[1] and introduce a more sophisticated database management system. The description contained is both descriptive and proscriptive for a similar effort can be made to customize databases to permit other agencies in their efforts to combat terrorism, white-collar crime, the financial industry, threats to individuals and the like. There exist in the literature references to business situations in which new systems failed regardless of the money thrust into the project; this is an account of a success that can be emulated.

[1] In this chapter, the terms 'drug trafficking' and 'drug trafficking offence' are used interchangeably. 'Drug trafficking offence' is defined in Section 1 of the Drug Trafficking Act 1994 and includes the production, supply, offer to supply, possession with intent to supply, importation, exportation, etc., of controlled drugs.

G. Monaghan (✉)
Drugs and HIV/AIDS Expert, United Nations Office on Drugs and Crime Russia and Viet Nam, CDTD Deputy Project Manager and Detective Sergeant, Drugs Directorate, New Scotland Yard, Metropolitan Police Service, London, UK
e-mail: Geoff.Monaghan@unodc.org

S. Terzi
Semeion Research Centre, Rome, Italy

M. Buscema and W.J. Tastle (eds.), *Intelligent Data Mining in Law Enforcement Analytics: New Neural Networks Applied to Real Problems*, DOI 10.1007/978-94-007-4914-6_6,

Concern about drug trafficking in London is not new. But during the last two decades, this concern has been heightened by the development of highly visible street drug markets and a rapid escalation in the level of violence associated with them. Many murders and other violent offences in London have been attributed to drug traffickers. In truth, the trafficking of heroin and cocaine, particularly in the form of crack, presents the MPS and communities in London with many difficulties. Crack trafficking in particular is posing acute problems and is increasing in severity. As noted by the Greater London Alcohol and Drug Alliance (GLADA) in its recent report *The GLADA Crack Cocaine Strategy 2005–08* between 1998/1999 and 2002/2003, the number of offences of supplying crack cocaine in London rose by 151%, and Drug Action Teams (DATs) have reported a proliferation of 'crack houses' in the same period. While organized criminal networks (OCNs) have been documented in London for hundreds of years, government officials, law enforcement officers and some researchers suggest that OCNs are now posing a more serious crime problem than in the past. In some parts of London, OCNs are credited with an alarming share of violent crime, especially murders. While reports conflict about the extent to which OCNs play in controlling drug trafficking in the capital, there is little doubt that Turkish OCNs play a leading role in terms of heroin trafficking while Jamaican OCNs and OCNs comprising black British nationals dominate the trade in crack. More recently, Vietnamese OCNs involved in the large-scale production of cannabis have been identified in a number of London boroughs.

Before the MPS can assess its ability and that of its partners to cope with the problem of drug trafficking, it needs to know much more about OCNs, the financial networks which support them and the structure of illicit drug markets. It needs to have a much better understanding of how controlled drugs (e.g. heroin, cocaine and amphetamine-type stimulants) get from their source to their points of retail sale, and it needs to know a good deal more about the ploys, tricks and stratagems traffickers use and the legal loopholes they exploit. However, despite the considerable amount of data generated by its enforcement efforts and a growing body of related academic research, the MPS still knows relatively little about these topics. It knows even less about the effectiveness of the tactics it uses in its efforts to disrupt and dismantle OCNs and drug markets. Drawing on research from a number of industrialized countries, Hough and Natarajan (2000) conclude that the relationship between the trafficking of controlled drugs, the demand for them and enforcement activities are poorly conceptualized by politicians and policymakers and are seriously under-researched. In terms of British research, Pearson and Hobbs (2001) make the point that the evidence base for drug trafficking in general and 'middle-market distribution' in particular is considerably underdeveloped, while Bean and Nimitz (2004) notes that little consideration has been given to the role foreign nationals play in terms of shaping or fuelling drug markets in Britain. As King (2004) points out, very little is known about the ways in which enforcement activity impacts prices and the ways in which prices are linked to demand. More importantly, he notes the difficulty in linking interdiction and intervention efforts by law enforcement agencies to fluctuations in the unit price of drugs.

While the so-called dark figure of crime – the mismatch between crime estimates produced by victimization surveys and those recorded by the police – is a well-known concept, scant regard has been paid to the accuracy of the published data on the number of drug-trafficking crimes reported by British police forces. This is interesting because the author has identified huge discrepancies in the MPS data.

Bean (2004) believes that these discrepancies are linked to two separate matters. First, there are limitations imposed by the methodology of data collection – that is, where data on drug traffickers provide only a partial picture. For example, a heroin trafficker may admit to supplying heroin on a dozen occasions but may only be prosecuted for two offences – 'specimen counts'. How best to record and interpret this? Or how to interpret the data on drug prices or the Forensic Science Service data on seizures? Bean (2004) notes that the 'counting rules' often muddy the waters. For example, he points out that Home Office statistics do not distinguish between production of cannabis (a drug-trafficking offence) and cultivation of cannabis (a non-drug-trafficking offence). There is also inconsistency in how the Crown Prosecution Service (CPS) and the police service interpret the 'rules'. Thus, a person growing a cannabis plant could be prosecuted for *producing* or *cultivating* or *possessing* cannabis. These limitations occur not as a result of defects in the data but because the reporting systems are not sufficiently comprehensive or adhered to.

The second matter relates to the validity and reliability of the data and directly to the collection of the data. Take, for example, the accuracy of the MPS Performance Information Bureau (PIB) data on Class A drug trafficking. Analyses of Crime Report Information System (CRIS)[2] data leave no doubt that they are seriously flawed and present a distorted picture of drug trafficking. In 1999, the author analyzed a sample of drug-trafficking reports on CRIS on behalf of the Independent Inquiry into the Misuse of Drugs Act. He found that nearly one third (31%) had been incorrectly classified. In another study on cannabis, the author found that a number of 'possession of cannabis' offences had been incorrectly classified as 'production of cannabis' (a drug-trafficking offence). He also found many instances where Class A controlled drugs (such as methadone) had been recorded as Class C controlled drugs, and in a number of cases, 'accused persons' had been charged or cautioned under the Misuse of Drugs Act 1971 with possessing ketamine (a dissociative anaesthetic), which, at the time, was not a controlled drug. Important fields in CRIS (e.g. nationality, ethnicity code and drug type and amount) were often left blank. All too often, even when the nationality field had been completed, the information was useless because the reporting officer has used terms such as 'West African', 'West Indian' or 'Eastern European'. In their study of CRIS records in 2000, Bean and Nemitz (2004) also found numerous errors, such as where 'drug-trafficking offences' were recorded and classified without any supporting evidence. The results of their study supported fully those found by the author.

[2]CRIS has been designed to provide an automated borough crime reporting and information system. It replaced the old-fashioned paper-based system and allows police officers to enter crime report data via computer terminals located in their own police stations. The database can be used for many types of enquiries, including crime analysis information and enquiries about specific or linked crimes.

The difficulty is compounded given the shortcomings of other databases. For example, the inconsistency in forces' records in getting data onto Phoenix (the Police National Computer (PNC)), both in terms of timeliness of input and the quality of information recorded, is well documented.

Given the problems identified, it is not surprising that MPS analysts struggle to produce quality reports on drug trafficking and OCNs. Obviously, before they can produce and disseminate quality information in the form of the various strategic and tactical assessments, they need to be able to access and retrieve reliable data quickly. Many analysts are well aware of the inherent weaknesses in the current systems. The following comments by two MPS senior analysts illustrate this point:

> A central Met-wide database of drug operations should be maintained to allow easily obtainable performance data (for use) against drug traffickers. It will then be possible to [gauge] the impact of anti-drug operations ... to inform future planning and coordination of such activity.
>
> Cook, J. (May 2004) *Drugs Desk Strategic Submission* Specialist Crime Directorate 10(4), MPS, p. 25
>
> We cannot expect to provide an effective response to the drugs problem when we don't truly understand the problem. We need to build up knowledge of all levels of street supply, from localized dealing, up to how ... importers distribute [drugs] throughout London.
>
> Metters, W. (July 2004) Drugs Strategic Committee: Minutes of meeting held on 14 July 2004 at New Scotland Yard

There is also widespread concern among MPS officers and analysts that they do not have access to the kind of reasonably complete and accurate data, which would allow them to determine (a) whether there had been any significant changes in drug-trafficking trends and (b) whether operational activity has had any discernible impact on OCNs and drug markets. They also recognize that comprehensive analyses cannot be done unless they have access to, and are trained to use, sophisticated data processing programs.

6.2 The MPS Central Drug-Trafficking Database (CDTD) Project

6.2.1 Evidence of Need

Over the years, researchers and police staff alike have experienced considerable difficulty in answering some highly complex questions relating to drug trafficking. The following list will give the readers some idea of the task they face:

- How much heroin was seized as a result of test purchase operations in 2004?
- What is the 'average street price' of heroin/crack in London in 2004?
- Based on current knowledge, how much is a gram of heroin likely to cost in 2020?
- How many foreign nationals were arrested for heroin trafficking in 2004?

- How many offenders charged with heroin trafficking in 2004 were in methadone programs?
- How much cannabis was produced (cultivated) on a large-scale commercial basis in 2004?
- How many offenders charged with cannabis trafficking in 2004 have convictions for theft, robbery or burglary?
- In terms of police tactics: 'What works? What doesn't work? What looks promising?'

To be able to answer these questions requires a considerable investment of time in terms of research. Even if the MPS analysts had the requisite knowledge,[3] and the relevant skills[4] and time, the authors doubt that the answers obtained would withstand academic scrutiny given the problems over the validity and reliability of CRIS data.

The coincidence of several recent and forthcoming developments strengthens the case for timely, accurate and complete recording of drug-trafficking information. Chief among these is the MPS's adoption of the National Intelligence Model (NIM). The NIM was produced by the NCIS on behalf of the ACPO Crime Committee to 'professionalise [sic] the intelligence discipline within law enforcement'. Briefly, the model comprises four components:

- Tasking and coordination (takes account of business planning needs in the context of government and local objectives)
- Four key intelligence products (see below)
- Knowledge products (i.e. rules and protocols for the conduct of business and the identification and promotion of best practice)
- System products (i.e. data systems, intelligence acquisition and operational security and effectiveness)

The four key intelligence products are (1) the strategic assessment, (2) the tactical assessment, (3) the target profile and (4) problem profile.

6.3 Aim of the Central Drug-Trafficking Database

The aim of the CDTD is to enable the MPS to produce reliable and objective data to help the MPS and its partners to (a) assess the extent of the drug problem in London and (b) devise appropriate responses to tackle the problem.

In addition to capturing the nominal details of offenders and their criminal histories, offences and drug seizures, the CDTD has been designed to store and

---

[3]For example, in-depth knowledge of OCNs, drugs legislation and drug trends. The authors also believe that operational experience would be advantageous.

[4]Knowledge of Artificial Adaptive Systems and Expert Systems, for example.

process a wealth of additional information relating to police tactics, unit drug prices and purity and the resolution of investigations. As such, it will provide the MPS with an effective and efficient information system which will make an important contribution to its strategic planning and will help to bring some clarity of thinking to the debate on drug-trafficking trends, the effectiveness of police tactics and the role organized crime networks (OCNs) play in shaping and influencing drug markets.

6.4 Objectives of the Central Drug-Trafficking Database

The objectives of the CDTD are to:

- Describe the nature and scale of drug-trafficking activity in London
- Help the MPS to identify drug misuse trends in London
- Help the MPS to identify and describe the structure of OCNs and describe the role they play in controlling drug markets
- Provide a description of MPS enforcement tactics (e.g. search warrants, test purchases, controlled deliveries) and evaluate their effectiveness
- Draw out any associations between drug trafficking and other types of offending (i.e. acquisitive and violent crime) and provide detailed profiles of offenders (e.g. sex, age, nationality)
- Provide profiles of operational police officers with the intention of monitoring and evaluating police performance
- Identify the legal lacunae and obstacles and logistical constraints, which frustrate or impede enforcement activity
- Identify and describe the ploys, stratagems and ruses used by drug traffickers with a view to introducing counter measures
- Help identify training needs for operational officers and support staff
- Provide reliable and objective data on the price of Class A drugs particularly heroin and cocaine (including crack) and the factors that may influence prices such as location, drug purity, trafficker profile, amount purchased and availability
- Determine how foreign nationals help shape and influence drug markets in London
- Help the MPS to gain a better understanding of drug-trafficking routes from the point of entry to onward distribution
- Help the MPS to identify those environmental factors associated with successful drug market locations
- Help the MPS to predict with a greater degree of certainty, drug-trafficking trends and drug market locations
- Provide the MPS and the Metropolitan Police Authority with the information they need to devise performance targets based on quantitative and qualitative data
- Help the MPS to assess the validity of its case disposal policy (e.g. alternatives to prosecution such as formal warnings, reprimands and cautions)

- Understand better the processes by which decisions are made by the Crown Prosecution Service to prosecute drug traffickers
- Assess sentencing practice in both Magistrates' Courts and the Crown Court (e.g. whether the judiciary is complying with the provisions of section 110 of the Powers of Criminal Courts (Sentencing) Act 2000 – minimum sentence of 7 years' imprisonment for offenders following a third conviction for a drug-trafficking offence involving a Class A drug)

6.5 The Central Drug-Trafficking Database: Description

The CDTD is in effect a composite database, the information for which is drawn from six MPS databases and six external databases including the Police National Computer (PNC). An important part of this process will be the distillation and refinement of these data.

The primary source of data will be drug-trafficking reports recorded on the MPS Crime Report Information System in 2004. In all, some 4,500 CRIS records will need to be reviewed. In turn, this means around 5,000 offenders will be identified. In order to capture all the important variables, each record will comprise over 400 fields.

As already noted, the limitations of the recording and retrieval systems inhibit robust performance assessment in several important areas. As Chatterton et al. (1998) and his colleagues point out, a key problem concerns the availability of good quality arrest data. They say that the crux of the matter is that the police make arrests using a range of tactics that are informed by different operational strategies. Treating these arrests as if they are identical and aggregating them masks important differences between them. This is unsatisfactory, especially if one is concerned with examining the impact of changes or the effectiveness of particular tactical options. It is widely known that certain types of tactics are used – stop and search, search warrants, strip searches, etc. – but there are others which are given less prominence in the literature, such as intimate searches, test purchases and controlled deliveries.[5] In order to rectify this, the authors have developed an interactive model that will help them identify the different arrest tactics used in an attempt to determine their effectiveness. Briefly, they categorize police tactics as primary, secondary, tertiary, etc.

Mention has already been made to the fact that the information on the CDTD will be drawn from a number of internal and external databases. Each of these data

[5]The term 'controlled delivery' is defined in Article 1(g) of the 1988 *Convention Against Illicit Traffic in Narcotic Drugs and Psychotropic Substances.* Article 11 of the convention provides 'for the use of controlled deliveries at international level and paves the way for "clean" deliveries'. This technique is frequently employed in drug importation investigations, and indeed this has been the case for many years.

sources provides only a partial view of drug-trafficking patterns and police tactics. However, when combined, they provide a wealth of information. So much so that the information will allow investigator to make detailed and informed statements about drug-trafficking trends, drug-trafficking groups and the effectiveness of enforcement tactics.

The databases used are set out below:

MPS Databases

- *Crime Report Information System* (CRIS)
- *Criminal intelligence* (CRIMINT) *database*
- *Custody Package database* (This database holds information on arrestees whether or not they are subsequently prosecuted.)
- *Stops database* (This database holds information on those suspects stopped and/or searched by police officers.)
- *Covert Drug Purchase database* (Designed by the author and funded by Home Office and piloted by the SCD 10 (4) Drugs Focus Desk. The database has since been modified by the author (see Appendix 1 to this chapter).)
- *Pharmacy database* (This database holds information on addicts who are prescribed controlled drugs. This information is gleaned from retail pharmacy records following routine inspection by SCD 3 (3) Controlled Drug Inspectors.)

External Databases

- *Police National Computer* (PNC)
- *Drugs database* (Forensic Science Service (FSS)) (This database holds information on drug seizures made by the MPS and other law enforcement agencies.)

6.6 The Relational Database: Architecture

6.6.1 The Database Design

The first step in designing this special type of database is to identify the main subjects (entities) that characterize and describe drug-trafficking activity and to define their relationships. The general model of the CDTD (Central Drug-Trafficking Database) was divided into five sub-models containing information on the five 'subject areas' identified during the process of designing the database. This type of subdivision of the activity has considerable relevance and influence on the activities described later, both in relation to the process of gathering and storing data and in the analysis which will subsequently be conducted. There are substantially two main reasons for this:

1. In order to prevent problems at the data entry stage, the pre-existing structure of the data to be collected (the format and organization of police reports) and the other source databases had necessarily to be taken into account.

2. The analyses required the data to be formulated in such a way that it could be accessed easily in accordance with the set objectives.

The five subject areas identified during the analysis process are as follows: *tactics*, *seizures*, *arrests*, *persons* and *locations*.

6.6.2 Tactics

In this subject area, all the steps followed by the police and other law enforcement officers when fighting drug-trafficking crimes have been collected and codified. In order to allow the data to be processed with the analysis tools, it was necessary to structure and classify them with a level of detail and indefiniteness sufficient to avoid unnecessary fragmentation and to provide an intelligible synthesis of the actual drug-trafficking activities in London. The fundamental entity in this subject area is tactic, meaning a single action performed by a police officer when confronting a suspect in order to obtain information on possible illegal activities he or she might be involved in.

The possible tactics have been divided into six generic categories:

- *Search of person* – a search of a person (including the clothes he or she is wearing) aimed at finding something on that person.
- *Search of object* – a search of an object (e.g. a briefcase, postal packet, container) aimed at finding an object.
- *Search of premises* – a search of a dwelling, office or motor vehicle aimed at finding something in the suspect's premises.
- *Controlled deliveries* – Article 11 of the 1988 United Nations Convention against Illicit Trafficking Narcotic Drugs and Psychotropic Substances endorses the investigative technique of controlled delivery at the international level. In Article 1, subparagraph (g), controlled delivery is defined as 'the technique of allowing illicit or suspect consignments of narcotic drugs, psychotropic substances, [precursor chemicals], or substances substituted for them, to pass out of, through or into the territory of one or more countries, with the knowledge and under the supervision of their competent authorities, with a view to identifying persons involved in the commission of offences established in accordance with Article 3, paragraph 1 of the Convention'. The majority of controlled deliveries carried out by MPS officers arise from drug seizures linked to postal packet interceptions such as checking an object (generally a postal package) containing drugs with the aim of tracing those involved in trafficking. For example, the post office intercepts a parcel containing drugs sent to some person and alerts the authorities who will not stop the delivery but, rather, will keep the addressee under surveillance to verify whether he is the final receiver or whether other people are involved.
- *Covert purchases* – an operation during which undercover police officers purchase drugs from trafficker dealers in order to obtain evidence of their illegal activity.

Table 6.1 Generic tactic typology

Generic tactic	Detailed tactic
Search of object	Other
	Not known
Search of person	Stop and search
	Other search
	Search following arrest
	Search in police detention
	Strip search in police detention
	Intimate search in police detention
	Search in prison
	Strip search in prison
	Intimate search in prison
	Not known
Search of premises	Warrant
	Without warrant
	Writ of assistance
	Other
	Not known
Covert purchase	Test purchase operation
	Undercover buy
	Not known
Controlled delivery	Controlled delivery
	Clean delivery
	Not known
Other	Other
	Not known

- *Other* – a general category covering all the tactics that do not fall into any of the above. For example, where the hospital authorities notify the police in a case where a drug smuggler is hospitalized as a result of having swallowed packages containing controlled drugs.

Each of these categories is then further divided into detailed tactics, the composition of which is given in Table 6.1.

Other items of information that distinguish the individual tactic are:

- The type of officer who carried out the tactic; this would be useful to know, for example, when assessing the relevance of the activities undertaken by particular officers, such as security guards at nightclubs.
- The type of legislation enforced by the officer in carrying out the tactic, the time, the date and place it was carried out, details of any objects or properties involved and also, in the case of a controlled delivery, information on the person acting as sender, intermediary and receiver.

Further information concerning the tactic is that related to the suspect's modus operandi, that is to say, where and how the drugs were taken, the means of transport used as well as the reaction to the officer on applying the tactic (attempt to run away, violent behaviour or attempt to dump or swallow the drugs). Each tactic is linked to detailed information concerning the officer who carried it out and the individuals on whom it was performed.

A police action may consist of one or more tactics applied within the same operation; this series of tactics that are logically linked to each other are gathered together to form a single entity, known as tactic sequence (TS), which often matches a police report. In the tactic sequence, information about the event that triggered the action is collected (such as information obtained by an informer, anonymous phone calls, chemical companies notifying suspicious transactions involving chemical substances, police officers on patrol reporting suspicious behaviour), and the possible link of the TS to an organized and structured activity is established: an operation.

A final piece of information linked to the TS, but interesting from a management point of view, is the classification of the report.

Summing up, the basic element of the database is the TS itself, the listing of a sequence of tactics; all other database components refer then to this element, inasmuch as each entity defined in the database is part of a TS.

6.6.3 Seizures

One of the results from adopting a tactic is the seizure of drugs, objects, weapons and money. All the detailed information regarding the 'goods' seized during the execution of a tactic is stored in this subject area. Given the nature of the database and its objectives, most of the details concern the seizure of drugs. Each seizure is linked to a tactic and to one or more persons.

The definition of seizure calls for clarification. In order to maintain the flexibility of the analysis at a very high level, it has been decided to consider 'seizure' in its most elementary unit possible: that is to say, a single type of drug found in the course of the same tactic in a single place.

To make this definition clearer, we will give a simple example:

A stops B and searches him; during the search, he finds 10 g of cannabis in B's right pocket and another 5 g of cannabis and 2 MDMA tablets under his jacket. Following the arrest, A takes B to the police station, and a further search leads to the seizure of another 10 g of cannabis and 30 MDMA tablets. This sequence of two tactics resulted in two seizures of cannabis and one of MDMA during the first tactic, and one seizure of cannabis and one of MDMA during the second.

The types of seizures taken into account are as follows:

- *Drugs* – with regard to drugs, the information taken into account relates to weight, number of units (e.g. tablets, stones or plants), the price paid and its form (i.e. powder, liquid); with regard to the classification of the drug type, three distinct periods of time were considered:
  - Primary, a classification made by the police officer at the time of seizure

Table 6.2 Drugs typology

Drug type	Drug
Class A	Cocaine
	Crack
	Heroin/diamorphine
	MDMA
	Opium
	Methadone
	LSD
	Other A
	Other MDMA-like drug
Class B	Amphetamine
	Methylamphetamine
	Other B
Class C	Cannabis
	Benzodiazepine
	Other C
Non-controlled drug	Ketamine
	Other non-controlled drug
Other substance	Other substance
NA	NA
	Multiple drug group
Unspecified	Unspecified

  - Secondary, a classification made by the police officer with the aid of a diagnostic kit or under supervision
  - Tertiary, a classification carried out at a test laboratory

In the case of a substance tested in a laboratory, further information is stored: its purity and possible links to other drug seizures.

The classification of the type of drug is summarized in Table 6.2.

- *Objects* – the only information available is the type of object.
- *Money* – the amount and type of currency.
- *Weapons* – manufacturer and model.

6.6.4 Arrests

Another of the results from adopting a tactic is the arrest of one or more of the persons involved. Information on the arrest procedure, type of crime committed, the officers who carried it out, the judicial outcome and details of the detention is stored in this subject area.

The basic element of this subject area is the arrest; in addition to the information concerning place and time of arrest, information on how it was carried out and on the violent or non-violent behaviour of the arrested person is now available. Detailed information on all police officers involved is provided, such as their role, age, sex, seniority and rank. Details of the detention also include important information about probation and about the subject's behaviour at the time.

The legal outcomes resulting from the arrest make up a relevant part of this subject area and are divided into two big categories: those which are sorted at the police station and those, on the other hand, which end up in front of a judge. In the latter case, detailed information is stored, such as type of court, charges, guilty-or-not-guilty pleas, sentences and penalties imposed.

Each arrest is linked to a tactic and to a person.

6.6.5 Persons

This subject area contains detailed information concerning people involved in drug-trafficking crimes (both charged and suspects). The recorded information specifies personal details (age, sex, nationality, ethnic origin) and the criminal record related to both drug-trafficking crimes (prior to 2004) and other types of offences. The person's criminal record is summarized with information such as number of charges and convictions and the dates of the first and the latest conviction. This information is then broken down according to type of offence.

Regarding the person, there is also a possibility of storing information concerning any past drug addiction of the subject.

6.6.6 Locations/Places

In this subject area, the information needed to geographically reference all the entities having among their features a location can also be stored. Data include different geographical localization systems, some of a general type such as GPS (Global Positioning System), while others are specific to London, with its subdivision into boroughs.

6.7 The Data Entry

Given the diversity of data sources, the data entry process is complex, both in terms of data storage support (electronic and paper based) and from the point of view of logical organization of data (the Crime Report Information System, or CRIS, is event (crime) oriented, whereas the Police National Computer, or PNC, is person oriented).

The process has been organized in three fundamental steps:

- Data retrieval and organization
- Understanding and codification of textual data
- Entering the data into the database

In case of doubt, at any of these stages, the operator always had the possibility of consulting a supervisor, a Scotland Yard officer experienced in drug trafficking, for help with the current data entry.

6.7.1 The Data Entry Operators Team (DEOs)

Apart from the manual data entry, the importance of the role of the DEOs in the CDTD project can be summed up in two activities: the interpretation and codification of the information available from the sources in free-text format and the verification of the validity and precision of the data by crossmatching, whenever possible, other sources containing the same information.

For example, when defining and classifying the tactics used by the police to fight drug trafficking, the interpretation of the descriptions of a given situation made by the officer who carried out the tactic assumes a particular relevance. This interpretation is important to attribute it to one of the tactics codified during the design stage of the database. In this case, the operator may refer to a text describing the whole police operation, in the description he or she must identify the individual tactics and recognize them as belonging to one of the categories used in the database.

To make the process clearer, while emphasizing the difficulties of the automation (which shows the need of a human interface), we give here as an example an imaginary report with its codification:

CRIS Report 1234567/04

Fibonacci is walking down the street carrying a briefcase (containing 30 grams of cocaine) and a newspaper. After reading the paper he throws it to the ground. Police officer Newton (in uniform) sees this and approaches him with the intention of cautioning against littering the street. At the sight of the officer Fibonacci panics, drops the briefcase and starts running. PC Newton becomes suspicious and stops him. Fibonacci refuses to answer any questions he is asked and this raises the officer's suspicions even more, who now thinks the briefcase may be stolen and therefore decides to inspect its contents. The briefcase is recovered, PC Newton searches it and finds the cocaine. Fibronacci is then arrested and charged with possession of cocaine with intent to supply.

From the analysis of this report, the operator creates a new sequence of tactics [TACTICS SEQUENCE] inserting the CRIS 1234567/04 code; during this event, no operation [OPERATION] is in progress, and the event which triggered the action [LEADING INFORMATION] is 'instigated by the police officer on patrol (e.g. witness to suspicious action (other crime))'. The tactic is identified by the operator as a 'search of object' [GENERIC TACTIC], 'stop and search' [DETAILED

TACTIC], and the Act of law [ACT] authorizing the officer to carry out the tactic is 'Section 1 of the Police and Criminal Evidence Act 1984'. The tactic is followed by a seizure of drugs [Drug Seizure], namely, 30 g of cocaine [DRUG], a Class A drug [DRUG TYPE]. The suspect is arrested [ARREST] for the offence of 'possession with intent to supply' [OFFENCE].

Fibonacci is then searched and the PC finds a British passport in the name of Alice Turing. However, the photograph has been carefully removed so Fibonacci is then arrested on suspicion of stealing the passport.

The arrest is followed by another tactic [TACTIC]: 'search of person' [GENERIC TACTIC] and 'search after arrest' [DETAILED TACTIC], and the Act of law [ACT] to be applied is 'Section 32 (2) (a) (ii) of the Police and Criminal Evidence Act 1984'. A stolen passport [PROPERTY] is seized [PROPERTY SEIZURE]. The subject is arrested [ARREST] with a theft allegation [OFFENCE].

6.7.2 Textual Data Preprocessing

As already mentioned, when entering data, it is particularly important that the data entry operators 'translate' the free text in the police reports and the test laboratory reports into structured information that can be used for subsequent data processing. In view of this, the DEOs were given complete training, ranging from aspects of law and police techniques allowing them to understand the events described in the texts and to classify them correctly, to basic knowledge of data processing and analysis.

The database aspects for which this type of operation has proved to be indispensable are practically the whole of the part dealing with police tactics, the codification of which was devised in the course of the project and which has proved to be efficient and exhaustive (in fact, in only 6% of the cases has a tactic had to be defined as 'Other'); in this case, the data source is the CRIS text report, while in the case of tactic sequences which are part of operations, the data source is the reports of the operation itself.

The detailed information regarding the results of drug tests carried out by the Forensic Science Service (FSS) is provided in a free-text format with quantitative information not systematically distributed within the text, and this does not allow for an easy identification of useful information.

6.7.3 The Data Quality: Crossing Different Data Sources

Manual data entry has also allowed a control on the quality of the data themselves; and even in the case of data already present on the sources in a structured form, the DEOs carried out cross-checks wherever the data were present in more than one

Table 6.3 Tactic sequence frequency x year

Year	Month	Frequency
2003	September	3
2003	October	5
2003	November	18
2003	December	38
2004	January	275
2004	February	284
2004	March	15
2004	April	4
2004	May	8
2004	June	24
2004	July	5
2004	August	34
2004	September	53
2004	October	95
2004	November	407
2004	December	286
2005	January	3
2005	February	1

source, by way of verification. For example, the information on a person's place of birth appears both in the CRIS and in the PNC: if they match, this information is entered directly; otherwise, the custody database is consulted. In the event of two sources having the same value, this value is entered; in the event of three discordant sources, the 'not known' value is entered.

Although this check slows down the data entry process, it ensures a high quality of the data entered and, as opposed to the extraction of data from free text, in the future, it will be possible to automate it.

6.8 The Database Data

6.8.1 Types of Data and Data Frequency

The information on drug trafficking in London relate to the year 2004. The total number of tactic sequences gathered is 1,558, and their temporary distribution is shown in Table 6.3.

It is evident from Table 6.3 that the data sampling throughout the year is not uniform. In particular, a clear concentration during the first 2 and the last 2 months of the year can be noticed, these being the only months for which the data collection and entry have been completed. This choice, imposed by the project's time restrictions which did not allow for a complete entry of all of the data for the year 2004, was taken for two reasons: on the one hand, it is preferable to complete

Table 6.4 Frequency of number of tactics

Number of tactics	Frequency
1	894
2	293
3	197
4	93
5	48
6	27
7	16
8	4
9	4
10	0
11	0
12	3

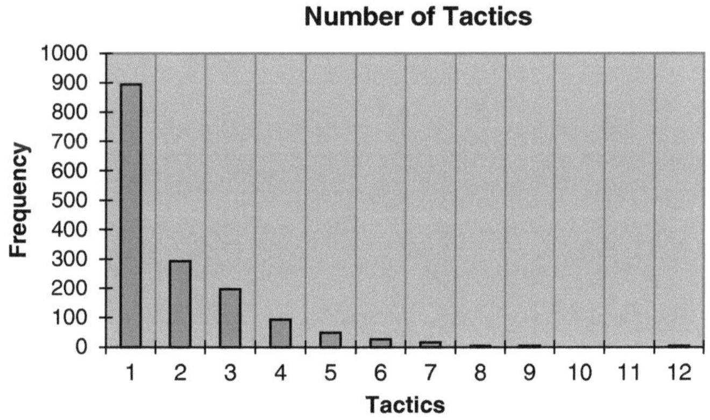

Fig. 6.1 Frequency of number of tactics

4 months out of 12 as opposed to 30% for each month in order to get a complete view of a given time period; on the other hand, it was difficult to establish an adequate random sampling strategy. The records distributed throughout the year have been entered to complete, for example, some logical sequences such as operations and to collect some particularly interesting tactic types such as the test purchase.

The total number of tactics is 3,061, and they are distributed within the sequences as shown in Table 6.4 and in Fig. 6.1.

The total number of arrests stored in the database, and therefore involved in the tactics, is 2,739, covering 1,523 people out of a total of 1,669 people entered in the database. The seizures of drugs as a result of the activities recorded on the CDTD are 2,873; the MPS Metropolitan Police officers involved in some form with tactics and arrests are 2,502.

Table 6.5 Persons – summary of missing data

Persons	Missing data	Percentage	Classes
Sex	22	1.32%	I
Birthdate	64	3.83%	I
Birthplace	18	1.08%	I
Trafficking level	1,317	78.91%	F
Education	1,628	97.54%	I
Nationality	231	13.84%	I
Ethnicity	56	3.36%	I
Self-ethnicity	87	5.21%	I

Table 6.6 Drugs – summary of missing data

Drug seizures	Missing data	Percentage	Classes
Weight	1,414	49.22%	I,C
Price paid	2,815	97.98%	E
Number	2,592	90.22%	I
Primary analysis	110	3.83%	I
Secondary analysis	2,857	99.44%	I
Tertiary analysis	847	29.48%	I

Table 6.7 Tactics – summary of missing data

Tactics	Missing data	Percentage	Classes
Sequence number	249	8.13%	F
Detailed tactics	7	0.23%	F
Acts	29	0.95%	I,F
Officers	0	0.00%	

6.8.2 Typology and Frequency of Missing Data

Among the particularly relevant data, where missing values are very frequent, the subject area which stands out the most is the one concerning seizure of drugs where the information on the price paid is available only in 2% of the cases. In this specific case, the lack of information is justified by the very nature of the data: in fact, the price paid can only be known if the purchase is made by a policeman, and therefore the values must in this case be left out of the equation. Conversely, in the same subject area, the information on weight, which shows a 50% of missing data, cannot be justified in the same way. The reason for the lack of data can be divided into four categories: **E** – endogenous (due to the same nature of the data, as in the case of price), **I** – unavailability (the data were missing at source), **C** – excessive cost (data were available, but the cost involving its retrieval in terms of resources was excessive given the project's timeframe), and **F** – lack of clarity (the operator's level of inference in retrieving the data is excessive and does not guarantee credibility).

For a general view on the situation of the missing data on the database, we provide tables (Tables 6.5, 6.6, 6.7, and 6.8) related to the situations for the most

Table 6.8 Arrests – summary of missing data

Arrests	Missing data	Percentage	Classes
Offence	9	0.33%	I
Place of offence	12	0.44%	I
Committed on from	14	0.51%	I
Arrest date	22	0.80%	I
Place of arrest	23	0.84%	I
Investigating officer 1	171	6.24%	I
Arrived at station	87	3.18%	I
Violent on arrest	2	0.07%	I
Arrest mode	651	23.77%	I
Arresting officer	409	14.93%	I
Officer I case	1,169	42.68%	I

relevant data in the five subject areas in which the database has been divided, adding a classification of the reasons for missing data and providing, where necessary, a more exhaustive analysis.

6.9 Automatic Inputting of New Data

For some of the data that were manually entered, being already electronically stored in a structured form, we can envisage an automatic loading procedure that would speed up the entering of new records. For this type of improvement, the most reliable subject area is the one linked to personal information since the main source, if not the only one, is the PNC (Police National Computer) and the quality of data is such that it does not require human intervention.

6.9.1 The Database Final Structure

6.9.1.1 Tactic Subject Area

Most of the information that feeds this subject area is from CRIS (Crime Report Investigation System). CRIS reports store information related to an arrest, with a description of the action in free text. The DEOs (data entry operators) will translate (codify) this free text in structured fields (tactics, acts of law, tools used, types of officers involved, etc.) and type in the CDTD to allow the following analysis. Every CRIS report about drug-trafficking offences will be translated (codified) as a tactic sequence in the CDTD, and each of the tactics described in the report is stored in the tactic table (Fig. 6.2, Table 6.9 and Fig. 6.3).

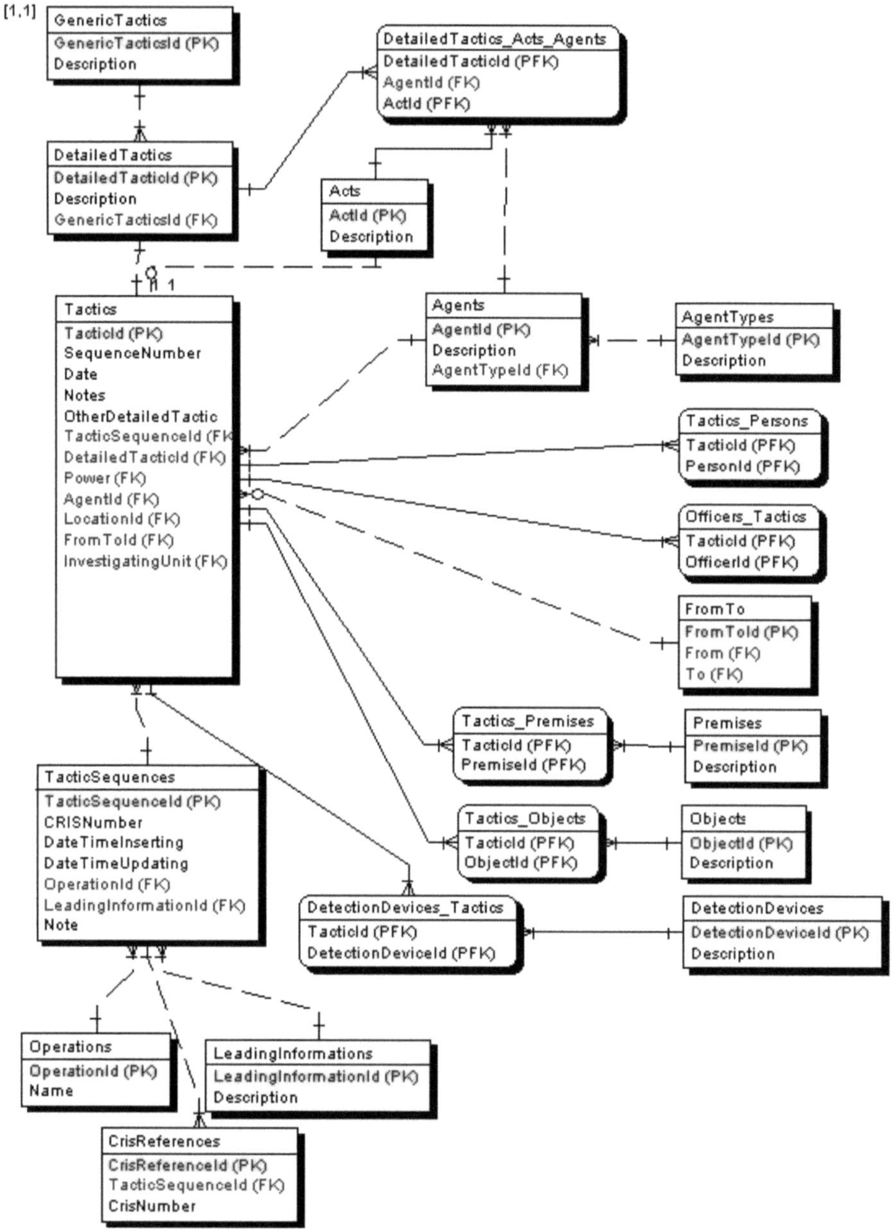

Fig. 6.2 The logic graph of the tactics

Table 6.9 Tactics description

Entity name	Description
Acts*[1]	Laws used by officers to apply the tactic
Agents*[1]	Type of officer performing the tactic
AgentTypes*[1]	Officer classes
CrisReferences	Reference to another CRIS report linked to this one
DetailedTactics*[1]	Tactic used
DetailedTactics_Acts_Agents	Many-to-many relationship table N:N:N
DetectionDevices*[1]	Devices used in tactic
Tactic_Premises	Many-to-many relationship table N:N
From/To	Source and destination of controlled delivery
GenericTactics*[1]	Tactic classes
LeadingInformations*[1]	Information that trigger the tactic sequence
Objects*[1]	Objects searched in tactic
Officers_Tactics	Many-to-many relationship table N:N (with officers entity in arrest subject area) – officers involved in tactic
Operations	Some tactics are performed during an operation
Premises*[1]	Premises searched in tactic
Tactics_Objects	Many-to-many relationship table N:N
Tactics_Persons	Many-to-many relationship table N:N (with Persons entity in Person subject area) – Persons involved in tactic
DetectionDevices_Tactics	Many-to-many relationship table N:N
TacticSequences	Sequence of linked tactics

*[1] – Tables with closed lists

Legenda

Seizures: each tactic can be followed by one or more seizures.
Persons: each tactic involves one or more persons (targets of the tactic).
Arrests: [Arrests entity] each tactic can be followed by the arrest of offenders – [Officer entity] each tactic is performed by one or more policemen.
Locations: geographical information about the place where tactic was performed and source and destination of controlled delivery packets.

6.9.1.2 Seizure Subject Area

Information that feeds this subject area is obtained from the CRIS report and the laboratory analysis report (see Fig. 6.4 and Table 6.10).

6.9.1.3 Arrest Subject Area

Information that feeds this subject area is from CRIS Report, Custody Database and from PNC (Fig. 6.5 and Table 6.11).

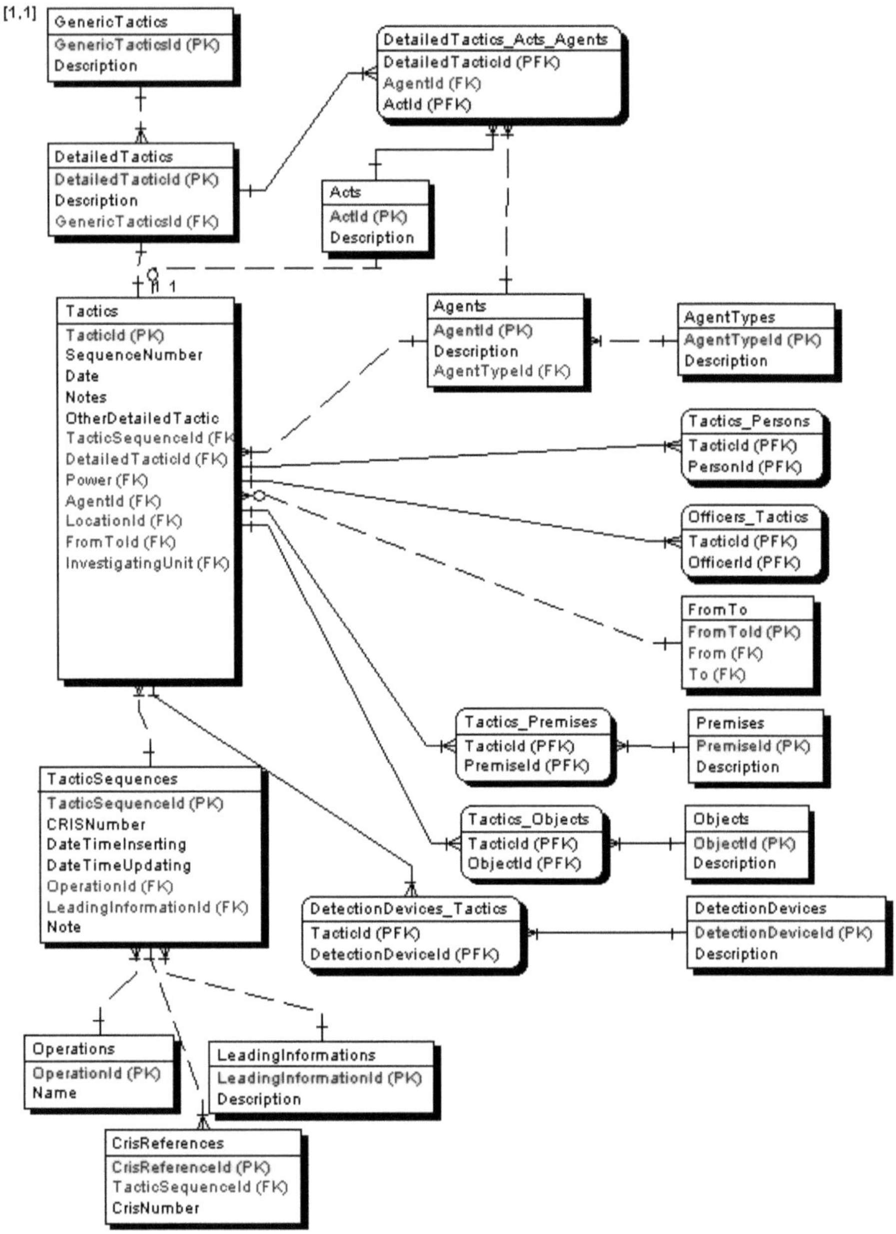

Fig. 6.3 Tactics Relationships with other entities

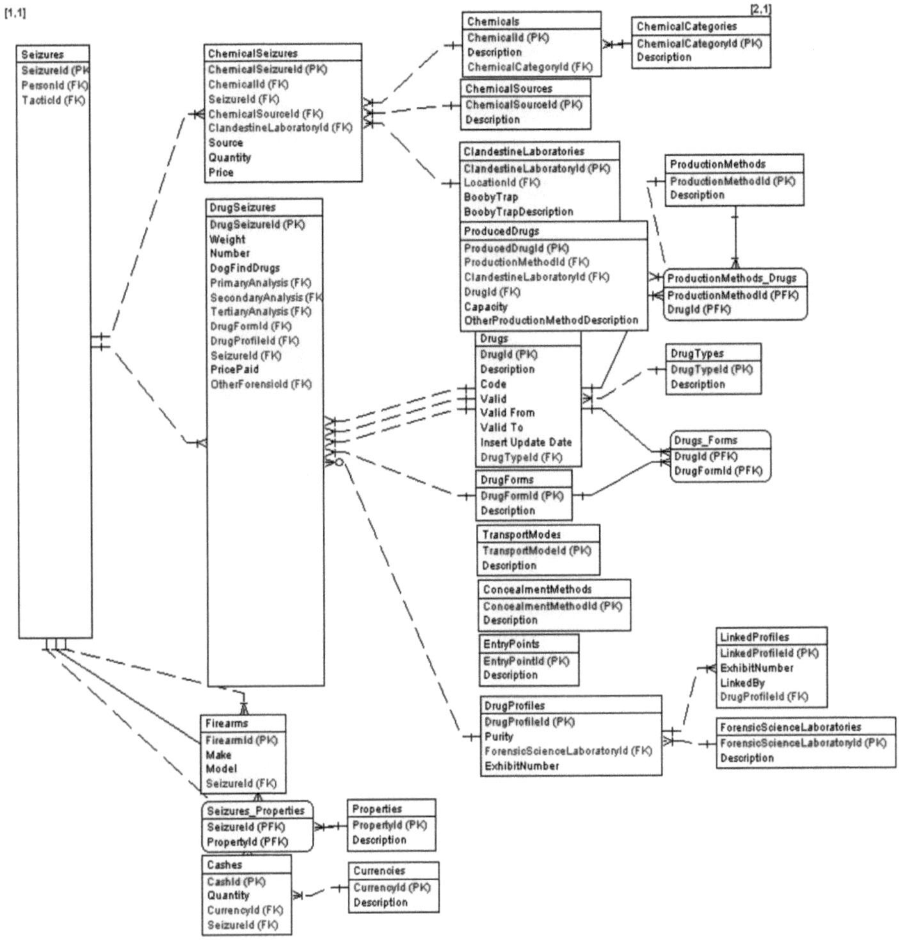

Fig. 6.4 The logic graph of seizures

6.9.1.4 Person Subject Area

Information that feeds this subject area is from PNC (Fig. 6.6 and Table 6.12).

Appendix

Notices10-06 Item 2 – The Metropolitan Police Service

The Metropolitan Police Service's (MPS) Covert Drug Purchases Database: Monitoring the price and purity of heroin, cocaine and other controlled drugs
(CR 216/04/17 and DP7/05/1)

Table 6.10 Seizures typology

Entity name	Description
Cashes	Money seized
ChemicalCategories*[1]	Type of chemicals used to produce drugs
Chemicals*[1]	Chemicals used to produce drugs
ChemicalSeizures	Chemicals seized
ChemicalSources*[1]	Chemicals producer
ClandestineLaboratories	Drug production laboratories discovered
ConcealmentMethods*[1]	Methods used to hide drug
Currencies*[1]	Currencies
DrugForms*[1]	Form of the seized drug (tablets, powder, liquid, etc.)
DrugProfiles	Information about drug purity
Drugs*[1]	Drug types
Drugs_Forms	Many-to-many relationship table N:N
DrugSeizures	Drug seized
DrugTypes*[1]	Drug classes
EntryPoints*[1]	Points where the seized drug enters into the UK
Firearms	Firearms seized
ForensicScienceLaboratories*[1]	Laboratory that performed the analysis on the seizure
LinkedProfiles	Other laboratory references linked to this
ProducedDrugs*[1]	Type of drug produced in the laboratory
ProductionMethods*[1]	Method of drug production
ProductionMethods_Drugs	Many-to-many relationship table N:N
Properties*[1]	Properties seized
Seizures	Seizure
Seizures_Properties	Many-to-many relationship table N:N
TransportModes*[1]	Methods used to transport drugs

*[1] – Tables with closed lists

Introduction

The purpose of this Notice is to set out the Specialist Crime Directorate's (SCD) criteria and procedure for monitoring the price and purity of heroin (crude diamorphine), cocaine (including crack) and other controlled drugs. Analyses of these data will:

- improve the understanding by MPS of the price and purity of specific drugs across London
- support Operation Paramount – a national drug price index that is collated by the National Criminal Intelligence Service and will be incorporated into the functions of the new Serious and Organised Crime Agency
- help MPS to have a greater understanding of drugs markets
- help the SCD Central Drug Trafficking Database (CDTD) Project team to monitor and evaluate the effectiveness of the tactic of covert drug purchases as a means for disrupting those organised crime networks involved in the trafficking of specific drugs
- assist the courts in determining appropriate sentences

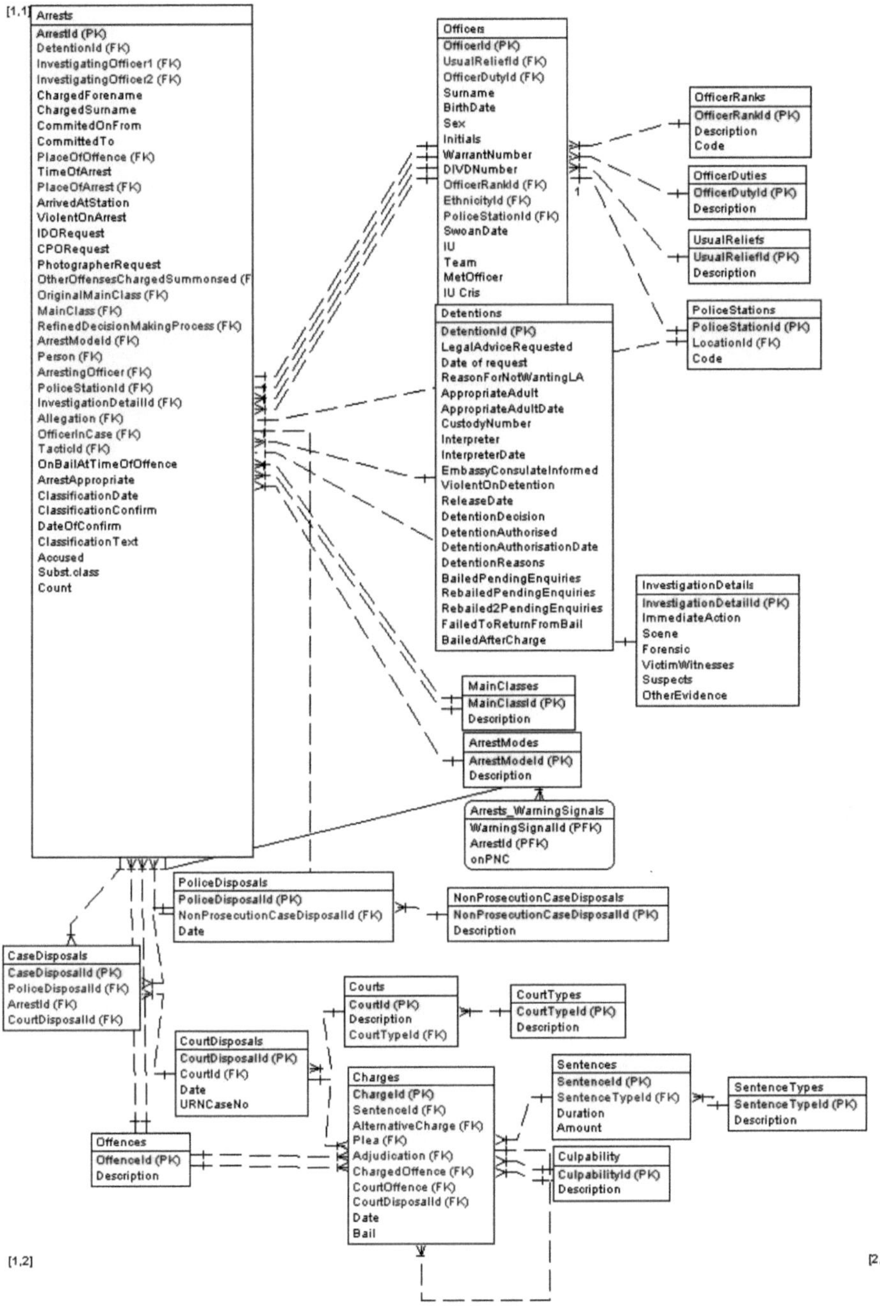

Fig. 6.5 The logic graph of the arrests

Table 6.11 Arrests typology

Entity name	Description
ArrestModes*[1]	How person was arrested
Arrests	Details on arrest
Arrests_WarningSignals	Many-to-many relationship table N:N
CaseDisposals	Legal consequences of the arrest
Charges	Details on offence charged to the arrested
CourtDisposals	Case disposal in front of a court
Courts*[1]	List of courts
CourtTypes*[1]	Court types
Detentions	Details on detentions
MainClass	Classification of the offence committed
NonProsecutionCaseDisposals*[1]	List of possible case disposals in police station
Offences*[1]	Offence committed
OfficerDuties*[1]	Type of officer activity during the arrest
OfficerRanks*[1]	Officer rank
Officers	Details on officers
PoliceDisposals	Case disposal in police station
PoliceStations	Police station
Sentences	Sentence in front of the court
SentenceTypes*[1]	Sentence types

*[1] – Tables with closed lists

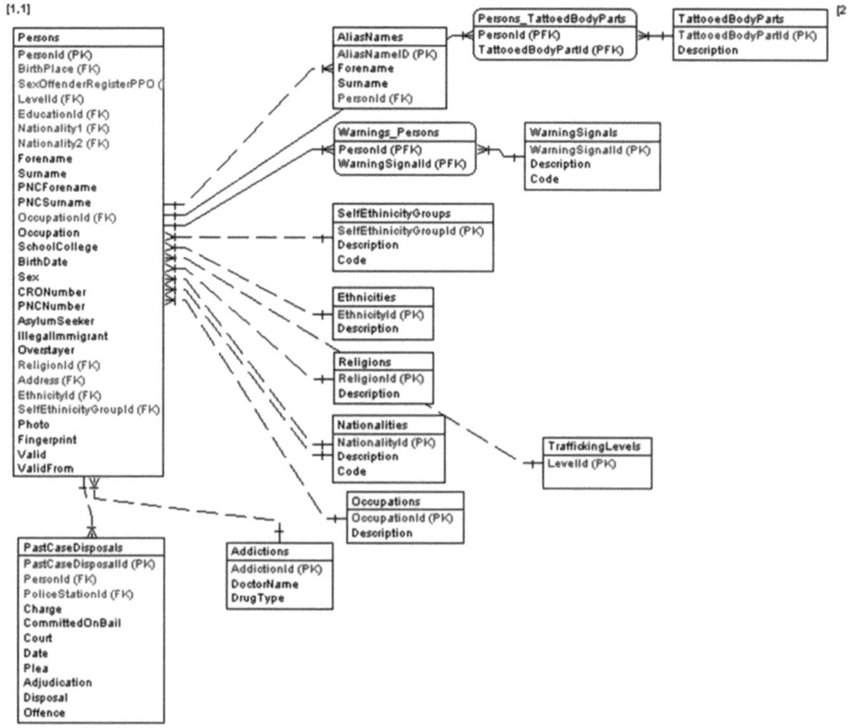

Fig. 6.6 The logic graph of the persons

Table 6.12 Persons typology

Entity name	Description
Addictions	Drug addiction
AliasNames	Known alias
Ethnicities*[1]	Ethnicity estimated by officer
Nationalities*[1]	Nationality
Occupations	Occupation
PastCaseDisposals	Drug offence committed before 2004 and other offences (not drugs)
Persons	Personal information
Persons_TattooedBodyParts	Many-to-many relationship table N:N
Religions*[1]	Religion
SelfEthnicitiesGroup*[1]	Self-defined ethnicity
TattooedBodyParts*[1]	Tattoos
TraffickingLevels*[1]	Type of drug trafficking (local, international ...)
Warnings_Persons	Many-to-many relationship table N:N
WarningSignals*[1]	Warning signals (addicted, violent, tried to commit suicide, etc.)

*[1] – Tables with closed lists

The results of this research will be incorporated into the National Intelligence Model (NIM) strategic and tactical assessments produced by the CDTD Project team, and made available to MPS staff. The research findings will also be available to statutory and non-statutory agencies working with the MPS.

Background

Drug price and purity information provides important intelligence. For intelligence to have true value, it must come from a reliable source and be as accurate as possible. As prices and purities vary considerably, the presentation of aggregate data must use the best statistical techniques available. Research has shown that information from police test purchase operations is the most reliable and accurate data source available. A methodology to collate and present this data has been developed after a pilot project in 2004. The procedure to be followed is outlined below, which has been agreed with Kevin O'Leary, Detective Chief Inspector of SCD11(10) Covert Operations.

Recording Price and Purity of Specified Controlled Drugs: Procedure

Details of **all (starting from Sunday 1 January 2006)** covert drug purchases will be recorded on the Covert Drug Purchase (CDP) form. Blank CDP forms can be found on the MPS Forms intranet site (form number 196). An example of a completed CPD form is at **Annex 1** of this Notice. The responsibility for completing the CDP form will rest with the police officer deputed to act as the Exhibits Officer for the duration of the Test Purchase Operation (TPO) or Undercover Buy Operation (UBO). The form is designed to be completed on screen so that it can be sent by e-mail to the Drugs Directorate (SCD3-CDP) for analysis.

One CDP form will be completed for each transaction. So, for example, in the case where a police officer carries out two test purchases from the same offender, at the same location in the course of one day, then the CDP form will be completed twice. In cases where the police officer buys two drugs (say crack and heroin) at the same time and from the same offender, then only one CDP form will be used, but the Exhibits Officer must record the details of both drugs on the form. When completing CDP form, the Exhibits Officer should make every effort to ensure that information is entered in the fields relating to drug type(s), amount(s), price(s), police exhibit reference(s) and exhibit bag number(s), weight(s) and purity, and that the details recorded match those on the laboratory form 1 and the witness statement provided by the forensic scientist. It is accepted that, in many cases, the police officers involved in the TPO or UBO will not know the identity of the person selling the drug. In these cases, the Exhibits Officer will not be able to complete the fields relating to the offender (for example name, age, date of birth, place of birth, nationality and CRO/Police National Computer number).

As soon as the result of the forensic analysis is known, the relevant fields on the CDP form will be completed. The Exhibits Officer will then send the completed CDP form to the group e-mail address SCD3-CDP using the MPS AWARE system.

The Senior Investigating Officer in charge of the TPO or UBO is responsible for ensuring that there is no undue delay in the completion of CPD form.

On receipt, the CPD form will be reviewed. If any errors or discrepancies are found the Exhibits Officer will be contacted by return e-mail.

The information relating to prices, purity, drug type and so on will be scrutinised, validated and entered onto the SCD Covert Drug Purchase Database. It will be analysed and the results will be incorporated into the NIM strategic and tactical assessments produced by the CDTD Project team.

Definitions

Covert Drug Purchase

For the purpose of this Notice, the term ‘covert drug purchase’ means:

- the purchase of any controlled drug
- the purchase of any non-controlled drug (for example ephedrine, caffeine, aspirin)

- the purchase of any innocuous substance (for example chalk, soap, plaster)
- by a Test Purchase Officer or an Undercover Officer whilst acting as such.

Controlled Drug

The term 'controlled drug' means any drug mentioned in Schedule 2, Part I, II or III to the Misuse of Drugs Act 1971.

Application

This Notice is of particular relevance to:

- Undercover Officers
- Test Purchase Officers
- Officers designated as Exhibits Officer in operations involving the covert purchasing of controlled drugs
- Senior Officers planning, supervising and managing operations involving the covert purchases of controlled drugs
- Criminal Justice Unit Managers and their deputies
- Staff in the Drugs Directorate
- Staff in the SCD11 (Covert Policing Unit)

For more information regarding this Notice please contact Paul Richards, Inspector on extension 64054 or by e-mail.

References

Bean, P., & Nemitz, T. (2004). *Drugs and crime* (2nd ed.). Devon: Willan Publishing.

Chatterton, M., Varley, M., & Langmead-Jones, P. (1998). *Testing performance indicators for local anti-drugs strategies* (Police Research Series Paper 97). Home Office, p. 2.

GLADA. (2003). *The GLADA Crack cocaine strategy 2005–08*. London: Greater London Authority, p. 12.

Hough, M., & Natarajan, M. (2000). Introduction: Illegal drug markets, research and policy. In *Illegal drug markets: From research to prevention policy, crime prevention studies* (Vol. 11, p. 8). Monsey/New York: Criminal Justice Press.

King, L. A. (2004). *Metropolitan police service specialist crime department test purchase database: Drug type, unit prices and purities in the period 1992 to 2003*. Final report to Home Office Drugs and Alcohol Research Unit, p. 16.

Pearson, G., & Hobbs, D. (2001). *Middle market drug distribution* (Home Office Research Study 227, p. 3). London: Home Office.

Chapter 7
Supervised Artificial Neural Networks: Backpropagation Neural Networks

Massimo Buscema

7.1 Introduction

Backpropagation (BP) refers to a broad family of artificial neural networks (ANNs), whose architecture consists of different interconnected layers (Werbos 1974; Rumelhart et al. 1986; Fahlman 1988; Jacobs 1988; Lapedes and Farber 1987; Tawel 1989; Minai and Williams 1990; Weigend et al. 1991; Chauvin and Rumelhart 1995). The BP ANNs represent a kind of ANN whose learning algorithm is based on the *deepest-descent* technique. If provided with an appropriate number of hidden units, they will also be able to minimize the error of nonlinear functions of high complexity. Theoretically, a BP provided with a simple layer of hidden units is sufficient to map any function $y = f(x)$.

Basically, it is often necessary to provide these ANNs with at least two layers of hidden units when the function to compute is particularly complex or when the data chosen to train the BP are not particularly reliable and a level filter is necessary for the input features.

The BP are networks whose learning function tends to "distribute itself" on the connections because of the specific correction algorithm of the weights used. In the case of BP, this means that these units, provided with at least a layer of hidden units, tend to *distribute among themselves* the codification of each feature of the input vector. This makes learning more compact and efficient, but it is more complex to know the "reasoning" which brings a BP to answer in a certain way in the testing process. In brief, it is difficult to explicate the *implicit knowledge* that these ANNs acquire in the training process.

A second theoretical and operative difficulty that BP poses concerns the *minimum number* of hidden units that are necessary in order for these ANNs to compute a

M. Buscema (✉)
Semeion Research Center of Sciences of Communication, via Sersale 117, Rome, Italy
e-mail: m.buscema@semeion.it

M. Buscema and W.J. Tastle (eds.), *Intelligent Data Mining in Law Enforcement Analytics: New Neural Networks Applied to Real Problems*, DOI 10.1007/978-94-007-4914-6_7,

function. In fact, it is known that if the function is not linear, at least one layer of hidden units will be necessary. However, at the moment, a method of how to exactly state the minimum number of hidden units needed to compute a nonlinear function is unknown. In these situations, the work is based on experience and on some heuristics.

Experience advises us to use a minimum number of hidden units in a first-time training of an ANN. If the training succeeds, an analysis of the sensitivity will normally allow us to understand the *singularity number* that each input node determines on the output, and consequently it will be able to deduce the degree of freedom needed by the ANN to resolve the equation and then to express these as hidden units.

This procedure is not guaranteed; during the training process, the BP can become trapped in local minima because of the relation between the morphological complexity of the hyperparaboloid characterizing the function and the weight values that are randomly set and placed before the training.

The dilemma of BP is that for a prior, unknown minimum number of hidden units, what numbers are useful to compute a function. If *too many* are created, then during some forms of training, the BP can create a condition of overfitting, causing a worsening of its generalization capacities in the testing process. If *an insufficient number* are created, the BP can have difficulties in learning either because the function is too complex or because the BP randomly falls into a local minimum.

The BP's family includes both feedforward ANN and feedback ANN (recurrent networks). In this section, we are going to examine only feedforward BP ANN, an understanding of which is essential prior to the study of feedback BP.

7.2 Standard Backpropagation

7.2.1 Theoretical Principles

A system functioning as *feedforward backpropagation* (from now on, BP) is theoretically based on the following principles:

(a) Through a series of trials, the system creates the *relations* between its units with respect to multiple tasks.
(b) After having learned the type of relations that are appropriate among its units, the system is able to exhibit through its structure the type of internal representation it has stabilized for the various tasks that it had to learn in order to carry out their multiplicity.
(c) The system can "easily" learn other tasks which are similar to the ones it has already learned, and then it can operate "generally."
(d) The relations becoming stabilized among the system's units during the learning of several tasks are the only *memory* of the system itself.

(e) Many of the system's units are of a *discriminant* or subconceptual type. They can belong to three logical types:

- *Input units*: the sensors through which the system receives the surrounding area's stimulation.
- *Output units*: the units through which the system expresses its behaviors as an *interpretation* of the received inputs.
- *Hidden units*: they are the internal units of the network's system. They receive the input either from the other hidden units or from the input units. Their behavior works as output for other hidden units or for the output units. They are the units providing the internal representation, through which the system interprets the received input, with respect to the output that it will produce.

(f) The relations between the system units are *uni-oriented*; that is, by activating itself in a certain way, unit A activates unit B with respect to the *strength* of their connection, but not vice versa.

This means that in BP, relations are *fuzzy rules*, for example,
if A activates itself in the manner of X {A's activation value},
then B will activate itself in the manner of $X \bullet W$ {where W is the strength of the connection between A and B}.

Since A's activation in an X way is also the outcome of a similar rule, unless A is an input unit, we can say that a BP is a continuously modifying *cascade of fuzzy rules*. Briefly, the *power of oriented connection* between one unit and another is the fuzzy rule by which one goes from the first unit to the second one. This strength of connection among the BP units is called *weight*, and it is indicated by w_{ij}, where i is the starting unit's identifier and j is the outcome unit's identifier:

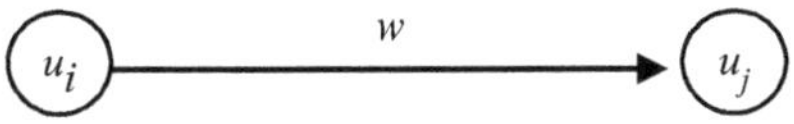

The weights among the units of a BP system are continuously modified in relation to the task that the BP has to carry out. In this sense, it is legitimate to say that the BP continuously adjusts its rules according to the experiences they carry out. This modification of weights continues until the BP individuates the weights, allowing them to handle the tasks it has learned up until that moment in the most appropriate way.

7.2.2 *Functioning and Learning*

Given these premises, it is better to thoroughly explain the functioning of a BP.

Let us imagine a very simple BP in which we have two input units (I_1, I_2), two hidden units (H_1, H_2), and two output units (O_1, O_2). Therefore, it is a BP made up of *three layers* (Fig. 7.1).

Fig. 7.1 The BP topology

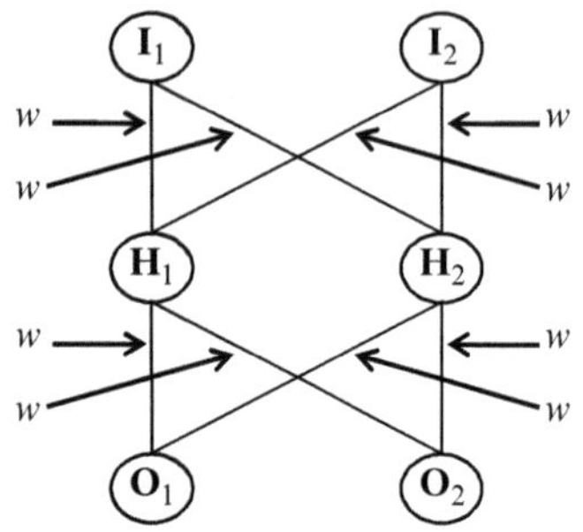

Furthermore, each layer level is connected to all units of the following layer through weights. The input units are connected to the hidden units through the weights $w_{I,H}$, practically, $w_{1,1}/w_{1,2}/w_{2,1}/w_{2,2}$. Similarly, the hidden units are connected to the output units through the weights $w_{H,O}$, practically, $w_{3,5}/w_{3,6}/w_{4,5}/w_{4,6}$.

In order to function, a BP needs to follow several steps. The first one consists of *activation conditions*:

1. The BP must be subjected to a certain type of input for at least a certain length of time.
2. It is necessary to imagine that the output units tend toward at least a certain type of objective, called the target, for the entire time the BP is subjected to a certain type of input.
3. It is necessary that the BP shows a value, even a random one, at the beginning of the relationship of connection among all its units, that is, its weights.

In brief, the activation connections suggest that the BP must have in the initial phase:

(a) At least one input
(b) At least one target to learn with respect to that input
(c) Random weights among its units

We emphasize that the BP is necessary in activation prequalification. We will express both values of the input units and those of the target units with numbers ranging between 0 and 1. We will use a similar criterion in order to stabilize the random weights among the different units (Fig. 7.2).

For example, we have planned a BP system with the following task: "giving an input of strength 1, determine which *weights* the system must possess in order to provide an answer for what should be the input for an output of strength 0,1."

In order *to function* and to resolve this task, it is necessary to explain the *functioning conditions* of a BP:

1. An algorithm able to calculate the activation value of each unit, except for the input layer ones, according to the *activation value* of the units connected to it and also according to the *strength* of *connection* through which these units are connected to it. We call this algorithm the *Forward Algorithm*.

Fig. 7.2 The BP trained

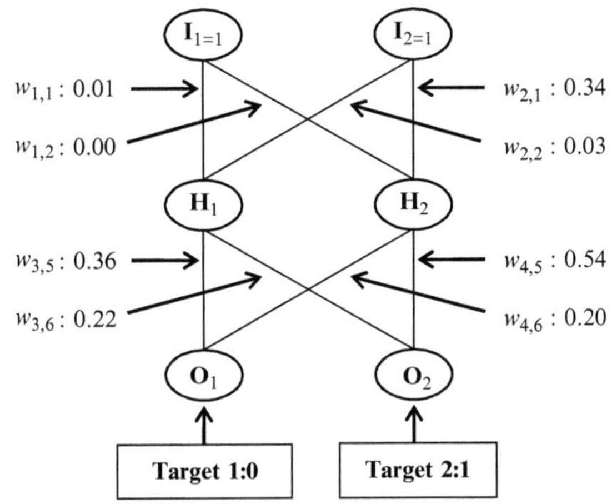

2. An algorithm able to gradually correct the *weights* among the different units, according to the difference between the output produced by the forward algorithm and the desired target. We call this algorithm the *Back Propagation*.

These functioning conditions presume that the BP carries out several trials in order to achieve an output that is more similar to the desired target. At each trial, the BP corrects its weights in order to bring the following trial as close as possible to the aim imposed from the outside.

We refer to an iteration of the *Forward Algorithm* and the consequent *Back Propagation Algorithm* as a *cycle*. We call the number of cycles needed, so that ANN will have experienced all couples of input and target at least one time in order to understand them an *epoch*.

The *epochs* constitute the lifetime of the BP. After a certain number of epochs during which the BP has been subjected to the same input and oriented toward the same target, it is expected to have *selected* the most adequate *weights* to attain this objective. It is also expected that the value of these last weights and of the hidden units will provide a good internal representation, at a subconceptual level, of the task that the BP has learned to carry out. This occurs if the *forward* and the *correction* algorithms are corrected.

Let us then attempt to define the *Forward Algorithm.*

The basic rule used to calculate the activation value of a unit with respect to other units connected to it with a strength w_{ji} is a function of the weighted sum of the inputs:

$$u_j = f\left(\sum_i u_i \cdot w_{ji}\right) = f(\mathrm{Net}_j) = \frac{1}{1 + e^{-\mathrm{Net}_j}}, \tag{7.1}$$

where Net_j = net input to the j level unit.

One must add to this equation the *threshold* of the unit, described as the *inclination* of the unit to activate or inhibit itself.

This means that

$$u_j = f(\text{Net}_j) = \frac{1}{1 + e^{-(\sum (i) w_{ji} \cdot u_i + \theta_j)}} \tag{7.1a}$$

where θ_j is the *bias* of unit u_j, the degree of *sensitivity*. The unit u_j is used as a response to the perturbation it receives from the net input. The bias is the opposite of the threshold, and it behaves as a weight generated by a fixed input of unitary value.

Equation (7.1a) represents the forward algorithm of BPs, assuming the sigmoidal function as default. It is necessary to calculate the dynamic of the backpropagation or correction algorithm.

The mathematical basis of this algorithm was already individuated by Rosemblatt in 1956 through the so-called *Delta Rule*, a procedure that allows for the correction of excess or deficit in the weights between the network units, basing itself on the difference between the actual output and the desired target.

Nevertheless, the delta rule allows adjustments to only those weights connecting the output units with those belonging to the immediately underlying layer. It does not allow one to know, in a three-layer network, how the weights connecting the input units with the hidden units should be modified at each cycle.

We shall in detail examine the *delta rule.*

The coefficient of error in this procedure is calculated by considering the difference between the actual output and the desired one (the target value) and relating this difference to the derivative between the activation state of the actual output and the net input of that output.

Therefore, if $\frac{\partial u_j}{\partial \text{Net}_j} = u_j \cdot (1 - u_j)$, then Δout_j (i.e., the error coefficient) will be

$$\Delta\text{out}_j = (t_j - u_j) \cdot u_j \cdot (1 - u_j), \tag{7.2}$$

where t_j = desired output (target), u_j = actual output, and u_j • $(1–u_j)$ = derivative between actual output and net input of unit u_j.

This is based on the fact that

$$\Delta w_{ji} = -\frac{\partial E_p}{\partial w_{ji}}; \tag{7.3}$$

(a) $E_p = \frac{1}{2} \sum_k (t_{pk} - u_{pk})^2 = \frac{1}{2} \sum_k (t_{pk} - f_k(\text{Net}_{pk}))^2$

$= \frac{1}{2} \sum_k (t_{pk} - f_k(\sum_j w_{kj} \cdot u_{pj} + \theta_k)),$

where E = error, p = model, t_k = target, u_k = output.

And then

(b) $\frac{\partial(\mathrm{Net}_{pk})}{\partial w_{kj}} = \left(\frac{\partial}{\partial w_{kj}} \cdot \sum_j w_{kj} \cdot u_{pj} + \theta_k\right) = u_{pj};$

(c) $-\frac{\partial E_p}{\partial w_{kj}} = (t_{pk} - u_{pk}) \cdot f_k'(\mathrm{Net}_{pk}) \cdot u_{pj}.$

At this point, one can say that the quantity of value to be added or subtracted from weight w_{ji} will be decided by value out_j, with respect to the activation state of unit u_i, namely, the activation with which u_j is connected to weight w_{ji} and it is in relation to the coefficient r. This is the correction rate to adopt (when $r = 1$, then the value of the adding or subtracting from weight w_{ji} is the one calculated by the whole procedure).

Equation (7.3) can be considered as follows:

$$\Delta w_{ji} = r \cdot \Delta\mathrm{out}_j \cdot u_j. \tag{7.4}$$

The value Δw_{ji} can be both negative and positive. It represents the "quantum amount" to be added or subtracted from the previous value of weight w_{ji}.

Then,

$$w_{ji(n+1)} = w_{ji(n)} + \Delta w_{ji}. \tag{7.5}$$

Nevertheless, Eq. (7.2) presupposes that each arriving unit of weight has an actual value which is comparable with an ideal value toward which it should tend (target). This presupposition, however, is valid only for the weights connecting a unit layer with the layer of the output units.

Therefore, the correction procedure discussed up to now relates only to BP provided with two layers (actually one layer if we think that the input unit cannot be considered as a layer of the network). Then, the Delta Rule represented by Eq. (7.2) permits the weight correction only for very limited networks. For *multilayer* BP, those with one or more layers of hidden units, the Delta Rule is insufficient as it is, for example:

In Fig. 7.3, it is evident that the correction of weights w_{ik} is not possible, although the weights w_{ji} can be adjusted through the Delta Rule because the value the units u_j should assume (t_j) is known. This can happen because there does not exist a value of ideal reference for the units u_i. In fact, the value that they are going to assume is one of the *results* of BP's learning work and therefore it cannot be constrained.

Rumelhart et al. (1986) and others solve this problem through a *generalization* of the traditional Delta Rule. The generalization of the Delta Rule consists of modifying Eq. (7.2) in those cases where the weight to modify is not connected to an output unit. Therefore, instead of computing the difference between the actual output and the desired one, a report summation will be computed between the error coefficient out_j, previously calculated, and the weights which that coefficient was referring to:

$$\Delta\mathrm{hidden}_i = u_i \cdot (1 - u_i) \cdot \sum \Delta\mathrm{out}_j \cdot w_{ji} \tag{7.3a}$$

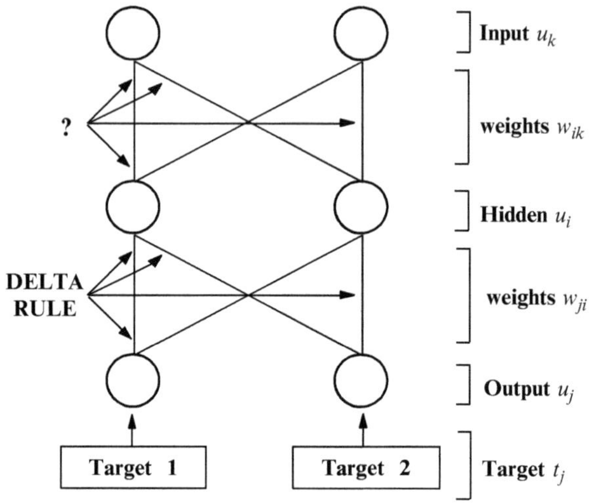

Fig. 7.3 Multilayer ANNs back propagation (weights details)

and, therefore,

$$\Delta w_{ik} = r \cdot \Delta \text{hidden}_i \cdot u_k; \tag{7.4a}$$

$$w_{ik(n+1)} = w_{ik(n)} + \Delta w_{ik}. \tag{7.5a}$$

In fact, starting from Eq. (a), we see that

(d)
$$\frac{\partial E_p}{\partial w_{kj}} = \frac{1}{2} \sum_k \frac{\partial}{\partial w_{kj}} \cdot (t_{pk} - u_{pk})$$
$$= -\sum_k (t_{pk} - u_{pk}) \cdot f_k(Net_{pk}) \cdot w_{kj} \cdot f_j(Net_{pj}) \cdot u_{pi}$$

where w_{pi} = hidden-inputs weights and w_{kj} = hidden-outputs weights.

Through the generalized Delta Rule, it is possible to create a backpropagation algorithm able to correct the weights of any BP's layer at every cycle.

We can now synthesize the two algorithms through which the BP would be able to work:

(a) Forward Algorithm

1. $\text{Net}_j = \sum_i u_i \cdot w_{ji} + \theta_j$
2. $u_j = f(\text{Net}_j)$

(b) Backpropagation Algorithm

(b1) *Correction calculation of the weights connected to the output*:

1. $\Delta \text{out}_j = (t_j - u_j) \cdot f'(u_j)$
2. $\Delta w_{ji} = r \cdot \Delta \text{out}_j \cdot u_i$

(b2) *Correction calculation of the weights not connected to the output*:

1. $\Delta \text{hidden}_i = f'(u_i) \cdot \sum_j \Delta \text{out}_j \cdot w_{ji}$
2. $\Delta w_{ik} = r \cdot \Delta \text{hidden}_i \cdot u_k$

(b3) *Putting into effect the corrections on the weights*:

1. $w_{ji(n+1)} = w_{ji(n)} + \Delta w_{ji}$
2. $w_{ik(n+1)} = w_{ik(n)} + \Delta w_{ik}$

At this point, both the *minimum conditions of activation* and those of *functioning*, or learning for BPs, have been explained.

7.2.3 The Self-Programming Bias

Both conceptually and mathematically, the bias is the unit's *threshold*. Nevertheless, from an arithmetical point of view, it is expressed through a sign value that is opposite to that of a threshold. For example, if unit u_i has a threshold of −2.0, then its bias will be +2.0. Conceptually, the bias is the inclination of a unit:

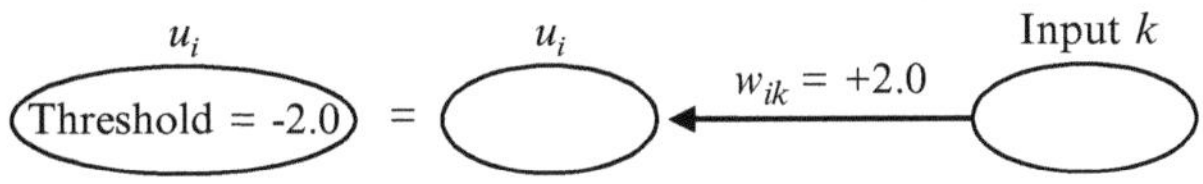

Then,

$$u_i = f(\text{Net}_i) = f\left(\sum_{j=1}^{N} u_j \cdot w_{ij} + \text{Bias}_i\right), \tag{7.6}$$

where $\text{Bias}_i = -\text{threshold}_i$.

Considered the bias' structural and functional nature, its dynamic for each network unit can be considered similar to that of any weight. Therefore, the weights' matrix is liable to the normal learning algorithm to which it is subjected.

This means that every BP can be provided with *dynamic threshold's* units. Each unit will dynamically learn its threshold in relation to the kind of experiences the whole ANN is carrying out.

This means that if the updating of the weights through the backpropagation is given by the following equations:

$$\Delta \text{out}_i = (t_i - u_i) \cdot f'(u_i) \tag{7.7}$$

$$w_{ij(n+1)} = w_{ij(n)} + \Delta \text{out}_i \cdot u_j \cdot \text{Rate} \tag{7.8}$$

for the output units, and

$$\Delta\text{hidden}_j = f'(u_j) \cdot \sum_{i=1}^{N} \Delta\text{out}_i \cdot w_{ij} \tag{7.9}$$

$$w_{jk(n+1)} = w_{jk(n)} + \Delta\text{hidden}_j \cdot u_k \cdot \text{Rate} \tag{7.10}$$

for hidden units, then Eqs. (7.11) and (7.12) will determine the bias update of the output layer and the one of any hidden layer:

$$\text{Bias}_{i(n+1)} = \text{Bias}_{i(n)} + \Delta\text{out}_i \cdot \text{Rate} \tag{7.11}$$

$$\text{Bias}_{j(n+1)} = \text{Bias}_{j(n)} + \Delta\text{hidden}_j \cdot \text{Rate} \tag{7.12}$$

It is correct to consider the bias of a unit as being a *dynamic threshold* subject to learning. From an algebraic point of view, it can be formulated as a dynamic weight generated from a fixed input of value 1 toward the unit.

Therefore, the bias represents the *historical and individual sensitivity* of each unit of ANN to the experiences of the whole network.

The self-programming bias has considerably increased the learning rapidity of BP ANNs, and of course it has decreased the possibility of entrapment of the network into local minima.

7.2.4 The Momentum

Experience has shown that the more the learning coefficients of Eqs. (7.7) and (7.8) are reduced, the less likely the probability of the network becoming stuck in local minima. Simultaneously, the smaller the learning coefficient, the correspondingly longer time it will take the network to learn. An attempt has been made to resolve this dilemma through the introduction of a new parameter, the *momentum* (McClelland and Rumelhart 1988). The *momentum* is a parameter through which the network reinforces the change of each of its connections in the descent's general direction of the paraboloid, which has already emerged during its previous updating process. The reason for this is the eventual deleting contingent upon oscillations produced by the steepest descent algorithm.

With the introduction of the *momentum*, Eq. (7.7) becomes

$$\Delta\text{out}_{i(n)} = (t_i - u_i) \cdot f'(u_i) \tag{7.7a}$$

$$\text{Momentum}_{ij(n)} = \Delta w_{ij(n-1)} \cdot k \quad 0 < k \leqslant 1 \tag{7.7b}$$

$$\Delta w_{ij(n)} = \text{Momentum}_{ij(n)} + \Delta\text{out}_{i(n)} \cdot u_j \cdot \text{Rate}, \tag{7.7c}$$

where rate_i is the learning rate of the ith level unit.

The introduction of the *momentum* also allows the ANN learning to speed up by using rather small learning coefficients. The experience of many authors recommends a value of $k = 0.9$, with a learning coefficient, rate $= 0.6$. Nevertheless, the major conclusion emerging from this research seems to offer other suggestions:

(a) It is not reasonable to suggest values for the *momentum* without knowing the problem's typology which the ANN must resolve.
(b) In this scientific field, it is good to be suspicious of every equation from which the researcher expects decisive constants: either they are useful only for certain experiments or they hide further relations that must be suitably explained.

Inserting arbitrary constants in these models is equivalent in a sense to adding *symbolic rules* to *subsymbolic procedures*. Moreover, the momentum does not eliminate the theoretical possibility that the ANN runs into local minima.

7.2.5 *The Transfer Equations*

Until now, the transfer function implicitly used in the previous equations has been the *sigmoidal* function.

Nevertheless, Eq. (7.6) was not specific on this matter:

$$u_i = f\,(\text{Net}_i) = f\left(\sum_{j=1}^{N} u_j \cdot w_{ij} + \text{Bias}_i\right). \tag{7.6}$$

If we mean that Net is the net input to a unit, and that $f\ (\ )$ is the equation of the sigmoid, Eq. (7.6) will be rewritten in the following way:

$$u_i = \frac{1}{1 + e^{-\text{Net}_i}} \quad 0 \leqslant u_i \leqslant 1, \quad e = 2.718281828459, \tag{7.13}$$

Through this equation, the activation value of u_i varies between 0 and 1 according to a semilinear function which has its flex point as 0.5. In fact, if $\text{Net}_i = 0$, then $u_i = 0.5$. Therefore, its derivative will be $f'(\text{Net}_i) = u_i \cdot (1 - u_i)$. This function is the most diffused in backpropagation ANNs, and some originators have tried to make it more complete through the introduction of other parameters, for example,

$$u_i = \left[1 + e^{\frac{-\text{Net}_i}{T}}\right]^{-1} \tag{7.13a}$$

where T is a parameter called temperature (see below).

For $T = 0$, the function is reduced to an exit *degree* of 1 or 0, while with the growing of T, the slope of the sigmoidal increases (for a closer examination, see below).

A second variation of Eq. (7.13) is the following:

$$u_i = \frac{\text{coef}}{\text{coef} + e^{-\text{Net}_i}} \tag{7.13b}$$

where for coef> 1, the unit sensitivity to the net input value increases, while for coef< 1, this sensitivity decreases.

This equation can be useful in order to avoid a problem that sometimes occurs within the hidden units in that all the units are overloaded with values that tend toward 0.0 or 1.0; the hidden unit layer is thus unable to codify the differences among the different input models (for a closer examination, see Buscema 1994)

A second transfer equation which is often used is the *hyperbolic tangent*:

$$u_i = \frac{e^{\text{Net}_i} - e^{-\text{Net}_i}}{e^{\text{Net}_i} + e^{-\text{Net}_i}} \qquad -1 \leqslant ui \leqslant +1 \tag{7.14}$$

In this case, the values of u_i vary between -1 and $+1$; therefore, at the moment of the weights' correction, the derivative of this function will have to be $(1.0 + u_i) \cdot (1.0 - u_i)$, replacing the derivative of the sigmoidal function $u_i \cdot (1 - u_i)$.

The hyperbolic tangent equation is not as soft as the sigmoidal function. This means it can be less useful in solving problems in which a fuzzy target is expected or in cases in which input vectors have no determinant differences.

If it is possible not to permit the ANN to fall into a local minima and strong corrections are desired for values near to the limits of -1 and $+1$, the hyperbolic tangent is preferable to the sigmoid for its learning velocity in binary problems (where the input and target vectors are laces of 0 and 1) is stronger.

A transfer equation exists that presents similar advantages but with less imprecision: it is the *archtangent equation*, and it has been shown to be efficient without causing any hindrance in the solution of complex problems:

$$u_i = \frac{1}{2} + \frac{1}{\Pi} \cdot \text{Arctg}(\text{Net}_i) \tag{7.15}$$

In Eq. (7.15), u_i varies between 0 and 1 and presents itself as a gentler sigmoid function than the traditional sigmoid equation (7.13).

There also exists a transfer equation that requires a different treatment called the *sinusoidal* function:

$$u_i = \sin(\text{Net}_i). \tag{7.16}$$

First, it is important to note that making use of this function requires modification of the derivative in the weight equations of correction: the sigmoid $u_i \cdot (1 - u_i)$

becomes cos(Net$_i$) in Eq. (7.16). Its characteristic consists in making clear a discrete class of couples (x, y), synthesizing the continuous function $y = g(x)$ that considers an ANN with only one output. In practice, it seems to decompose the principal components of the function that is described in a discrete way in the input-output examples.

In many experiments, we noticed that by selecting a half unit with the sigmoid and a half with the sine for the layer of hidden unit, *two* split halves of the input models are self-created. The first one, constituted by the sigmoidal units, tends to codify the determinant differences of the details among input models, while the second one, constituted by sinusoidal units, tends to codify the similarity of the details between the different models. This peculiarity, also noticed by other authors (NeuralWare 1995), often guarantees a better generalization capacity to the whole ANN.

Actually, the choice of the transfer functions is just one of the choices determining the functioning of an ANN. Moreover, this choice should be made while considering all the other choices: kind of problem, topology of ANN, learning rates, type of momentum, type of backpropagation, etc. Therefore, it is useless to assert *beforehand* which of those equations is more effective, excepting rare cases (e.g., the weakness of the linear transfer in hidden units). Further on, we will see that their effectiveness changes according to the kind of problems and to the global architecture of ANN.

On the basis of research implemented since 1985 for backpropagation ANNs, maybe it can be asserted that the sigmoidal function has always shown a good behavior in very different cases and problems.

7.3 Methodological Developments

7.3.1 The SoftMax

This is a function used in the learning process of BP networks for the classification of problems.

In order to resolve this kind of problem, the desired output (target) is usually represented through the code *1 of N* classes. Each class is represented by an output unit; then, in the case of N classes, the target will be constituted by a *vector* composed by N output nodes, and it will have the following form:

$$\underline{d} = (0, \ldots, 0, 1, 0, \ldots, 0)$$

where only the unit corresponding to the desired class has a target value different from 0 (i.e., 1).

There are two problems with this formula:

1. The codified *1 of N* is dependent on the values of the output units. When one of the units has a value equal to 1, the other units have a value equal to 0. Nevertheless, there is no assumption of dependence in the error made in a trained backpropagation network.
2. If there are more classes, the network can find a reasonable solution through a mapping of the output vector. This will give an RMS of $\frac{1}{\sqrt{k}}$ (where k is equal to the class number).

This solution is a very simple one to find. For example, if we use the sigmoid as the transfer function and all the weights of the output layer are large and negative, the outputs of all the output units will be equal to 0. If the class number is not too large, and a very low learning rate is set, the network will have difficulty converging on this artificial solution. In order to have a satisfying solution for these two problems, it has been recommended to use the "SoftMax activation function" for the layer of output units (Bridle 1989).

The SoftMax function is a refined version of the competitive function *1 of N* and has some convenient mathematical properties:

$$y_k = \frac{e^{I_k}}{\sum_{l=1}^{k} e^{I_l}}$$

Bridle (1989) proposed to use the SoftMax function combining it with the function that follows that resulted from the relative entropy measure between the target and the actual output:

$$J = -\sum_{j=i}^{k} d_j \ln(y_j)$$

We can calculate the value that will return from the output unit j in the following way:

$$\frac{-\partial J}{\partial I_j} = -\sum_{k=1}^{k} \frac{\partial J}{\partial y_k} \cdot \frac{\partial y_k}{\partial I_j}$$

We must begin with the quotient rule:

$$\frac{\partial y_k}{\partial I_j} = y_k(\partial k_j - y_j)$$

Therefore,

$$\frac{-\partial J}{\partial J} = -\sum_k \left(\left(\frac{-d_j}{y_k} \right) \cdot y_k \cdot \left(\delta k_j - y_j \right) \right) = d_j - y_j \sum_k d_k = d_j - y_j$$

where $\delta k_j = \begin{cases} 1 \text{if} k = j \\ 0 \text{if} k \neq j \end{cases}$

Thus, the backpropagation's standard algorithm can be used with the SoftMax function for the backpropagation of the real error.

7.3.2 *The Fast Propagation*

In order to speed up the learning of backpropagation ANNs, Tariq Samad (1988) changed Eq. (7.8) in the following way:

$$w_{ij(n+1)} = w_{ij(n)} + \text{Rate} \cdot \Delta\text{out}_i \cdot (u_j + k \cdot \Delta\text{hidden}_j) \tag{7.8a}$$

The error registered in the previous cycles in proportion k ($k > 0$; for $k = 0$, it is the normal backpropagation of Eq. (7.8)) has to be added to the value of the rising unit u_i of the connection. NeuralWare (1995) implemented this technique by defining it as "fast propagation" and sustained its advantage in terms of rapidity (for comparisons, see Buscema 1994; for a closer examination of this technique, see Samad 1988 and 1989).

7.3.3 *Semeion's Self-Momentum*

In 1989, Semeion attempted to resolve the problem concerning the achievement of the rapid rise in learning in networks. This effort resulted in a similar success but achieved in a slightly different way (December: Tests at Semeion Institute in Buscema and Sacco 2000). The hypothesis led to the conclusion that it was necessary to reinforce the descending direction of the paraboloid as a function of the error that each node was generating exactly in that moment. In other words, it is good to spank the child in order to remind him about the good examples when he does wrong, but it is silly to spank him every time he moves.[1]

[1]Old Italian proverb.

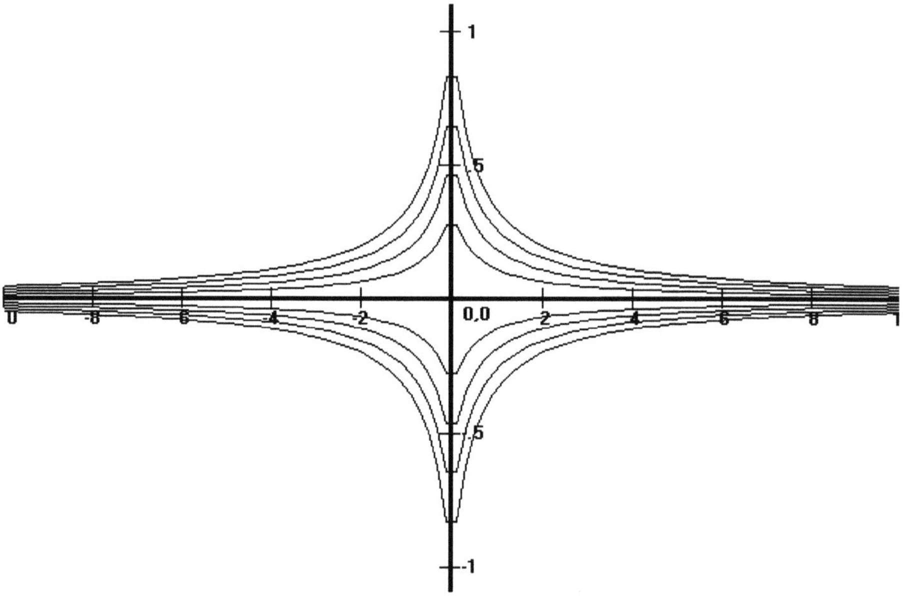

Fig. 7.4 Progress of the self-momentum on the ordinate, value of the weights in the interval [−1.0, 1.0] on the abscissa

More formally, starting from Eq. (7.7a), the *self-momentum equation* appears in the following way:

$$\Delta \mathrm{out}_{i(n)} = (t_i - u_i) \cdot f'(u_i) \tag{7.7a}$$

$$\mathrm{Self-Momentum}_{ij} = \Delta w_{ij(n-1)} \cdot \left|\Delta \mathrm{out}_{i(n)}\right| \cdot \frac{1}{0.5 + \left|w_{ij}\right|} \tag{7.17}$$

$$\Delta w_{ij(n)} = \mathrm{Self-Momentum}_{ij(n)} + \Delta \mathrm{out}_{i(n)} \cdot u_j \cdot \mathrm{Rate} \tag{7.18}$$

where $\left|w_{ij}\right|$ = absolute value of connection w_{ij}.

The *self-momentum* eliminates the arbitrary parameter k of the *momentum* and allows ANN to be able to stabilize, in an independent way, the strength with which the direction of the weights' correction is reinforced. On a practical level, the *self-momentum* equation allows all the problems to be resolved through the *momentum* by maintaining the coefficient of unitary learning (rate = 1). Figure 7.4 shows the progress of the self-momentum formulating $\Delta \mathrm{out}_{i(n)} = 1$, varying the $\Delta w_{ij(n-1)}$ in the interval [−0.5, 0.5] and the w_{ij} weights in the interval [−1.0, 1.0].

The self-momentum technique has shown its advantages in different applications, especially in nonlinear regression problems and when the dataset is a temporal series (Buscema and Sacco 2000).

References

Bridle, J. S. (1989). Probabilistic interpretation of feedforward classification network outputs, with relationships to statistical pattern recognition. In F. Fogelman-Soulié & J. Hérault (Eds.), *Neuro-computing: Algorithms, architectures*. New York: Springer.

Buscema, M. (1994). *Squashing Theory. Modello a Reti Neurali per la Previsione dei Sistemi Complessi* [Squashing theory: A neural network model for prediction of complex systems] Rome: Armando, Semeion Collection.

Buscema, M. (1996). *SQUASH. Shell for program feed forward neural networks*, Semeion Software n. 5, Rome, 1992–1996.

Buscema, M., & Sacco, P. L. (2000). Feedforward networks in financial prediction: The future that modifies the present. *Expert System, 17*(3), 149–170.

Chauvin, Y., & Rumelhart, D. E. (Eds.). (1995). *Backpropagation: Theory, architectures, and applications*. Hillsdale: Lawrence Erlbaum.

Fahlman, S. E. (1988). *An empirical study of learning speed in back-propagation networks* (CMU Technical Report, CMU-CS-88-162).

Jacobs, R. A. (1988). Increased rates of convergence through learning rate adaptation. *Neural Network, 1*, 295–307.

Lapedes, A., & Farber, R. (1987). *Nonlinear signal processing using neural networks: Prediction and system modeling* (Los Alamos National Laboratory Report LA-UR-87-2662).

McClelland, J. L., & Rumelhart, D. E. (1988). *Explorations in parallel distributed processing*. Cambridge, MA: MIT Press.

Minai, A. A., & Williams, R. D. (1990). Acceleration of backpropagation through learning rate and momentum adaptation. *International Joint Conference on Neural Networks, 1*, 676–679.

Neuralware. (1995). *Neural computing*. Pittsburgh: NeuralWare Inc.

Rumelhart, D. E., Hinton, G. E., & Williams, R. J. (1986). Learning internal representations by error propagation. *Nature, 323*, 533–536.

Samad, T. (1988). *Back-propagation is significantly* . . . . International Neural Network Society Conference Abstracts.

Samad, T. (1989). *Back-propagation extension* (Honeywell SSDC Technical Report).

Tawel, R. (1989). Does neuron learn like the synapse? In D. S. Touretzky (Ed.), *Advances in neural information processing systems* (Vol. 1). San Mateo: Morgan Kaufman.

Weigend, A. S., Rumelhart, D. E., & Huberman, B. A. (1991). Generalization by weight-elimination with applications to forecasting. In R. P. Lippmann et al. (Ed.), *Advances in neural information processing systems 3* (pp. 875–882). Morgan Kaufmann, San Mateo, CA.

Werbos, P. (1974). *Beyond regression: New tools for prediction and analysis in behavioral sciences*, Ph.D. thesis, Harvard, Cambridge, MA.

Chapter 8
Preprocessing Tools for Nonlinear Datasets

Massimo Buscema, Alessandra Mancini, and Marco Breda

8.1 Introduction

One of the most difficult problems for training an artificial neural network (ANN) is to establish the size and quality of the training and testing sets. The subset quality is related to the assessment of its representativeness with respect to the complete dataset. Classic statistical resampling strategies used in the literature turn a random distribution of samples into two or more subsamples (i.e., cross validation and bootstrapping techniques). These procedures are based on the premise that the subsets extracted from the total dataset are a good approximation of the density function of the measured variables.

However, when the amount of data is limited or the dataset is particularly complex (when the data consists of discrete hyperpoints of some unknown nonlinear function), the application of resampling techniques based on random procedures requires some caution; one of the limits of random resampling techniques is that of splitting data into a training and a testing set, though this method does not take into account the problem of outliers. When treating "noisy data" during the preprocessing phase, typically outliers are not eliminated from the database because they could reveal significant information. Nevertheless, when dealing with the subsets that are going to be used to train the ANN, no particular care is taken toward the distribution of outlier data. The evaluation of the actual performance of the ANNs can therefore be associated to the importance of quality analysis based on the representativeness of the real data and of the training and testing subsets generated from the global database. Moreover, when evaluating the success of the ANN, one must consider that, since the learning algorithm for training these

M. Buscema (✉) • A. Mancini • M. Breda
Semeion Research Center of Sciences of Commuication, via Sersale 117, Rome, Italy
e-mail: m.buscema@semeion.it

M. Buscema and W.J. Tastle (eds.), *Intelligent Data Mining in Law Enforcement Analytics: New Neural Networks Applied to Real Problems*, DOI 10.1007/978-94-007-4914-6_8,

models is usually based on a random initialization of weights (characterized by the presence of a random starting rate in the feedforward ANNs), an internal source of variance is generated which will inevitably influence its performance (Kolen and Pollack 1990).

To manage these problems, we need to "optimize" the data sampling strategy in order to be able to improve both the representativeness of the data subsets used in training and, at the same time, the assessment of the performance accuracy of the ANN model.

The artificial organisms proposed in this work, which we call *T&T* (training and testing) and *T&Tr* (training and testing reverse), can be considered data preprocessing systems which permit one to obtain more effective procedures for the training, testing, and validation of ANN models. These are evolutionary organisms whose populations are composed of ANNs which carry out the sampling procedure in an inductive, rather than random, way. The primary objective consists of the arrangement of the source sample into *n* subsamples all possessing a similar probability density function to each other.

Another significant methodological problem related to the application of ANN to real databases comes to light when these are comprised of a large number of variables which, apparently, seem to provide the largest possible amount of information. Under these conditions, the *input space*, determined by all the possible combinations of the values of the observed variables, becomes so large that any research strategy to find the best solutions becomes very cumbersome when specific problems are tackled, for example, classification tasks.

It is necessary to carry out a preliminary analysis of the variables of the dataset since these can have a different relevance with respect to the data mining that one intends to conduct. Some of the attributes may contain redundant information which is included in other variables, confusing information (noise), or may not even contain any significant information at all and thus be completely irrelevant. Therefore, a procedure that will identify and select, from the global set, a subset consisting of those variables that are most informative toward the representation of input patterns is necessary when dealing with classification problems solved with induction algorithms. Moreover, the accuracy of the procedure, learning time, and the number of examples necessary are all dependent upon the choice of variables.

Among the methods used to reduce the dimensionality of the data, the *feature selection* techniques (also known as *subset attribute selection* or *input selection*) were developed to determine which variables (or attributes) are most relevant in the representation of an input pattern, derived from a large dataset. When dealing with a database with a large number of variables, feature selection is used to reduce the number of variables used in the classification while maintaining an acceptable level of accuracy in the procedure. By extracting the most relevant attributes, the dimensions of the input space are reduced; thus, it is easier to find the best solutions, and if the extracted attributes are actually the most significant, the definition of a predictive pattern in the global data is effective.

On the whole, feature selection extracts, from a given dataset D_Γ of N characteristics, the best subset consisting of K characteristics. The number of possible subsets of K characteristics is given by $\sum_{K=0}^{N} \binom{N}{K} = 2^N$, and among these subsets, the best one is the one that maximizes some cost function $F(\cdot)$. Research into the attribute space to determine the pool of characteristics is carried out with a search function by measuring the discrimination capacity of each of the subsets. This evaluation is carried out on the possible subsets D'_Γ using each variable subset as training samples ($D'_\Gamma{}^{[tr]}$) and testing samples ($D'_\Gamma{}^{[ts]}$) for an inductor $\Omega_{D'_\Gamma{}^{[tr]},A,F,Z}(\cdot)$, using a fixed induction algorithm A, the configuration parameters F, and the installation parameters Z.

Excluding the exhaustive search strategy on the global set of characteristics, which is not applicable to a dataset with a high number of variables, the techniques that can be used are a blind search (e.g., depth first) or heuristic search (hill climbing, best first), but in the literature, evolutionary search techniques have also been proposed (Kudo and Sklansky 2000; Siedlecki and Slansky 1989). Genetic algorithms have been shown to be very effective as global search strategies when dealing with nonlinear and large problems.

Feature selection techniques can be developed using two different general approaches based on whether the selection of the variables is carried out dependently or independently of the learning algorithm used to build the inductor. The filter approach attempts to select the best attribute subset by evaluating its relevance based on the data. The "wrapper" approach, instead, requires that the selection of the best attribute subset takes place considering as relevant those attributes that allow the induction algorithm to generate a more accurate performance (John et al. 1994).

Input selection (*IS*) operates as a specific evolutionary wrapper system that responds to the need to reduce the dimensionality of the data by extracting the minimum number of variables necessary to control the "peaking" phenomenon and, at the same time, conserving the most information available.

8.2 Artificial Organisms: The Models

We now introduce a new concept called the artificial organism (AO). We define AO as a *group* of dynamic systems (ANN, evolutionary algorithms, etc.) that use the same sensors and effectors of a process, working in synergy without explicit supervision. T&T, T&Tr, and IS systems satisfy these characteristics and are therefore considered models of AO.

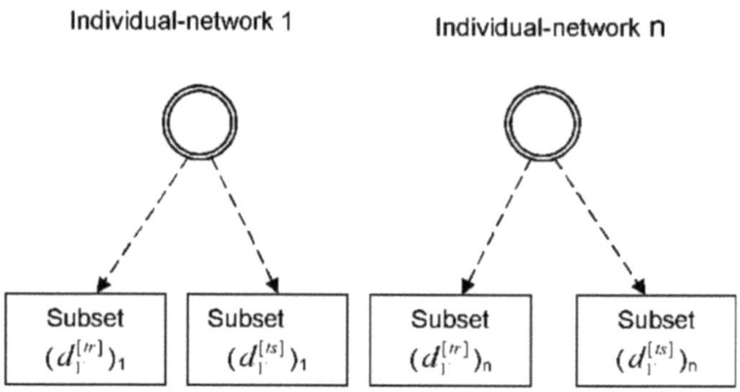

Fig. 8.1 The T&T algorithm: each individual-network of the population distribute the complete dataset D_Γ in two subsets, $d_\Gamma^{[tr]}$ (training) and $d_\Gamma^{[ts]}$ (testing)

8.2.1 *The Training and Testing Algorithm*

The "training and testing" algorithm (T&T) is based on a population of *n* artificial neural networks (ANNs) managed by an evolutionary system. In its simplest form, this algorithm reproduces several distribution models of the complete dataset D_Γ (one for every ANN of the population) in two subsets ($d_\Gamma{}^{[\mathrm{tr}]}$, the training set, and $d_\Gamma{}^{[\mathrm{ts}]}$, the testing set). During the learning process, each ANN, according to its own data distribution model, is trained on the subsample $d_\Gamma{}^{[\mathrm{tr}]}$ and blind validated on the subsample $d_\Gamma{}^{[\mathrm{ts}]}$ (see Fig. 8.1).

The performance score reached by each ANN in the testing phase represents its "*fitness*" value (i.e., the individual probability of evolution). The genome (the full content of all information) of each "network individual" thus codifies a data distribution model with an associated validation strategy. The *n* data distribution models are combined according to their fitness criteria using an evolutionary algorithm. The selection of "network individuals" based on fitness determines the evolution of the population, that is, the progressive improvement of performance of each network until the optimal performance is reached, which is equivalent to the better division of the global dataset into subsets.

The evolutionary algorithm mastering this process, named "genetic doping algorithm" (GenD for short), was created at Semeion Research Center (Buscema 2004). GenD has similar characteristics to a genetic algorithm, but (1) the criteria of evolution and the mathematics of the crossover are completely new and different from classical models; (2) a species-health-aware evolutionary law and genetic operators are used; and (3) the individuals are organized into a structure (Buscema 2004).

In T&T systems, the solution space of GenD is constituted by all the possible partitions of records between the training and testing sets.

Given a dataset D_Γ of N records, the number of samples d_Γ which are comprised of K possible records is given by

$$\binom{N}{K}$$

Varying K, you have

$$\sum_{K=0}^{N} \binom{N}{K} = 2^N \quad (*)$$

(*) the search space is 2^N, but the acceptable space is 2^N-2.

However, the possible useful partitions of records between training and testing sets are limited to

$$\sum_{K=\lfloor r\cdot N\rfloor}^{\lfloor(1-r)\cdot N\rfloor} \binom{N}{K} < 2^N$$ with r typically having a value between 0.4 and 0.5.

The evolutionary algorithm codes those partitions according to a two-symbol alphabet:

$$\Phi_{\mathrm{T\&T}} = \{\phi_{\mathrm{tr}}, \phi_{\mathrm{ts}}\}$$

where

(ϕ_{tr}) represents a record belonging to the *training* set $d_\Gamma^{[\mathrm{tr}]}$ and
(ϕ_{ts}) represents a record belonging to the *testing* set $d_\Gamma^{[\mathrm{ts}]}$.

Therefore, a pair of training and testing sets represents, in the solution space, a possible solution $x = \left(D_\Gamma^{[\mathrm{tr}]}, D_\Gamma^{[\mathrm{ts}]}\right)$, given by the vector

$$x = \left(D_\Gamma^{[\mathrm{tr}]}, D_\Gamma^{[\mathrm{ts}]}\right) = [x_1, x_2, \cdots, x_N] \in \Phi_{\mathrm{T\&T}}^N \qquad x_i \in \Phi_{\mathrm{T\&T}}$$

The elaboration of T&T is articulated in two phases:

1. Preliminary phase: the parameters of the fitness function that will be used on the global dataset are evaluated. During this phase, an inductor $\Omega_{D_\Gamma^{[\mathrm{tr}]},A,F,Z}(\cdot)$ is configured which consists of an artificial neural network with a standard algorithm (A) back propagation. For this inductor, the optimal configuration to reach convergence is stabilized at the end of different training trials on the global dataset D_Γ; in this way, the configuration that most "suits" the available dataset is determined: the number of layers and hidden units and some possible generalizations of the standard learning law. The parameters thus determined that define the configuration (F) and the initialization (Z) of the population's individual networks will then stay fixed in the following computational phase. Basically, during this preliminary phase, there is a fine-tuning of the ANN

that defines the fitness values of the population's individuals during evolution. Additionally, a value E_0 of epochs is necessary to give an adequate evaluation of the fitness of the individuals. The selection of the individuals is carried out on the basis of the fitness value defined according to a cost function that is deemed useful to the optimal interpolation of the set.

2. Computational phase: the system extracts from the global dataset the best training and testing sets. During this phase, the individual network of the population is running, according to the established configuration and the initialization parameters. From the evolution of the population, managed by the GenD algorithm, the best distribution of the global dataset D_Γ into two subsets is generated, starting from the initial population of possible solutions $x = \left(D_\Gamma^{[tr]}, D_\Gamma^{[ts]}\right)$. For each GenD epoch, each individual of the initial population is trained on the training set $D_\Gamma^{[tr]}$ for a number of epochs E_0 and is tested on the corresponding testing set $D_\Gamma^{[ts]}$. For the evolutionary system, the following options are fixed: only one tribe and the two global genetic operators of crossover and mutation. This allows the algorithm to converge on the desired evolution in minimum time.

8.2.2 T&Tr (Training and Testing Reverse) Algorithm

The T&T algorithm can be enhanced by introducing a "reverse" procedure in order to achieve a better measure of the accuracy of the performance on the global dataset, when the representativeness of the dataset is not completely satisfactory.

In the T&Tr evolutionary system, every individual of the population is composed of a pair of ANNs. Each pair represents a distribution model of the global dataset D_Γ in two subsets: $d_\Gamma^{[tr]}$ (*tr*aining set) and $d_\Gamma^{[ts]}$ (*tes*ting set). For each pair, the first ANN is trained on subsample $d_\Gamma^{[tr]}$, and it is validated blind on the subsample $d_\Gamma^{[ts]}$. For the second ANN, completely independent from the first, the subset $d_\Gamma^{[ts]}$ is used as a training set, and the subset $d_\Gamma^{[tr]}$ is used as a testing set (see Fig. 8.2).

The average value of the performance reached by the two ANNs during the testing phase is the fitness of the individual. T&Tr optimizes the procedure that splits the global set into training and testing subsets where

$$f_1(d_\Gamma^{[tr]}) \cong f_2(d_\Gamma^{[ts]}) \cong f_0(d_\Gamma^{[global]})$$

$f_1(d_\Gamma^{[tr]})$ and f_2 $(d_\Gamma^{[ts]})$ = probability density function of the testing and training subset, respectively; f_0 $(d_\Gamma^{[global]})$ = probability density function of the global dataset.

The goal of such optimization is to achieve the best performance with a single ANN trained on one of these subsets and tested on the other. In T&Tr, performance overestimation is avoided by training the inductor on the testing set $D_\Gamma^{[ts]}$ for E_0 epochs and testing it on the corresponding training set $D_\Gamma^{[tr]}$ ("reverse" procedure), that is, exchanging the subsets in the training and testing phase of the pair's second ANN. For each individual in the population, we obtain a different model of data

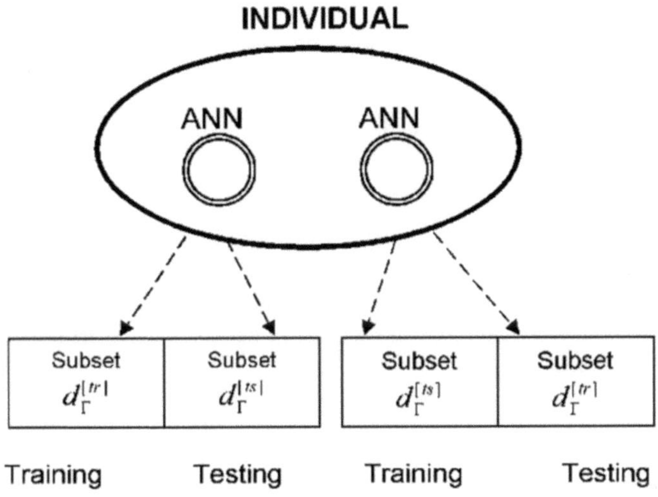

Fig. 8.2 The T&Tr algorithm: every individual of the population is composed of a pair of ANNs, each representing a distribution model of the global dataset D_Γ in two subsets $d_\Gamma^{[tr]}$ (training) and $d_\Gamma^{[ts]}$ (testing). The first network of each pair is trained on subset $d_\Gamma^{[tr]}$ and it is blind-validated on the subset $d_\Gamma^{[ts]}$; instead, the second network is trained on subset $d_\Gamma^{[ts]}$ and tested on the subset $d_\Gamma^{[tr]}$

distribution which, at every generation, is combined by the GenD evolutionary algorithm according to the fitness criterion. In this way, the best distribution of the overall dataset into training and testing subsets is reached after a finite number of generations.

8.2.3 Input Selection

Input selection (IS) is an adaptive system based on the evolutionary algorithm GenD and able to evaluate the relevance of the different variables of the available dataset in an intelligent way. Therefore, it can be considered on the same level as a feature selection technique.

For a pair of training and testing subsets evaluated by the inductor in a classification/prediction problem, IS is able to determine which variables are relevant for the considered problem; the inductor is therefore trained on this pool of variables using the variation in its performance as feedback. It is possible to assume that, if the selection of the input variables has some influence on the performance of the inductor, the goodness of the results obtained in the classification/prediction problem depends mainly on the relevance of the selected variables.

From a formal point of view, IS is an artificial organism based on the GenD algorithm and consists of a population of ANNs, in which each ANN carries out a selection of independent variables from the available database.

In a specific domino problem, the solution space is determined by $\sum_{H=0}^{M}\binom{M}{H} = 2^M$ (*) possible combinations of H variables which describe the data.

(*) the acceptable solution space is $(2^M - 1)$.

Given the following two-symbol alphabet,

$$\Phi_{\text{IS}} = \{\phi_{rel}, \phi_{irrel}\}$$

in which

ϕ_{rel} represents membership of a variable in the set V_{rel} of relevant variables and

ϕ_{irrel} represents membership of a variable in the set V_{irrel} of irrelevant variables. Therefore, the vector

$$x = (V_{rel}, V_{irrel}) = [x_1, x_2, \ldots, x_M] \in \Phi_{\text{IS}}^M \qquad x_i \in \Phi_{\text{IS}}$$

represents a single possible solution $x = (V_{rel}, V_{irrel})$, given a pair of sets of relevant and irrelevant variables.

The elaboration of IS, as for T&T, is developed in two phases:

1. Preliminary phase: during this phase, an inductor $\Omega_{D_{\Gamma}^{[\text{tr}]},A,F,Z}(\cdot)$ is configured to evaluate the parameters of the fitness function. This inductor is a standard back-propagation ANN. The parameters configuration and the initialization of the ANNs are carried out with particular care to avoid possible overfitting problems that can surface when the database is characterized by a high number of variables that describe a low quantity of data. The number of epochs E_0 necessary to train the inductor is determined through preliminary experimental tests.
2. Computational phase: the inductor is active, according to the stabilized configuration and the fixed initialization parameters, to extract the most relevant variables of the training and testing subsets. Each individual network of the population is trained on the training set $D_{\Gamma}'^{[\text{tr}]}$ and tested on the testing set $D_{\Gamma}'^{[\text{ts}]}$. The evolution of the individual network of the population is based on the algorithm GenD and leads to the selection of the best combination of input variables, that is, the combination that produces the best performance (maximum accuracy) of the inductor $\Omega_{D'_{\Gamma}{}^{[\text{tr}*]},A,F,Z}(\cdot)$ in the testing phase with the least number of input variables:

$$Acc_X\left(\Omega_{D'_{\Gamma}{}^{[\text{tr}**]},A,F,Z}(\cdot)\right) = \underset{D'_{\Gamma}{}^{[\text{tr}*]}\text{variables}}{\text{Max}} \left\{Acc_X\left(\Omega_{D'_{\Gamma}{}^{[\text{tr}*]},A,F,Z}(\cdot)\right)\right\}$$

8.3 Experimental Procedure

8.3.1 *Definition of the Baseline*

In the experiments conducted, it was possible to compare the performance obtained in the classification of artificial organisms developed at the Semeion Research

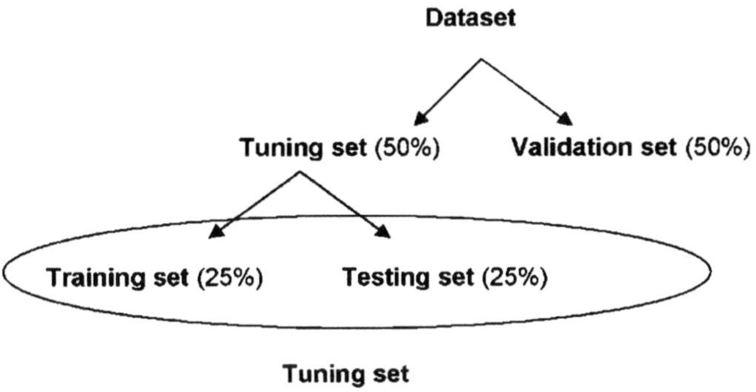

Fig. 8.3 The T&T algorithm: the dataset distribution schema

Center (see paragraph 2) to a baseline derived from the results of simulations carried out using a back-propagation network model, complete with an output layer and the neural transfer function softmax (Sutton and Barto 1998), on subsets extracted with random procedure, to measure the relative increase in performance.

8.3.2 Software

The artificial organisms T&T and T&Tr and the IS system were carried out with the Semeion S.W. n. 17© software, developed in the Semeion Research Center. The back-propagation networks described above were created with the Supervised 6.53© (Semeion Software number 12) software.

8.3.3 Training and Testing

During the first step of the procedure, the T&T system is applied to the total dataset; in this way, the data are divided into two different subsets: the tuning set with approximately 50% of the records and the validation set with the other 50%. T&T is then reapplied to the tuning set only, which, in turn, is divided into two new subsets (the training set with approximately 50% of the records used for tuning, equal to 25% of the total records, and the testing set with the other 50%, equivalent to another 25% of the total amount). In this way, the complete dataset is distributed into three subsets according to the schema showed in Fig. 8.3.

After splitting, an ANN is trained using only the training and the testing subsets. This is called a tuning phase and it allows the ANN to modify its weights configuration until the correct interpolation function is reached. Then, in the generalization phase, the system is validated on the validation set. As mentioned earlier, the experimental phase is carried out using a standard back-propagation ANN.

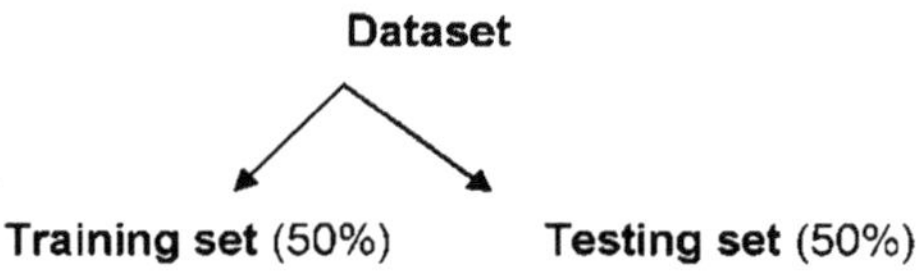

Fig. 8.4 The T&Tr algorithm: the dataset distribution schema

8.3.4 Testing and Training Reverse

The experimental procedure requires only one application of the T&Tr system to the total dataset. In this way, the data is divided into two different subsets: the training set with approximately the 50% of records and the testing set with the other 50%. Thus, the dataset is distributed according to the schema showed in Fig. 8.4.

In this case as well, the experimental phase is carried out *training* a standard back-propagation ANN on the first subset and *testing* it on the second subset.

8.3.5 Input Selection

In a preliminary phase, the T&Tr algorithm is used on the global dataset which is divided into two subsets that will be used to train the back-propagation network models, and then the IS system is applied. In this way, a subset of variables from the global dataset is selected and extracted following the procedure previously described.

8.4 Datasets

To test the T&T, T&Tr, and IS systems' effectiveness, we used two different datasets, the first chosen from the UC Irvine Machine Learning Repository (Merz and Murphy 1998) and the second from the StatLog database (a subset of the datasets used in the European StatLog project):

1. *Wisconsin Breast Cancer dataset (1991):* from Dr. W. H. Wolberg at University of Wisconsin Hospitals (for short, Breast Cancer). The donor is Olvi Mangasarian (Mangasarian and Wolberg 1990; Wolberg and Mangasarian 1990; Mangasarian et al. 1990; Bennett and Mangasarian 1992).
2. *Australian Credit Approval dataset (1992)*: this database is included in the StatLog datasets; the source of dataset is confidential. The donor is Ross D. King, Department of Statistic and Modeling Science, University of Strathclyde, Glasgow (UK) (Quinlan 1987, 1993).

On these datasets, we compared the results obtained applying the T&T, T&Tr, and IS systems to those of a standard back-propagation network.

Table 8.1 Breast cancer dataset – T&T subsets

Subset	Number of cases (%)
Training	197 (28%)
Testing	168 (24%)
Validation	334 (48%)

Table 8.2 Breast cancer dataset – T&Tr subsets

Subset	Number of cases (%)
A	330 (48%)
B	369 (52%)

8.5 Results of the Experiments with T&T and T&Tr

8.5.1 *Breast Cancer Dataset*

The records in this dataset are the results of breast cytology tests and represent a diagnosis of benign or malignant tumor. The format is a data frame with 699 observations on nine independent variables (each variable is measured by an ordinal scale on ten levels):

1. Clump thickness
2. Uniformity of cell size
3. Uniformity of cell shape
4. Marginal adhesion
5. Single epithelial cell size
6. Bare nuclei
7. Bland chromatin
8. Normal nucleoli
9. Mitoses

Class distribution: 458 cases are benign (65.5%), and 241 cases are malignant (34.5%).

The application of the T&T system on the global dataset required that the data be divided into three subsets; the tuning set contains the 52% of the complete data (28% in the training set and 24% in the testing set) and the remaining 48% of the data (see Table 8.1) for validation.

A back-propagation ANN with four hidden units was trained and tested on the first two subsets, and at the end, the same ANN was tested again on the validation set.

On the other hand, the application of T&Tr splits the dataset into a subset A with 48% and a subset B with 52% of cases (see Table 8.2); the first subset is used as training, the other as a testing set, and then reversed on a standard back-propagation ANN.

Table 8.3 Breast cancer dataset – results – *% mean; maximum performance = 98.79%; minimum performance = 97.56%

Method	% Recognition
Back-prop + T&T	99.4
Back-prop + T&Tr	98.17*
Back-prop	95.0

Table 8.4 Australian credit approval dataset – T&T subsets

Subset	Num of cases (%)
Training	229 (33%)
Testing	111 (16%)
Validation	350 (51%)

Table 8.3 shows our results obtained training the ANNs on the subsets created by T&T and T&Tr, as percentage of accuracy reached during classification.

The standard back-propagation ANN model trained on subsets extracted from the global dataset using the T&T and T&Tr systems performed better than other classifier systems when compared with results obtained on the same dataset. Therefore, the data preprocessing obtained with the T&T and T&Tr systems was successful on this illustration.

8.5.2 Australian Credit Approval Dataset

This dataset is composed of credit card application data. All attribute names and values have been coded to keep the data confidential. The goal is to distinguish between two classes of applicants for credit ("good risk" and "bad risk"). The dependent variable represents final judgment about the client's behavior after they have received the credit. There are 14 independent variables recorded at time t_0, and they are an interesting mix of attributes: continuous, nominal with a small number of values, and nominal with larger numbers of values. The sample consists of 690 people or cases: 307 "good risk" clients (44.5%) and 383 "bad risk" clients (55.5%). T&T divided the entire sample into three subsets; the tuning set contains 49% of the data (33% in the training set and 16% in the testing set), and the validation set contains the remaining 51% (see Table 8.4).

T&Tr divided the total dataset in a subset A containing 52% of cases and a subset B containing the remaining 48% (Table 8.5); a standard back-propagation ANN model was trained using first the subset A for the training phase and subset B for the testing phase and then reversed. When training the ANN models on the

Table 8.5 Australian credit approval dataset – T&Tr subsets

Subset	Num of cases (%)
A	362 (52%)
B	328 (48%)

Table 8.6 Australian credit approval dataset – results – *% mean; maximum performance = 90.24%; minimum performance = 89.50%

Method	% Recognition
Back-prop + T&T	96.00
Back-prop + T&Tr	89.87*
Back-prop	86.66

subsets extracted by T&T and T&Tr, the results obtained are shown in Table 8.6; the accuracy percentage reached is superior compared to the simple back-propagation network.

8.6 Results Analysis

An analysis of the effectiveness of the T&T and T&Tr systems as tools for preprocessing the databases subjected to elaboration with RNA can be carried out comparing the subsets extracted from the overall sample by the above techniques and those generated using a criterion of random distribution. This test was carried out based on the quality of the performance obtained on the databases considered in our experiments.

The analysis was carried out by comparing, for each of the resampling techniques (T&T, T&Tr, and random), the two subsets of training and testing in order to verify the nature of the distribution of the cases within each of them. The test was carried out between the testing and training subsets based on a criterion of comparison that calculates an index $\boldsymbol{S_i}$, given by the *sum of the squares of the differences of the linear correlations* over every variable with respect to all the others. This index, which constitutes an indicator of the uniformity of the distribution of the information between the training and testing sets, is given by

$$S_i = \sum_{i=1}^{N} \sum_{J=1,\, j \neq i,}^{N} \left(R_{ij}^{(t)} - R_{ij}^{(v)} \right)^2 \tag{8.1}$$

where

N = number of variables
${R_{ij}}^{(t)}$ = linear correlation between two variables of the training subset
${R_{ij}}^{(v)}$ = linear correlation between two variables of the testing subset

Table 8.7 Breast cancer – comparison of the *S* indices calculated between the subsets of training and testing generated by T&T, T&Tr, and the random distribution

Variables	S T&T	T&Tr	Random
Var 1	0.0176	0.0190	0.0221
Var 2	0.0809	0.0083	0.0140
Var 3	0.0730	0.0129	0.0134
Var 4	0.1125	0.0438	0.0125
Var 5	0.0327	0.0311	0.0632
Var 6	0.0434	0.0352	0.623
Var 7	0.0617	0.0104	0.0265
Var 8	0.0755	0.0216	0.0297
Var 9	0.0428	0.0214	0.0176
Target 1	0.0729	0.0149	0.0077
Target 2	0.0729	0.0149	0.0077
Mean$\overline{S}$	0.0623	0.0212	0.0252

The coefficient of linear correlation is calculated according to the

$$R_{ij} = \frac{\sum_{p=1}^{M} \left(v_{pj} - \bar{v}_j\right) \cdot \left(v_{pi} - \bar{v}_i\right)}{\sqrt{\sum_{p=1}^{M} \left(v_{pj} - \bar{v}_j\right)^2 \cdot \sum_{p=1}^{M} \left(v_{pj} - \bar{v}_i\right)^2}} \tag{8.2}$$

where $M =$ number of cases; $v_{pi} = j^{th}$ variable of the pth case; and $\bar{v}=$ mean.

Tables 8.7 and 8.8 show the index $\boldsymbol{S_i}$ calculated between the subsets generated by T&T, T&Tr, and the random distribution in the two databases Breast Cancer and Australian Credit.

In T&T, the comparison was carried out between the training set of the tuning phase and the validation set. The tables report the mean value of the indices $\boldsymbol{S}$, calculated according to (8.2), which is the estimate of similarity of the two subsets generated through the three different techniques:

$$\bar{S}^{(\mathrm{T\&T})} = \frac{1}{N}\sum_{i=1}^{N} S_i^{(\mathrm{T\&T})} \quad \bar{S}^{(\mathrm{T\&Tr})} = \frac{1}{N}\sum_{i=1}^{N} S_i^{(\mathrm{T\&Tr})} \quad \bar{S}^{(\mathrm{Random})} = \frac{1}{N}\sum_{i=1}^{N} S_i^{(\mathrm{Random})} \tag{8.3}$$

A greater similarity between the two subsets of training and testing, in terms of distribution of information, is obtained as $\overline{S} \rightarrow \mathbf{0}$. In each of the datasets, T&T seems to generate the subsets of training and testing that are most different from each other, in conformity with a strategy of assigning the more relevant information present in the global database to the training subset.

In this way, for a classifier (ANN) trained on data preprocessed with T&T, the learning phase is optimal, though there is a loss in the sampling carried out on the complementary testing set. Therefore, the outcome of the variance in the linear correlations between the variables in the training/testing subset pairs of the T&T is "emphasized." The improvement in the performance obtained through the classifier is just determined by the inclusion of the *outlier* data, and of complex

Table 8.8 Australian credit – comparison of the indices *S* calculated between the training and testing subsets generated by T&T, T&Tr, and the random distribution

	S		
Variables	T&T	T&Tr	Random
Var 1	0.0979	0.0448	0.0788
Var 2	0.0738	0.0539	0.0524
Var 3	0.2127	0.1130	0.0847
Var 4	0.0544	0.1500	0.0843
Var 5	0.5424	0.0627	0.0794
Var 6	0.1960	0.0587	0.0908
Var 7	0.2626	0.0571	0.0634
Var 8	1.0195	0.0629	0.0859
Var 9	0.7258	0.0717	0.0615
Var 10	0.2887	0.1004	0.0831
Var 11	0.0822	0.0724	0.0429
Var 12	0.0909	0.1526	0.0459
Var 13	0.2033	0.0467	0.1483
Var 14	0.2896	0.0976	0.1671
Target 1	0.8241	0.0505	0.0557
Target 2	0.8241	0.0505	0.0557
Mean$\overline{S}$	0.3618	0.0779	0.0800

sections of the dataset, during the training phase, which give it the ability to operate efficiently even on this type of data in the subsequent test. However, this condition is unfavorable during the validation phase of the classifier; because the most significant information, including the outlier data, was directed into the training subsets, there is no guarantee that the testing subsets generated using this strategy are an acceptable approximation of the function of density of probability of each of the variables measured in the database.

The value of the index S_i when calculated on the training and testing subsets generated by T&Tr and by the random procedure is very similar in both databases. This means that the subsets obtained with the application of T&Tr, besides having a similar distribution of information, appear to be very similar to the ones obtained through the random subdivision procedure of the global set. Therefore, the improvement in the performance of networks trained on subsets generated through this procedure can be explained by a significant difference in the distribution of the probability density function of the two subsets that is not revealed through the linear indices calculation.

8.7 Results of the Experiments with IS

8.7.1 Breast Cancer Dataset

Table 8.9 shows the results obtained on the classification task (as a percentage of accuracy) on training the ANNs on subsets extracted by T&Tr with the pool of variables selected by IS.

Table 8.9 Breast cancer dataset – results – *% mean; max performance = 98.79%; min performance = 97.56%; **% mean; max performance = 99.09%; min performance = 96.48%

Method	% recognition
Back-prop + T&T	99.7
Back-prop + T&Tr	98.17*
Back-prop + T&Tr + IS	97.78**
Back-prop	95.0

Table 8.10 Breast cancer – in gray are the correlation coefficients between the variables selected by IS and the target in the T&Tr testing

Variables	Target
1. Clump thickness	−0.7309
2. Uniformity of cell size	−0.8068
3. Uniformity of cell shape	−0.7966
4. Marginal adhesion	−0.6530
5. Single epithelial cell size	−0.5744
6. Bare nuclei	−0.8027
7. Bland chromatin	−0.7439
8. Normal nucleoli	−0.7000
9. Mitoses	−0.4277

From the nine variables that constitute the database, IS selects a subset of six variables; these are:

1. Uniformity of cell size
2. Uniformity of cell shape
3. Single epithelial cell size
4. Bare nuclei
5. Normal nucleoli
6. Mitoses

The correlation coefficients of all variables selected by IS related to the target are shown in Table 8.10.

8.7.2 Australian Credit Approval Dataset

The accuracy percentage that was obtained training the ANNs on the subsets extracted with T&Tr within the pool of variables selected with IS is shown in Table 8.11.

IS selected 12 of the 14 independent variables that comprise the global dataset, shown below according to their progressive number: 1, 3, 4, 5, 6, 7, 8, 9, 10, 11, 12, and 13.

Table 8.11 Australian credit approval dataset – results – *% mean; max performance = 90.24%; min performance = 89.50%; **% mean; max performance = 91.77%; min performance = 89.50%

Method	% Recognition
Back-prop + T&T	96.00
Back-prop + T&Tr + IS	90.63**
Back-prop + T&Tr	89.87*
Back-prop	86.66

Table 8.12 Australian credit – in gray are the correlation coefficients between the variables selected by IS and the target in the T&Tr testing subset

Variables	Target
Var 1	−0.0523
Var 2	0.1336
Var 3	0.1677
Var 4	0.2354
Var 5	0.4051
Var 6	0.2580
Var 7	0.3463
Var 8	0.7369
Var 9	0.4863
Var 10	0.4794
Var 11	0.0747
Var 12	0.1311
Var 13	−0.1091
Var 14	0.1869

The correlation coefficients of all the variables selected by IS related to the target are shown in Table 8.12.

8.8 Discussion

The application of the data preprocessing strategies developed by the T&T, T&Tr, and IS systems to real databases is shown to be very effective in enhancing the performance of the standard ANN models for classification/prediction problems. This result is evident independently from the complexity of the available data sample; the two databases analyzed in this chapter have very different complexity levels. While the Breast Cancer database can be adequately represented through linear functions, the variables measured in the Australian Credit database are characterized by more complex and highly nonlinear relations. On either of these databases, the training and testing process of the ANN models, carried out using the training and testing subsets selected by T&T and T&Tr, reaches a high level of accuracy, such as to determine a significant improvement in the performance of the networks on the classification problem. In our experiments, the choice of a simple ANN model like the back-propagation, rather than a more complex inductor, is mainly due to the advantage given by the speed of convergence and the possibility of being able to treat polarization problems effectively. The use of more complex inductors could actually lead to the side effect of polarization on the subsets, and as a consequence, it could negatively influence data processing. Simple inductors have shown to be less accurate but are able to avoid these side effects.

In general, based on the results obtained, it is possible to state that the *probability of errors* occurring in the classification of new inputs in networks, that we can call $S_{T\&T}$ and $S_{T\&Tr,}$ trained and tested, respectively, on the subsets selected with T&T and T&Tr systems is *inferior* when compared to the probability obtained applying other methods of resampling. Therefore, these systems offer the possibility of handling classification problems whose reference dataset size could possibly be enlarged by adding new cases.

The real behavior of the networks $S_{T\&T}$ and $S_{T\&Tr}$ when faced with the addition of a new record, which is not present in training and testing, can produce the following situations:

1. If the new record is similar to a record present in the training subset, it will probably be classified correctly by both $S_{T\&T}$ and $S_{T\&Tr}$.
2. If it is similar to an outlier present in the training subset, it will again be probably classified correctly by both networks.
3. If it is similar to an outlier that is only present in the testing subset, it will probably be correctly recognized by $S_{T\&Tr}$, but it is likely that $S_{T\&T}$ will fail.
4. If it is similar to an outlier that is not present in either the training or the testing sets, it will probably not be recognized by either networks.

The choice criterion in the application of the two systems is determined by the representativeness of the dataset analyzed. T&T can be conveniently used when the dataset exhibits a good representativeness of the real universe of data; on the other hand, if the dataset is not sufficiently representative, through the application of T&Tr, one can obtain a higher level of accuracy in subsample extraction. Under the same conditions, T&T has the advantage of a faster execution.

By applying the input selection system to the subsets that were previously selected with the artificial organisms, T&T and T&Tr obtained an advantage that has to be interpreted in relation to the lower computational costs brought about by the reduction in the dimensionality of the data. The profit from the accuracy of the prediction is actually fairly modest when compared to what can be gained by optimizing the subsets through T&T and T&Tr, and in addition, with this system, it is possible to eliminate certain variables that are costly to acquire without diminishing the informative power of the entire dataset.

Another particularly interesting aspect that arises from our results is that IS has certain characteristics that make it a valid alternative to the traditional machine learning methods. In particular, the application of the GenD algorithm to the IS system is useful because the selection of the optimal set of variables can take place while also considering those individuals of the ANN population whose fitness value is lower or equal to the value obtained for the best individual. These sets can compose an archive of input selections, which can be used in cases in which it is not possible or convenient to carry out an optimization because of economic reasons or other factors. Nonetheless, the selection of fitness higher than a certain threshold of predictive capability which is considered adequate according to certain criteria, for example, clinical or industrial, is in itself a great advantage because it provides the option for a wide range of alternative applications.

IS seems to be a very reliable system when treating databases characterized by the presence of a large number of irrelevant variables and when dealing with correlated characteristics, even when these are relevant. IS effectively inherits the solidity of gradient-based neural networks, like the back propagation, and the flexibility of evolutionary algorithms in the exploration of the solution space of the problem in study.

References

Bennett, K. P., & Mangasarian, O. L. (1992). Robust linear programming discrimination of two linearly inseparable sets. *Optimization Methods and Software, 1*, 23–34.

Buscema, M. (2004). Genetic Doping Algorithm (GenD): Theory and applications. *Expert Systems, 21*(2), 63–79.

John, G., Kohavi, R., & Pfleger, K. (1994). Irrelevant features and the subset selection problems. In *11th International Conference on Machine Learning*, pp. 121–129.

Kolen, J. F., & Pollack, J. B. (1990). Back-propagation is sensitive to initial conditions. *Complex Systems, 4*(3), 269–280.

Kudo, M., & Sklansky, J. (2000). Comparison of algorithms that select features for pattern classifiers. *Pattern Recognition, 33*(1), 25–41.

Mangasarian, O. L., & Wolberg, W. H. (1990). Cancer diagnosis via linear programming. *SIAM News, 23*(5), 1–18.

Mangasarian, O. L., Setiono, R., & Wolberg, W. H. (1990). Pattern recognition via linear programming: Theory and application to medical diagnosis. In T. F. Coleman & Y. Li (Eds.), *Large-scale numerical optimization* (pp. 22–30). Philadelphia: Siam Publications.

Merz, C. J., & Murphy, P. M. (1998). *UCI repository of machine learning databases*. Irvine: University of California, Department of Information and Computer Science.

Quinlan, J. R. (1987). *Generating production rules from decision tree*. In International Joint Conference on Artificial Intelligence (pp. 304–307). Milan.

Quinlan, J. R. (1993). *C 4.5: Programs for machine learning*. San Mateo: Morgan Kaufmann.

Siedlecki, W., & Slansky, J. (1989). A note on genetic algorithms for large scale on feature selection. *Pattern Recognition Letters, 10*, 335–347.

Sutton, R. S., & Barto, A. G. (1998). *Reinforcement learning: An introduction*. Cambridge: MIT Press.

Wolberg, W. H., & Mangasarian, O. L. (1990). Multisurface method of pattern separation for medical diagnosis applied to breast cytology. *Proceedings of the National Academy of Sciences of the United States of America, 87*, 9193–9196.

Chapter 9
Metaclassifiers

Massimo Buscema and Stefano Terzi

9.1 The Problem

A classification system is responsible for creating a set of classes based on some common relations or affinities that are present in the data. It is obvious to the typical neural network researcher that there does not currently exist a single algorithm that can correctly process every set of data regardless of application, and this problem has led to the development of a vast library of available instruments in a vain attempt to have available at least some kind of classifier. Each classifier is trained on a particular set of data to capture the underlying constraints that might be substantially different from the next set of data even if that data were derived from a similar application.

Hence, when facing a real problem, the standard process of classification is the training or the network, with an appropriate validation protocol, based on the many different typologies of classifiers and selecting one from the set of classifiers that contains the better characteristics.

If we consider the classification process as a data-mining exercise, we discover that each classifier can classify the same inputs into different classes. This means that the quantity and quality of extracted information change from classifier to classifier; in particular, some typologies of classifiers like neural networks and decisional trees present a great internal variability, producing sensibly different models even when applied to the same problem. In the standard process, we briefly described above, this diversity of models that are typically produced is not exploited; rather, the aim is to choose a single classifier while excluding all the others.

M. Buscema (✉) • S. Terzi
Semeion Research Center of Sciences of Commuication, via Sersale 117, Rome, Italy
e-mail: m.buscema@semeion.it

M. Buscema and W.J. Tastle (eds.), *Intelligent Data Mining in Law Enforcement Analytics: New Neural Networks Applied to Real Problems*, DOI 10.1007/978-94-007-4914-6_9,

This chapter offers a different kind of strategy altogether: to use all of the classifiers as collaborators to exploit the possible complementarities of information extracted from each classifier.

Dietterich (2002) suggests three motivations to explain why the fusion of the single classifiers should produce a more efficient one: one is statistic, one computational, and the last is representational. From a statistic point of view, the "fusion" of more classifiers reduces the probability of choosing a single inadequate one, so even if we have no guarantee that the set of classifiers is better than all the single classifiers, the fusion implies a reduction of the risk of a wrong choice. Conversely, the strongest classifier might be diluted with the inclusion of a very weak classification. From a computational point of view, much of the classifiers use heuristic optimization algorithms to define the optimum parameters able to stop at a local optimum, the aggregation of which can lead to solutions closer to a global optimum even when starting from a local optimum. The last motivation concerns the space of solutions of the single classifiers that may not contain the optimum solution for the classification problem; in this case, the set of classifiers can expand this space of solutions obtaining better results (e.g., a set of linear classifiers cannot singularly reach the solution of a nonlinearly separable problem, but a combination of them can do so).

The problem central to the construction of a metaclassifier is quite complex: a schematization of possible alternatives and the definition of a terminology and a taxonomy can be useful, also taking into account that due to the nature of the complexity problem, this schematization will likely be subject to various exceptions.

There are four project dimensions in the development of a metaclassifier:

1. The database for the training and the validation
2. The selection of the "significant" variables
3. The choice and training of the single classifiers
4. The definition of combination strategy

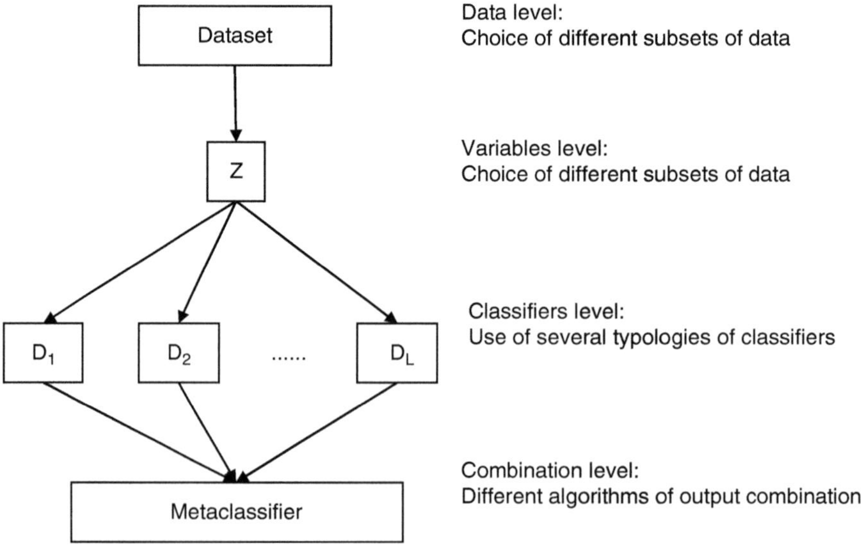

The first three levels represent a forward propagation chain of variations that generate the final classifier:

1. If two classifiers are trained with two different training sets, the two classifiers will develop *different* data models.
2. If two classifiers are trained with the same records, but different variables, the two classifiers will develop *different* data models.
3. If two classifiers are trained with the same training set, but the mathematics of the two classifiers is different (topology, learning rule, signal dynamics, or cost function), the two classifiers will develop *different* data models.
4. If two classifiers are trained with the same training set and if they have the same algorithm, but they begin the learning session with initial random weights e/o parameters, then the two classifiers will develop *different* data models.

In any case, we shall give the greater attention to level four, the development of particular strategies of combination of single classifiers. It has been pointed out that levels three and four are often strictly connected (Kuncheva 2004). The choices made for the first one have consequences on the possible choices for the other and vice versa.

The possible choices regarding the combination strategies belong to two wide classes: fusion and selection. In the first case, each single classifier contributes to the definition of the label of the class to which each record belongs, and in the second case, the responsibility of producing the classification for a specific region of entry space is given to each classifier. Obviously there are metaclassifiers, classification systems that are composed of classification systems, which belong to both of these categories. Until now, the majority of development has been dedicated to fusion, even if selective fusion was to be the most efficient choice.

Another analysis dimension in projecting a metaclassifier concerns the objective of an optimization process in which two conditions are possible. On one side, the optimization is concerned with the kinds and parameters of combination strategy (decision optimization) for a fixed set of classifiers; on the other side, the optimization concerns the parameters of classifier data on a given fixed combination strategy.

The classifiers used in metaclassifiers can produce crisp outputs, the label of predicted class, or fuzzy outputs, a vector with the same dimension of one of the classes of the problem representing the "plausibility" of a record belonging to each of the examined classes. One could further distinguish the metaclassifiers depending on the kind of output of base classifiers used.

Some key distinctions can articulate the metaclassifier world in different logical categories:

1. *Algorithmic* category—each metaclassifier can define its characteristics in two ways:

   (a) *Statically* through a calculation of characteristics and resulting classifiers executed in a noniterative way. The algorithm does not plan an iterative analysis of composing characteristics of classifiers to optimally define its parameters. A static algorithm can be:

    i. *Flexible*: when a vector of parameters emerges from a calculation
    ii. *Strict*: when only one parameter emerges from a calculation

   (b) *Dynamically* with an iterative calculation made on characteristics and results of composing classifiers in order to optimize a vector of parameters. A dynamic algorithm can be:

    i. *Trainable*: when the iterative algorithm tends to define the data entry continuous function parameters
    ii. *Optimizable*: when the iterative algorithm tends to optimize whatever cost function

2. *Extensional* category *(scope)*—each metaclassifier can define its characteristics depending on whether or not it is:

   (a) *Local* in which each composing classifier, in an independent way, provides the metaclassifier with some characteristics.
   (b) *Global* in that characteristics and results of all composing classifiers interact, thereby globally defining metaclassifier characteristics.

3. *Teleological* category—each metaclassifier can define its characteristics depending on these bases:

   (a) *Supervised* in that the relevance of each composing classifier is weighted on the basis of the right/wrong results it produced.
   (b) *Autopoietic* in that the relevance of each composing classifier is weighted on the basis of the produced results, without considering its mistakes or its successes. Autopoietic metaclassifiers, obviously, offer interesting performances when all composing classifiers have a confusion matrix which respects the following condition:

$$\text{Target}_i = Err_{i,i} - \sum_{j=1,j\neq i}^{N} Err_{i,j} < 0;$$

   that is, for each output, the number of errors is lower than the number of the sum of false attributions.

4. *Functional* category—each metaclassifier can evaluate each new entry input in this way:

   (a) *Feed forward* by which the metaclassifier provides only one response for each new entry input.

(b) *Recursive* by which the metaclassifier generates more responses, each considering the previous ones, until the process optimizes a specific cost function (providing the same classification response occurs). During the recall process, this kind of metaclassifiers works as a dynamic system. When a new input is presented, each one of its components hypothesizes a class for it, and then all components negotiate their different hypothesis until they dynamically reach an agreement.

9.1.1 The MetaNet as Metaclassifiers

Between 1994 and 2008, Semeion created a series of metaclassifiers based on some common traits, and for this reason, they have been named MetaNets (see Buscema 1998; Buscema et al. 2010):

- All MetaNets have a typical neural network architecture, that is to say, which input nodes are most involved in determining the composing classifiers and which output nodes are the output classes of classification problems.
- The connections between MetaNet inputs and outputs always consist of a complete grid structure and are defined by specific algorithms characterizing MetaNet peculiarities.
- The MetaNet output vector is calculated from a specific probabilistic equation called softmax (Bridle 1989):

$$Net_j = \sum_k^P \sum_i^N Out_i^k \cdot w_{i,j}^k; \qquad i, j \in \{1, 2, ..., N\}; \text{ Number of output classes.}$$

$$MetaOut_j = \frac{e^{Net_j}}{\sum_z^N e^{Net_z}}. \qquad k \in \{1, 2, ..., P\}; \text{ Number of Classifiers.}$$

- All MetaNets are unsupervised, and each one evaluates its own output without knowledge of the errors of its composing classifiers but only the statistic of their responses. Thus, MetaNets are strongly sensitive to the quality of classifiers to be optimized. This means that each MetaNet, in order to be considered excellent, should be composed of classifiers whose confusion matrices present these properties:

Classifier(k)		j Output 1	Output 2	Output ...	Output N
i	Target 1	MC 1,1	MC 1,2	MC 1, ...	MC 1, N
	Target 2	MC 2,1	MC 2,2	MC 2, ...	MC 2, N
	Target ...	MC...,1	MC ...,2	MC ...,...	MC ..., N
	Target N	MC N,1	MC N,2	MC N, ...	MC N, N

$$\forall k, k \in P : MC_{i,i}^{k} - \sum_{j \neq i}^{N-1} MC_{i,j}^{k} > 0; \quad \begin{array}{r} i,j \in \{1,2,...,N\}; \\ \text{Number of output classes.} \\ k \in \{1,2,...,P\}; \\ \text{Number of Classifiers.} \end{array}$$

However, from the following testing, we will verify this condition that if not respected, produces a "very smooth" fall of MetaNet capacities according to the typical characteristics of ANNs.

We can summarize the common features of MetaNet algorithms in this way:

1. To apply a MetaNet algorithm, the dataset must be divided randomly into three subsamples: training set, testing set, and prediction or validation set; training and testing set constitute the tuning set, and the prediction or the validation set represents the sole subsample for MetaNet performances evaluation.
2. The number of input nodes (M) is equal to the number of the original output (N) times the number of the classifiers working as components of the MetaNet (P): $M = P \times N$.
3. The number of output nodes is the same of the original classes (N).
4. The connections, C, between input and output are a complete matrix $C = M \times N$.
5. Each connection value represents the degree with which every component classifier supports every classification node of the MetaNet.
6. The numerical value of each MetaNet connection can belong to the interval between −inf (lowest degree) and +inf (highest degree).
7. The plausibility and the implausibility of each connection are functions of the probability of each MetaNet component during the testing phase.
8. The final classification of a MetaNet algorithm is the probabilistic composition of the input weighted to each output node (softmax algorithm, see Bridle 1989).

To explain this procedure, we need to start from the analysis of the confusion matrix of a single classifier:

In this matrix, it is necessary to distinguish four terms for each cell, $v_{i,j}^{k}$:

1. The "Rights": the degree through which the kth classifier considered correct in the classification of the records in the cell $v_{i,j}^{k}$, in relation to the summation of each row cell:

		j			
Classifier(*k*)		O(utput) 1	O(utput) 2	O(utput) ...	O(utput) N
i	T(arget) 1	v(1, 1)	v(1, 2)	v(1, ...)	v(1, N)
	T(arget) 2	v(2, 1)	v(2, 2)	v(2, ...)	v(2, N)
	T(arget) ...	v(..., 1)	v(..., 2)	v(..., ...)	v(..., N)
	T(arget) N	v(N, 1)	v(N, 2)	v(N, ...)	v(N, N)

$$R_{i,j}^{k} = \frac{v_{i,j}^{k}}{\sum_{j}^{N} v_{i,j}^{k}}; \tag{9.1}$$

2. The "Missed": the degree through which the *k*th classifier considered as incorrectly different. You must explain to me what you mean and classifying the records in the cell $v_{i,j}^{k}$, in relation to the summation of each row cell:

$$M_{i,j}^{k} = \frac{\sum_{j}^{N} v_{i,j}^{k} - v_{i,j}^{k}}{\sum_{j}^{N} v_{i.j}^{k}}; \tag{9.2}$$

3. The "Corrects": the degree through which the *k*th classifier considers "corrects" the records classified in the cell $v_{i,j}^{k}$, in relation to the summation of each column cell:

$$C_{j,i}^{k} = \frac{v_{i,j}^{k}}{\sum_{i}^{N} v_{i,j}^{k}}. \tag{9.3}$$

4. The "False": the degree through which the *k*th classifier considers "false" the records classified in the cell $v_{i,j}^{k}$, in relation to the summation of each column cell:

$$F_{j,i}^{k} = \frac{\sum_{i}^{N} v_{i,j}^{k} - v_{i,j}^{k}}{\sum_{i}^{N} v_{i,j}^{k}}. \tag{9.4}$$

The "Rights" and the "Missed" are correlated to the probability that any specific output depends from a specific target:

$$p_{j,i}^{k} = p(O_{j}^{k}|T_{i}^{k}).$$

The "Corrects" and the "False," instead, are correlated to the probability that any specific target comes from a specific output:

$$p^k_{i,j} = p(T^k_i | O^k_j).$$

The strength of the connection between any output of each classifier and any output of the metaclassifier (MetaNet weights) is given by this simple equation:

$$w^k_{i,j} = f\left(R^k_{i,j}, C^k_{i,j}, 1 - R^k_{i,j}, 1 - C^k_{i,j}\right)$$

$$k = \text{any specific classifier.}$$

There are several algorithms that will calculate the value of the weights of MetaNet, and experience has demonstrated that many of them can be defined into this paradigm. From simple algorithms of probability that are calculations of co-occurrence couples of output nodes in composing classifiers, through the continuum of real ANNs that are globally or locally defining the MetaNet connections, all are part of complete grid coverage. The closing diagram below illustrates the role of the existing ANNs as inputs and the resulting ANN that produces a response based on the best inputs as determined by the individual ANNs. Each ANN possesses its own algorithm and works on the data in its own way, and it is sometimes in opposition to the output of other algorithms. These methods take the best from all the other ANNs and merge them to produce an output that was at least partially driven by their results. The output has been shown to be better than the individual ANNs.

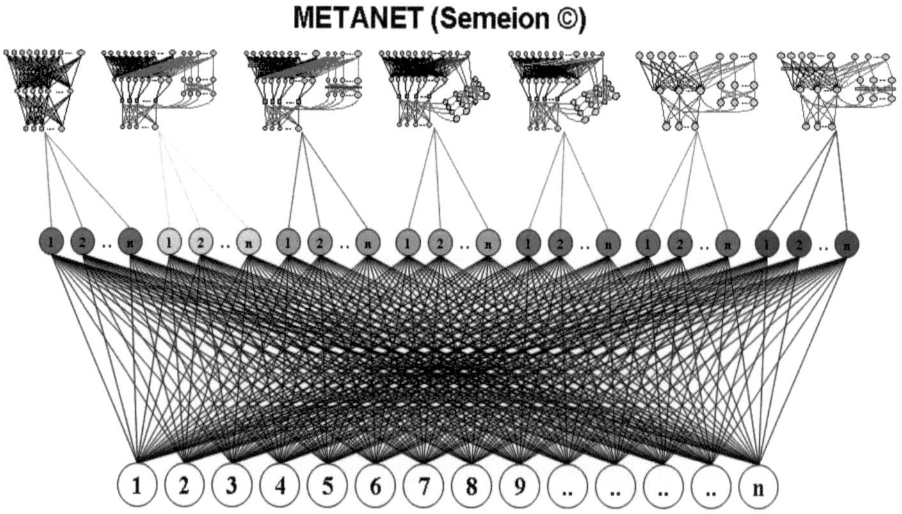

References

Bridle, J. S. (1989). Probabilistic interpretation of feedforward classification network outputs, with relationships to statistical pattern recognition. In F. Fogelman-Soulié & J. Hérault (Eds.), *Neuro-computing: Algorithms, architectures*. New York: Springer.

Buscema, M. (1998). MetaNet: The theory of independent judges. In *Substance use and misuse* (Vol. 33, n. 2 (Models), pp. 439–461). New York: Marcel Dekker, Inc.

Buscema, M., Terzi, S., & Tastle, W. (2010). *A new meta-classifier*. In NAFIPS 2010, 12–14 July, Toronto, Canada.

Dietterich, T. (2002). Ensemble learning. In M. A Arbib (Ed.), *The handbook of brain theory and neural networks* (2nd ed.). Cambridge, MA: The MIT Press.

Kuncheva, L. I. (2004). *Combining pattern classifiers. Methods and algorithms*. Hoboken: Wiley.

Bibliography

Asuncion, A., & Newman, D. J. (2007). *UCI machine learning repository*.http://www.ics.uci.edu/~mlearn/MLRepository.html. Irvine: University of California, School of Information and Computer Science.

Breiman, L., Friedman, J. H., Olshen, R. A., & Stone, C. J. (1993). *Classification and regression trees*. Boca Raton: Chapman & Hall.

Duda, R. O., Hart, P. E., & Stork, D. G. (2001). *Pattern classification*. New York: Wiley.

Hastie, T., Tibshirani, R., & Friedman, J. (2009). *The elements of statistical learning. Data mining, inference and prediction* (Springer series in statistics). New York: Springer.

Huang, Y. S., & Suen, C. Y. (1995). A method for combining multiple expert for the recognition of unconstrained handwritten numerals. *IEEE Transactions on Pattern Analysis and Machine Intelligence, 17*, 90–93.

Kuncheva, L. I. (2000). Clustering-and-selection model for classifier combination. In *Proceedings of knowledge-based intelligent engineering systems and allied technologies* (pp. 185–188). Brighton, UK.

Quinlan, J. R. (1993). *C.45: Programs for machine learning*. San Mateo: Morgan Kaufmann Publishers.

Rogova, G. (1994). Combining the results of several neural network classifiers. *Neural Networks, 7*, 777–781.

Witten, I. H., & Eibe, F. (2005). *Data mining. Practical machine learning tools and techniques*. Amsterdam: Elsevier.

Woods, K., Kegelmeyer, W. P., & Bowyer, K. (1997). Combination of multiple classifiers using local accuracy estimates. *IEEE Transactions on Pattern Analysis and Machine Intelligence, 19*, 405–410.

Chapter 10
Auto-Identification of a Drug Seller Utilizing a Specialized Supervised Neural Network

Massimo Buscema and Marco Intraligi

10.1 Database Description

This chapter describes the application of supervised neural networks on a dataset composed of 139 variables for 1,120 cases[1] which represents the arrests for different kind of drugs carried out in the neighborhoods of London. The testing's aim is to train a supervised neural network system to distinguish crack dealer subjects from all other kinds of subjects which are included in the dataset (other kinds of drug dealers). In order to accomplish this task, dataset subjects have been divided into two classes:

- Crack dealer, 252 subjects (22.50%)
- Generic dealer, 868 subjects (77.50%)

The 139 variables, shown in Table 10.1, refer to:

- Sex (male, female, not defined)
- Residence district of the arrested subject (32 districts + 1 not available)
- District where each subject was arrested (32 districts + 1 not available)
- Nationality of the arrested subject
- Somatic stock of the arrested subject
- Age (expressed in classes)
- Number of previous convictions (convictions)
- Number of previous offences (offences)

[1]This dataset was extracted in June 2006 when the situation of the CDTD database was composed of 1,590 tactic sequences, 1,667 persons, and 1,190 accused persons (70 incomplete cases).

M. Buscema (✉) • M. Intraligi
Semeion Research Center of Sciences of Commuication, via Sersale 117, Rome, Italy
e-mail: m.buscema@semeion.it

M. Buscema and W.J. Tastle (eds.), *Intelligent Data Mining in Law Enforcement Analytics: New Neural Networks Applied to Real Problems*, DOI 10.1007/978-94-007-4914-6_10,

Table 10.1 List of the 139 variables of the DB

List of 139 variables		
1 Sex_Male	48 ConvictionsNumber	95 AR_PL_Newham
2 Sex_Female	49 OffencesNumber	96 AR_PL_Redbridge
3 Sex_notknown	50 Off_FirstConvAge	97 AR_PL_Richmond_ upon_ Thames
4 Barking_ and_ Dagenham	51 Off_LastConvAge	98 AR_PL_Southwark
5 Barnet	52 Off_Drug	99 AR_PL_Sutton
6 Bexley	53 Off_TheftKindred	100 AR_PL_Tower_Hamlets
7 Brent	54 Off_AgainstPerson	101 AR_PL_Waltham_Forest
8 Bromley	55 Off_OffensiveWeapons	102 AR_PL_Wandsworth
9 Camden	56 Off_Sexual	103 AR_PL_Westminster
10 Croydon	57 Off_RelatedToPolice	104 AR_PL_NA
11 Ealing	58 Off_Fraud	105 NumOfDrugSeizures
12 Enfield	59 Off_Total	106 NumOfCashSeizures
13 Greenwich	60 Off_AgainstProperty	107 Pounds
14 Hackney	61 NumOfArrests	108 NumOfTactics
15 Hammersmith_ and_ Fulham	62 AR_OFF_Theft_and_ Kindred_ Offences	109 Non-Law_ Enforcement_ Agent
16 Haringey	63 AR_OFF_Offences_the_ Person_ Offences	110 Other_Law_Enforcement_ Agent
17 Harrow	64 AR_OFF_Drug_trafficking_ Offences	111 Police
18 Havering	65 AR_OFF_Drug_Possession_ Offences	112 Search_of_Object
19 Hillingdon	66 AR_OFF_Other_ Drug_ Offences	113 Search_of_Person
20 Hounslow	67 AR_OFF_Offensive_ Weapon_ Offences	114 Search_of_Premises
21 Islington	68 AR_OFF_Firearms_Offences	115 Covert_Purchase
22 Kensington_ and_ Chelsea	69 AR_OFF_Kidnapping_and_ Abduction_ offences	116 Controlled_Delivery
23 Kingston_ upon_ Thames	70 AR_OFF_Other_violent_ offences	117 Other_Generic_Tactic
24 Lambeth	71 AR_OFF_Other_offences	118 NumOfTacticSequences
25 Lewisham	72 AR_PL_Barking_and_ Dagenham	119 InOperation
26 Merton	73 AR_PL_Barnet	120 ViolentOnArrest
27 Newham	74 AR_PL_Bexley	121 ArrMode_NA
28 Redbridge	75 AR_PL_Brent	122 ArrMode_Direct
29 Rich- mond_upon_Thames	76 AR_PL_Bromley	123 ArrMode_Result_ of_ Enquiries
30 Southwark	77 AR_PL_Camden	124 ArrMode_Given_ into_ custody
31 Sutton	78 AR_PL_Croydon	125 ArrMode_Other
32 Tower_Hamlets	79 AR_PL_Ealing	126 OnBailAtTimeOfOffence
33 Waltham_Forest	80 AR_PL_Enfield	127 NA
34 Wandsworth	81 AR_PL_Greenwich	128 AFR

(continued)

Table 10.1 (continued)

List of 139 variables		
35 Westminster	82 AR_PL_Hackney	129 ASIA
36 (EA1)_White_European	83 AR_PL_Hammersmith_and_Fulham	130 EASTEU
37 (EA2)_Dark_European	84 AR_PL_Haringey	131 EU
38 (EA3)_Afro-Caribbean	85 AR_PL_Harrow	132 IRE
39 (EA4)_Asia	86 AR_PL_Havering	133 JAM
40 (EA5)_Oriental	87 AR_PL_Hillingdon	134 ME
41 (EA6)_Arab	88 AR_PL_Hounslow	135 NK
42 Age(<18)	89 AR_PL_Islington	136 SAME
43 Age(18–21)	90 AR_PL_Kensington_and_Chelsea	137 TU-CY
44 Age(21–25)	91 AR_PL_Kingston_upon_Thames	138 UK
45 Age(25–35)	92 AR_PL_Lambeth	139 VTN
46 Age(35–45)	93 AR_PL_Lewisham	
47 Age(>45)	94 AR_PL_Merton	

- Specifications about previous offences
- Number of previous arrests
- Specifications about previous arrests
- Kinds of tactics used by the police
- Aims of police operations
- Behavior of the person when he was arrested (violent or nonviolent)

10.2 Testing Protocol

Our testing allows the application of a protocol developed through different phases shown to be able to attain excellent results. The protocol we used, in addition to a *random* procedure, is based on a specific process of *optimization* producing results that are firmly better than the random ones.

10.2.1 Random Procedure

In Fig. 10.1, the random protocol 5×2 cross validation (Dietterich 1997) is shown. It is obtained by dividing five times the global sample into five pairs of subsamples equally randomly distributed according to the class to which the patients belong (crack dealer/generic dealer). Each subsample (form sub1a to 5a and from sub1b to 5b – see Fig. 10.1) is used independently by the different models both for the training phase and for the testing phase. In this way, for each type of neural network, there will be ten processings and respective results.

The provisional models used in this procedure are the back-propagation neural network created with feed-forward (FF_BP) architecture and the linear discriminant

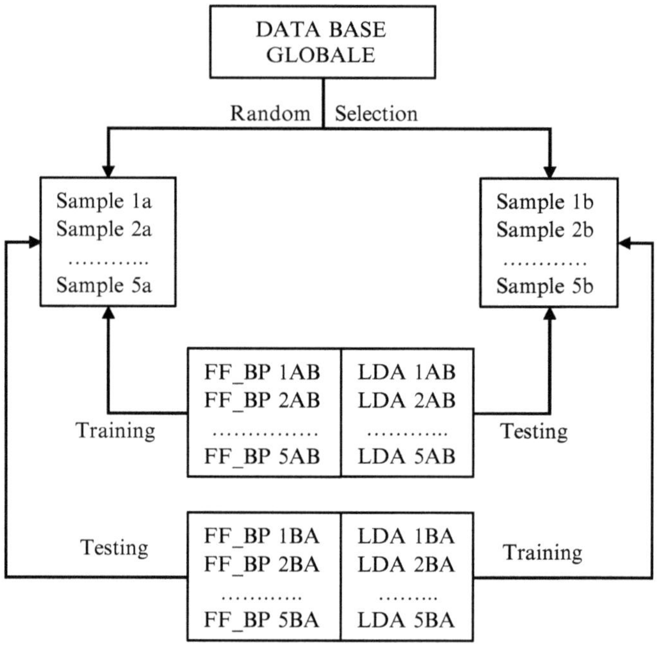

Fig. 10.1 Research protocol

Table 10.2 Results obtained from the ten linear discriminant models on random distribution

Random – Results of linear discriminant analysis

LDA	Crack dealer (%)	Generic dealer (%)	A.Mean Acc (%)	W.Mean Acc (%)	Errors
LDA1AB	50.79	85.25	68.02	77.50	126
LDA1BA	59.52	78.57	69.05	74.29	144
LDA2AB	46.03	84.56	65.30	75.89	135
LDA2BA	60.32	82.95	71.63	77.86	124
LDA3AB	53.17	82.03	67.60	75.54	137
LDA3BA	57.14	79.95	68.55	74.82	141
LDA4AB	61.11	81.11	71.11	76.61	131
LDA4BA	44.44	84.56	64.50	75.54	137
LDA5AB	55.56	81.11	68.33	75.36	138
LDA5BA	59.52	80.65	70.08	75.89	135
Mean	*54.76*	*82.07*	*68.42*	*75.93*	*135*

analysis (LDA), linear statistic model in baseline form. The results obtained from the two models are shown in Tables 10.2 and 10.3.

The results of Tables 10.2 and 10.3 show the big difficulty for both provisional models in correctly distinguishing the "crack dealer" class from the "generic dealer" class. In particular, concerning the "crack dealer" class, the BP model has on average

Table 10.3 Results obtained from the ten models of back-propagation neural networks on random distribution

Random – results of back propagation ANN					
BP	Crack dealer (%)	Generic dealer (%)	A.Mean Acc (%)	W.Mean Acc (%)	Errors
FF_Bp1AB	61.11	83.64	72.38	78.57	120
FF_Bp1BA	64.29	80.65	72.47	76.96	129
FF_Bp2AB	63.49	78.80	71.15	75.36	138
FF_Bp2BA	61.11	85.25	73.18	79.82	113
FF_Bp3AB	69.05	74.19	71.62	73.04	151
FF_Bp3BA	63.49	81.11	72.30	77.14	128
FF_Bp4AB	61.90	82.26	72.08	77.68	125
FF_Bp4BA	65.87	74.88	70.38	72.86	152
FF_Bp5AB	63.49	80.18	71.84	76.43	132
FF_Bp5BA	67.46	81.34	74.40	78.21	122
Mean	*64.13*	*80.23*	*72.18*	*76.61*	*131*

10 percentage points more than the LDA, while the "generic dealer" class, also on average in this case, has the LDA at more than two percentage points over the BP.

10.2.2 Twist Optimization Procedure

The Semeion Twist (2007) model permits us to optimize new training and testing (T&T) samples and to reduce the input variables, thereby obtaining better results than the ones obtained with the random procedure (see references below for Twist applications in medical field).

Twist is a model combining the application of two artificial beings, T&T and input selection (IS) based on the evolutive algorithm GenD created at Semeion (Buscema 2004). The objective of T&T is to preprocess the dataset such that the best distribution of data subsets is available for both training and testing. This is accomplished through iterative procedures by which each subsample has a probability density function equal to the global sample. A set of ANNs is used to form T&T by learning on the training set data and being validated on the testing set. The result obtained by each ANN in the testing phase represents its fitness that is the level of distribution optimization. In substance, it represents how the ANN is reliable and how the optimization contributes to the process of generalizing the results obtained.

Concerning the optimized samples, then, a selection is made by the input selection algorithm (IS) to identify the most significant variables. As T&T does, IS can be considered a data preprocessor able to select useful information to optimize the relations between input and output variables.

Also, IS is formed by a population of ANNs, but in this case, each one is based on a single selection of independent variables. After having been training and validated, the evolutive process chooses the best combination of input variables. Therefore, the GenD fitness rule creates the best performance in testing the lower number of input values associated with each generation (for previous applications in medical field, see Buscema et al. 2005; Grossi and Buscema 2007; Grossi et al. 2007; Penco et al. 2005; Buscema et al. 2010; and Rotondano et al. 2011).

At the end of this double process in our testing, the Twist system optimizes the five couples of training and testing samples, thereby reducing the number of input variables from 139 to 80 (Table 10.4).

The results obtained from the Twist ANNs are shown in the Table 10.5.

Table 10.4 In bold typeset, the 59 variables eliminated by the Twist system, among the 139 ones of the global dataset, are shown

Twist – list of 80 variables selected from 139		
1 Sex_Male	48 ConvictionsNumber	95 AR_PL_Newham
2 Sex_Female	49 OffencesNumber	96 AR_PL_Redbridge
3 Sex_notknown	50 Off_FirstConvAge	97 AR_PL_Richmond_ upon_ Thames
4 Bark-ing_and_Dagenham	51 Off_LastConvAge	98 AR_PL_Southwark
5 Barnet	52 Off_Drug	99 AR_PL_Sutton
6 Bexley	53 Off_TheftKindred	100 AR_PL_Tower_Hamlets
7 Brent	54 Off_AgainstPerson	101 AR_PL_Waltham_Forest
8 Bromley	55 Off_OffensiveWeapons	102 AR_PL_Wandsworth
9 Camden	56 Off_Sexual	103 AR_PL_Westminster
10 Croydon	57 Off_RelatedToPolice	104 AR_PL_NA
11 Ealing	58 Off_Fraud	105 NumOfDrugSeizures
12 Enfield	59 Off_Total	106 NumOfCashSeizures
13 Greenwich	60 Off_AgainstProperty	107 Pounds
14 Hackney	61 NumOfArrests	108 NumOfTactics
15 Hammersmith_and_ Fulham	62 AR_OFF_Theft_and_ Kindred_ Offences	109 Non-Law_ Enforcement_ Agent
16 Haringey	63 AR_OFF_Offences_the_ Person_ Offences	110 Other_Law_Enforcement_ Agent
17 Harrow	64 AR_OFF_Drug_ trafficking_ Offences	111 Police
18 Havering	65 AR_OFF_Drug_ Possession_ Offences	112 Search_of_Object
19 Hillingdon	66 AR_OFF_Other_ Drug_ Offences	113 Search_of_Person
20 Hounslow	67 AR_OFF_Offensive_ Weapon_ Offences	114 Search_of_Premises
21 Islington	68 AR_OFF_ Firearms_ Offences	115 Covert_Purchase

(continued)

Table 10.4 (continued)

Twist – list of 80 variables selected from ?139		
22 Kensington_and_ Chelsea	69 AR_OFF_Kidnapping_ and_ Abduction_ offences	116 Controlled_Delivery
23 Kingston_upon_ Thames	70 AR_OFF_Other_ violent_ offences	117 Other_Generic_Tactic
24 Lambeth	71 AR_OFF_Other_offences	118 NumOfTacticSequences
25 Lewisham	72 AR_PL_Barking_and_ Dagenham	119 InOperation
26 Merton	73 AR_PL_Barnet	120 ViolentOnArrest
27 Newham	74 AR_PL_Bexley	121 ArrMode_NA
28 Redbridge	75 AR_PL_Brent	122 ArrMode_Direct
29 Rich- mond_upon_Thames	76 AR_PL_Bromley	123 ArrMode_Result_ of_ Enquiries
30 Southwark	77 AR_PL_Camden	124 ArrMode_ Given_ into_ custody
31 Sutton	78 AR_PL_Croydon	125 ArrMode_Other
32 Tower_Hamlets	79 AR_PL_Ealing	126 OnBailAtTimeOfOffence
33 Waltham_Forest	80 AR_PL_Enfield	127 NA
34 Wandsworth	81 AR_PL_Greenwich	128 AFR
35 Westminster	82 AR_PL_Hackney	129 ASIA
36 (EA1)_White_ European	83 AR_PL_Hammersmith_ and_ Fulham	130 EASTEU
37 (EA2)_Dark_ European	84 AR_PL_Haringey	131 EU
38 (EA3)_Afro- Caribbean	85 AR_PL_Harrow	132 IRE
39 (EA4)_Asia	86 AR_PL_Havering	133 JAM
40 (EA5)_Oriental	87 AR_PL_Hillingdon	134 ME
41 (EA6)_Arab	88 AR_PL_Hounslow	135 NK
42 Age(<18)	89 AR_PL_Islington	136 SAME
43 Age(18–21)	90 AR_PL_Kensington_ and_ Chelsea	137 TU-CY
44 Age(21–25)	91 AR_PL_Kingston_ upon_ Thames	138 UK
45 Age(25–35)	92 AR_PL_Lambeth	139 VTN
46 Age(35–45)	93 AR_PL_Lewisham	
47 Age(>45)	94 AR_PL_Merton	

Finally, a MetaNet is used (see preceding chapter) to take the outputs from the various ANNs and use them as inputs to the MetaNet (Lahner et al. 2008). Each ANN analyzes data from its own unique perspective utilizing its own particular algorithms and displays a view of the underlying relationships. One cannot say that any one particular view is incorrect, but one can say that each view shows a different view. By using these various views as input to another ANN, one generalized view

Table 10.5 Results obtained from the ten models by Twist system

Twist ANNs	Crack dealer (%)	Generic dealer (%)	A.Mean Acc (%)	W.Mean Acc (%)	Errors
Twist 1	77.17	76.22	76.69	76.44	131
Twist 2	81.89	65.50	73.70	69.24	171
Twist 3	70.08	86.48	78.28	82.73	96
Twist 4	70.08	86.25	78.16	82.55	97
Twist 5	76.38	77.86	77.12	77.52	125
Twist 6	80.31	72.73	76.52	74.46	142
Twist 7	75.59	75.52	75.56	75.54	136
Twist 8	74.80	77.16	75.98	76.62	130
Twist 9	77.17	79.49	78.33	78.96	117
Twist 10	82.68	72.03	77.35	74.46	142
Twist mean	*76.62*	*76.92*	*76.77*	*76.85*	*128.70*

Table 10.6 Results obtained by MetaNet

MetaNets

	Crack dealer (%)	Generic dealer (%)	A.Mean Acc (%)	W.Mean Acc (%)	Errors
Twist ANNs fusion	75.59	86.48	81.04	83.99	89

Table 10.7 Summary of the average results obtained by the different systems and procedures

Summary	Crack dealer (%)	Generic dealer (%)	A.Mean Acc (%)	W.Mean Acc (%)	Errors
LDA (mean of 10)	54.76	82.07	68.42	75.93	135
Back prop (mean of 10)	64.13	80.23	72.18	76.61	131
Twist mean (mean of 10)	75.10	80.72	77.91	79.47	115
MetaNet (fusion of 10)	75.59	86.48	81.04	83.99	89

is produced that is better than any individual view. Thus, using the parameters and results of every previously trained neural network, performance has been improved (Table 10.6).

Table 10.7 shows the huge vantages provided by the Twist procedure and by MetaNet fusion in this pattern recognition problem. In fact, if we consider the LDA results, we have to conclude that the predictability of the crack dealer is not good enough to engage the London Metropolitan Police in the creation of an automatic pattern recognition technology. All the same, we have to arrive, more or less, at the same conclusion if we analyze the results generated by a classic artificial neural network (back propagation). But the results we have gotten on the same data using the Twist ANNs and their fusion (MetaNet) are really interesting from a technological point of view.

References

Buscema, M. (2004). Genetic doping algorithm (GenD): Theory and applications. *Expert Systems, 2*(2), 63–79.

Buscema, M., Grossi, E., Intraligi, M., Garbagna, N., Andriulli, A., & Breda, M. (2005). An optimized experimental protocol based on neuro-evolutionary algorithms. Application to the classification of dyspeptic patients and to the prediction of the effectiveness of their treatment. *Artificial Intelligence in Medicine, 34*, 279–305.

Buscema, M., Grossi, E., Capriotti, M., Babiloni, C., & Rossini, P. M. (2010). The I.F.A.S.T. model allows the prediction of conversion to Alzheimer Disease in patients with mild cognitive impairment with high degree of accuracy. *Current Alzheimer Research, 7*, 173–187.

Dietterich, T. G. (1997). Approximate statistical tests for comparing supervised classification learning algorithms. *Neural Computation, 10*(7), 1895–1923.

Grossi, E., & Buscema, M. (2007). Introduction to artificial neural networks. *European Journal of Gastroenterology & Hepatology, 19*, 1046–1054.

Grossi, E., Mancini, A., & Buscema, M. (2007). International experience on the use of artificial neural networks in gastroenterology. *Digestive and Liver Disease, 39*, 278–285.

Lahner, E., Intraligi, M., Buscema, M., Centanni, M., Vannella, L., Grossi, E., & Annibale, B. (2008). Artificial neural networks in the recognition of the presence of thyroid disease in patients with atrophic body gastritis. *World Journal of Gastroenterology, 14*(4), 563–568.

Penco, S., Grossi, E., Cheng, S., Intraligi, M., Maurelli, G., Patrosso, M. C., Marocchi, A., & Buscema, M. (2005). Assessment of the role of genetic polymorphism in venous thrombosis through artificial neural networks. *Annals of Human Genetics, 69*, 693–706.

Rotondano, G., Cipolletta, L., & Grossi, E. (2011). Artificial neural networks accurately predict mortality in patients with nonvariceal upper GI bleeding. *Gastrointestinal Endoscopy, 73*(2), 218–226.

Software

Buscema, M. (1999–2011). Supervised ANNs and Organisms, Semeion Software #12, ver. 18.0, Rome (Italy).

Buscema, M. (2006–2007). Twist – Input Search and Training & Testing Reverse, Semeion Software #39, ver. 2.0, Rome (Italy).

Buscema, M. (2008–2010). Meta Net, Semeion Software #44, ver. 8.0, Rome (Italy).

Chapter 11
Visualization and Clustering of Self-Organizing Maps

Giulia Massini

11.1 Introduction

The Self-organizing map (SOM) is a neural network developed mainly by Teuvo Kohonen (Kohonen 1972, 1995b) between 1972 and 1982. It is an unsupervised type of network which allows for the classification of the input vectors creating a prototype of the classes and a projection of the prototypes on a two-dimensional map (but n-dimensional maps are also possible) able to record the relative proximity (or neighborhood) between the classes. Therefore, the network imposes important synthetic information on the input:

1. It creates a classification of the input vectors on the basis of their vector similarity and assigns them to a class.
2. It creates a prototypical model of the classes with the same cardinality (number of variables) as the input vector.
3. It provides a measurement, expressed as a numerical value, of the distance/proximity of the various classes.
4. It creates a relational map of the various classes, placing each class on the map itself.
5. It provides a measurement of the distance/proximity existing between the input vectors from the class to which they belong and between the input vectors and other classes.

The relative simplicity of the SOM architecture is also the key to its popularity in scientific applications.

G. Massini (✉)
Semeion Research Center of Sciences of Commuication, via Sersale 117, Rome, Italy
e-mail: g.massini@semeion.it

M. Buscema and W.J. Tastle (eds.), *Intelligent Data Mining in Law Enforcement Analytics: New Neural Networks Applied to Real Problems*, DOI 10.1007/978-94-007-4914-6_11,

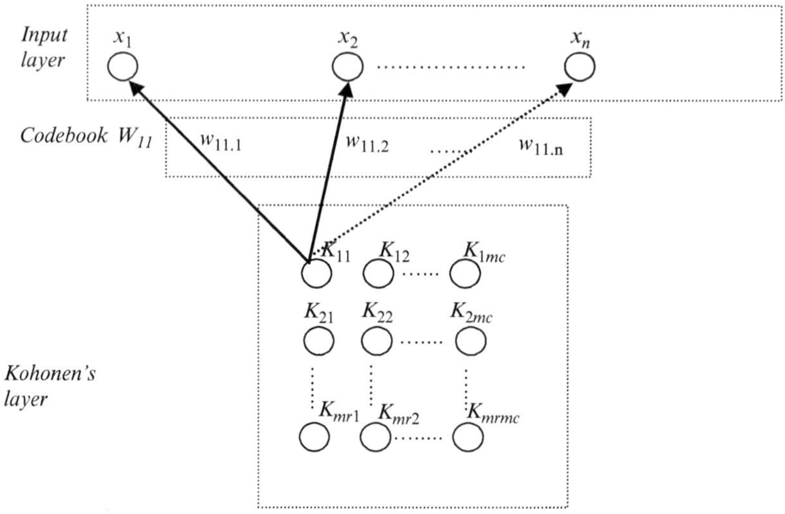

Fig. 11.1 SOM with n-nodes of input, with ($m_r{\cdot}m_c$) units of Kohonen's layer. This architecture allows the inputs to be classified into m^2 classes, each being a subclass represented by a codebook

11.2 Architecture

A typical SOM network is made up of two layers of units: a one-dimensional input (n-cardinality vector) and a two-dimensional output layer (lines (r) × columns (c)), also known as Kohonen's map (M matrix of mr × mc dimensions). A matrix of the weights records the relation between each unit of the output layer and each unit of the input layer (W matrix of (mr × mc × n) dimensions). The weight vector connecting each output unit to an input unit is called a "codebook" (vector *wrc* of n-cardinality) (see Fig. 11.1). Within the SOM network, each output unit can be interpreted as a class whose codebook represents the prototype.

11.3 Base Algorithm

The SOM algorithm is based on a competitive algorithm founded on the vector quantification principle: at each cycle of life in the network, the unit from Kohonen's layer whose codebook is most similar to the input wins. This unit is given the name of winner unit (WU). Consequently, the WU codebook is modified to get it even closer to the input. The codebooks belonging to the units that are physically near the WU (which are part of the neighborhood) are also put closer to the input of a given delta.

The algorithm calculates a first stage during which the parameters of the neighborhood and corrections of weights are set and the codebook initialization is carried out; this stage is followed by the cyclic stage of codebook adjustment. In this stage, the codebooks are modified for the network to classify the input records.

In short, the SOM algorithm is organized as follows:

Initialization stage

- Layering of the input vectors.
- Definition of the dimensions (rows × columns) of the matrix which, in its turn, determines the number of classes and therefore of prototypes (codebook).
- Initialization of the codebooks: the value of the vectors of each codebook is random.
- Definition of the function (Gaussian, Mexican hat, etc.) and of the parameters regulating the neighborhood of the winner unit and of the weight correction delta.

Cyclic calibration stage

- Presentation of the input vectors (pattern) in a random and cyclic way.
- Calculation of the *d*-activation of the K units of Kohonen's layer: the activation is calculated as vector distance between the input vector X and the weight vector Wj (mj codebook) which links the K unit to the input nodes.

  The classic way to calculate the Euclidean distance between the vectors is:

$$d_j = \|X - W_j\| = \sqrt{\sum_{i=1}^{N} (x_i - w_{ij})^2}$$

- Determination of the winning unit WU: the node of the K layer whose activation is least:

$$\mathrm{WU} : d_w = \min_{j \in [1..M]} \left\{ d_j \ \|X - W_j\| = \sqrt{\sum_{i=1}^{N} (x_i - w_{ij})^2} \right\}$$

- Correction of the codebook (matrix of the W_{ij} weights) of the winning unit and the units adjacent to the winning unit in relation to the function set to determine the level of weight correction according to the input and the proximity to the WU.
- Updating of the factors determining the proximity and layering of the delta correction of the codebooks.

The distinctive characteristic of the SOM is mainly related to the updating of the weights, carried out not only on those related to the WU but also, according to the chosen function, on the weights belonging to the units which are physically close to it. This characteristic also allows the SOM to show the position occupied by the class within the matrix in relation to the position occupied by the other classes. This type of topological mapping, able to organize the classes through spatial relations, has been given the name of *feature mapping*.

11.3.1 Topology of the Neighborhood

The neighborhood of a WU is defined by the degree of physical proximity (v) existing between the WU and the other K units. Each unit of Kohonen's layer occupies a position on the matrix of the coordinates (r, c) for which the neighborhood is indexed with a scalar degree from 1 to the maximum line and column dimension:

$$v_i = \pm r \quad \text{OR} \quad v_i = \pm c \qquad \text{where} \quad \max i = \max r \ \ \text{OR} \ \ \max c$$

Function $h(v)$ regulates the size of the neighborhood and the extent of the corrections which need to be made on the codebooks of the units close to the WU. With the passing of time (cycles during which all the training set models are viewed), the neighborhood is reduced until it disappears; in this case, the only unit to which the codebook is corrected is the WU. Since the codebooks are set during the initialization stage with random values within the layering range, the proximity of the WU at the beginning of the learning stage is regulated with a maximum size in order to allow for all the codebooks to be modified and put closer to the input vectors. The reduced proximity with wide matrices can determine the fact that some areas of the K matrix remain isolated because the codebooks are too different from the input vectors. Function $h(v)$ must also allow for the extent of the correction to be bigger for the units close to the WU and therefore to decrease when v is larger. The Gaussian function has been shown to meet these needs remarkably well:

$$h(v) = e^{-\frac{v^2}{\sigma}}$$

$$h(v) : \text{EXP}(-(\text{SQR}(v)/\sigma))$$

where d is the physical proximity of the unit to the WU and σ is a parameter which linearly decreases by a Δ as time increases, thereby modifying the width of the curve (bell), thus the extent of the neighborhood. Figures 11.2 and 11.3 show examples of neighborhood space topologies:

11.3.2 Correction of the Codebooks

The rate of correction a codebook undergoes is determined by various factors:

1. Difference (d) existing between the vector codebook and the input vector
2. Physical distance to the WU (v)
3. Function of the neighborhood $h(v)$ which determines a $\Delta\sigma$
4. Function of weight layering in relation to the period of life of the network which determines a $\Delta\alpha$

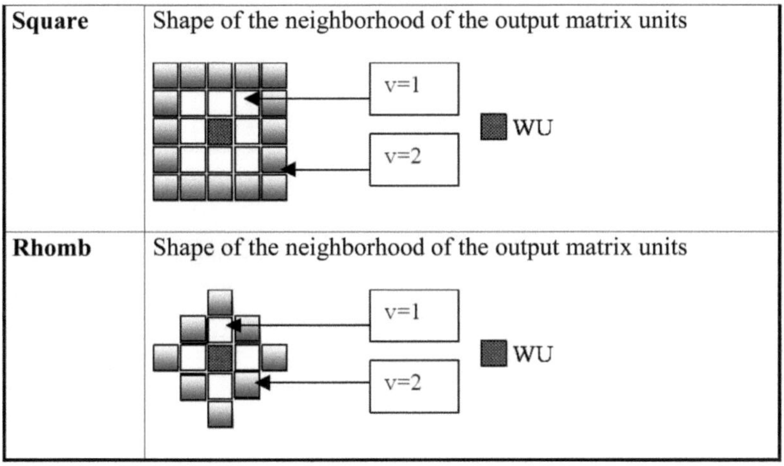

Fig. 11.2 Topology of the neighborhood space of a winner unit in a square and in a rhomb; in the illustration, *v* is the degree of proximity of the K units to the WU

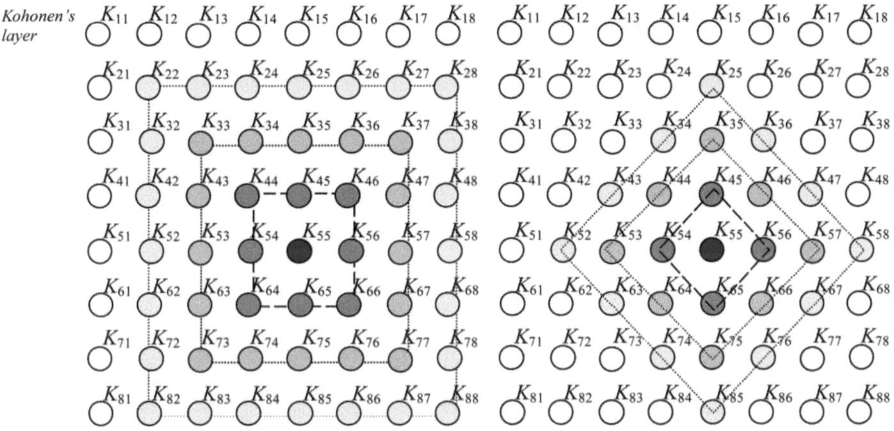

Fig. 11.3 Example of the topology of the neighborhood space with matrix *K* (8r × 8c) where the WU is the K_{55} unit (The first matrix shows a neighborhood in a square while the second a neighborhood in a rhomb. We can notice from the illustration that, for example, while in the matrix to the *left* the *v* distance of the K_{66} unit to the WU is 1, in the matrix to the *right* the *v* distance of the K_{66} unit to the WU is 2)

In a SOM, the *codebooks* are moved closer to the input vector; therefore, for each generic *codebook W*, the distance existing between the corresponding weights w_{ij} and the variables x_i of the generic input vector *X* is calculated:

$$d_j = \|X - W_j\| = \sqrt{\sum_{i=1}^{N} (x_i - w_{ij})^2}$$

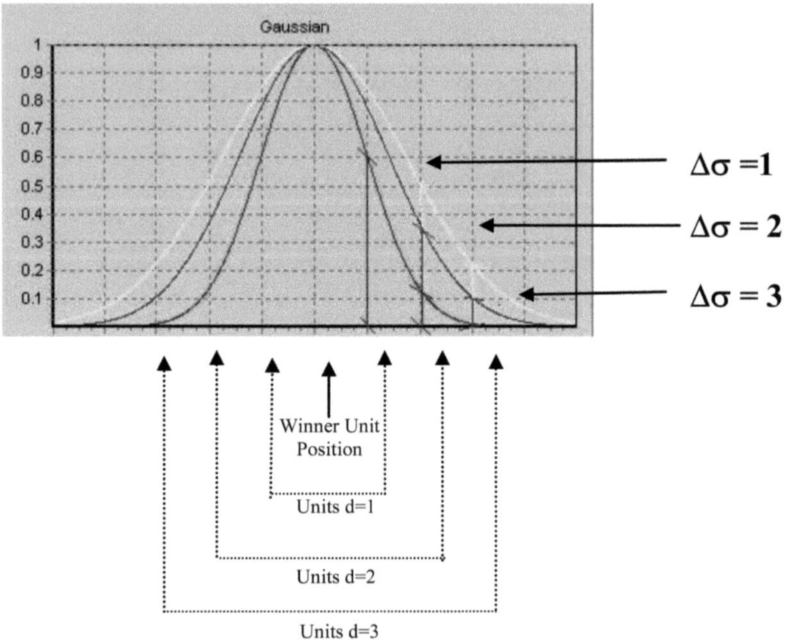

Fig. 11.4 The illustration shows how, when the parameter σ (1, 2, 3) changes, that is, when the parameter that determines the correction curve of the neighborhood function, the number of units that are part of the neighborhood and the extent of the correction ($\Delta\sigma$) made on the weights also change

On the basis of the function $h(v)$ of the neighborhood, the $\Delta\sigma$ is therefore calculated in relation to the value of the parameter σ, and the proximity (v) of the unit K to the *WU*. $\Delta\sigma$ is the measure which assumes *the peak of the bell* in the function $h(v)$, when $x = v$. In the case in which function $h(v)$ is the *Gaussian curve*, then the $\Delta\sigma$ will be calculated in the following way (see Fig. 11.4):

$$\Delta\sigma = e^{\frac{-d^2}{\sigma}}$$

The $\Delta\alpha$ is calculated as a factor of a linear function decreasing in relation to the time the network is alive. Therefore, the function of correction of the *codebooks* is as follows:

$$f(w) = \alpha \cdot e^{-\frac{v^2}{\sigma}} \sqrt{\sum_{i=1}^{N} (x_i - w_{ri})^2}$$

$$w_{ij} = w_{ij} + \alpha(x_i - w_{ij}).$$

11.4 Classes and Their Interpretation

The number of classes is determined in advance, when the network architecture is set and the number of units of Kohonen's matrix is defined.

Each class is thus characterized by:

- A codebook (model of the input vectors which have been attributed to that class)
- Input vectors which belong to that class
- Numerosity (number of input vectors belonging to that class)
- Position on the matrix
- Vector distance with each of the other codebooks of the other classes
- Variance within its own class (to what extent input vectors belonging to that class are similar to or different from one another)

The best representation of the classes and the information they provide enables one to understand the work of classification the SOM makes on a given sample of data. Let us give an example (which we call A) of a classification worked out by the SOM on a training sample with 690 input vectors formed by six variables in which Kohonen's matrix measures 10×10.

To view the numerical values of the classes in which the sample of the input vectors is distributed on the matrix, we have chosen to use circles of analogue dimensions to the numerical values (Fig. 11.5).

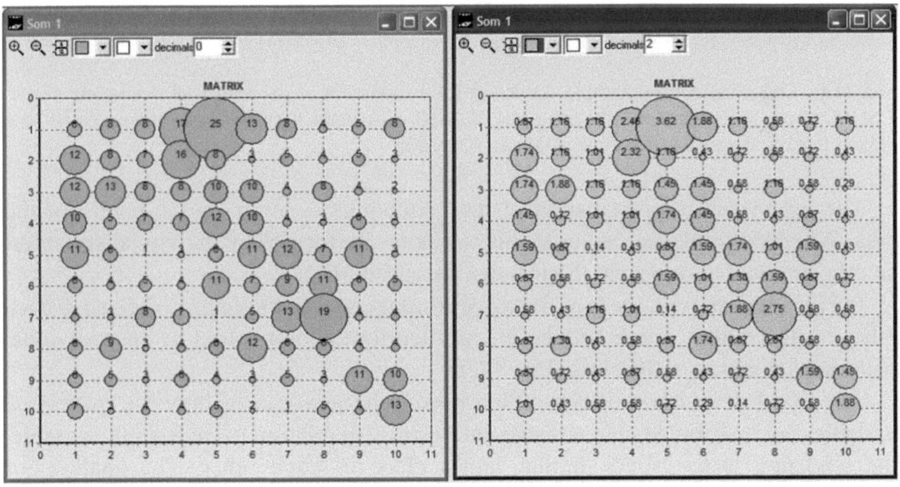

Fig. 11.5 The illustration shows two representations of the number of the classes on the matrix of example A. In the matrix to the *left*, the values correspond to the frequency, while to the *right* the values correspond to the percentage of the frequency in comparison to the number of the sample (Taken from Software of Semeion: SOM ver. 0.5, 2005)

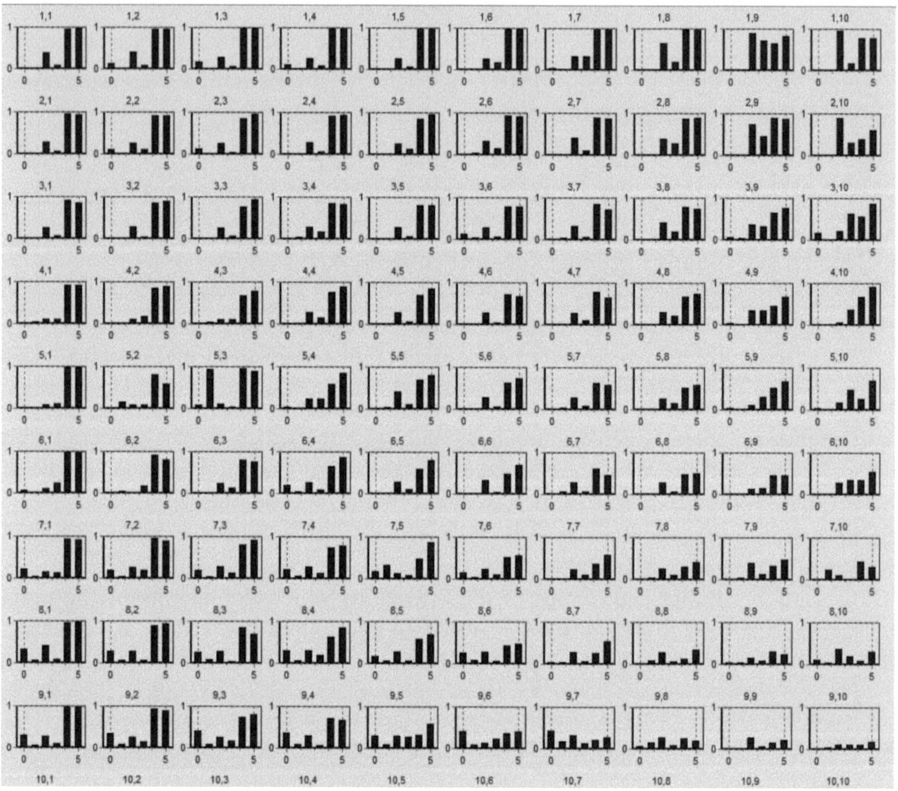

Fig. 11.6 The illustration (example A) shows the histograms of the values of the codebooks of a SOM whose Kohonen's matrix measures 10×10 with input vectors of cardinality equal to 6 (Taken from Software of Semeion: SOM ver. 0.5, 2005)

The best visual representation of the codebooks depends on the nature of the data. When the input vectors, as in this example, are formed by individual variables, an efficient way to represent them is the histogram (Fig. 11.6).

In Fig. 11.6, one can see how the more similar codebooks are placed close to each other on the matrix. Each codebook can therefore be compared to the individual patterns that make up the class (Fig. 11.7).

The SOM neighborhood logic also makes the analysis of how the individual variables are distributed on the matrix interesting. In relation to each single variable, the tendency is to group similar values close to each other. The analysis of how the individual variables are distributed highlights the linear and nonlinear correlations between the variables, where the linear correlation would generate similar distributions. In the following illustrations (Figs. 11.8 and 11.9), also taken from example A, one can see how variables 5 and 6 have between them a relation of linear tendency: when the values of one increase, the values of the other also increase, while between the others, nonlinear relations occur.

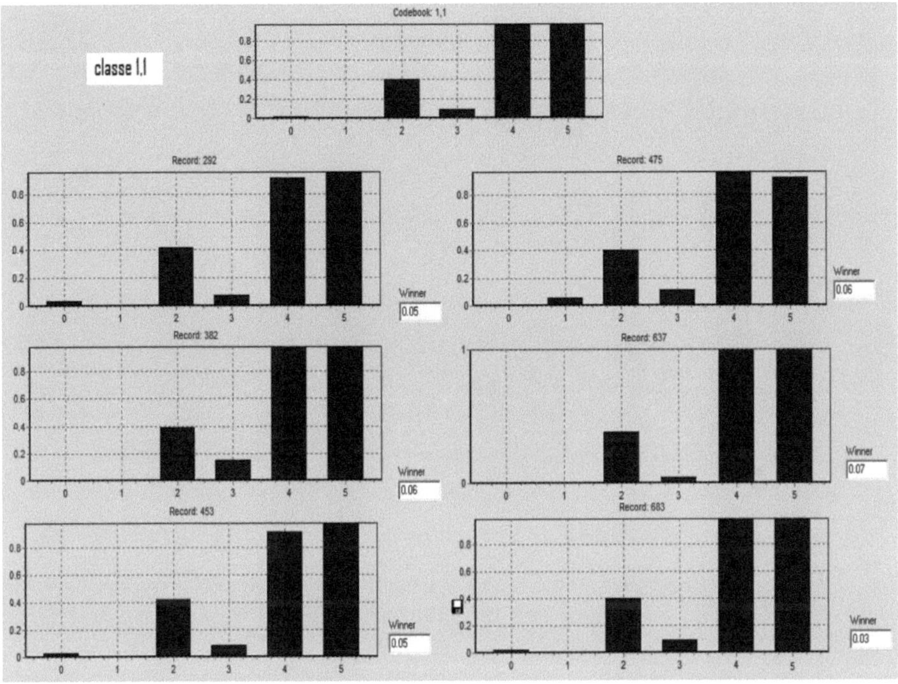

Fig. 11.7 The illustration compares the codebook of class 1.1 to the individual vectors of six input records (which in this example are records 292, 382, 453, 475, 637, and 683) which are part of this class. The "winner" value to the side of each record shows the vector distance existing between the record and the codebook. It should be noted that the minimum distance is from record 683 (0.03) (Taken from SW of the Semeion: SOM ver. 0.5, 2005)

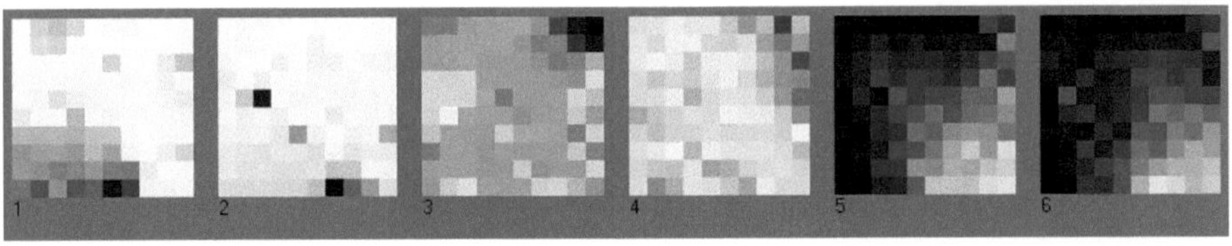

Fig. 11.8 The illustration compares the value distribution of the six variables of example A on the matrix. The values are represented on the *gray scale* (the variable with the lowest value corresponds to *white*, that with the highest value corresponds to *black*) (Taken from Software of Semeion: SOM ver. 0.5, 2005)

When the input models are images, the codebooks offer a synthesis of these images. We show an example (which we call example B) of a database of XR mammographs taken from Progetto CALMA (*computer-aided library for mammography*) of the National Institute of Nuclear Physics. From each mammography,

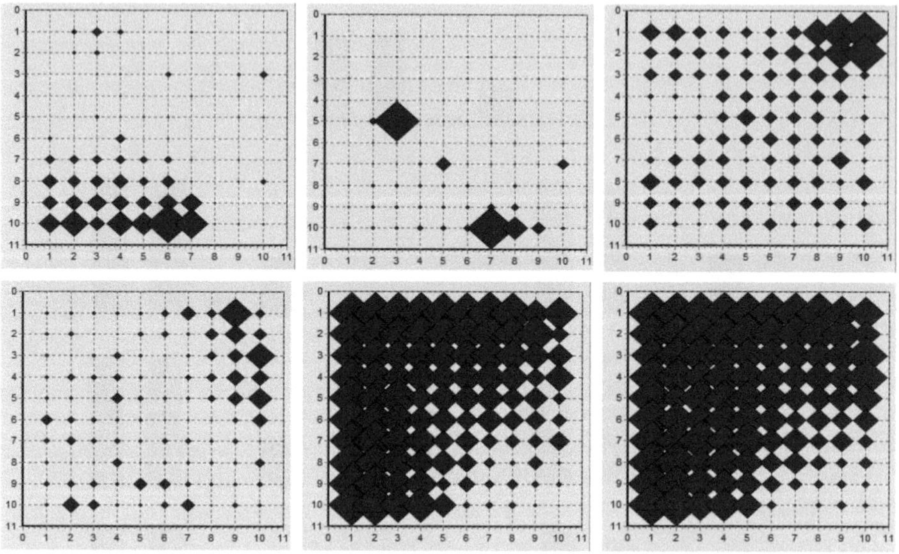

Fig. 11.9 The illustration compares the value distribution of the six variables of example A on the matrix. The values are represented by rhombs of analogue dimensions to the value of the variable in the individual codebook (Taken from Software of Semeion: SOM ver. 0.5, 2005)

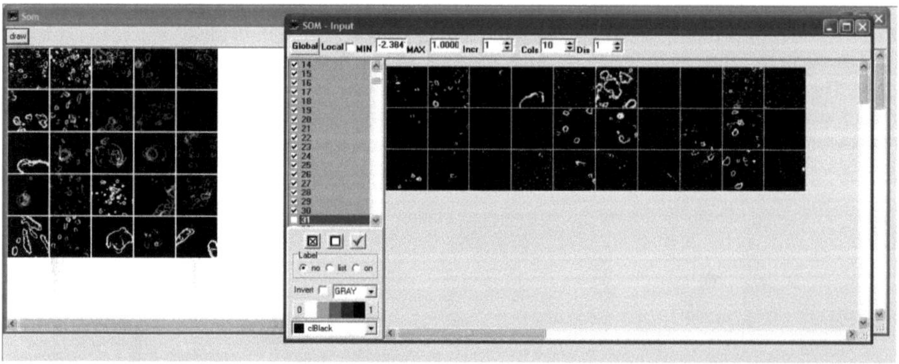

Fig. 11.10 The illustration (*left*) compares the image of the 25 codebooks of a SOM measuring 5×5, which has processed 384 input models, of which (*right*) the first 30 are shown. The inputs, measuring 60×60 pixels, are portions of images from X-ray mammographs containing micro-calcifications or healthy tissue. Before being processed by the SOMs, the inputs were processed with the ACM systems (*active connection matrices*, Semeion's patent). The images were taken from the CALMA database of the National Institute of Nuclear Physics (Taken from Software of Semeion: SOM ver. 0.5, 2005)

portions of images containing micro-calcifications and images with healthy tissue were taken, both measuring 60×60 pixels. The inputs were then processed with ACM systems (*active connection matrices*) of the Semeion before being processed by the SOMs (Fig. 11.10).

11.5 Macro-classes and Their Interpretation

When contiguous classes on the matrices are very similar to one another due to the vector neighborhood between their codebooks, the problem arises of grouping these classes together in order to reduce the number of classes. To this end, various procedures have been identified to determine, measure, and view the grouping of classes.

One possible procedure is to force the data to be processed with SOM matrices of reduced dimensions and then to compare the two distributions: that is, to verify which classes of the matrix with the biggest dimensions have clustered with the matrix with the smallest dimensions.

On the other hand, another procedure consists of widely expanding the matrix dimensions in relation to the numerosity of the sample in order to see how the records are distributed on the matrix. We show an example here (which we call example C), in which the matrix dimensions have been widely expanded (50 × 50) while keeping the neighborhood dimensions contained and regulated by the *alpha* function (Fig. 11.11).

In this example, the population was made up of 18,355 cases of which the following variables were known:

- Age
- Total bilirubin
- Total cholesterol
- Creatinine (serum)
- AST (SGOT)
- ALT (SGPT)

Fig. 11.11 The illustration shows a 50 × 50 matrix and, inside it, the neighborhood, which in cycle 1 involves 21 × 21 units of the matrix

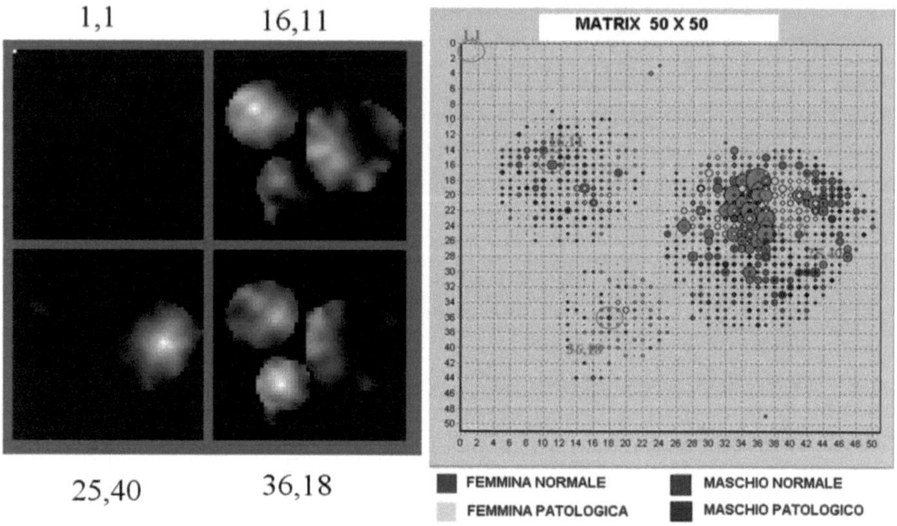

Fig. 11.12 The illustration shows a SOM with a 50 × 50 matrix. The sample, made up of 18,355 cases, was placed on the matrix by the SOMs based on three groups (*image to the right*). The *image to the left* shows the vector distances in relation to four codebooks: r1, c1 – r16, c11– r25, c40 – r36, and c18; the *light areas* represent short distances, the *dark* ones long distances. The positions of the four codebooks are also highlighted on the matrix to the *left* (Taken from Software of Semeion: SOM ver. 0.5, 2005)

- HDL cholesterol
- Triglycerides
- Glucose (serum)

The subjects have been distributed by the SOMs in three macro-groups as shown in Fig. 11.12:

Another procedure consists of globally verifying which of the codebooks are most similar to one another by measuring their vector distance. In this case, various types of calculations and representations are also possible according to their complexity as follows:

- *Distance of each codebook from the others* (Fig. 11.13): for an $n \times n$ matrix, $n \times n$ matrices are rewritten, each representing the distance of each codebook from the others.
- *Local distance*: distance of each codebook from its neighborhood formed by a set of eight codebooks in which the mean value is placed at the center (Fig. 11.14).
- *Minimum local distance*: minimum distance that separates each codebook from one of the eight codebooks of its neighborhood (Fig. 11.14). In this case, a graph is drawn with the matrix units placed at the vertices, and the units of the codebooks that possess a minimum vector distance with respect to the neighborhood formed by eight codebooks are linked to each other (Figs. 11.15 and 11.16).

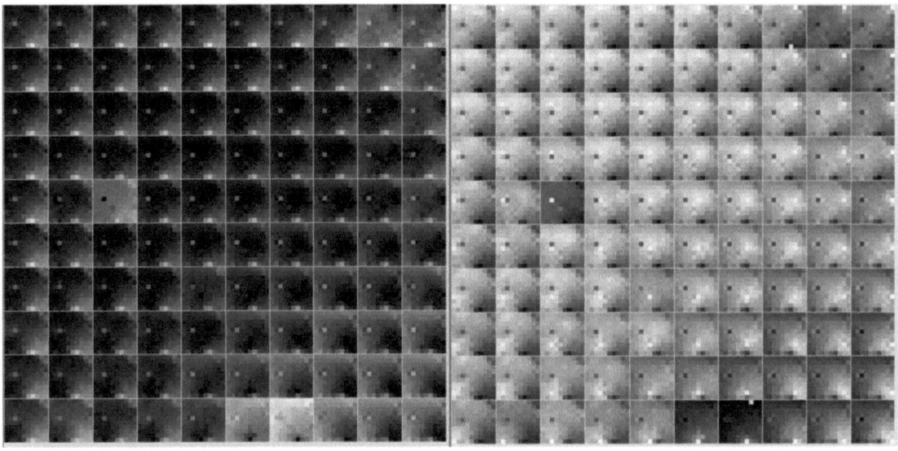

Fig. 11.13 The illustration shows the visualization of the vector distance between each codebook and all the others in a SOM measuring 10 × 10. In the matrix to the *right*, the *light areas* represent short distances, while the *dark* ones long distances; in the matrix to the left, the relation is the opposite (Taken from Software of Semeion: SOM ver. 0.5, 2005)

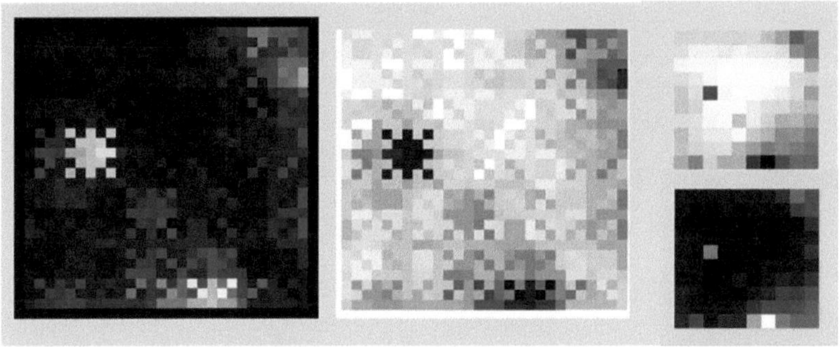

Fig. 11.14 The two big figures show the distance between one codebook and the others (with two different color codes), while the small figures show the absolute value of the same codebook in comparison with the absolute values of other codebooks (always with two color codes) (Taken from Software of Semeion: SOM ver. 0.5, 2005)

- *Minimum global distance*: minimum distance that separates each codebook from all the others (Figs. 11.15 and 11.16). In this case, a graph is also drawn with the matrix units placed at the vertices.
- *Minimum spanning tree* of the vector distances between all the codebooks (Figs. 11.15 and 11.16).

Following the rule of identifying the minimum distances between the units in order to link their vertices, the minimum spanning tree graph suggests a spatial positioning which makes the identification of the macro-classes easier. If, for example, all the units resting on the same vertex are linked to one another, it is possible to rewrite Fig. 11.16 as shown in Fig. 11.17.

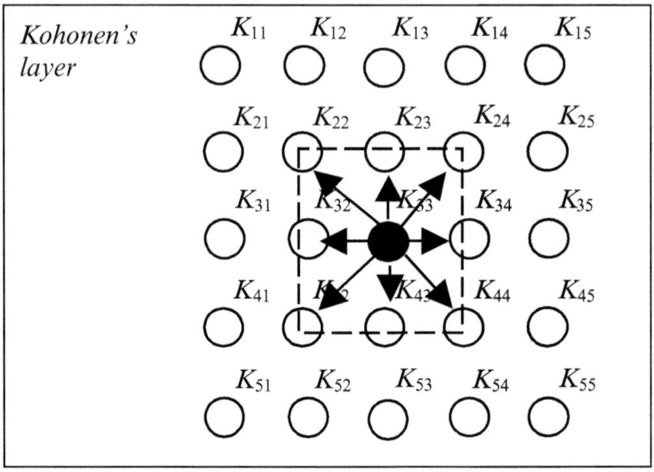

Fig. 11.15 The illustration shows the neighborhood of the K_{33} units formed by the eight units K22, K23, K24, K32, K34, K42, K43, and K44

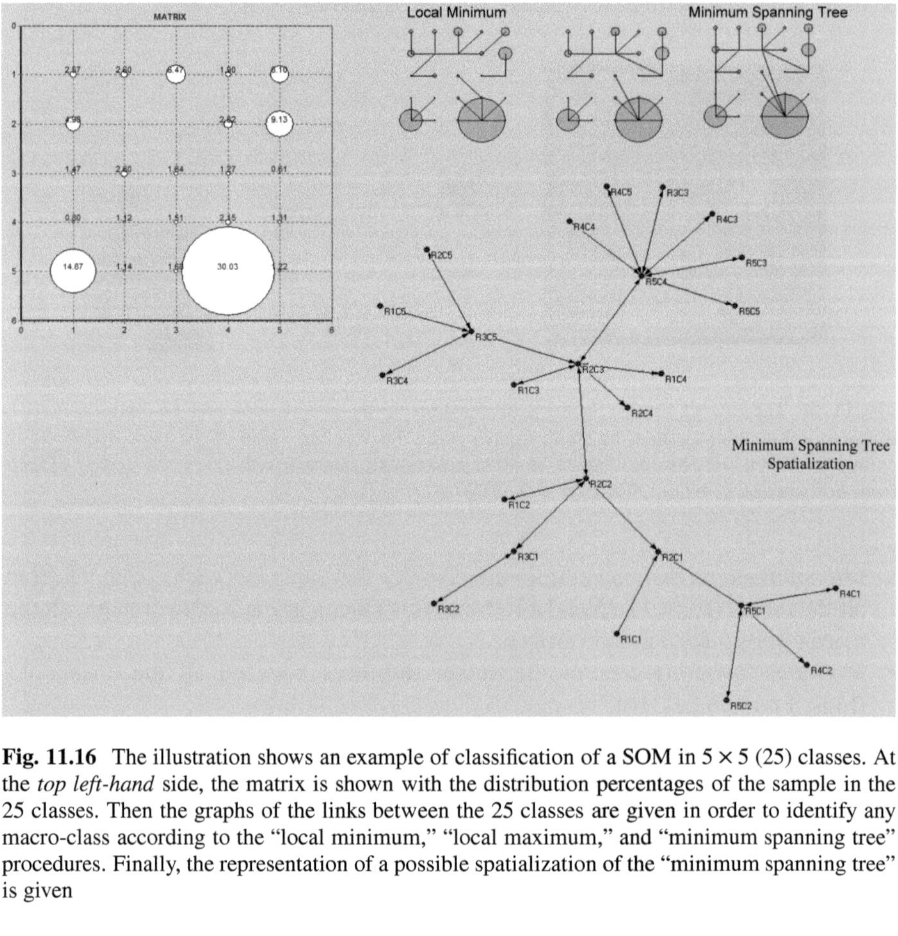

Fig. 11.16 The illustration shows an example of classification of a SOM in 5 × 5 (25) classes. At the *top left-hand* side, the matrix is shown with the distribution percentages of the sample in the 25 classes. Then the graphs of the links between the 25 classes are given in order to identify any macro-class according to the "local minimum," "local maximum," and "minimum spanning tree" procedures. Finally, the representation of a possible spatialization of the "minimum spanning tree" is given

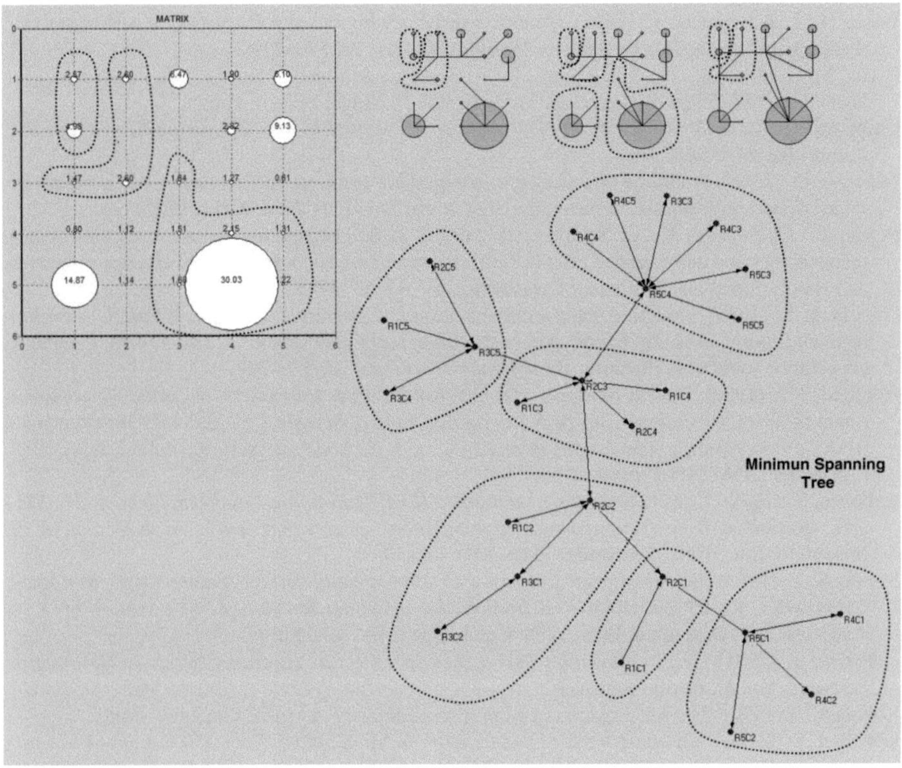

Fig. 11.17 The illustration shows a processing of Fig. 11.16 in which at the base of the vertices that possess more than two links, the "A, B, C, D, E, and F" macro-classes are defined. Such macro-classes have been represented on the matrix of the distribution percentages of the sample and on the "local minimum," "local maximum," and "minimum spanning tree" graphs

More in-depth analyses are possible on the *minimum spanning tree* principle where, in order to identify the macro-classes, not only is the position of the vertices on the graph taken into account but also the actual distances. The procedure can then be repeated in such a way that in more complex graphs, macro-classes become vertices.

References

Amari, S., & Arbib, M. A. (1977). Competition and cooperation in neural nets. In J. Metzler (Ed.), *System neuroscience* (pp. 119–165). New York: Academic.

Anderson, J. A., & Rosenfeld, E. (1988). *Neurocomputing foundations of research.* Cambridge, MA/London: MIT Press.

Arbib, M. A. (Ed.). (1995). *The handbook of brain theory and neural networks.* Cambridge, MA/London: The MIT Press, A Bradford Book.

Batchelor, B. G. (1974). *Practical approach to pattern recognition.* New York: Plenum Press.

Bauer, H. U., & Pawelzik, K. (1992). Quantifying the neighborhood preservation of self-organizing feature maps. *IEEE Transactions on Neural Networks, 30*, 570–579.
Breda, M. (1999). Self-organizing maps. In M. Buscema & Semeion Group (Eds.), *Reti Neurali e Sistemi Sociali Complessi* (Teoria e Modelli, Vol. I). Milan: Franco Angeli.
Carpenter, G. A., & Grossberg, S. (1991). *Pattern recognition by self-organizing neural network.* Cambridge, MA: MIT Press.
DeSieno, D. (1988). Adding a conscience to competitive learning. In *Proceedings of the International Conference on Neural Networks* (Vol. 1, pp. 117–124). New York: IEEE Press.
Erwin, E., Obermayer, K., & Schulten, K. (1992). I: Self-organizing maps: Stationary states, metastability and convergence rate; II: Self-organizing maps: Ordering, convergence properties and energy functions. *Biological Cybernetics, 67*, 35–45, 47–55.
Grossberg, S. (1976). Adaptive pattern classification and universal recoding, I: Parallel development and coding of neural feature detectors. *Biological Cybernetics, 33*, 121–134; II: Feedback, expectation, olfaction, illusions. *Biological Cybernetics, 23*, 187–202.
Hopfield, J. J. (1982). Neural networks and physical systems with emergent collective computational abilities. *Proceedings of the National Academy of Sciences, 79*, 2554–2558. (Reprinted from *Neurocomputing foundations of research*, by J. A. Anderson & E. Rosenfeld, Eds., 1988, Cambridge, MA: MIT Press.)
Kohonen, T. (1972). Correlation matrix memories. *IEEE Transactions on Computers, C-21*, 353–359. (Reprinted from *Neurocomputing foundations of research*, by J. A. Anderson & E. Rosenfeld, Eds., 1988, Cambridge, MA: MIT Press.)
Kohonen, T. (1982). Self-organized formation of topologically correct feature maps. *Biological Cybernetics, 43*, 59–69. (Reprinted from *Neurocomputing foundations of research*, by J. A. Anderson & E. Rosenfeld, Eds., 1988, Cambridge, MA: MIT Press.)
Kohonen, T. (1984). *Self-organization and associative memory* (Springer Series in Information Sciences, Vol. 8). Berlin: Springer.
Kohonen, T. (1990). The self-organizing Map. *Proceedings of the IEEE, 78*, 1464–1480.
Kohonen, T. (1995a). Learning Vector Quantization. In M. A. Arbib (Ed.), *The handbook of brain theory and neural networks*. Cambridge, MA/London: The MIT Press, A Bradford Book.
Kohonen, T. (1995b). *Self-organizing maps*. Berlin/Heidelberg: Springer.
Obermayer, K., Blasdel, G., & Schulten, K. (1992). A statistical mechanical analysis of self-organization and pattern formation during the development of visual maps. *Physical Review A, 45*, 7568–7589.
Ritter, H. (1991). Asymptotic level density for a class of vector quantization processes. *IEEE Transactions on Neural Networks, 2*, 173–175.
Ritter, H. (1995). Self-organizing feature maps: Kohonen maps. In M. A. Arbib (Ed.), *The handbook of brain theory and neural networks*. Cambridge, MA/London: The MIT Press, A Bradford Book.
Ritter, H., & Kohonen, T. (1989). Self-organizing semantic maps. *Biological Cybernetics, 61*, 241–254.
Ritter, H., Martinetz, T., & Schulten, K. (1992). *Neural computation and self-organizing maps.* Reading: Addison Wesley.
Widrow, B., & Winter, R. (1988). Neural nets for adaptive filtering and adaptive pattern recognition. *Computer, 21*, 25–39.
Willshaw, D. J., & von der Malsburg, C. (1976). How patterned neural connections can be set up by self-organization. *Proceedings of the Royal Society B, 194*, 431–445.
Yuille, A. L., & Geiger, D. (1995). Winner-take-all mechanisms. In M. A. Arbib (Ed.), *The handbook of brain theory and neural networks*. Cambridge, MA/London: The MIT Press, A Bradford Book.

Software

MASSINI G. (2007). SOM (Self Organizing Maps), Semeion Software #19, v. 7.

Chapter 12
Self-Organizing Maps: Identifying Nonlinear Relationships in Massive Drug Enforcement Databases

Giulia Massini

12.1 Introduction

The classification of the seizures affected by Scotland Yard in the fight against drug dealing is based on a database (DB) consisting of 848 cases of seizure,[1] each of which is broken down into 93 variables (see Appendix A).

A self-organizing map (SOM) (Kohonen 1972, 1982, 1984, 1990, 1995a, b) is constructed with a matrix of size 10×10 capable of supporting 100 classes.

The construction parameters of the SOM are as follows:

Input nodes	93
Pattern	848
Kohonen layer	100
Columns	10
Rows	10
Neighborhood topology	Square
View of the patterns	Random
ALPHA function	Gaussian
ALPHA MAX	1
ALPHA MIN	0
ALPHA INC	0.01
Set W	0–1
ALPHAW function	Linear
ALPHAW MAX	1
ALPHAW MIN	0
ALPHAW INC	0.01
Number of epochs	100

[1]Dataset of seizures was extracted in March 2006 when the situation of the CDTD database was 954 tactic sequences, 1,084 persons, and 888 persons seized (40 incomplete cases).

G. Massini (✉)
Semeion Research Center of Sciences of Commuication, via Sersale 117, Rome, Italy
e-mail: g.massini@semeion.it

M. Buscema and W.J. Tastle (eds.), *Intelligent Data Mining in Law Enforcement Analytics: New Neural Networks Applied to Real Problems*, DOI 10.1007/978-94-007-4914-6_12,

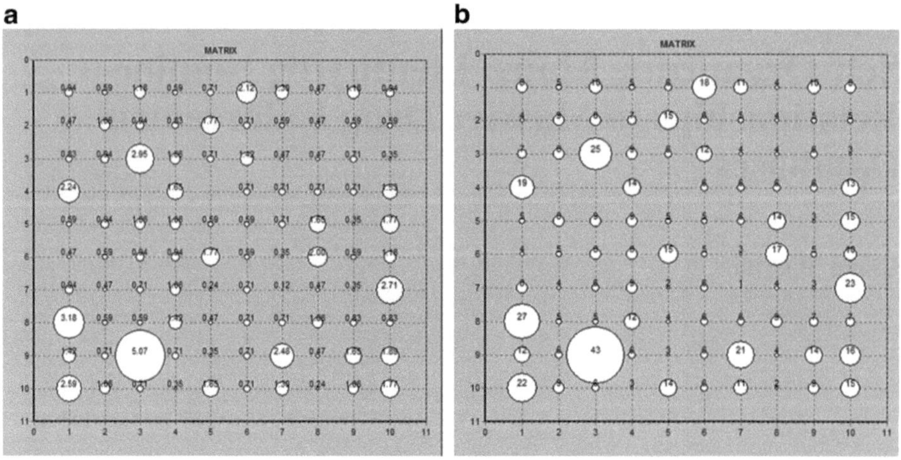

Fig. 12.1 Distribution of the 848 records onto the 10 × 10 matrix of an SOM. In the matrix on the *left*, (**a**) the percentages are shown, while the matrix on the *right* (**b**) shows the corresponding real frequencies

where

- *Input nodes*: number of units comprising the input layer.
- *Pattern*: number of patterns.
- *Kohonen layer, columns, rows*: the number of units of the Kohonen layer matrix is determined by the product of the number of columns and of rows.
- *Neighborhood topology*: topology of winning unit's neighborhood, either square or diamond shaped.
- *View of the patterns*: for the ordering for the input to the patterns.
- *ALPHA function, ALPHA MAX, ALPHA MIN, ALPHA INC*: parameters for the determination of the winning unit's neighborhood.
- *Set W*: range of the initialization of the weights.
- *ALPHAW function, ALPHAW MAX, ALPHAW MIN, ALPHAW INC*: parameters for the determination of the correction of the weights.
- *Number of epochs*: determined by the AlphaMax/AphaInc.

At the end of the training phase, the records are distributed by the SOMs onto the matrix yielding 97 classes out of a possible 100; three classes remained empty (row 4, column 2; R4C3; R4C5); see the distribution of the records in classes on the matrix in Fig. 12.1.

12.2 Potential Problems

The first problem to be addressed was the grouping of the 100 classes into broader macroclasses in order to be able to interpret the classification obtained. To define the macroclasses, the Euclidean vectorial distance between each codebook vector to each of the 100 classes and all the others was calculated:

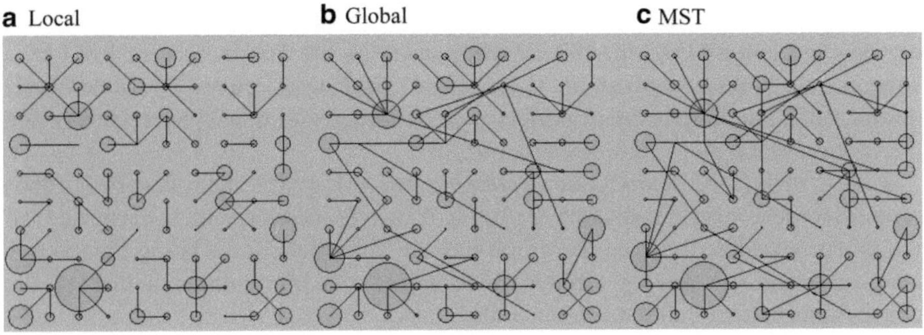

Fig. 12.2 The figure shows three images of the relationships between the classes of the SOM network. The *circles* represent the frequency, while the *straight lines* represent the nearness relationships between the classes. (**a**) Local shows the smaller vectorial distance between the codebook of a class and the eight ones that are near it; (**b**) Global shows the smaller vectorial distance between the codebook of a class and all the codebooks of the matrix; (**c**) MST shows the minimum spanning tree calculated on the matrix of the distances between all the codebooks

Fig. 12.3 Shows the neighborhood of the K_{33} unit consisting of the eight units K22, K23, K24, K32, K34, K42, K43, and K44

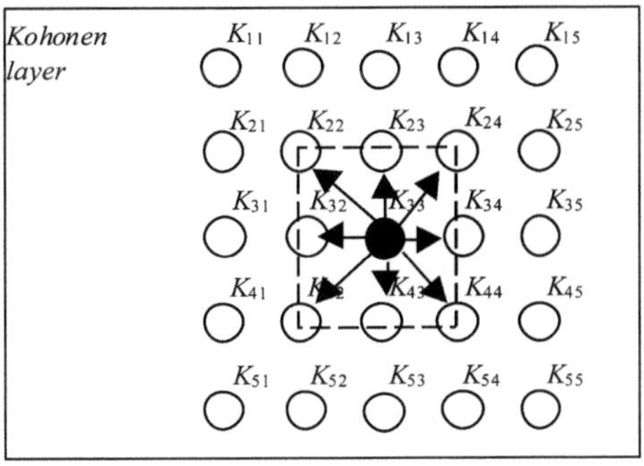

$$d_{i\times j} = \left\| X_i - X_j \right\| = \sqrt{\sum_{z=1}^{N} \left(x_{iz} - x_{jz}\right)^2}$$

where x_{ik}andx_{jk} are, respectively, the *k*th value of the codebook *i* and the *k*th value of the codebook *j*; *N* is the number of variables.

Then, a distance matrix of dimension 100×100 was constructed (symmetrical with respect to the diagonal); on these values, it was then possible to visualize some relationships. Figure 12.2a, b, and c shows three procedures of increasing complexity, which define some macroclasses on the SOM matrix beginning with the vectorial distance between classes.

In Fig. 12.2a, *Local* displays a straight line as the smaller vectorial distance between the codebook of a class and one of the eight adjacent classes constituting its outline. In this way, the matrix local macroclasses are revealed to be made up of classes that have physical proximity and are connected with one another (Fig. 12.3).

In Fig. 12.2b, *Global,* just the smaller vectorial distance between the codebook of one class and all the other 99 classes present on the matrix, is visualized by a straight line. In this visualization, the adjacent connections remain prevalent on the matrix, but some macroclasses project their nearness with nonadjacent classes and therefore reveal a more complex structure.

Lastly, in Fig. 12.2c, *minimum spanning tree* (*MST*) (Graham and Hell 1985) represents not only the smaller vectorial distances between each class and all the others visualized, but as the graph has the characteristic of being completely connected and not having circuits, it identifies the minimum structure of the connections between all the classes of the matrix.

12.3 Visualization of the Data

A spatialized representation of the MST graph (c) is shown in Fig. 12.4 where generally speaking the vertices represent the individual classes of the SOM matrix, while in particular, the yellow vertices represent the empty classes of the matrix (R4C2; R4C3; R4C5).

A study of the MST graph reveals some important characteristics:

1. The strategic nodes that are central in the graph (in white) coincide with the three empty classes in the SOM matrix.
2. The hubs that have a greater number of connections in the graph coincide with some of the classes of greater frequency on the matrix.

Based on the projection of the branchings of the MST graph's hubs, the macroclasses were identified on the SOM matrix as indicated in Fig. 12.5.

To understand the structure of the SOM macroclasses identified through the MST, it is necessary to compare it with the way the variables are distributed (the value that they assume) in the different codebooks. It was noted that the variables that determine the structure of the macroclasses in a more significant way are shown in Fig. 12.6.

12.4 Giving Meaning to the Analyses

With regard to the first group of variables, it is then possible to subdivide the MST graph more thoroughly, identifying subdivisions in the original macrogroups and then transferring them onto the SOM matrix (Fig. 12.7).

At this level of analysis, it is possible to understand how the strategies of the seizures relating to the different drugs are structured, namely,

- Cannabis
- Cocaine
- Crack

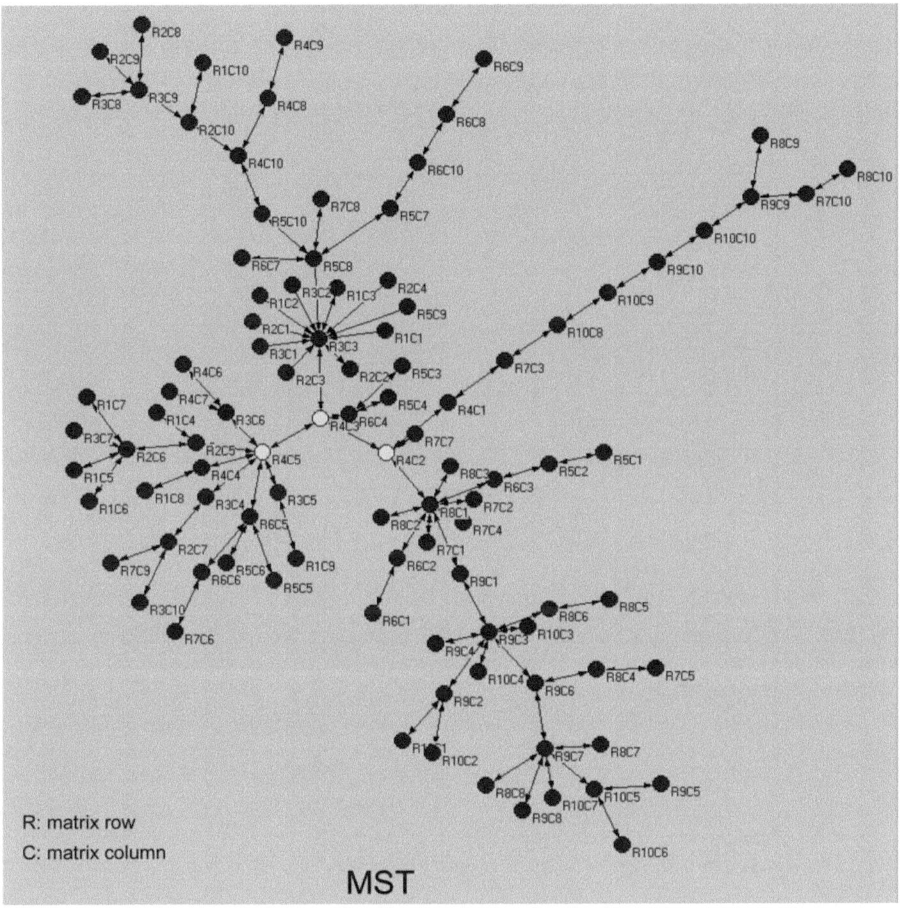

Fig. 12.4 Spatialization of the MST graph shown in Fig. 12.2c: the vertices are made up of the SOM classes, and the arcs are made up of the vectorial nearness relationship between these classes. The three *white* nodes are the empty classes (frequency 0.00) on the SOMs

- Heroin/diamorphine
- MDMA
- Other drugs

The seizures of *Cannabis* on the matrix are subdivided mainly into three macroclasses defined, respectively, by the variables "Search of Person" and "Search of Premises." Two macroclasses are in strategic positions on the MST graph, both have a hub node relating to a class with high frequency, and the grade is defined as the number of links with each node: the class R3C3 (a hub with grade 12) for "Search of Person" and the class R8C1 (grade 9 hub) for "Search of Premises." Another hub (grade 7) consists of the node R9C3 that represents the class with greater frequency in the matrix (5.07% of the cases, i.e., 43 records) (Fig. 12.8).

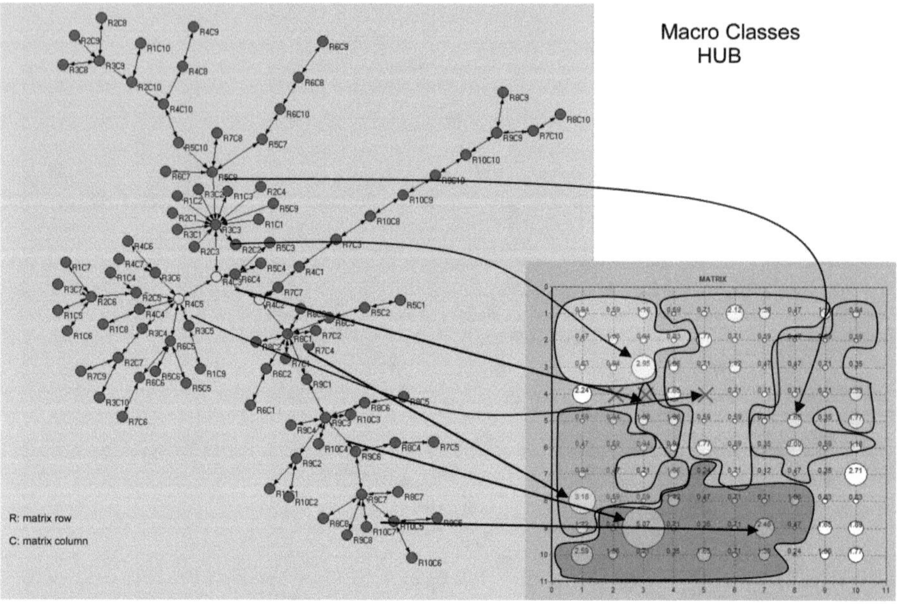

Fig. 12.5 The branchings of the hubs on the MST graph have been marked out with different colors and then transferred onto the SOM matrix. Notice that the *black* and *white* color has been replaced with *red* and *yellow*

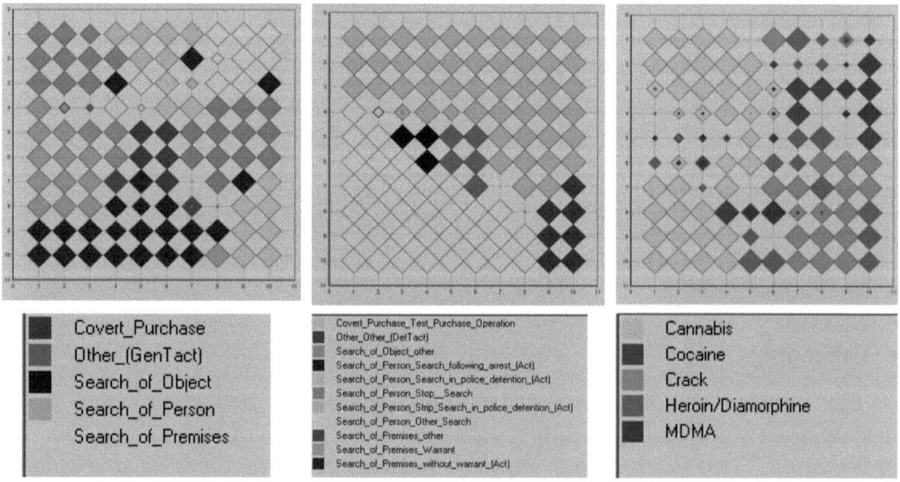

Fig. 12.6 The distribution of the values of some of the variables present in the DB in the different codebooks of the SOM matrix and the size of the individual rhombi relate to the value assumed by the variables that may vary between the value 1 (maximum value) and 0 (minimum value). Each variable has been standardized between 0.00 and 1.00

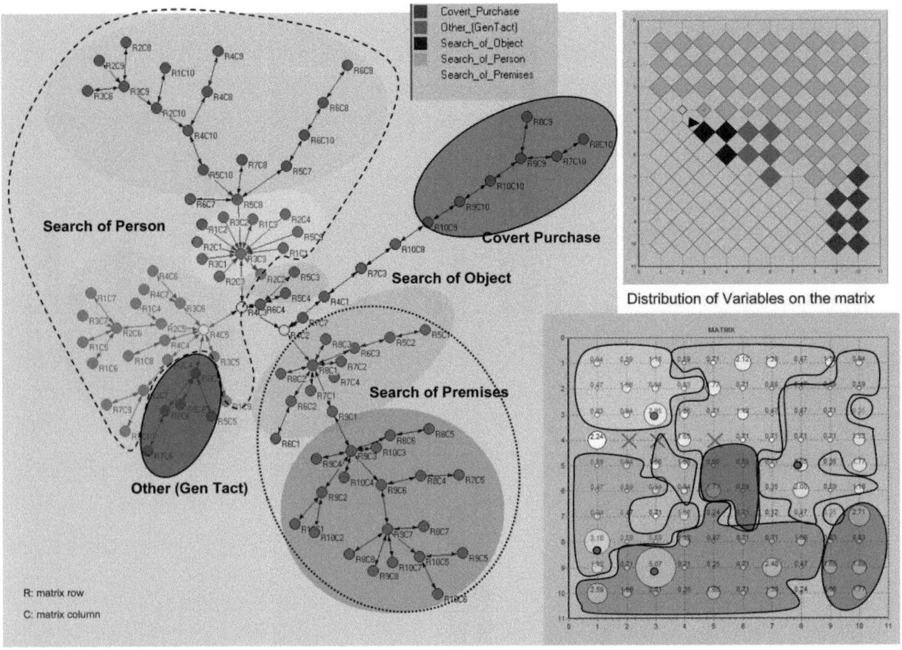

Fig. 12.7 Identification of the variables having a significant impact on the structure of the macroclasses' breakdown

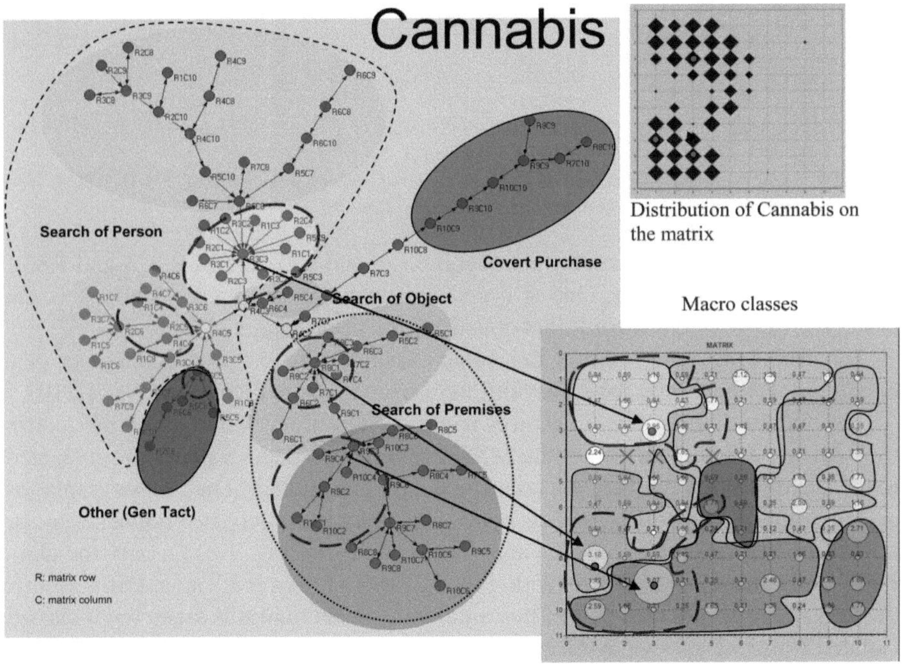

Fig. 12.8 Identification of the variable "Cannabis" in the classes of the SOMs and then in the nodes of the MST graph

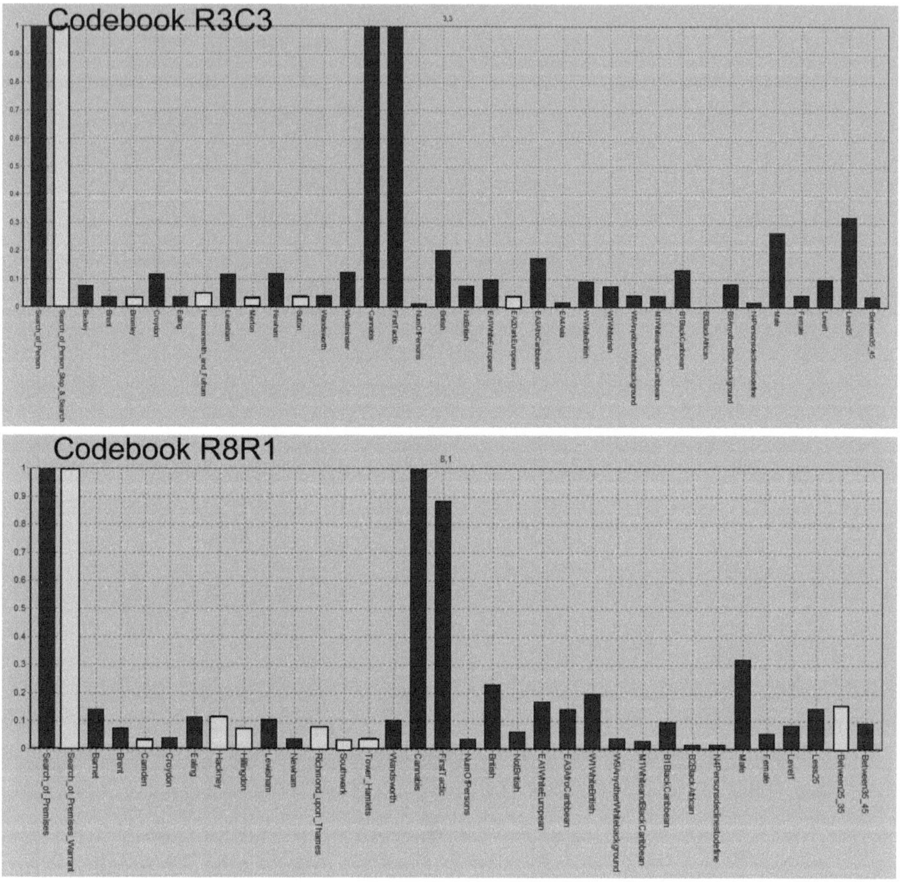

Fig. 12.9 Histogram of the values not equal to 0 of the codebooks R3C3 and R8C8. The similarities are shown in *black* and the differences in *yellow*

In the following figures are the codebooks of the two classes R3C3 and R8C1 in order to compare them. One is urged to remember that each codebook is the prototype image of the records that constitute the class (Fig. 12.9).

The following figure shows the values assumed by the two codebooks in a histogram in which only the values different from zero are displayed (Fig. 12.10).

The situation of the *Crack* seizures is very different: even though on the matrix they occupy a definite area, on the MST graph, the respective nodes are relegated to marginal positions and are not connected with one another. The classes displayed on the graph are mostly *connector* nodes, i.e., they have only two connections, or *terminal* nodes, i.e., they have a single connection (see Fig. 14.11); only the class R9C7 constitutes a grade 6 hub within the "Search of Premises" area. Furthermore, the next figure shows how part of this area of the SOM matrix is made up of classes isolated from one another and from the other areas. In this area (see Figs. 12.10 and 12.12), the variables "Not British" and "Afro Caribbean" are particularly prevalent (Fig. 12.13).

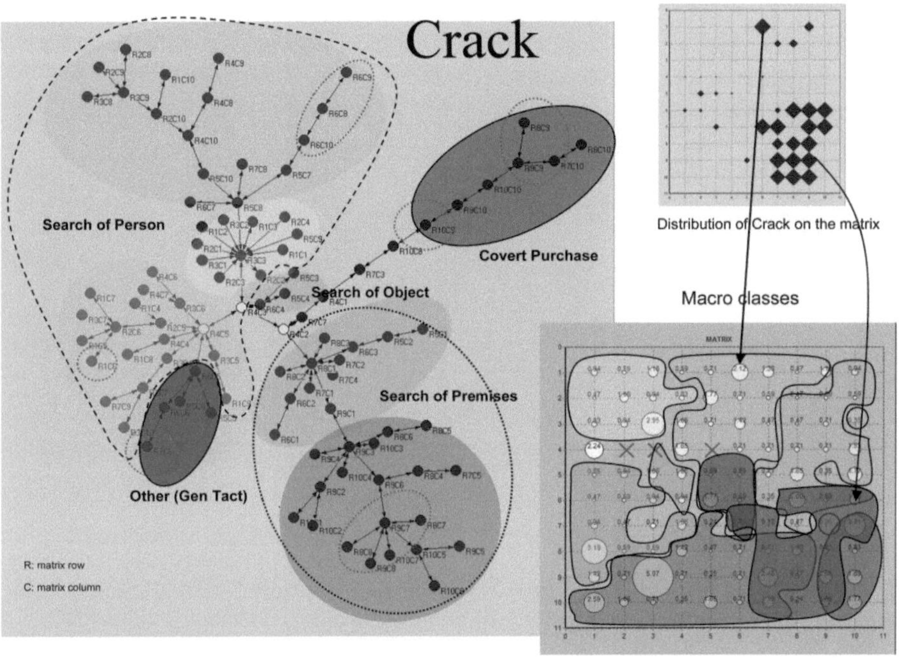

Fig. 12.10 This shows the distribution of the classes in whose codebooks, the "Crack" variable is present. On the matrix, the variable is present in codebooks that, for the most part, are adjacent, while on the MST graph, the respective classes occupy marginal positions that are very distant from each other (*black circles*)

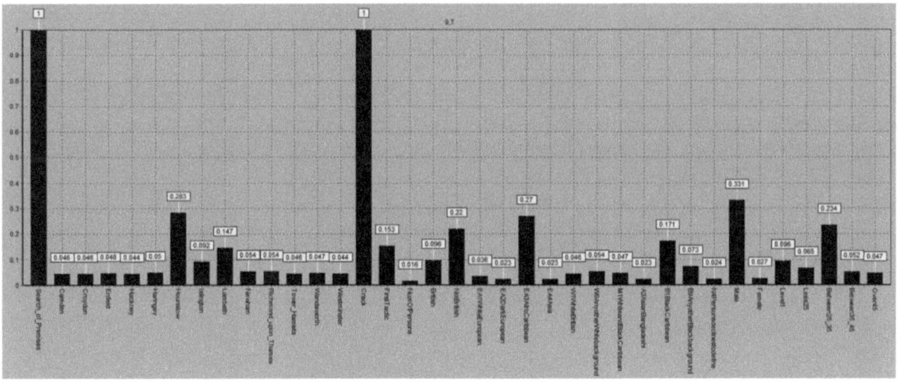

Fig. 12.11 R9C7 codebooks relating to drug Crack, grade 6 hub on the MST

The Heroin seizures both on the matrix and on the MST graph are very much connected with the Crack seizures occupying close positions (Fig. 12.14).

Next, we show the overall distribution of the different types of drug seizures on the MST graph which reveals once again how the strategies linked with the seizures of "Cannabis" are the strategic ones in relation to the other drugs (Fig. 12.15).

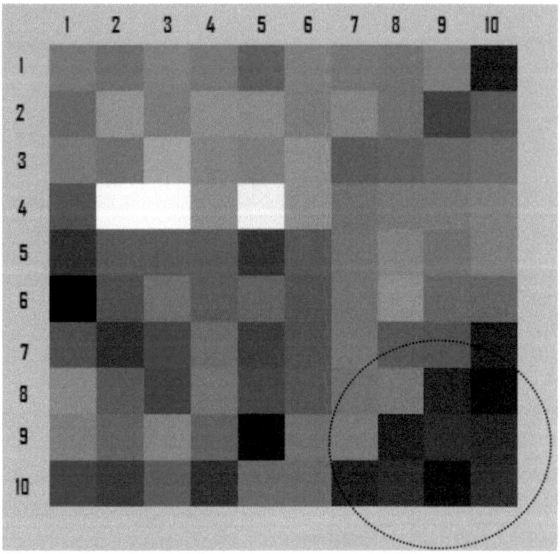

Fig. 12.12 This shows the mean of the distance between the codebooks of the respective neighborhood for each cell of the matrix. *Black*: *maximum difference*. *White*: *maximum similarity*

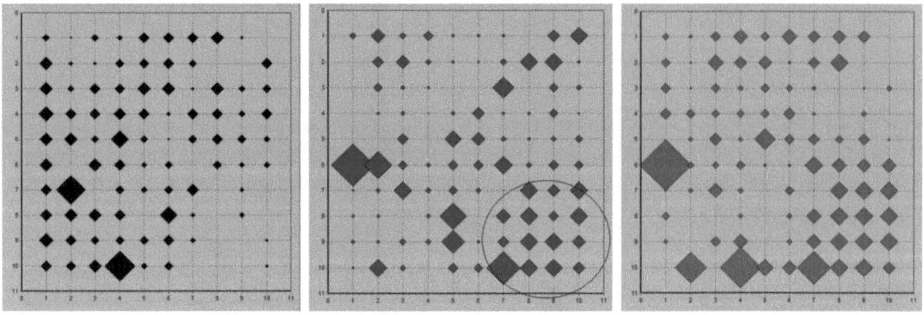

Fig. 12.13 This shows the distribution of the variables "British" (the matrix on the *left*) and "Not British" (the matrix in the *center*) and "Afro Caribbean" (the matrix on the *right*). As it can be seen, the bottom right area of the center matrix is characterized by the presence of the variables "Not British," and in the right matrix, the ethnicity "Afro Caribbean"

The DB of the arrests carried out by Scotland Yard as part of the fight against drug dealing consists of 269 cases of arrest,[2] each of them broken down into 78 variables (see Appendix B).

A SOM network was constructed with a matrix of dimensions 10×10, i.e., capable of creating 100 classes.

---

[2]Dataset of arrest was extracted in January 2006 when the situation of the CDTD database was 338 tactic sequences, 513 persons, 351 accused persons (only 260 of whom were completed for processing).

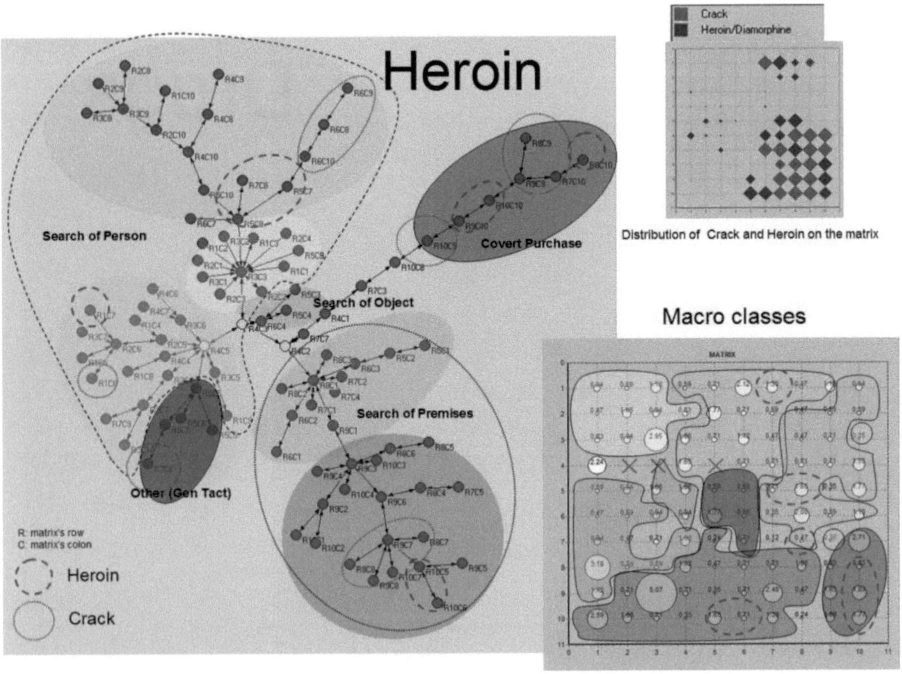

Fig. 12.14 This shows the distribution of the classes in whose codebooks, the "Heroin" variable is present, in comparison with the "Crack" variable previously analyzed. It can be noted how close they are, both on the MST graph and on the matrix

The construction parameters of the SOM were as follows (notice the change in the number of input nodes and patters):

Input nodes	78
Pattern	269
Kohonen layer	100
Columns	10
Rows	10
Neighborhood topology	Square
View of the patterns	Random
ALPHA function	Gaussian
ALPHA MAX	1
ALPHA MIN	0
ALPHA INC	0.01
Set W	0–1
ALPHAW function	Linear
ALPHAW MAX	1
ALPHAW MIN	0
ALPHAW INC	0.01
Number of epochs	100

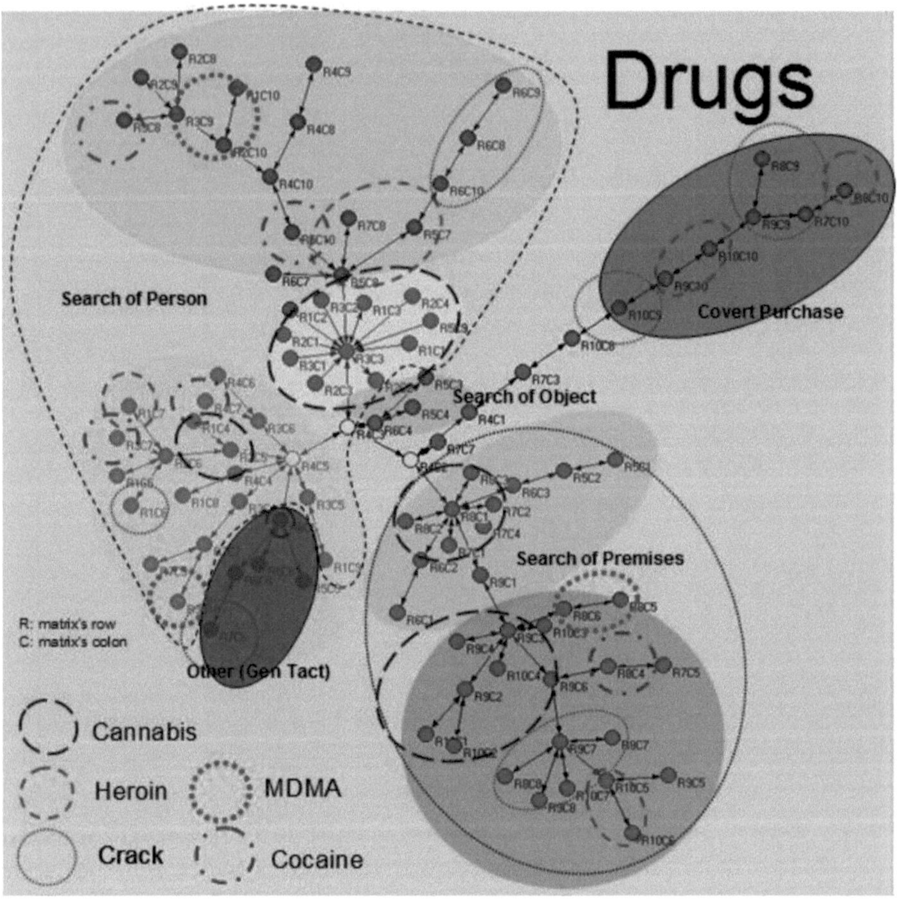

Fig. 12.15 The distribution of the variables relating to the different drug types on the MST graph

At the end of the training phase, the records were distributed by the SOMs onto the matrix creating 83 classes out of the 100 possible, since 17 classes remained empty; see the distribution of the records in the classes on the matrix in Fig. 12.16.

In order to identify the macroclasses, it was necessary to define the vectorial distance of each codebook among all the others of the matrix. Then the relationships between these distances were analyzed in accordance with the procedure previously described using the analysis of the global local nearness and the breakdown of the latter on the MST graph (*minimum spanning tree),* as illustrated in Fig. 12.17.

A spatialized representation of the MST (c) graph is shown in the following figure where, generally speaking, the vertices represent the individual classes of the SOM matrix, while the yellow vertices represent in detail the empty classes of

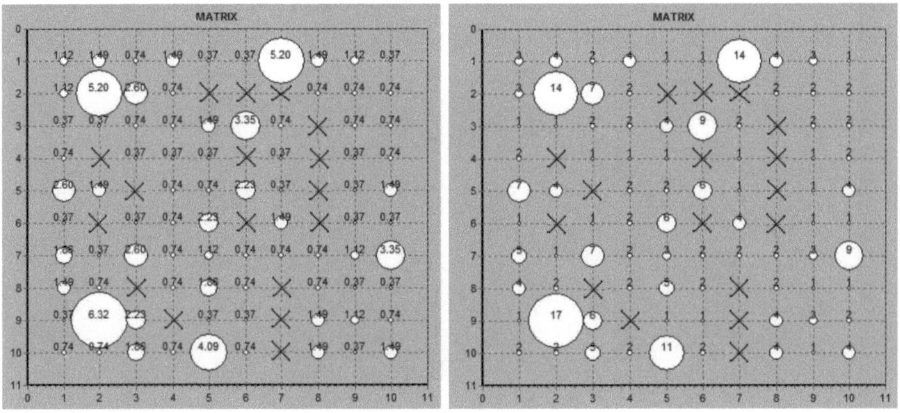

Fig. 12.16 Distribution of the 269 records on the 10 × 10 matrix of a SOM. The matrix on the *left* shows the percentages, while the one on the *right* shows the corresponding real frequencies; *X* represents the empty classes

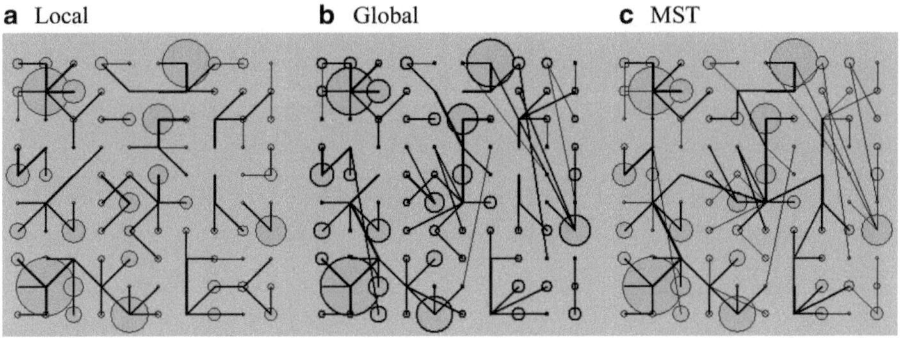

Fig. 12.17 Analysis of the distances between the codebooks of the matrix. The *circles* represent the distribution of the records in the different classes of the matrix, while the *straight lines* represent the vectorial nearness relationship between the codebooks. The thickness of the *straight lines* relates to the nearness degree of the codebooks. The three figures show analyses of increasing complexity between the nearness relationships between the codebooks

the matrix (Fig. 12.18). A study of the MST graph reveals that the central nodes in the graph coincide with the empty classes (in white) in the SOM matrix. Going into the detail in this graph, the empty classes seem to form the "backbone" of the graph itself, while the proper classes form its "branchings." This is an experiment to attempt to reduce the dimensions of the SOM matrix by a sufficient number of cells equal to the number of empty classes. However, the new classification tended to keep the number of empty classes constant and unite the records in the

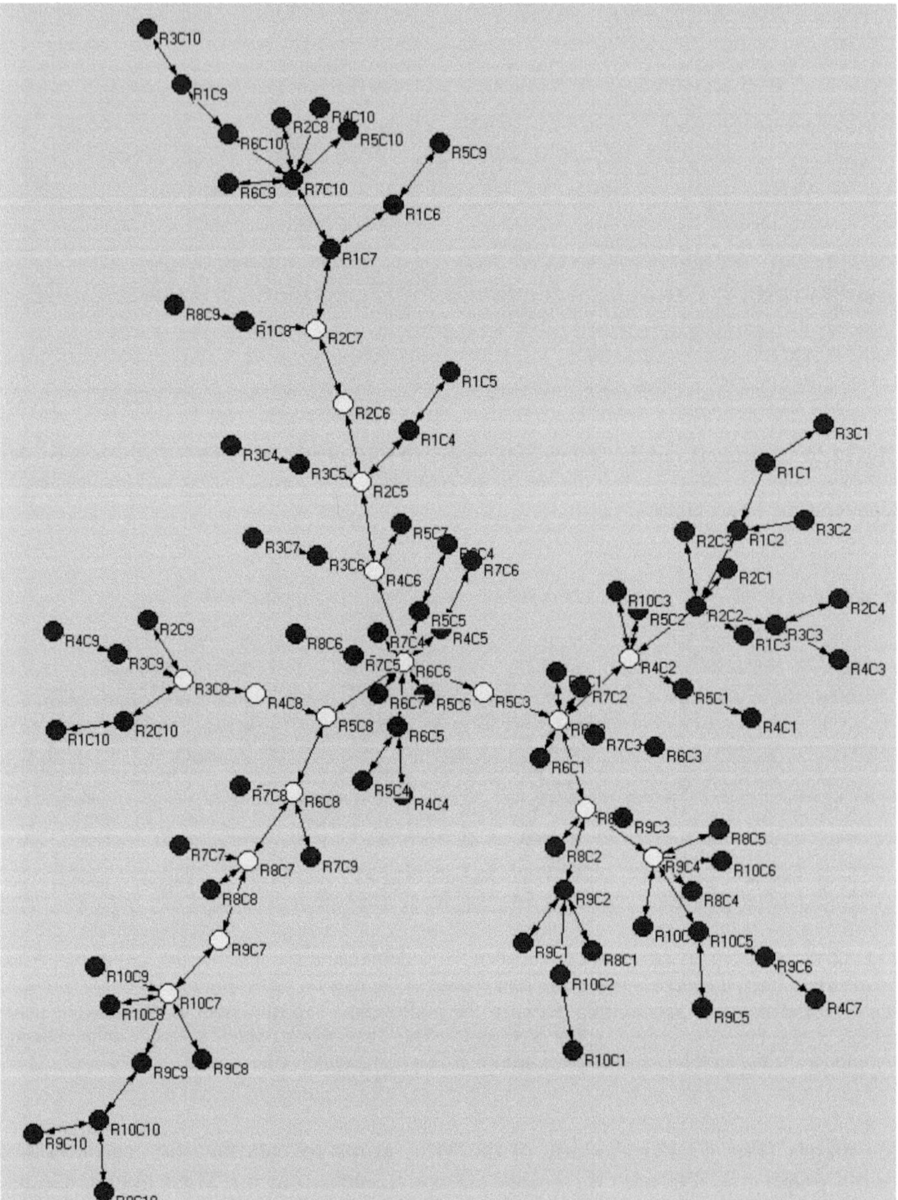

Fig. 12.18 Spatialization of the MST graph shown in Fig. 12.16c. The vertices are made up of the SOM classes, and the arcs are made up of the vectorial nearness relationship between these classes. The white nodes are the empty classes (frequency 0.00) in the SOM

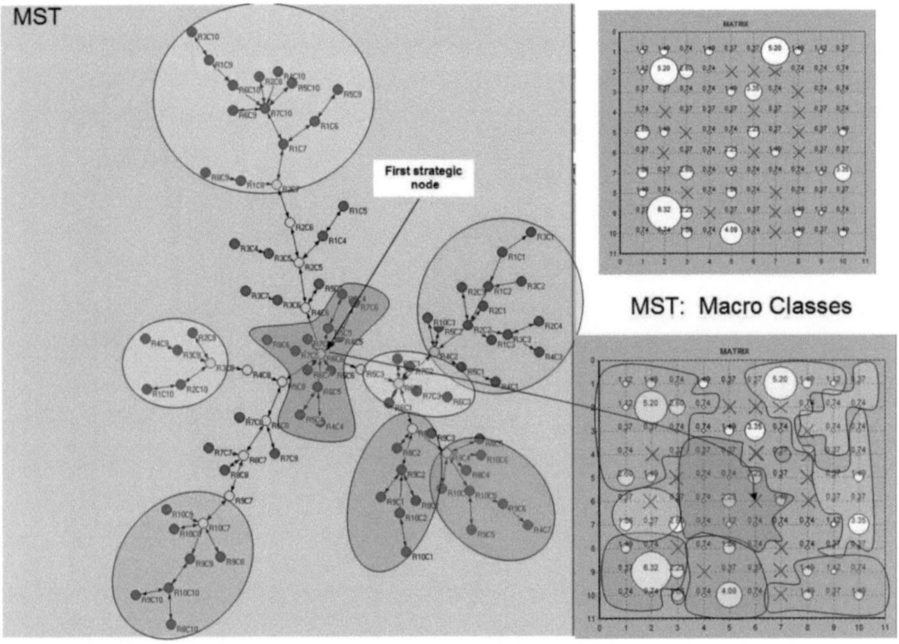

Fig. 12.19 Identification of the macroclasses on the basis of the MST graph's hubs through projection in the SOM matrix of the nodes that form part of the branchings

remaining classes. This has been interpreted as meaning that the distance between the codebooks belonging to different macroclasses is significant. The empty classes have codebooks for mediation between the different classes adjacent to them.

Based on the projection of the branchings of the MST graph's hubs, the macroclasses on the SOM matrix were identified as indicated in Fig. 12.19.

Figure 12.19 shows the definition of the macroclasses through the identification of the hub nodes on the graph which in this case coincide with some empty classes of the SOM, namely, R6C6 (grade 10), R6C2 (grade 7), R9C4 (grade 6), R4C2 (grade 5), R8C3 (grade 5), R10C7 (grade 5), R3C8 (grade 4), and R2C7 (grade 4). It can be noted that the class R6C6 of grade 10 occupies the central position and it is the strategic node of the MST graph (the one whose branchings have the most distant peripheral nodes). This node occupies a central position also on the matrix. It may also be noted that three of its branchings are articulated, and, therefore, four areas of the matrix can be identified, as we see in Fig. 12.20.

The areas can then be semanticized by identifying the distribution of the variables in the codebooks. Figure 12.21 shows the variables that have determined the structure of the area to the greatest extent.

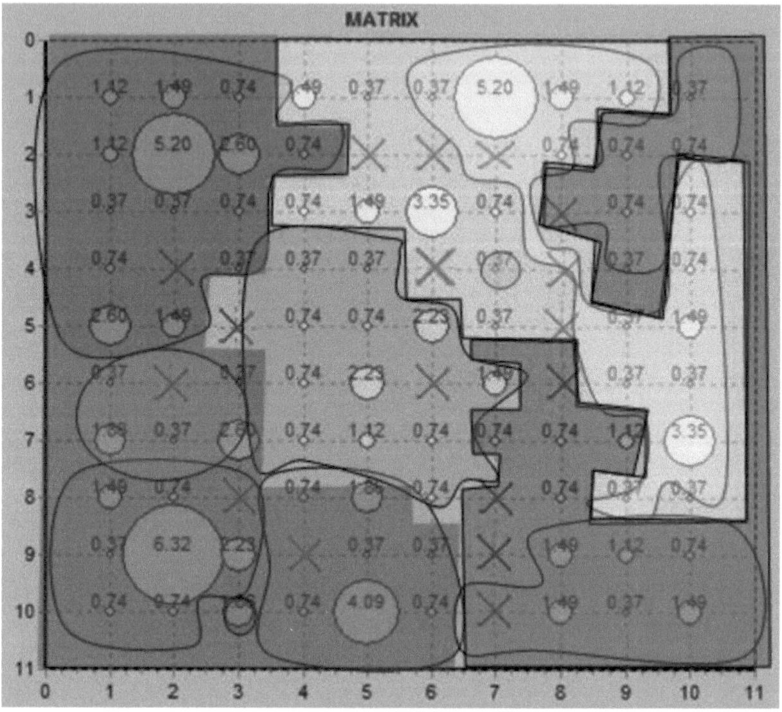

Fig. 12.20 Subdivision of the SOM matrix into four areas starting from the strategic hub R6C6 that constitutes the central area and then the other three areas defined on the basis of its three branchings

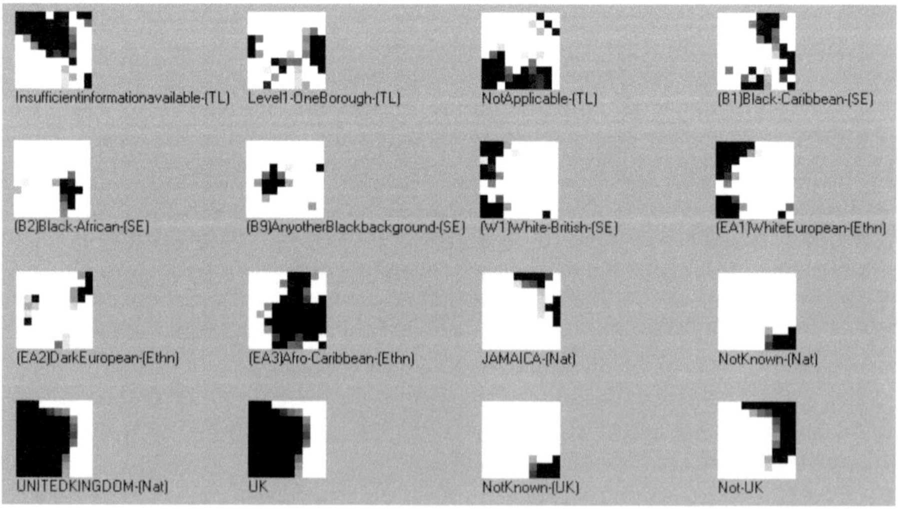

Fig. 12.21 Distribution of the variables in the codebooks that have contributed most significantly to the subdivision of the macroclasses on the matrix. In *black* are the higher values close to 1, and in *white* are the lower values close to 0

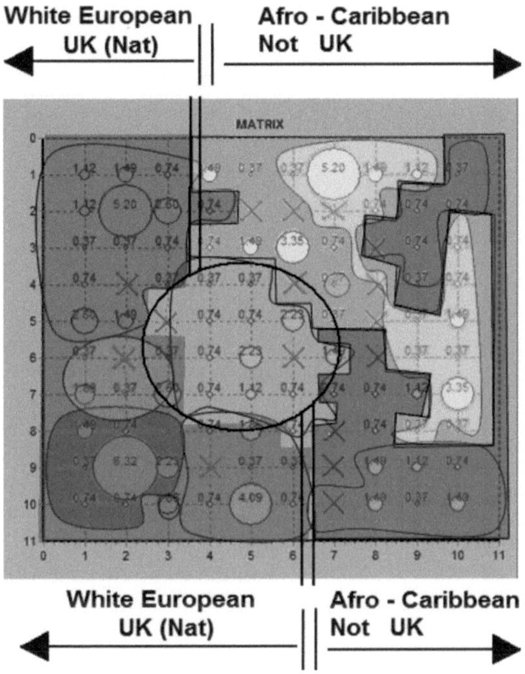

Fig. 12.22 Identification of four areas on the SOM matrix and their semanticization on the basis of the variables "place of birth" and "ethnicity"

From an analysis of the variables, it is then possible to identify (see Fig. 12.22), at the center of the MST graph and of the SOM matrix, a macroclass characterized by arrests of persons *born in the UK* of *Afro-Caribbean* ethnicity; on the left, the arrests involving persons *born in the UK of White European* ethnicity; and lastly, on the right, the arrests of persons born in other states, predominantly of *Afro-Caribbean* ethnicity.

12.5 Conclusion

Self-organizing maps in combination with minimal spanning trees created using an analysis of codebooks used to structure the data can produce interesting and meaningful visualizations. Even when the outcome of certain datasets is known to police officials, their graphical representation, as shown in this chapter, can produce knowledge that can be brought to bear by focusing limited forces in areas known to produce certain kinds of crime. This chapter explains how certain kinds of illegal drugs can be identified with specific ethnic groups in certain neighborhoods, in which certain search methods are found to be more successful than others and even areas that appear to be absent drug crime still have an involvement.

Appendix A: Variables of the Seizures

1.	Covert Purchase
2.	Other (GenTact)
3.	Search of Object
4.	Search of Person
5.	Search of Premises
6.	Covert Purchase Test Purchase Operation
7.	Other (DetTact)
8.	Search of Object other
9.	Search of Person Search following arrest (Act)
10.	Search of Person Search in police detention (Act)
11.	Search of Person Stop & Search
12.	Search of Person Strip Search in police detention (Act)
13.	Search of Person Other Search
14.	Search of Premises other
15.	Search of Premises Warrant
16.	Search of Premises without warrant (Act)
17.	Barking and Dagenham
18.	Barnet
19.	Bexley
20.	Brent
21.	Bromley
22.	Camden
23.	City of London
24.	Croydon
25.	Ealing
26.	Enfield
27.	Greenwich
28.	Hackney
29.	Hammersmith and Fulham
30.	Haringey
31.	Harrow
32.	Havering
33.	Hillingdon
34.	Hounslow
35.	Islington
36.	Kensington and Chelsea
37.	Lambeth
38.	Lewisham
39.	Merton
40.	Newham
41.	Redbridge
42.	Richmond upon Thames
43.	Southwark
44.	Sutton
45.	Tower Hamlets
46.	Waltham Forest
47.	Wandsworth

(continued)

(continued)

48.	Westminster
49.	Cannabis
50.	Cocaine
51.	Crack
52.	Heroin/Diamorphine
53.	MDMA
54.	Other Drugs
55.	Other Not Controlled
56.	Other Substance
57.	First Tactic
58.	In Operation
59.	Num Of Persons
60.	British
61.	Not British
62.	EA1WhiteEuropean
63.	EA2DarkEuropean
64.	EA3AfroCaribbean
65.	EA4Asia
66.	EA5Oriental
67.	EA6Arab
68.	EA7Unknown
69.	W1WhiteBritish
70.	W1WhiteIrish
71.	W9AnyotherWhitebackground
72.	M1WhiteandBlackCaribbean
73.	M3WhiteandAsian
74.	M9AnyotherMixedbackground
75.	A1AsianIndian
76.	A2AsianPakistani
77.	A3AsianBangladeshi
78.	A9AnyotherAsianbackground
79.	B1BlackCaribbean
80.	B2BlackAfrican
81.	B9AnyotherBlackbackground
82.	O1Chinese
83.	O9AnyOther
84.	N4Personsdeclinestodefine
85.	Male
86.	Female
87.	Level1
88.	Level2
89.	Level3
90.	Less25
91.	Between25 35
92.	Between35 45
93.	Over45

(continued)

Appendix B: Variables of the Arrests

1.	Insufficient information available (TL)
2.	Level1 OneBorough (TL)
3.	Level2a Region (TL)
4.	Level2 More Boroughs (TL)
5.	Level3 International (TL)
6.	Not Applicable (TL)
7.	(A1)Asian-Indian (SE)
8.	(A2)Asian-Pakistani (SE)
9.	(A3)Asian-Bangladeshi (SE)
10.	(A9)Any other Asian background (SE)
11.	(B1)Black-Caribbean (SE)
12.	(B2)Black-African (SE)
13.	(B9)Any other Black background (SE)
14.	(M1)White and Black Caribbean (SE)
15.	(M3)White and Asian (SE)
16.	(M9)Any other Mixed background (SE)
17.	(N4)Persons declines to define (SE)
18.	(NS)Not Stated (SE)
19.	(O1)Chinese (SE)
20.	(O9)Any Other (SE)
21.	(W1)White British (SE)
22.	(W2)White Irish (SE)
23.	(W9)Any other White background (SE)
24.	(EA1)White-European (Ethn)
25.	(EA2)Dark-European (Ethn)
26.	(EA3)Afro-Caribbean (Ethn)
27.	(EA4)Asia (Ethn)
28.	(EA5)Oriental (Ethn)
29.	(EA6)Arab (Ethn)
30.	ALGERIA (Nat)
31.	FRANCE (Nat)
32.	GAMBIA (Nat)
33.	GHANA (Nat)
34.	GREECE (Nat)
35.	GRENADA (Nat)
36.	IRELAND (Nat)
37.	ITALY (Nat)
38.	JAMAICA (Nat)
39.	NIGERIA (Nat)
40.	Not Known (Nat)
41.	PHILIPPINES (Nat)
42.	PORTUGAL (Nat)
43.	SOMALIA (Nat)

(continued)

(continued)

44.	SPAIN (Nat)
45.	SRI LANKA (Nat)
46.	UNITED KINGDOM (Nat)
47.	VIETNAM (Nat)
48.	UK
49.	Not Known (UK)
50.	Not UK
51.	Female
52.	Male
53.	Sex Undefined
54.	Age
55.	Convictions Number
56.	Offenses Number
57.	Age At First Conviction
58.	Theft and kindred offenses
59.	Offenses against the person
60.	Old Drug offenses
61.	Firearms Shotguns Offensive weapons
62.	Public disorder and rioting
63.	Sexual offenses
64.	Offenses against the person2
65.	Offenses related to police/courts/prisons
66.	New offenses
67.	Offenses against property
68.	Fraud and kindred offenses
69.	Miscellaneous offenses
70.	Num Of Arrests
71.	Num Of Tactic Sequences
72.	Num Of Tactics
73.	Num Of Drug Seizures
74.	Cannabis
75.	Cocaine
76.	Crack
77.	Heroin Diamorphine
78.	Other Drugs

References

Graham, R. L., & Hell, P. (1985). On the history of the minimum spanning tree. *Annals of the History of Computing, 7*(1), 43–57.

Kohonen, T. (1972). *Correlation matrix memories. IEEE Transactions on Computers, C-21*, 353–359. (Reprinted from *Neurocomputing foundations of research*, by J. A. Anderson & E. Rosenfeld, Eds., 1988, Cambridge, MA: MIT Press.)

Kohonen, T. (1982). *Self-organized formation of topologically correct feature maps. Biological Cybernetics, 43*, 59–69. (Reprinted from *Neurocomputing foundations of research*, by J. A. Anderson & E. Rosenfeld, Eds., 1988, Cambridge, MA: MIT Press.)

Kohonen, T. (1984). *Self-organization and associative memory* (Springer series in information sciences, Vol. 8). Berlin: Springer.
Kohonen, T. (1990). The self-organizing map. *Proceedings of the IEEE, 78*, 1464–1480.
Kohonen, T. (1995a). Learning vector quantization. In M. A. Arbib (Ed.), *The handbook of brain theory and neural networks*. Cambridge, MA/London: The MIT Press, A Bradford Book.
Kohonen, T. (1995b). *Self-organizing maps*. Berlin/Heidelberg: Springer.

Software

Massini G. (2007). SOM (Self Organizing Maps), Semeion Software #19, v. 7.

Chapter 13
Theory of Constraint Satisfaction Neural Networks

Massimo Buscema

13.1 Introduction

One kind of particularly interesting problem involves finding a solution to a set of constraints that impose a series of conditions on the solution that the variables must satisfy. One type of neural network that is used in solving this kind of problem is called the constraint satisfaction (CS) artificial neural network (ANN). It can be used to consider and analyze very different and sometimes unconventional problems. The concepts and theory necessary to understand the operation of this specialized neural network are explained in this chapter; a detailed example of its use in law enforcement is provided in a later chapter.

The way a CS ANN addresses a solution to different problems becomes clear by knowing its structural and functional characteristics. It is a one-layer ANN, and, therefore, each unit or node is similar to any other, and it is not characterized by a specific geography. The connections or weights among the different nodes are symmetric; therefore, $w_{ij} = w_{ji}$. Furthermore, their reflexive connections do not exist: $w_{ii} = 0$.

Each node can have its own *Bias*. This generally means that a constraint satisfaction provided with N nodes will have M number of connections (weights and bias) equal to

$$M = \underbrace{\frac{N \cdot (N-1)}{2}}_{w_{ij}\,\text{weights}} + \underbrace{N}_{\text{Bias}\ \theta_i} = \frac{N \cdot (N+1)}{2} \tag{13.1}$$

M. Buscema (✉)
Semeion Research Center of Sciences of Commuication, via Sersale 117, Rome, Italy
e-mail: m.buscema@semeion.it

M. Buscema and W.J. Tastle (eds.), *Intelligent Data Mining in Law Enforcement Analytics: New Neural Networks Applied to Real Problems*, DOI 10.1007/978-94-007-4914-6_13,

A constraint satisfaction. Constraint satisfaction is an ANN starting with a trained weights matrix and updates the values of its own units on the basis of

(a) The external inputs to which it is subjected
(b) The relational constraints imposed by the weights that characterize it

13.2 Learning Through Backpropagation

The values of the weights matrix, *W*, that characterize a constraint satisfaction can be generated in different ways. A certain kind of ANN can be charged with learning the weights matrix that characterizes all patterns specifying the values of the database about which the constraint satisfaction is questioned. Otherwise, it is possible to resort to some traditional Bayesian equations about the probabilities that characterize the positive and/or negative co-occurrence relationship between each node's couplet in the constraint satisfaction or, alternatively, to a reformulation of equations based on Hebb's (1949; also Hopfield 1982, 1984; Buscema 1995) hypotheses about the connections among neurons.

The use of a backpropagation ANN in order to learn the constraint satisfaction weights has shown to be a fairly effective method (see the classic Rumelhart et al. 1986b and Buscema 1998b, especially for Eq. 13.6). The procedure is computationally simple.

An *auto-associative* backpropagation ANN at a maximum gradient of *only 2 layers* is designed with the main diagonal fixed to zero: One layer is composed of Input units and the other layer of Output units. "Auto-associative" means that in this ANN the *target* vector will be the same with respect to that of every pattern of input.

The number of Input and Output nodes, which are equal, can be defined through two different strategies:

- *Strategy A*: Each field of the database (DB) under investigation is a node whose specific value will vary in the lattice {0,1} according to the variety of the field.
- *Strategy B*: Each field option is a node that can be active or passive if, in each record, that option is or is not present. In this case, the ANN will be constituted exclusively by binary Inputs [0,1].

In this second strategy, the total number of ANN's Input and Output nodes is given by the sum of all options of every field expected in the whole DB. Strategy B is advisable. In fact, this option is more computationally expensive than the first; it will be an ANN with more nodes and consequently more complex modeling. Nevertheless, this permits us to clearly consider the dynamic of each DB option through the constraint satisfaction. The two options are synthesizable as follows:

- *Codification a* [values {0,1}]: $\text{NumInput} = \text{NumOutput} = \sum_{i}^{N_f} \text{field}_i$
- *Codification b* [values [0,1]]: $\text{NumInput} = \text{NumOutput} = \sum_{j}^{N_{\text{op}}} \text{op}_j$

where NumOutput = number of outputs, NumInput = number of inputs, field = vector of DB's fields, op = vector of all options of each DB's field, N_f = total number of the fields, and N_{op} = total number of the options.

Once designed, the ANN will have, as a learning pattern, all DB records on which it intends to operate. The learning algorithm will be ANN's backpropagation classical algorithm, but it will be provided with some heuristic suggestions deduced from the experimentations carried out at Semeion's Research Center.

13.2.1 Forward Algorithm

$$u_i = f(\text{Net}_i) = f\left(\sum_j^N u_j \cdot w_{ij} + \text{Bias}_i\right) \tag{13.2}$$

Suggestion 1: It has been verified that in order to generate a useful weights matrix for the constraint satisfaction network, it is advisable to make the random initialization of space R at the beginning of the learning process very small. In practice,

$$\pm R = \frac{1}{\sqrt{\text{NumInput}}} \tag{13.3}$$

Furthermore, it is suggested to put all Bias = 0.0 and not to randomize them.

Suggestion 2: It has been further verified that the most effective transfer function f (Net_i) is the classic sigmoid; therefore,

$$f(\text{Net}_i) = \frac{1}{1 + e^{-\text{Net}_i}} \tag{13.4}$$

The function of the sine makes the quantitative relations among the nodes ambiguous. The function of the Hyperbolic Tangent excessively stresses the weights matrix. The Arctangent is too soft on the strong differences among records.

13.2.2 Backward Algorithm

$$\Delta\text{out}_{i(n)} = (t_i - u_i) \cdot f'(u_i) \tag{13.5}$$

$$\text{SelfMomentum}_{ij(n)} = \Delta w_{ij(n-1)} \cdot \left|\Delta\text{out}_{i(n)}\right| \cdot \frac{1}{0.5 + |w_{ij}|} \tag{13.6}$$

where $|w_{ij}|$ = absolute value of connection w_{ij}.

$$\Delta w_{ij(n)} = \text{SelfMomentum}_{ij(n)} + \Delta \text{out}_{i(n)} \cdot u_j \cdot \text{Rate}_i \tag{13.7}$$

$$\Delta b_{i(n)} = \text{Bias}_{i(n-1)} \cdot \left|\Delta \text{out}_{i(n)}\right| \cdot \frac{1}{1 + \left|\text{Bias}_{i(n)}\right|} + \Delta \text{out}_{i(n)} \cdot \text{Rate}_i \tag{13.8}$$

$$\text{Bias}_{i(n+1)} = \text{Bias}_{i(n)} + \Delta b_{i(n)} \tag{13.9}$$

$$w_{ij(n+1)} = w_{ij(n)} + \Delta w_{ij(n)} \cdot u_j \tag{13.10}$$

Suggestion 3: It is useful to initiate the *Rate* at very low values ($Rate < 1$). The learning process will be longer but more precise, and the measure's weights will be smaller.

Suggestion 4: It is useful not to allow an ANN to correct the reflexive weights ($i = j$); the learning will be longer and more complex, but the generated connections matrix will be more "refined" and then more efficacious when it is adapted as a constraint satisfaction weights matrix.

After having concluded the learning step, it is necessary to translate the weights matrix W of backpropagation ANN into a new matrix, New W, of constraint satisfaction. So, the bidirectional connections of the backpropagation will be reduced to symmetric connections by calculating the medium value:

$$\text{New } w_{ij} = \frac{w_{ij} + w_{ji}}{2} \tag{13.11}$$

$$\text{New } w_{ji} = \text{New } w_{ij} \tag{13.12}$$

$$\text{Bias}_i = \text{Bias}_i \tag{13.13}$$

At this point, the weights of the constraint satisfaction are defined.

13.3 Learning Through Symmetric Backpropagation

This learning technique is similar to BP with the previously described ΔRule. The only difference between the two techniques consists in the correction of their connections values. In classic BP, the correction takes place without considering the *symmetry* existing among the weights. In this case, indeed, the new weights are directly updated in the learning process by considering the error variation with respect to the couples of symmetric weights (Buscema 1998a):

$$\Delta w_{ij} = \frac{1}{2}\left(\Delta \mathrm{out}_i \cdot u_j + \Delta \mathrm{out}_j \cdot u_i\right) \tag{13.14}$$

$$\Delta w_{ij} = \Delta w_{ji} \tag{13.15}$$

13.4 The Prior Probability Algorithm (PPA)

A more rapid and sometimes more robust system for generating constraint satisfaction's weights matrix consists of using Bayesian's equations concerned with the probability of the positive and/or negative co-occurrence of all constraint satisfaction node couples in all the records (Rumelhart et al. 1986a). The reference equation for generating the weights matrix is the following:

$$w_{ij} = w_{ji} = -\log_{(n)} \frac{p\left(x_i = 0 \text{ and } x_j = 1\right) \cdot p\left(x_i = 1 \text{ and } x_j = 0\right)}{p\left(x_i = 1 \text{ and } x_j = 1\right) \cdot p\left(x_i = 0 \text{ and } x_j = 0\right)} \tag{13.16}$$

where x_i and x_j are the *i*th and *j*th node of constraint satisfaction and p is the co-occurrence probability of a certain event.

The calculation for Bias is done in the usual way:

$$\mathrm{Bias}_i = -\log \frac{p(x_i = 0)}{p(x_j = 1)} \tag{13.17}$$

These equations are utilizable for ANNs whose nodes have been designed both with Strategy A (*nodes number = summation of DB's record fields*) and with Strategy B (*nodes number = summation of all options of every DB's record field*). In fact, if we indicate the four different co-occurrence probabilities defined in the weight's equations in the following way:

- $p1 : (x_i = 0 \text{ and } x_j = 1)$
- $p2 : (x_i = 1 \text{ and } x_j = 0)$
- $p3 : (x_i = 1 \text{ and } x_j = 1)$
- $p4 : (x_i = 0 \text{ and } x_j = 0)$

then the specific probability of each node's couple could be calculated in the following way:

$$p1_{ij} = \frac{\sum_{m=1}^{M} \left(1 - x_{m_i}\right) \cdot x_{m_j}}{M} \tag{13.18}$$

$$p2_{ij} = \frac{\sum_{m=1}^{M} \left(1 - x_{m_j}\right) \cdot x_{m_i}}{M} \tag{13.19}$$

$$p3_{ij} = \frac{\sum_{m=1}^{M} x_{m_i} \cdot x_{m_j}}{M} \tag{13.20}$$

$$p4_{ij} = \frac{\sum_{m=1}^{M} (1 - x_{m_i}) \cdot \left(1 - x_{m_j}\right)}{M} \tag{13.21}$$

At this point, the weights matrix W is

$$w_{ij} = w_{ji} = -\log_{(n)} \frac{p1_{ij} \cdot p2_{ij}}{p3_{ij} \cdot p4_{ij}} \quad \text{if}\,(i = j) \text{ then } w_{ij} = 0. \tag{13.22}$$

In the same way, the Bias calculation can be carried out:

$$p5_i = \frac{\sum_{m=1}^{M} (1 - x_i)}{M} \tag{13.23}$$

$$p6_i = \frac{\sum_{m=1}^{M} x_i}{M} \tag{13.24}$$

$$\text{Bias}_i = -\log_{(n)} \frac{p5_i}{p6_i} \tag{13.25}$$

(For practical reasons, when a certain co-occurrence probability is 0, it is better to assume an artificial value of type 0.00001 for purposes of computability.)

Through this procedure, the whole weights matrix and the constraint satisfaction's bias vector are generated. This is also the case in which the nodes value of each model is a fuzzy value, closed in the space {0,1} (Buscema 1998a).

13.5 Algorithm of Constraint Satisfaction Network

At each cycle, the updating algorithm of the constraint satisfaction node is very elementary. Its philosophy is the following: It is assumed that each constraint satisfaction node is equivalent to a *hypothesis*. The weights connecting the nodes are then the *solidarity*, *contradictory*, and *fuzzy indifference* relations between every possible couple of hypotheses. Consequently, the bias of each node represents the *fuzzy inclination* of every hypothesis to contract, *generally*, solidarity or exclusion relations with the other hypotheses.

The constraint satisfaction aims to *maximize* the activation grade of each of its *hypotheses* (node) with respect to the constraints that the *relations* between each hypothesis and any other (weights and bias) impose upon them. This means that a constraint satisfaction provided with three nodes presents:

- Three hypotheses
- Three relations among different hypotheses
- Three thresholds, one for each hypothesis
- 2^3 combinations of different answers, if each hypothesis could assume only the values 0 or 1

If each original combination (disposition) of a binary answer (0 or 1) was a cube's edge, all the infinite solutions of constraint satisfaction would be included in the volume of a tridimensional cube (in fact, each node of the constraint satisfaction can assume values included between 0 and 1, limits included):

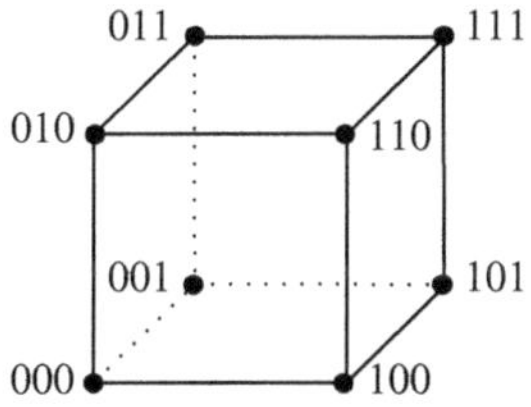

The updating algorithm of the constraint satisfaction units tends to find a closer solution to the edge "111," but it considers the weights connecting each node to any other.

Of course, if the nodes' number of the constraint satisfaction is 50, the solution space is equal to a 50th dimension hypercube space with 2^{50} different edges.

A practical use of the constraint satisfaction consists of assuming all its nodes are units directly externally manipulatable, as is an input unit. This gives an *arbitrary value* to one or more hypotheses to the external. This value indicates the *consistency* that is intended to give to that hypothesis. This has the aim to consider which other hypotheses will be able to activate themselves having fixed certain conditions from the external.

In this case, the updating algorithm of the constraint satisfaction units will always try to *maximize* the activation degree of each of its nodes. But this time, its work would not be simply constrained by the weights connecting the different hypotheses but from one or more external inputs that arbitrarily were activated and maintained actively during the work of the constraint satisfaction. This procedure allows for the testing of how different hypotheses groups are *optimized*, considering the whole context where they live. The updating algorithm of the constraint satisfaction units is composed of four steps:

- Calculation of the NetInput arriving to each unit
- Calculation of the updating Delta of each unit
- Updating of each unit
- Calculation of the reached maximization grade (goodness)

The NetInput calculation to each unit is calculated in the following way:

$$\mathrm{Net}_i = \sum_j^N u_j \cdot w_{ij} + \mathrm{Bias}_i + \mathrm{InputExt}_i \tag{13.26}$$

Actually, the most used parameters are included between 0 and 1 in order to scale the strength, both of the internal and external NetInput. In fact, if we define as *intr* the internal NetInput scaling and as *estr* the external NetInput scaling, then the previous equation will become

$$\mathrm{Net}_i = \mathrm{intr} \cdot \left(\sum_j^N u_j \cdot w_{ij} + \mathrm{Bias}_i \right) + \mathrm{estr} \cdot (\mathrm{InputExt}_i) \tag{13.26a}$$

Experience has currently taught us that good values for these two parameters are given by the two following equations:

$$\mathrm{intr} = \frac{1}{N}; \quad \mathrm{estr} = \frac{1}{\sqrt{N}} \tag{13.27}$$

where N = number of nodes.

Nevertheless, at the moment, there is a questionable area.

In order to modulate and/or contain the minimum values of the NetInput, other systems can be used in combination or as alternative to these two parameters. The most elementary method consists in normalizing the maximum and the minimum of its values in a linear way within predefined limits.

There is also the possibility to manipulate the NetInput through a semilinear function. A transfer function of the constraint satisfaction NetInput that was used for experiments giving satisfactory results is the *hyperbolic tangent* (Buscema 1998a):

$$\mathrm{Net'}_i = \frac{e^{\mathrm{Net}_i} - e^{-\mathrm{Net}_i}}{e^{\mathrm{Net}_i} + e^{-\mathrm{Net}_i}} \tag{13.28}$$

In this case, the values of each NetInput vary in a logistic way between −1 and +1. Obviously, these function's results can be normalized with some of the already considered equations.

The equation for the calculation of each unit's updating Delta is the following:

$$\begin{array}{ll} \mathrm{if}(\mathrm{Net}_i > 0) & \Delta_i = \mathrm{Net}_i \cdot (1 - u_i) \\ \mathrm{else} & \Delta_i = \mathrm{Net}_i \cdot u_i \end{array} \tag{13.29}$$

At this point, it is possible to update the units:

$$u_{i(n+1)} = u_{i(n)} + \Delta_i \tag{13.30}$$

The double branch of the penultimate equation must not be a "shock." Practically, it is necessary so that the units do not exceed the one limit of the lattice [0, 1].

The degree of goodness for a solution the constraint satisfaction method finds at every cycle is defined *Goodness*, $G_{(n)}$, where n is the actual cycle:

$$G_{(n)} = \sum_i \sum_j w_{ij} \cdot u_{i(n)} \cdot u_{j(n)} + \sum_i \text{Bias}_i - \sum_i \text{InputExt}_i \cdot u_{i(n)} \tag{13.31}$$

The question to consider is how much each node contributes to the maximization of the constraint satisfaction with respect to the external constraints (InputExt) established in the simulation.

It is evident that the Goodness values will be a floating point value because the constraint satisfaction moves within the n-dimensional volume designed by its nodes (Rumelhart et al. 1986a, b).

13.6 The Hidden Units of the Constraint Satisfaction Network

In a constraint satisfaction, the hidden units are not directly externally manipulatable. That is, though the hidden units cannot directly receive input values from the investigator, they can easily react to the NetInput produced on them from the other constraint satisfaction units. For example,

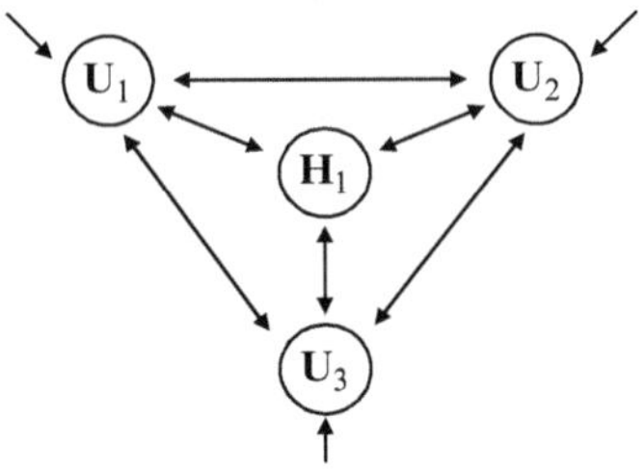

where U_i = input/output units of CS; H_i = hidden units of CS.

The hidden units of the constraint satisfaction have a *similar structure* but a *different function* with respect to the hidden units of a feedforward ANN.

The reference equation for their NetInput calculation is the following:

$$\text{Net}_i = \left(\sum_j^N u_j \cdot w_{ij} + \text{Bias}_j \right) \cdot \text{intr} \tag{13.32}$$

The only difference between these units and the units provided with inputs in a constraint satisfaction is that the first parameter of the external input cannot be considered (InputExt = 0).

Often it is useful to construct a constraint satisfaction possessing hidden units in order to codify the records of a DB. This requires the necessity of assigning a hidden unit to each record. The problem then consists of the way in which to connect the hidden units *among them* and *with all the other units* of the constraint satisfaction.

It was noticed that the solution of the second point also offers a solution to the first. There are more procedures for the choice of the weights' value between each *visible unit* of the constraint satisfaction and *each hidden unit* (Buscema 1998a). These are two of them:

13.6.1 Procedure 1: The Simple Transposition

In this case, each hidden unit that represents a record is connected with all the visible units by a weight whose value is similar to the one each visible unit had in the pattern corresponding to that particular record. For example,

Table R

		Fields				
		Male	*Female*	*20 years old*	*30 years old*	*40 years old*
Records	*Subject* 1	1	0	1	0	0
	Subject 2	0	1	0	1	0
	Subject 3	0	1	0	0	1

With such an elementary case, it is possible to proceed in the following manner:

(a) Construction of an auto-associative feedforward ANN with two layers in order to define the weights connecting the five options each other. It would be an ANN provided with five Inputs and five Outputs and charged with learning three Patterns; the learning process will conclude after generating 20 weights (w_{ij}) and five bias ($Bias_i$).

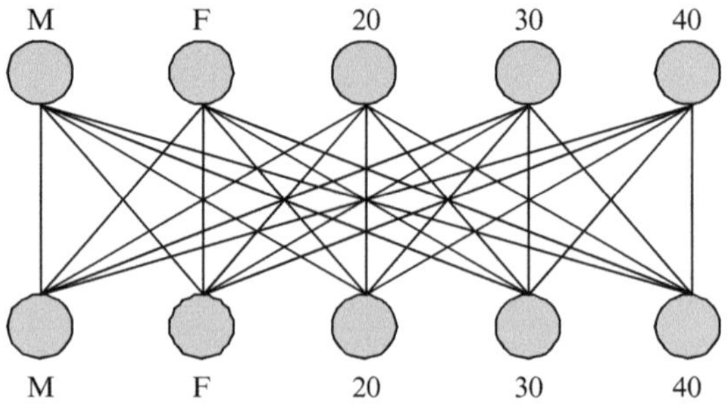

(b) Reduce the overall number of weights by

$$\text{if } (i \neq j) \quad \text{then} \quad \left\{ w'_{ij} = \frac{w_{ij} + w_{ji}}{2}; w'_{ji} = w'_{ij} \right\} \tag{13.33}$$

while the bias of the constraint satisfaction remains the same as the feedforward bias. In this way, the number of real weights of the constraint satisfaction becomes 10, plus five bias.

(c) Construction of the hidden units through *simple transposition.*

As there are only three DB records, the hidden units to be added to constraint satisfaction will also be three. This is defined as a weights matrix W^H that connects each hidden unit with each of the five visible units. The "Table R" becomes the content of the weights matrix W^H with the sole nominal difference that each table row contains the weights value W^H retained through the five visible units of the constraint satisfaction and each hidden unit represents a specific record. The bias of the three hidden units either are put equal to zero or can be computed through one of the previously analyzed co-occurrence equations.

13.6.2 Procedure 2: The Weighted Transposition

In order to activate this procedure, it is necessary to follow the same steps in the previous procedure. However, at the last step, *Table R* is not directly identified with the weights matrix W^H. If *Table R* is identified with a matrix called T^R, then the following equation regulates the values passage from the matrix T^R to the weights W^H:

$$w_{ij}^H = f^m\left(T_{ij}^R\right) \tag{13.34}$$

The function f^m rewrites the matrix values T^R according to the value of the weights matrix W, which interconnects the visible units of the constraint satisfaction among them. Then,

$$w_{ij}^H = f^m\left(T_{ij}^R\right) = \text{Scale} \cdot T_{ij}^R + \text{Offset} \tag{13.35}$$

where

$$\text{Scale} = \frac{\text{high} - \text{low}}{\text{Max} - \text{Min}} \tag{13.36}$$

$$\text{Offest} = \frac{\text{Max} \cdot \text{low} - \text{Min} \cdot \text{high}}{\text{Max} - \text{Min}} \tag{13.37}$$

Parameters "Max" and "Min" indicate the maximum and minimum input values of function f^n, while parameters "high" and "low" indicate the maximum and minimum Output values of function f^n. The problem consists in how to determine these four parameters.

The problem is already solved for the Input values Min and Max. Each value of T^R can vary between 0 and 1 and then Max = 1.0 and Min = 0.0.

There are at least three options for the values of the parameters high and low:

$$\begin{aligned} \textit{Maximization} \text{ of weights } W: \quad & \text{high} = \text{Max}\left[\left|w_{ij}\right|\right] \\ & \text{low} = -\text{high} \end{aligned} \tag{13.38}$$

In this case, the absolute value is chosen as it is higher than the weights matrix W connecting the visible units and assumed to be the maximum border, while its inverse is considered to be the minimum border of the new weights W^H.

$$\begin{aligned} \textit{Minimization} \text{ of weights } W: \quad & \text{high} = \text{Min}\left[\left|w_{ij}\right|\right] \quad w_{ij} \neq 0.0 \\ & \text{low} = -\text{high} \end{aligned} \tag{13.39}$$

The procedure is similar to the previous one, the only difference being the exit borders of function f^n which are represented by the weights matrix W that is smaller and different from zero.

$$\begin{aligned} \text{Mean } \textit{weighting} \text{ of weights } W: \quad & \text{high} = \overline{x} \\ & \text{low} = -\text{high} \end{aligned} \tag{13.40}$$

where $\overline{x} = \frac{\sum_i^N \sum_i^N \left|w_{ij}\right|}{N^2}$ with $i \neq j$

In this instance, the mean of the absolute values of the weights matrix W is calculated leaving the principal diagonal of the matrix W out of this calculation.

Hundreds of experiments were carried out with both of these procedures at Semeion. Because of our experiments, we are convinced that the *weighed transposition* procedure has been shown to be the best strategy aimed at varying the number of fields and the number of records for any DB.

The reason for this could be the following: In the simple transposition, the hidden units function as *passive units* with respect to the visible nodes corresponding to properties that do not characterize the record represented by the hidden unit, or they exercise a moderately excitatory strength over those visible units characterizing the record they represent.

In contrast with this, the hidden units in the *weighted transposition* act more or less *inhibitively*. This should allow a greater filter capacity on the hidden and visible units when the constraint satisfaction has to manage particularly complex DBs. This is of great dimension and contains records presenting very fuzzy diversifications.

It is not necessary to give an opinion about the efficacy of the three proposed options aimed at making the weighted transposition. From our point of view, they represent three real options. Each one can be useful in analyzing the answers that the constraint satisfaction provides to its hidden units' different intensities of influence on the other visible units.

The analysis of these two procedures permits the deduction of different ways to connect each of the hidden units with their pertinent weights. In choosing the *simple transposition*, it is appropriate to conceptualize the hidden units as a unique units pool that is in competition among them. In this case, the weights matrix W^{HH} interconnecting them should be filled entirely with values corresponding to the inverse of the maximum value, each visible node having a possible input. Therefore,

$$w_{kp}^{HH} = -1 \tag{13.41}$$

where $w_{kp} = w_{pk}$ and $k \neq p$.

A similar solution was adopted by McClelland and Rumelhart for IAC ANNs. Otherwise, if a *weighted transposition* is adopted, four options can be utilized; the first three are similar to those previously described: maximization, minimization, and mean weighting of the weights matrix W:

$$w_{kp}^{HH} = -\left(\mathrm{Max}_{kp}\left[\left|w_{ij}\right|\right]\right) \quad \{\text{maximized competition}\} \tag{13.42}$$

$$w_{kp}^{HH} = -\left(\mathrm{Min}_{kp}\left[\left|w_{ij}\right|\right]\right) \quad \{\text{minimized competition}\} \tag{13.43}$$

$$w_{kp}^{HH} = -\left(\frac{\sum_i^N \sum_j^N \left|w_{ij}\right|}{N^2}\right) \quad \text{with } i \neq j \quad \{\text{weighted competition}\} \tag{13.44}$$

The fourth option consists of annulling the competition among the hidden units of the constraint satisfaction:

$$w_{kp}^{HH} = 0.0 \quad \{\text{Null competition}\} \tag{13.45}$$

These four options are also useful as a filter system through the distribution of the constraint satisfaction answers to the different questions.

13.7 Final Considerations

The constraint satisfaction are ANNs that try to maximize their node activations beginning with the entire set of constraints. The first step consists in understanding which problems represent a suitable fit for being treated with this kind of ANN; more precisely, it is necessary to understand how any problem must be treated so that it can be analyzed with a constraint satisfaction.

It could be asserted that each problem of *resource optimization* and *data profiling* has to be explored with a constraint satisfaction. The effectiveness of the solutions an ANN is able to produce depends on a series of factors. The *first* of these *factors* is the *data representativity* in the simulation model; in order to start this process, it is necessary *to atomize* the original problem into the smallest component that will represent the whole of the atomic hypotheses of the problem itself. In addition to the *atomization* principle, already known in other contexts, it is also useful to take into consideration the *principle of data variety*. The hypotheses that will be part of the constraint satisfaction must not only be those considered "more incisive" in defining the problem but also all those hypotheses appearing in the problem space-time even if they are considered not to be determined. The ANN utilized to identify the weights of the constraint satisfaction will establish the, more or less, strong significance of each hypothesis with respect to any other. In sum, the data representativity consists in predisposing the *most complete nodes model* of the atomic variables in the constraint satisfaction able to define the real problem.

The *second factor* is the effectiveness of constraint satisfaction that is dependent on the implementation of the *weight generation system*. In this sense, the learning that occurs with the weights matrix, like that of Bayesian equations, has been shown to be simple and robust (Eqs. 13.16, 13.17, 13.18, 13.19, 13.20, 13.21, 13.22, 13.23, 13.24, 13.25).

The *third* fundamental *factor* for the functioning of a constraint satisfaction is the choice of the *updating algorithm* of the units. Here we have presented a modified version (Eqs. 13.29, 13.30, 13.31, 13.32, 13.33, 13.34) of the classic algorithm proposed by Rumelhart and his collaborators.

The *fourth factor* that is critical in deciding the performance quality of a constraint satisfaction concerns the weights matrix connecting the constraint satisfaction hidden units with each other and with the other visible units. In this respect, the new algorithm has been shown to produce interesting results (Eqs. 13.35, 13.36, 13.37, 13.38, 13.39, 13.40, 13.41, 13.42, 13.43, 13.44, 13.45, 13.46, 13.47, 13.48).

References

Buscema, M. (1995). Self-reflexive networks, theory, topology, applications. *Quality & Quantity, 29*(4), 339–403.

Buscema, M. (1998a). Constraint satisfaction neural networks. *Substance Use & Misuse, 33*(2), 389–408.

Buscema, M. (1998b). Back propagation neural networks. *Substance Use & Misuse, 33*(2), 233–270.

Hebb, D. O. (1949). *The organization of behavior*. New York: Wiley.

Hopfield, J. J. (1982). Neural networks and physical systems with emergent collective computational abilities. *Proceedings of the National Academy of Sciences, USA, 79*, 2554–2558.

Hopfield, J. J. (1984). Neurons with graded response have collective computational properties like those of two-state neurons. *Proceedings of the National Academy of Sciences, USA, 81*, 3088–3092.

Rumelhart, D. E., Hinton, G. E., & Williams, R. J. (1986a). Learning internal representations by error propagation. In D. E. Rumelhart & J. L. McClelland (Eds.), *Parallel distributed processing* (Foundations, explorations in the microstructure of cognition, Vol. 1). Cambridge, MA/London: MIT Press.

Rumelhart, D. E., Smolensky, P., McClelland, J. L., & Hinton, G. E. (1986b). Schemata and sequential thought processes in PDP models. In J. L. McClelland & D. E. Rumelhart (Eds.), *PDP, exploration in the microstructure of cognition* (Vol. II, pp. 7–57). Cambridge, MA: MIT Press.

Chapter 14
Application of the Constraint Satisfaction Network

Marco Intraligi and Massimo Buscema

14.1 Introduction

The previous chapter described the theoretical underpinnings of the constraint satisfaction network (Buscema 1998 and Rumelhart et al. 1986). In this chapter, we describe the application of the constraint satisfaction (CS) network to a dataset comprising 144 variables for 1,120 cases representing arrests for drug trafficking offenses in the 32 boroughs of London. This dataset was extracted in June 2006 when the situation of the CDTD (Central Drugs Trafficking Database) was 1,590 tactic sequences, 1,667 individuals, 1,190 accused individuals, and 70 incomplete cases.

The 144 variables described in Table 14.1 include:

- Sex (male, female, unknown)
- Arrestee's borough of residence (32 boroughs + 1 not available)
- Borough where each individual was arrested (32 boroughs + 1 not available)
- Nationality of the arrestee
- Ethnic appearance of the arrestee (as determined by the police)
- Self-defined ethnic classification of the arrestee
- Age (expressed in six categories)
- Number of convictions[1]

[1]In English law, the word conviction refers to the outcome of a criminal prosecution which concludes in a judgment or finding that the defendant is guilty of the crime charged. The term

M. Intraligi (✉)
Semeion Research Centre, Rome, Italy
e-mail: m.intraligi@semeion.it

M. Buscema
Department of Mathematical and Statistical Sciences, University of Colorado at Denver, Denver, CO, USA
e-mail: m.buscema@semeion.it

M. Buscema and W.J. Tastle (eds.), *Intelligent Data Mining in Law Enforcement Analytics: New Neural Networks Applied to Real Problems*, DOI 10.1007/978-94-007-4914-6_14,

Table 14.1 List of the 144 variables of the dataset

List of 144 variables		
1 Sex_Male	49 (EA1)_White_European	97 AR_PL_Haringey
2 Sex_Female	50 (EA2)_Dark_European	98 AR_PL_Harrow
3 Sex_notknown	51 (EA3)_Afro-Caribbean	99 AR_PL_Havering
4 Barking_and_Dagenham	52 (EA4)_Asia	100 AR_PL_Hillingdon
5 Barnet	53 (EA5)_Oriental	101 AR_PL_Hounslow
6 Bexley	54 (EA6)_Arab	102 AR_PL_Islington
7 Brent	55 Age(<18)	103 AR_PL_Kensington_and_Chelsea
8 Bromley	56 Age(18–21)	104 AR_PL_Kingston_upon_Thames
9 Cam den	57 Age(21–25)	105 AR_PL_Lambeth
10 Croydon	58 Age(25–35)	106 AR_PL_Lewisham
11 Ealing	59 Age(35–45)	107 AR_PL_Merton
12 Enfield	60 Age(>45)	108 AR_PL_Newham
13 Greenwich	61 ConvictionsNumber	109 AR_PL_Redbridge
14 Hackney	62 OffensesNumber	110 AR_PL_Richmond_upon_Thames
15 Hammersmith_and_Fulham	63 Off_FirstConvAge	111 AR_PL_Southwark
16 Haringey	64 Off_LastConvAge	112 AR_PL_Sutton
17 Harrow	65 Off_Drug	113 AR_PL_Tower_Hamlets
18 Havering	66 Off_TheftKindred	114 AR_PL_Waltham_Forest
19 Hillingdon	67 Off-AgainstPerson	115 AR_PL_Wandsworth
20 Hounslow	68 Off-OffensiveWeapons	116 AR_PL_Westminster
21 Islington	69 Off-Sexual	117 AR_PL_NA
22 Kensington_and_Chelsea	70 Off-RelatedToPolice	118 NumOfDrugSeizures
23 Kingston_upon_Thames	71 Off-Fraud	119 Cannabis
24 Lambeth	72 Off-Total	120 Cocaine
25 Lewisham	73 Off-AgainstProperty	121Crack
26 Merton	74 NumOfArrests	122 Heroin_Diamorphine

27 Newham	75 AR_OFF_Theft_and_Kindred_Offenses	123 MDMA
28 Redbridge	76 AR_OFF_Offenses_the_Person_Offenses	124 NumOfCashSeizures
29 Richmon_upon_Thames	77 AR_OFF_Drug_trafficking_Offenses	125 Pounds
30 Southward	78 AR_OFF_Drug_Possession_Offenses	126 NumOfTactics
31 Sutton	79 AR_OFF_Other_Drug_Offenses	127 Non-Law_Enforcement_Agent
32 Tower_Hamlets	80 AR_OFF_Offensive_Weapon_Offenses	128 Other_Law_Enforcement_Agent
33 Waltham_Forest	81 AR_OFF_Firearms_Offenses	129 Police
34 Wandsworth	82 AR_OFF_Kidnapping_and_Abduction_Offenses	130 Search_of_Object
35 Westminster	83 AR_OFF_Other_violent_Offenses	131 Search_of_Person
36 Boroughs_NA	84 AR_OFF_Other_Offenses	132 Search_of_Premises
37 AFR	85 AR_PL_Barking_and_Dagenham	133 Covert_Purchase
38 ASIA	86 AR_PL_Barnet	134 Controlled_Delivery
39 EASTEU	87 AR_PL_Bexley	135 Other_Generic_Tactic
40 EU	88 AR_PL_Brent	136 NumOfTacticSequences
41 IRE	89 AR_PL_Bromley	137 InOperation
42 JAM	90 AR_PL_Camden	138 ViolentOnArrest
43 ME	91 AR_PL_Croydon	139 ArrMode_NA
44 NK	92 AR_PL_Ealing	140 ArrMode_Direct
45 SAME	93 AR_PL_Enfield	141 ArrMode_Result_of_Enquiries
46 TU-CY	94 AR_PL_Greenwich	142 ArrMode_Given_into_custody
47 UK	95 AR_PL_Hackney	143 ArrMode_Other
48 VTN	96 AR_PL_Hammersmith_and_Fulham	144 OnBailAtTimeOfOffence

- Number of offenses[2]
- Specific details of offenses (e.g., theft, burglary, unlawful possession of a controlled drug)
- Number of arrests
- Mode of arrest (e.g., direct, as a result of enquiries, given into custody)
- Kinds of generic tactics used by the police (e.g., search of person, search of premises, covert drug purchase)
- Aims of the police operations
- Behavior of the person at the time of the arrest (e.g., tried to escape, attempted to discard drugs, resisted arrest)
- Type(s) of drug(s) seized from the arrestee
- Number of drug seizures

14.2 Variables and Analysis

The database was used to generate a *connections matrix* between all the variables through the *prior probability algorithm* (Buscema 1998). Once generated, the connections matrix was queried through a CS network.

Each query produces an output vector equal to the number of the variables which can be considered as the prototype resulting from the query formulated. Therefore, for each query, there will be a certain number of variables activated with values between 0 and 1.

Overall, 77 different queries were asked:

- One for each borough of residence of the arrestee (32)
- One for each borough where the arrest occurred (32)
- One for each type of drug seized (5)
  - Cannabis
  - Cocaine
  - Crack
  - Heroin
  - Ecstasy-type drugs (e.g., MDMA, MDA, MDEA)
- One for each type of generic tactic used by the police (6)
  - Search of object
  - Search of person
  - Search of premises
  - Covert purchase

---

summary conviction refers to the consequence of a trial before a court or magistrate, without a jury, which generally involves a minor offense.

[2]The word *offense is* synonymous with the word *crime* in English law.

Table 14.2 Variables never active

List of variables *never* activated in the 77 queries			
36	Boroughs_NA	53	(EA5)_Oriental
37	AFR	54	(EA6)_Arab
38	ASIA	55	Age(<18)
40	EU	64	Off_LastConvAge
41	IRE	79	AR_OFF_Other_Drug_Offenses
43	ME	117	AR_PL_NA
44	NK	127	Non-Law_Enforcement_Agent
45	SAME	128	Other-Law_Enforcement_Agent
48	VTN		

  – Controlled delivery
  – Other generic tactic

• One for each sex (2)

The results obtained are shown in the tables below, one for each interrogation, where the activated variables having a value greater than 0 are listed.

The 77 answers of the CS were organized into a single dataset and processed with the AutoCM network (Buscema and Sacco 2010) in order to obtain a distance matrix and generate a minimum spanning tree (MST) graph (Kruskal 1956). Then, by applying the evolutionary GenD algorithm (Buscema 2004), the 77 answers were organized into clusters. Following this process, the MST graph enables us to see which boroughs have similar characteristics.

Seventeen of the 144 variables in the dataset were never activated following the 77 queries. These are shown in Table 14.2.

Figure 14.1 shows 5 clusters identified by the GenD algorithm among all the 77 answers of the CS network to the same number of queries. It is important to remember that the MST graph was calculated on the distance matrix obtained from the weights matrix of an AutoCM network trained on the dataset of the CS network's answers. Therefore, it is assumed that when the variables belonging to each cluster are formulated as an external input for the CS network, they activate more or less the same variables of the dataset. This datum was highlighted, creating a table for each cluster that lists all the variables activated with relative value.

Next, we show the tables relating to each cluster with the activated variables and respective values (Table 14.3) (Fig. 14.2).

14.2.1 Comments on Cluster 1

Cluster 1 consists of 11 queries that overall activate 29 variables (see Table 14.4).

The queries activating the greatest number of variables are *AR PL* (arrest place) *Greenwich* and *Barking and Dagenham* (14 out of 29) followed by the queries *Bexley*, *Greenwich*, *Cannabis,* and *AR PL Bexley* (13 out of 29). The queries activating the least number of variables are *Camden* (5) and *AR PL Hillingdon* (6) (Table 14.5).

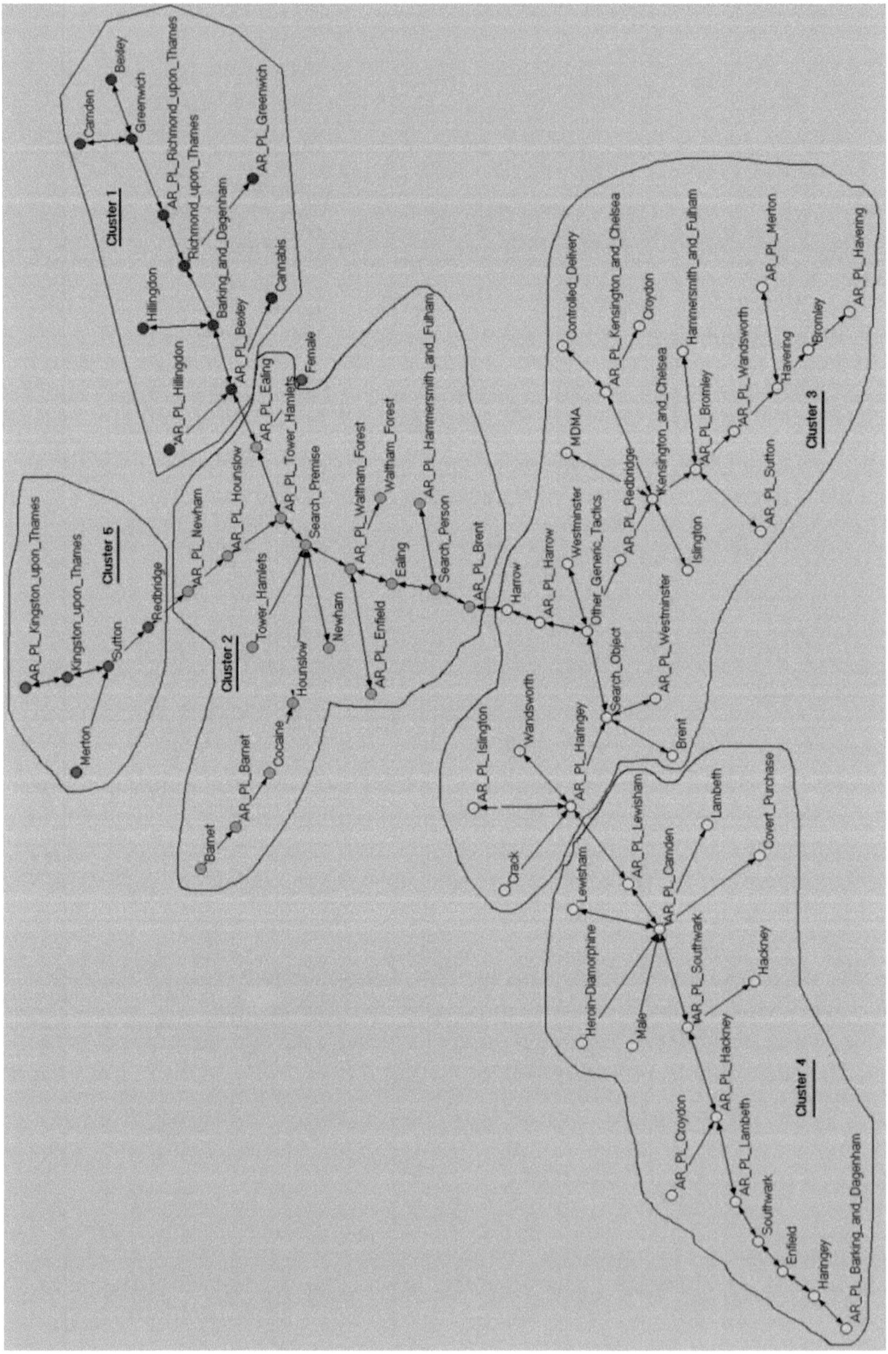

Fig. 14.1 MST graph of the 77 answers of the CS network arranged in 5 clusters

Table 14.3 Table with the activated variables in the 11 queries of cluster 1

Num	Cluster 1 – variables activated	Camden	Bexley	Green-wich	AR_PL_ Richmond_ upon_Thames	Richmond_ upon_Thames	AR_PL_ Greenwich	Barking_and_ Dagenham	Hilling-don	AR_PL_ Hillingdon	Cann-abis	AR_PL_ Bexley
1	Sex_Male	1	1	1	1	1	1	1	1	1	1	1
4	Barking_and_ Dagenham							1				
6	Bexley		1	1	1	1	1				1	1
9	Camden	1										
13	Greenwich			1								
17	Harrow							1			1	1
19	Hillingdon								1	1		
29	Richmond_upon_ Thames		1	1	1	1	1	1	1			
39	EASTEU	1										
47	UK		1	1	1	1	1	1	1	1	1	1
49	(EA1)_White_ European	1	1	1	1	1	1	1	1			
52	(EA4)_Asia									1	1	1
56	Age(18–21)		1	1	1	1	1	1	1	1	1	1
58	Age(25–35)	1										
75	AR_OFF_Theft_and_ Kin- dred_Offenses		1	1	1	1	1					
80	AR_OFF_Offensive_ Weapon_Offenses		1								1	1
84	AR_OFF_Other_ Offenses		1	1			1					
85	AR_PL_Barking_ and_Dagenham							1				

(continued)

Table 14.3 (continued)

Num	Cluster 1 – variables activated	Camden	Bexley	Green-wich	AR_PL_Richmond_upon_Thames	Richmond_upon_Thames	AR_PL_Greenwich	Barking_and_Dagenham	Hilling-don	AR_PL_Hillingdon	Cann-abis	AR_PL_Bexley
87	AR_PL_Bexley		1	1	1	1	1				1	1
94	AR_PL_Greenwich						1					
98	AR_PL_Harrow							1			1	1
100	AR_PL_Hellingdon								1	1		
110	AR_PL_Richmond_upon_Thames		1	1	1	1	1	1	1			
119	Cannabis		1	1	1	1	1	1			1	1
124	NumOfCashSeizures							1	1		1	1
125	Pounds								1			
131	Search_of_Person										1	1
132	Search_of_Premises		1	1	1	1	1	1	1		1	1
139	ArrMode_NA				1	1	1	1	1			

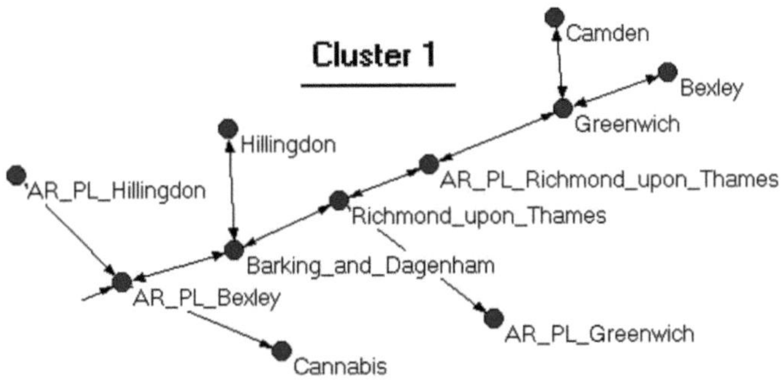

Fig. 14.2 Detail taken from the global MST graph – cluster 1

Table 14.4 Number and % of variables activated by each query

Cluster 1 – queries	Variables activ.	%
AR_PL_Greenwich	14	48.3
Barking_and_Dagenham	14	48.3
Bexley	13	44.8
Greenwich	13	44.8
Cannabis	13	44.8
AR_PL_Bexley	13	44.8
AR_PL_Richmond_upon_Thames	12	41.4
Richmond_upon_Thames	12	41.4
Hillingdon	12	41.4
AR_PL_Hillingdon	6	20.7
Camden	5	17.2

The variables, in order of magnitude, which are activated on more than six occasions, are *Sex Male*, *UK*, *Age (18–21)*, *Search of Premises*, *EA1-White_European*, *Cannabis*, *Bexley*, *Richmond upon Thames*, *AR PL Bexley*, and *AR PL_Richmond upon-Thames*. It is possible to consider these variables as the most prototypical of this cluster.

It is interesting to note that the activated variables following the inclusion of *Cannabis* are the same as the activated ones following the inclusion of *AR PL Bexley* (see Tables 14.3 and 14.6) (Fig. 14.3).

14.2.2 Comments on Cluster 2

Cluster 2 is formed by 18 queries that overall activate 53 variables (Table 14.7).

In this case, all the queries activate more than 50 % of the variables, except *AR PL Ealing* which activates 47 % (25 out of 53). Among others, *AR PL Ealing* is the variable-query in this cluster which connects to cluster 1. The query that

Table 14.5 Number and % of activations of the 29 variables

Cluster 1 – variables activated	Num activ.	%
Sex_Male	11	100
UK	10	91
Age(18–21)	10	91
Search_of_Premises	9	82
(EA1)_White_European	8	73
Cannabis	8	73
Bexley	7	64
Richmond_upon_Thames	7	64
AR_PL_Bexley	7	64
AR_PL_ Richmond_upon_Thames	7	64
AR_OFF_Theft_and_Kindred_Offenses	5	45
ArrMode_NA	5	45
NumOfCashSeizures	4	36
Harrow	3	27
(EA4)_Asia	3	27
AR_OFF_Offensive_Weapon_Offenses	3	27
AR_OFF_Other_Offenses	3	27
AR_PL_Harrow	3	27
Hillington	2	18
AR_PL_Hillingdon	2	18
Search_of_Person	2	18
Barking_and_Dagenham	1	9
Camden	1	9
Greenwich	1	9
EASTEU	1	9
Age(25–35)	1	9
AR_PL_Barking_and_Dagenham	1	9
AR_PL_Greenwich	1	9
Pounds	1	9

activates the most variables is *Barnet* followed by *Waltham Forest* and *Hounslow* (Table 14.8).

There are 14 variables always activated (to the 18 queries). In all, 32 variables activated over 50 % of the queries.

In contrast to cluster 1, the activated variables relating to convictions and offenses are important in terms of describing the characteristics of the prototypical arrestee found in cluster 2.

Harrow is the borough most activated, both as a place of arrest and of residence, together with Tower_Hamlets (mostly as the place of arrest). The interesting datum is that Harrow is not part of the queries of this cluster; rather, it belongs to cluster 3, and in the MST graph, it is the vertex closest to cluster 2. It can be said that the prototypical variables of cluster 2 are very similar to those activated by the Harrow and AR_PL_Harrow queries (see Tables 14.7 and 14.9).

Table 14.6 Table with the variables activated in the 18 queries of cluster 2

Num	Cluster 2 – variables activated	AR_PL_Ealing	AR_PL_Tower_Hamlets	AR_PL_Hounslow	AR_PL_Newham	Search Premise	Tower_Hamlets	Newham	Hounslow	Cocaine	AR_PL_Barnet	Barnet	AR_PL_Waltham_Forest	AR_PL_Enfield	Waltham_Forest	Ealing	Search Person	AR_PL_Hammersmith_and_	AR_PL_Brent
1	Sex Male	1	1	1	1	1	1	1	1	1	1	1	1	1	1	1	1	1	1
5	Barnet											1							
11	Ealing															1			
17	Harrow	1	1			1	1	1	1	1	1	1	1	1	0.8	1	1	1	1
20	Hounslow								1										
27	Newham							1											
32	Tower_ Hamlets		1	1	1	1	1												
33	Waltham_ Forest														1				
47	UK	1	1	1	1	1	1	1	1	1	1	1	1	1	1	1	1	1	1
51	(EA3)_Afro-Caribbean												1	1	1	1	1	1	1
52	(EA4)_Asia	1	1	1	1	1	1	1	1	1	1	1							
57	Age(21–25)	1	0.5	1	1		0.5		0.1			0.4			0.8				
61	Convictions Number		1	1	1	1	1	1	1	1	1	1	1	1	1	1	1	1	1
62	Offenses Number		1	1	1	1	1	1	1	1	1	1	1	1	1	1	1	1	1
65	Off_Drug	0.2	1	1	1	1	1	1	1	1	1	1	1	1	1	1	1	1	1
66	Off_ Theft Kindred		1	1	1	1	1	1	1	1	1	1	1	1	1	1	1	1	1

(continued)

Table 14.6 (continued)

Num	Cluster 2 – variables activated	AR_PL_Ealing	AR_PL_Tower_Hamlets	AR_PL_Hounslow	AR_PL_Newham	Search Premise	Tower_Hamlets	Newham	Hounslow	Cocaine	AR_PL_Barnet	Barnet	AR_PL_Waltham_Forest	AR_PL_Enfield	Waltham_Forest	Ealing	Search Person	AR_PL_Hammersmith_and_	AR_PL_Brent
67	Off_ Against Person		1	1	1	1	1	1	1	1	1	1	1	1	1	1	1	1	1
68	Off_Offensive Weapons		1	1	1	1	1	1	1	1	1	1	1	1	1	1	1	1	1
70	Off_RelatedToPolice		1	1	1	1	1	1	1	1	1	1	1	1	1	1	1	1	1
71	Off_Fraud		1	1	1	1	1	1	1	1	1	1	1	1	1	1	1	1	1
72	Off_Total		1	1	1	1	1	1	1	1	1	1	1	1	1	1	1	1	1
73	Off_AgainstProperty		1	1	1	1	1	1	1	1	1	1	1	1	1	1	1	1	1
74	NumOfArrests	1	1	1	1	1	1	1	1	1	1	1	1	1	1	1	1	1	1
76	AR_OFF_Offenses_the_Person_ Offenses																1	1	1
77	AR_OFF_ Drug_trafficking_Offenses		1	0.3		1	1	1	1	1	1	1	1	1	1	1	1	1	1
78	AR_OFF_ Drug_Possession_Offenses	1	1	1	1	1	1	1	1	1	1	1	1	1	1	1			

80	AR_OFF_ Offensive_ Weapon_ Offenses	1	1	1	1	1	1	1	1	1	1	1	1	1	1	1	1	1	1
81	AR_OFF_ Firearms_ Offenses	1	1	1	1	1	1	1	1	1	1	1	1	1	1	1	1	1	1
82	AR_OFF_ Kidnapping_ and_Abduction_ Offenses	1													1				
83	AR_OFF_ Other_ violent_ Offenses	1	1	1	1	1	1	1	1	1	1	1	1	1	1	1	1	1	1
86	AR_PL_Barnet	1									1	0.1							
88	AR_PL_Brent								1	1	1	1	0.1				0.2		1
92	AR_PL_Ealing	1																	
93	AR_PL_Enfield													1					
96	AR_PL_ Hammersmith_ and_Fulham																	1	
98	AR_PL_Harrow	1	1	1	1	1	1	1	1	1	1	1	1	1	1	1	1	1	1
101	AR_PL_Hounslow			1															
108	AR_PL_Newham				1														
113	AR_PL_ Tower_ Hamlets	0.2	1	1	1	1	1	1	1	1	1	1	1	1	1		1		
114	AR_PL_ Waltham_ Forest	1									1		1						

(continued)

Table 14.6 (continued)

Num	Cluster 2 – variables activated	AR_PL_Ealing	AR_PL_Tower_Hamlets	AR_PL_Hounslow	AR_PL_Newham	Search Premise	Tower_Hamlets	Newham	Hounslow	Cocaine	AR_PL_Barnet	Barnet	AR_PL_Waltham_Forest	AR_PL_Enfield	Waltham_Forest	Ealing	Search Person	AR_PL_Hammersmith_and_	AR_PL_Brent
118	NumOfDrugSeizures						1	1	1	1									
119	Cannabis	1								0.5			0.5						
120	Cocaine							1											
124	NumOfCashSeizures	1	1	1	1	1	1	1	1	1	1	1	1	1	1	1	1	1	1
126	NumOfTactics	1	1	1	1	1	1	1	1	1	1	1	1	1	1	1	1	1	1
129	Police	1	1	1	1	1	1	1	1	1	1	1	1	1	1	1	1	1	1
131	Search_of_Person	1	1	1	1	1	1	1	1	1	1	1	1	1	1	1	1	1	1
132	Search_of_Premises	1	1	1	1	1	1	1	1	1	1	1	1	1	1	1			
135	Other_Generic_Tactic		1	1	1	1	1	1	1	1	1	1	1	1	1	1	1	1	1
138	ViolentOnArrest		1	1		1	1	1	1			0.4	1	1	1	1	1	1	1
139	ArrMode_NA		1	1	1	1	1	1	1	1	1	1	1	1	1	1	1	1	1
140	ArrMode_Direct	1	1	1	1	1	1	1	1	1	1	1	1	1	1	1	1	1	1
144	OnBailAtTimeOfOffence	1	1	1	1	1	1	1	1	1	1	1	1	1	1	1	1	1	1

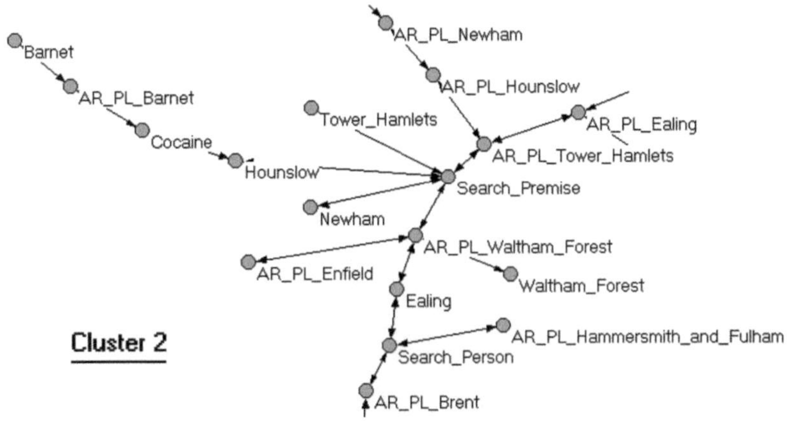

Fig. 14.3 Detail taken from the global MST graph – cluster 2

Table 14.7 Number and % of variables activated by each query

Cluster 2 – queries	Variables activ.	%
Barnet	38	72
Waltham_Forest	37	70
Hounslow	36	68
AR_PL_Tower_Hamlets	34	64
AR_PL_Hounslow	34	64
Tower_Hamlets	34	64
Cocaine	34	64
AR_PL_Barnet	34	64
AR_PL_Waltham_Forest	34	64
Search Premise	33	62
Newham	33	62
AR_PL_Enfield	33	62
Ealing	33	62
AR_PL_Newham	32	60
Search Person	32	60
AR_PL_Brent	32	60
AR_PL_Hammersmith_and_Fulham	31	58
AR_PL_Ealing	25	47

The tactics activated are *Search of Person*, *Other Generic*, and *Search of Premises*. Asian and Afro-Caribbean are the most activated ethnicities, though little. It is interesting to note that in this cluster the activated variables following the inclusion of the variable *Cocaine* are the same as those activated following the AR_PL_Barnet input (Table 14.7). This does not mean that the two variables are directly connected, and in fact, no one of them activates the other. However, this detail shows that the characteristics of the persons arrested and of the tactics used in the borough of Barnet are the same emerging when cocaine is seized in the other boroughs (Fig. 14.4).

Table 14.8 Number and % of activations of the 53 variables

Cluster 2 – variables activated	Num activ.	%	Cluster 2 – variables activated	Num activ.	%
Sex_Male	18	100	AR_OFF_Drug_trafficking_Offenses	16	89
UK	18	100	AR_OFF_Drug_Possession_Offenses	15	83
Off_Drug	18	100	Search_of_Premises	15	83
NumOfArrests	18	100	ViolentOnArrest	14	78
AR_OFF_Offensive_Weapon_Offenses	18	100	(EA4)_Asia	11	61
AR_OFF_Firearms_Offenses	18	100	Age(21–25)	8	44
AR_OFF_Other_violent_Offenses	18	100	(EA3)_Afro-Caribbean	7	39
AR_PL_Harrow	18	100	AR_PL_Brent	7	39
NumOfCashSeizures	18	100	Tower_Hamlets	5	28
NumOfTactics	18	100	NumOfDrugSeizures	4	22
Police	18	100	AR_OFF_Offenses_the_Person_Offenses	3	17
Search_of_Person	18	100	AR_PL_Barnet	3	17
ArrMode_Direct	18	100	AR_PL_Waltham_Forest	3	17
OnBailAtTimeOfOffence	18	100	Cannabis	3	17
ConvictionsNumber	17	94	AR_OFF_Kidnapping_and_Abduction_Offenses	2	11
OffensesNumber	17	94	Barnet	1	6
Off_TheftKindred	17	94	Ealing	1	6
Off_AgainstPerson	17	94	Hounslow	1	6
Off_OffensiveWeapons	17	94	Newham	1	6
Off_RelatedToPolice	17	94	Waltham_Forest	1	6
Off_Fraud	17	94	AR_PL_Ealing	1	6
Off_Total	17	94	AR_PL_Enfield	1	6
Off_AgainstProperty	17	94	AR_PL_Hammersmith_and_Fulham	1	6
AR_PL_Tower_Hamlets	17	94	AR_PL_Hounslow	1	6
Other_Generic_Tactic	17	94	AR_PL_Newham	1	6
ArrMode_NA	17	94	Cocaine	1	6
Harrow	16	89			

Table 14.9 Table with the variables activated in the 26 queries of cluster 3

Num	Cluster 3 – variables activated	Harrow	AR_PL_Harrow	Other Generic Tactics	Westminster	AR_PL_Red-bridge	Kensington_and_Chelsea	MDMA	Islington	AR_PL_Kens-ington_and_Chelsea	Controlled Delivery	Croydon	AR_PL_Bromley	AR_PL_Sutton
1	Sex Male	1	1	1	1	1	1	1	1	1	1	1	1	1
7	Brent													
8	Bromley													
9	Camden		1	1	1									
10	Croydon											1		
15	Hammersmith_and_Fulham													
17	Harrow	1												
18	Havering													
21	Islington								1					
22	Kensington_and_Chelsea					1	1	1	1	1	1	1	1	1
34	Wandsworth													
35	Westminster				1									
42	JAM													
47	UK	1	1	1	1	1	1	1	1	1	1	1	1	1
51	(EA3)_Afro-Caribbean	1	1	1	1	1	1	1	1	1	1	1	1	1
59	Age(35–45)	1	1	1	1	1	1	1	1	1	1	1	1	1
60	Age(>45)													
61	Convictions Number	1	1	1	1	1	1	1	1	1	1	1	1	1
62	OffensesNumber	1	1	1	1	1	1	1	1	1	1	1	1	1
65	Off-Drug	1	1	1	1	1	1	1	1	1	1	1	1	1
66	Off_TheftKindred	1	1	1	1	1	1	1	1	1	1	1	1	1

(continued)

Table 14.9 (continued)

Num	Hammer-smith_and_Fulham	AR_PL_Wands-worth	Havering	AR_PL_Merton	Bromley	AR_PL_Haver-ing	Search Object	AR_PL_West-minster	Brent	AR_PL_Haringey	Crack	AR_PL_Isling-ton	Wandsworth
1	1	1	1	1	1	1	1	1	1	1	1	1	1
7									1				
8					1								
9							1	1		1	1	1	
10													
15	1												
17													
18			1										
21													
22	1	1	1	1	1	1			0.3				
34													1
35													
42										1	1	1	1
47	1	1	1	1	1	1	1	1	1				
51	1	1	1	1	1	1	1	1	1	1	1	1	1
59	1	1	1	1	1		1	1	1	1	1	1	1
60						1							
61	1	1	1	1	1	1	1	1	1	1	1	1	1
62	1	1	1	1	1	1	1	1	1	1	1	1	1
65	1	1	1	1	1	1	1	1	1	1	1	1	1
66	1	1	1	1	1	1	1	1	1	1	1	1	1

(continued)

Table 14.9 (continued)

Num	Cluster 3 – variables activated	Harrow	AR_ PL_ Harrow	Other Generic Tactics	West-minster	AR_ PL_ Red-bridge	Kensington_ and_ Chelsea	MDMA	Islington	AR_ PL_ Kens-ington_ and_ Chelsea	Controlled Delivery	Croydon	AR_ PL_ Bromley	AR_ PL_ Sutton
67	Off_AgainstPerson	1	1	1	1	1	1	1	1	1	1	1	1	1
68	Off_OffensiveWeapons	1	1	1	1	1	1	1	1	1	1	1	1	
69	Off_Sexual	1	1	1	1	1	1	1	1	1	1	1	1	1
70	Off_RelatedToPolice	1	1	1	1	1	1	1	1	1	1	1	1	1
71	Off_Fraud	1	1	1	1	1	1	1	1	1	1	1	1	1
72	Off_Total	1	1	1	1	1	1	1	1	1	1	1	1	1
73	Off_AgainstProperty	1	1	1	1	1	1	1	1	1	1	1	1	1
74	NumOfArrests	1	1	1	1	1	1	1	1	1	1	1	1	1
77	AR_OFF_ Drug_ trafficking_ Offenses	1	1	1	1	1	1	1	1	1	1	1	1	1
88	AR_PL_ Brent	1	1	1	1	1	1	1	1	1	1	1	1	1
89	AR_PL_ Bromley	1	1	1	1	1	1	1	1	1	1	1	1	1
90	AR_PL_ Camden	1	1	1	1	1	1	1	1	1	1	1	1	1
97	AR_PL_ Haringey	1	1	1	1	1	1	1	1	1	1	1	1	1
98	AR_PL_ Harrow	1	1											
99	AR_PL_Havering													
102	AR_PL_Islington													
103	AR_ PL_ Kensington_ and_ Chelsea	1	1	1	1	1	1	1	1	1	1	1	1	1
107	AR_PL_Merton													
109	AR_PL_Redbridge					1								
112	AR_PL_Sutton													1
115	AR_PL_Wandsworth													
116	AR_PL_Westminster													

(continued)

Table 14.9 (continued)

Num	Hammersmith_and_Fulham	AR_PL_Wandsworth	Havering	AR_PL_Merton	Bromley	AR_PL_Havering	Search Object	AR_PL_Westminster	Brent	AR_PL_Haringey	Crack	AR_PL_Islington	Wandsworth
67	1	1	1	1	1	1	1	1	1	1	1	1	1
68	1	1	1	1	1	1	1	1	1	1	1	1	1
69	1	1	1	1	1	1	1	1	1	1	1	1	1
70	1	1	1	1	1	1	1	1	1	1	1	1	1
71	1	1	1	1	1	1	1	1	1	1	1	1	1
72	1	1	1	1	1	1	1	1	1	1	1	1	1
73	1	1	1	1	1	1	1	1	1	1	1	1	1
74	1	1	1	1	1	1	1	1	1	1	1	1	1
77	1	1	1	1	1	1	1	1	1	1	1	1	1
88	1	1	1	1	1	1	1	1	1	1	1	1	1
89	1	1	1	1	1	1	1	1	1	1	1	1	1
90	1						1	1	1	1	1	1	1
97	1	1	1	1	1	1	1	1	1	1	1	1	1
98													
99						1							
102												1	
103	1	1	1	1	1	1	1	1	1	1	1	1	1
107				1									
109													
112													
115		1											
116								1					

(continued)

Table 14.9 (continued)

Num	Cluster 3 – variables activated	Harrow	AR_ PL_ Harrow	Other Generic Tactics	Westminster	AR_ PL_ Red-bridge	Kensington_ and_ Chelsea	MDMA	Islington	AR_ PL_ Kens-ington_ and_ Chelsea	Controlled Delivery	Croydon	AR_ PL_ Bromley	AR_ PL_ Sutton
118	NumOfDrugSeizures	1	1	1	1	1	1	1	1	1	1	1	1	1
121	Crack	1	1	1	1	1	1	1	1	1	1	1	1	1
122	Heroin_ Diamorphine	1	1	1	1	1	1	1	1	1	1	1	1	1
123	MDMA							1						
126	NumOfTactics	1	1	1	1	1	1	1	1	1	1	1	1	1
129	Police	1	1	1	1	1	1	1	1	1	1	1	1	1
130	Search_ of_ Object													
133	Covert_ Purchase	1	1	1	1	1	1	1	1	1	1	1	1	1
134	Controlled_ Delivery										1			
135	Other_Generic_ Tactic	1	1	1	0.3									
136	NumOfTacticSequences	1	1	1	1	1	1	1	1	1	1	1	1	1
137	InOperation	1	1	1	1	1	1	1	1	1	1	1	1	1
138	ViolentOnArrest	1												
139	ArrMode_NA		1	1	0.3	1								
140	ArrMode_Direct	1	1	1	1	1	1	1	1	1	1	1	1	1
141	ArrMode_Result_ of_Enquiries	1	1	1	1	1	1	1	1	1	1	1	1	1
143	ArrMode_Other	1	1	1	1	1	1	1	1	1	1	1	1	1

(continued)

Table 14.9 (continued)

Num	Hammer-smith_and_Fulham	AR_PL_Wands-worth	Havering	AR_PL_Merton	Bromley	AR_PL_Haver-ing	Search Object	AR_PL_West-minster	Brent	AR_PL_Haringey	Crack	AR_PL_Isling-ton	Wandsworth
118	1	1	1	1	1	1	1	1	1	1	1	1	1
121	1	1	1	1	1	1	1	1	1	1	1	1	1
122	1	1	1	1	1	1	1	1	1	1	1	1	1
123													
126	1	1	1	1	1	1	1	1	1	1	1	1	1
129	1	1	1	1	1	1	1	1	1	1	1	1	1
130							1						
133	1	1	1	1	1	1	1	1	1	1	1	1	1
134													
135							1	1	1	1	1	1	1
136	1	1	1	1	1	1	1	1	1	1	1	1	1
137	1	1	1	1	1	1	1	1	1	1	1	1	1
138									0.5				
139													
140	1	1	1	1	1	1	1	1	1	1	1	1	1
141	1	1	1	1	1	1	1	1	1	1	1	1	1
143	1	1	1	1	1	1	1	1	1	1	1	1	1

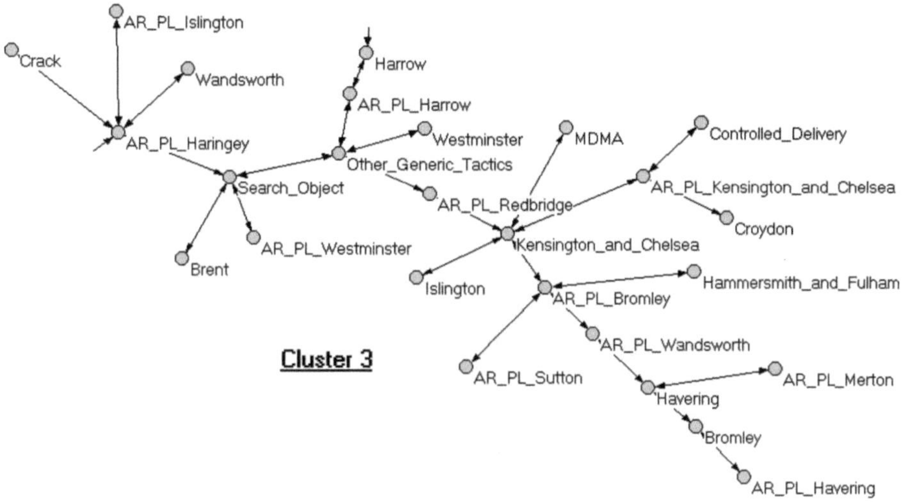

Fig. 14.4 Detail taken from the global MST graph – cluster 3

14.2.3 Comments on Cluster 3

Cluster 3 is formed by 26 queries which activated 61 variables (Table 14.10).

In this cluster, all the queries activate more than 50 % of the 61 variables listed. *Harrow*, *AR PL Harrow*, *Westminster*, and *Brent* are the queries activating the majority of variables (38 out of 61) (Table 14.11).

Thirty-one variables are always activated (out of the 26 queries). In all, 35 activated over 50 % of the queries.

As in cluster 2, many variables are activated relating to previous convictions and offenses by the subjects. A common characteristic among the 26 queries is the seizure of crack and heroin, even though the latter is not present as a query in the cluster. The variable, *MDMA-type drug*, on the other hand, is present in the cluster and activated only when it is used as a query.

The boroughs most activated are *Brent*, *Bromley*, *Haringey*, *Kensington and Chelsea*, and *Camden*, all representing a place of arrest. *Kensington and Chelsea* and *Camden* as boroughs of residence are activated to a lesser extent.

Afro-Caribbean is the most activated ethnicity, while *Covert Purchase* is the tactic most used (Table 14.12) (Fig. 14.5).

14.2.4 Comments on Cluster 4

Cluster 4 consists of 16 queries which resulted in the activation of 53 variables (see Table 14.13).

Table 14.10 Number and % of variables activated by each query

Cluster 3 – queries	Variables activ.	%
Harrow	38	62
AR_PL_Harrow	38	62
Westminster	38	62
Brent	38	62
Other Generic Tactics	37	61
AR_PL_Redbridge	37	61
Search Object	37	61
AR_PL_Westminster	37	61
AR_PL_Islington	37	61
MDMA	36	59
Islington	36	59
Controlled Delivery	36	59
Croydon	36	59
AR_PL_Sutton	36	59
Hammersmith_and_Fulham	36	59
AR_PL_Haringey	36	59
Crack	36	59
Wandsworth	36	59
Kensington_and_Chelsea	35	57
AR_PL_ Kensington_and_Chelsea	35	57
AR_PL_Bromley	35	57
AR_PL_Wandsworth	35	57
Havering	35	57
AR_PL_Merton	35	57
Bromley	35	57
AR_PL_Havering	35	57

In this case, all the queries activate more than 50 % of the variables, except *AR PL Barking and Dagenham* which activates 43 % (23 out of 53). A particular point of interest here is that the variable *AR PL Barking and Dagenham* is the query that is at the end of the MST graph.

Among the queries that activate most variables (38 out of 53) are *AR PL Lewisham*, *AR PL Camden*, *AR PL Croydon*, *AR PL Lambeth*, *Lewisham*, *Lambeth*, *Covert Purchase,* and *Heroin* (Table 14.14).

The number of variables that are always activated in response to the 16 queries is 17. In all, over 50 % of the queries activated some 37 variables.

As in the two previous clusters, the *Convictions* and *Offenses* variables are activated. In addition, the *Crack* and *Heroin* variables are always activated. The boroughs most activated are *Bromley*, *Haringey*, *Kensington and Chelsea*, *Brent*, *Camden*, and *Lewisham,* all as places of arrest. *Camden* and *Haringey* are somewhat activated as boroughs of residence, while Jamaican is the nationality most activated. Once again, *Covert Purchase* is the tactic most used (Table 14.15) (Fig. 14.6).

Table 14.11 Number and % of activations of the 61 variables

Cluster 3 – variables activated	Num activ.	%	Cluster 3 – variables activated	Num activ.	%
Sex_Male	26	100	Age(35–45)	25	96
(EA3)_Afro-Caribbean	26	100	UK	22	85
ConvictionsNumber	26	100	AR_PL_Camden	21	81
OffensesNumber	26	100	Kensington_and_Chelsea	16	62
Off_Drug	26	100	Other_Generic_Tactic	11	42
Off_TheftKindred	26	100	Camden	8	31
Off_AgainstPerson	26	100	JAM	4	15
Off_OffensiveWeapons	26	100	ArrMode_NA	4	15
Off_Sexual	26	100	AR_PL_Harrow	2	8
Off_RelatedToPolice	26	100	ViolentOnArrest	2	8
Off_Fraud	26	100	Brent	1	4
Off_Total	26	100	Bromley	1	4
Off_AgainstProperty	26	100	Croydon	1	4
NumOfArrests	26	100	Hammersmith_and_Fulham	1	4
AR_OFF_Drug_trafficking_Offenses	26	100	Harrow	1	4
AR_PL_Brent	26	100	Havering	1	4
AR_PL_Bromley	26	100	Islington	1	4
AR_PL_Haringey	26	100	Wandsworth	1	4
AR_PL_Kensington_and_Chelsea	26	100	Westminster	1	4
NumOfDrugSeizures	26	100	Age(>45)	1	4
Crack	26	100	AR_PL_Havering	1	4
Heroin_Diamorphine	26	100	AR_PL_Islington	1	4
NumOfTactics	26	100	AR_PL_Merton	1	4
Police	26	100	AR_PL_Redbridge	1	4
Covert_Purchase	26	100	AR_PL_Sutton	1	4
NumOfTacticSequences	26	100	AR_PL_Wandsworth	1	4
InOperation	26	100	AR_PL_Westminster	1	4
ArrMode_Direct	26	100	MDMA	1	4
ArrMode_Result_of_Enquiries	26	100	Search_of_Object	1	4
ArrMode_Other	26	100	Controlled_Delivery	1	4
OnBailAtTimeOfOffence	26	100			

14.2.5 Comments on Cluster 5

Cluster 5 is formed by 6 queries that overall activate 47 variables (Table 14.16).

In this particular cluster, a few queries including *Female* are present that in all the other queries were never activated. *Merton*, *Redbridge*, and *Sutton* are the queries that activate most variables (Table 14.17).

Table 14.12 Table with the variables activated in the 16 queries of cluster 4

Num	Cluster 4 – variables activated	AR_PL_Lewisham	AR_PL_Camden	Lewisham	Lambeth	Covert Purchase	Heroin-Diamorphine	Male	AR_PL_Southwark	Hackney	AR_PL_Hackney	AR_PL_Croydon	AR_PL_Lambeth	Southwark	Enfield	Haringey	AR_PL_Barking_and_Dagenham
1	Sex_Male	1	1	1	1	1	1	1	1	1	1	1	1	1	1	1	1
9	Camden	0.9	0.5			0.5	0.5	0.5									
12	Enfield														1		
14	Hackney									1							
16	Haringey															1	1
24	Lambeth				1												
25	Lewisham			1													
30	Southwark													1			
42	JAM	1	1	1	1	1	1	1	1	1	1	1	1	1	1	1	
46	TU-CY																1
50	(EA2)_Dark_European																1
51	(EA3)_Afro-Caribbean	1	1	1	1	1	1	1	1	1	1	1	1	1	1	1	
58	Age(25–35)										1	0.6	1	1	1	1	1
59	Age(35–45)	1	1	1	1	1	1	1	1	1							
61	ConvictionsNumber	1	1	1	1	1	1	1	1	1	1	1	1	1	1	1	
62	OffensesNumber	1	1	1	1	1	1	1	1	1	1	1	1	1	1	1	
63	Off_FirstConvAge																1
65	Off_Drug	1	1	1	1	1	1	1	1	1	1	1	1	1	1	1	1
66	Off_TheftKindred	1	1	1	1	1	1	1	1	1	1	1	1	1	1	1	
67	Off_AgainstPerson	1	1	1	1	1	1	1	1	1	1	1	1	1	1	1	
68	Off_OffensiveWeapons	1	1	1	1	1	1	1	1	1	1	1	1	1	1	1	
69	Off_Sexual	1	1	1	1	1	1	1	1	1	1	1	1	1	1	1	
70	Off_RelatedToPolice	1	1	1	1	1	1	1	1	1	1	1	1	1	1	1	

71	Off_Fraud	1	1	1	1	1	1	1	1	0.4	1	0.9	0.4				
72	Off_Total	1	1	1	1	1	1	1	1	1	1	1	1	1	1	1	
73	Off_AgainstProperty	1	1	1	1	1	1	1	1	1	1	1	1	1	1	1	
74	NumOfArrests	1	1	1	1	1	1	1	1	1	1	1	1	1	1	1	1
77	AR_OFF_Drug_trafficking_Offenses	1	1	1	1	1	1	1	1	1	1	1	1	1	1	1	1
85	AR_PL_Barking_and_Dagenham																1
88	AR_PL_Brent	1	1	1	1	1	1	1	1	1	1	1	1	1	1	1	
89	AR_PL_Bromley	1	1	1	1	1	1	1	1	1	1	1	1	1	1	1	1
90	AR_PL_Camden	1	1	1	1	1	1	1	1	1	1	1	1	1	1	1	
91	AR_PL_Croyden											1					
95	AR_PL_Hackney										1						
97	AR_PL_Haringey	1	1	1	1	1	1	1	1	1	1	1	1	1	1	1	1
103	AR_PL_Kensington_and_Chelsea	1	1	1	1	1	1	1	1	1	1	1	1	1	1	1	1
105	AR_PL_Lambeth												1				
106	AR_PL_Lewisham	1	1	1	1	1	1	1				1	1	1			
111	AR_PL_Southwark								1								
118	NumOfDrugSeizures	1	1	1	1	1	1	1	1	1	1	1	1	1	1	1	1
121	Crack	1	1	1	1	1	1	1	1	1	1	1	1	1	1	1	1
122	Heroin_Diamorphine	1	1	1	1	1	1	1	1	1	1	1	1	1	1	1	1
126	NumOfTactics	1	1	1	1	1	1	1	1	1	1	1	1	1	1	1	1
129	Police	1	1	1	1	1	1	1	1	1	1	1	1	1	1	1	1
133	Covert_Purchase	1	1	1	1	1	1	1	1	1	1	1	1	1	1	1	1
135	Other_Generic_Tactic	1	1	1	1	1	1	1	1	1	1	0.7	1	1	1	1	
136	NumOfTacticSequences	1	1	1	1	1	1	1	1	1	1	1	1	1	1	1	1
137	InOperation	1	1	1	1	1	1	1	1	1	1	1	1	1	1	1	1
138	ViolentOnArrest	0.9	1	1	1	1	1	1	1	0.9	1	1	1	1	1		
140	ArrMode_Direct	1	1	1	1	1	1	1	1	1	1	1	1	1	1	1	1
141	ArrMode_Result_of_Enquiries	1	1	1	1	1	1	1	1	1	1	1	1	1	1	1	
143	ArrMode_Other	1	1	1	1	1	1	1	1	1	1	1	1	1	1	1	1
144	OnBailAtTimeOfOffence	1	1	1	1	1	1	1	1	1	1	1	1	1	1	1	

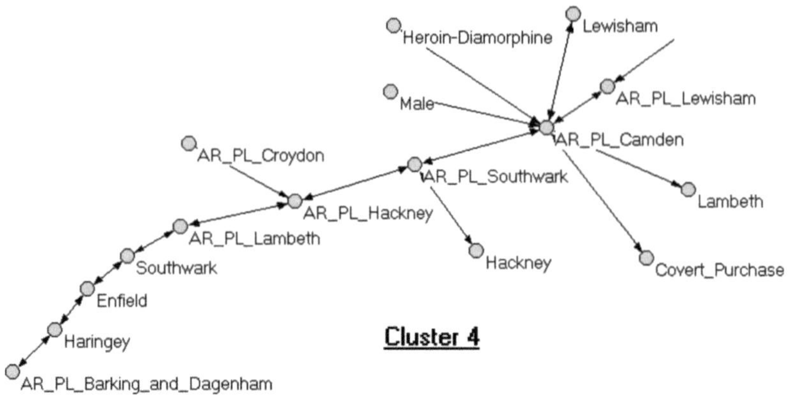

Fig. 14.5 Detail taken from the global MST graph – cluster 4.

Table 14.13 Number and % of variables activated by each query

Cluster 4 – queries	Variables activ.	%
AR_PL_Lewisham	38	72
AR_PL_Camden	38	72
Lewisham	38	72
Lambeth	38	72
Covert Purchase	38	72
Heroin-Diamorphine	38	72
Male	38	72
AR_PL_Croydon	38	72
AR_PL_Lambeth	38	72
AR_PL_Southwark	37	70
Hackney	37	70
AR_PL_Hackney	37	70
Southwark	37	70
Enfield	36	68
Haringey	35	66
AR_PL_Barking_and_Dagenham	23	43

Two variables are always activated in response to the six queries, namely, *UK* and *EA 1- White European*. In all, over 50 % of the queries activated some 27 variables.

In addition, the following two variables are activated: *Convictions* and *Offenses*, but, unlike clusters 2, 3, and 4, the drugs *Heroin* and *Crack* are not activated. The *MDMA-type drugs* variable is activated in response to the *Kingston upon Thames* query, *Cannabis* to the *Redbridge* query, and *Cocaine* to the *Female* query. The boroughs most activated are *Kingston upon Thames*, *Sutton*, *Merton*, and *Westminster*, as places of arrest and places of residence. *White European* is the ethnicity most activated, while *Search of Premises* and *Search of Person* are the tactics most used.

The following tables show the profile of the variables in four typical boroughs (Barking & Dagenham, Barnet, Camden, and Sutton). It is important to compare

Table 14.14 Number and % of activations of the 53 variables

Cluster 4 – variables activated	Num activ.	%	Cluster 4 – variables activated	Num activ.	%
Sex_Male	16	100	Off_AgainstProperty	15	94
Off_Drug	16	100	AR_PL_Brent	15	94
NumOfArrests	16	100	AR_PL_Camden	15	94
AR_OFF_Drug_trafficking _Offenses	16	100	ArrMode_Result_of _Enquiries	15	94
AR_PL_Bromley	16	100	OnBailAtTimeOfOffence	15	94
AR_PL_Haringey	16	100	Other_Generic_Tactic	15	94
AR_PL_Kensington_and _Chelsea	16	100	ViolentOnArrest	14	85
NumOfDrugSeizures	16	100	Off_Fraud	11	67
Crack	16	100	AR_PL_Lewisham	10	62
Heroin_Diamorphine	16	100	Age(35–45)	9	56
NumOfTactics	16	100	Age(25–35)	7	41
Police	16	100	Camden	3	18
Covert_Purchase	16	100	Haringey	2	13
NumOfTacticSequences	16	100	Enfield	1	6
InOperation	16	100	Hackney	1	6
ArrMode_Direct	16	100	Lambeth	1	6
ArrMode_Other	16	100	Lewisham	1	6
JAM	15	94	Southwark	1	6
(EA3)_Afro-Caribbean	15	94	TU-CY	1	6
ConvictionsNumber	15	94	(EA2)_Dark_European	1	6
OffensesNumber	15	94	AR_PL_Barking_and _Dagenham	1	6
Off_TheftKindred	15	94	AR_PL_Croydon	1	6
Off_AgainstPerson	15	94	AR_PL_Hackney	1	6
Off_OffensiveWeapons	15	94	AR_PL_Lambeth	1	6
Off_Sexual	15	94	AR_PL_Southwark	1	6
Off_RelatedToPolice	15	94	Off_FirstConvAge	1	6
Off_Total	15	94			

how the profile of each borough changes when the boroughs are considered as "place of residence" of the arrestees or when it is considered as specific place where the persons are arrested.

14.3 Profiles

14.3.1 Barking and Dagenham

Looking at the results derived from the CS network, some major differences can be seen between the two prototypes described in Table 14.18.

Table 14.15 Table with the variables activated in the 6 queries of cluster 5

Num	Cluster 5 – variables activated	Female	Redbridge	Sutton	Merton	Kingston_ upon_ Thames	AR_ PL_ Kingston_ upon_ Thames
1	Sex_Male		1	1	1		
2	Sex_Female	1					
3	Sex_notknown					1	1
23	Kingston_upon_Thames			1	0.1	1	1
26	Merton	1		1	1		
28	Redbridge		1				
31	Sutton	1	0.9	1	0.4		
35	Westminster					1	1
47	UK	1	1	1	1	1	1
49	(EA1)_White_European	1	1	1	1	1	1
57	Age(21–25)	1					
59	Age(35–45)			1	1	1	1
60	Age(>45)		1				
61	ConvictionsNumber		1	1	1	1	1
62	OffensesNumber		1	1	1	1	1
65	Off_Drug		1	1	1	1	1
66	Off_TheftKindred		1	1	1	1	1
67	Off_AgainstPerson		1	1	1	1	1
68	Off_OffensiveWeapons		1	1	1	1	1
69	Off_Sexual			1	1	1	1
70	Off_RelatedToPolice		0.9	1	1	1	1
71	Off_Fraud		1	1	1	1	1
72	Off_Total		1	1	1	1	1
73	Off_AgainstProperty		1	1	1	1	1
74	NumOfArrests		0.7	1	1		
75	AR_OFF_Theft_and_ Kindred_Offenses		0.9				
78	AR_OFF_Drug_ Possession_Offenses	1		1	1	1	1
80	AR_OFF_Offensive_ Weapon_Offenses		1	1	1		
81	AR_OFF_Firearms_ Offenses	1	1	1	1		
104	AR_PL_Kingston_upon_ Thames					1	1
107	AR_PL_Merton	1		1	1		
109	AR_PL_Redbridge		0.3				
112	AR_PL_Sutton	1	1	1	1		
116	AR_PL_Westminster					1	1
118	NumOfDrugSeizures	1					
119	Cannabis		0.6				
120	Cocaine	1					
123	MDMA					1	1

(continued)

Table 14.15 (continued)

Num	Cluster 5 – variables activated	Female	Redbridge	Sutton	Merton	Kingston_ upon_ Thames	AR_ PL_ Kingston_ upon_ Thames
124	NumOfCashSeizures	1	1		0.1		
126	NumOfTactics						
129	Police				0.2		
131	Search_of_Person	1			0.9		
132	Search_of_Premises	1	1		0.2		
135	Other_Generic_Tactic		0.6				
139	ArrMode_NA		1	1	1	1	1
142	ArrMode_Given_into_ custody					1	1
143	ArrMode_Other				0.2		

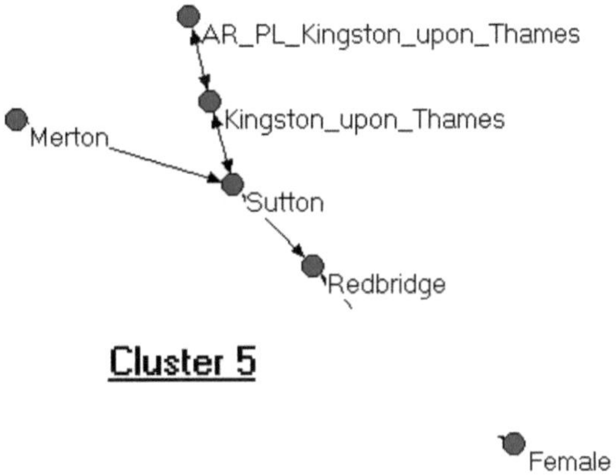

Fig. 14.6 Detail taken from the global MST graph – cluster 5

Table 14.16 Number and % of variables activated by each query

Cluster 5 – queries	Variables activ.	%
Merton	30	64
Redbridge	27	57
Sutton	25	53
Kingston_upon_Thames	23	49
AR_PL_Kingston_upon_Thames	23	49
Female	15	32

The first table refers to the borough in which the arrestee resides and describes the characteristics of the prototypical arrestee as:

- Male.
- Aged between 18 and 21.
- British.
- White (ethnic appearance (EA 1) – White European).

Table 14.17 Number and % of activations of the 47 variables

Cluster 5 – variables activated	Num activ.	%	Cluster 5 – variables activated	Num Activ.	%
UK	6	100	AR_PL_Merton	3	50
(EA1)_White_European	6	100	NumOfCashSeizures	3	50
ConvictionsNumber	5	83	Search_of_Premises	3	50
OffensesNumber	5	83	Sex_notknown	2	33
Off_Drug	5	83	Westminster	2	33
Off_TheftKindred	5	83	AR_PL_Kingston_upon_Thames	2	33
Off_AgainstPerson	5	83	AR_PL_Westminster	2	33
Off_OffensiveWeapons	5	83	MDMA	2	33
Off_RelatedToPolice	5	83	Search_of_Person	2	33
Off_Fraud	5	83	ArrMode_Given_into_custody	2	33
Off_Total	5	83	Sex_Female	1	17
Off_AgainstProperty	5	83	Redbridge	1	17
AR_OFF_Drug_Possession_Offenses	5	83	Age(21–25)	1	17
ArrMode_NA	5	83	Age(>45)	1	17
Kingston_upon_Thames	4	67	AR_OFF_Theft_and_Kindred_Offenses	1	17
Sutton	4	67	AR_PL_Redbridge	1	17
Age (35–45)	4	67	NumOfDrugSeizures	1	17
Off_Sexual	4	67	Cannabis	1	17
AR_OFF_Firearms_Offenses	4	67	Cocaine	1	17
AR_PL_Sutton	4	67	Police	1	17
Sex_Male	3	50	Other_Generic_Tactic	1	17
Merton	3	50	ArrMode_Other	1	17
NumOfArrests	3	50	NumOfTactics	0	0
AR_OFF_Offensive_Weapon_Offenses	3	50			

- Cannabis accounts for the majority of seizures that is also confirmed by the activation of the same boroughs as place of arrest, and the predominant generic tactic used by the police is *Search of Premises.*

For those *arrested* in the boroughs of Barking and Dagenham, their prototypical characteristics are as follows:

- Male.
- Aged between 25 and 35.
- A resident of Haringey borough.
- Turkish/Cypriot nationals (Jamaican nationals also feature in this category).
- Jamaican (ethnic appearance (EA 2) – Dark European).
- With a criminal record for drug trafficking.
- Crack and heroin account for the majority of drug seizures, while the predominant generic used by the police is *Covert Purchase.*

Table 14.18 Profile of the variables in Barking and Dagenham: first table refers to the borough as "place of residence"; second table refers to the borough as "place of arrest"

Query: Barking & Dagenham		
Variables activated		**Values**
1	Sex_Male	1
4	Barking_and_Dagenham	1
17	Harrow	1
29	Richmond_upon_Thames	1
47	UK	1
49	(EA1)_White_European	1
56	Age(18–21)	1
85	AR_PL_Barking_and_Dagenham	1
98	AR_PL_Harrow	1
110	AR_PL_Richmond_upon_Thames	1
124	NumOfCashSeizures	1
139	ArrMode_NA	1
Drugs activated		
119	*Cannabis*	1
Tactics activated		
132	*Search_of_Premises*	1
Query: Barking & Dagenham		
Variables activated		**Values**
1	Sex_Male	1
16	Haringey	1
42	JAM	0.02
46	TU-CY	1
50	(EA2)_Dark_European	1
58	Age(25–35)	1
63	Off_FirstConvAge	0.98
65	Off_Drug	1
74	NumOfArrests	1
77	AR_OFF_Drug_trafficking_Offenses	1
85	AR_PL_Barking_and_Dagenham	1
89	AR_PL_Bromley	1
97	AR_PL_Haringey	1
103	AR_PL_Kensington_and_Chelsea	1
118	NumOfDrugSeizures	1
126	NumOfTactics	1
129	Police	1
136	NumOfTacticSequences	1
137	InOperation	1
140	ArrMode_Direct	1
143	ArrMode_Other	1
Drugs activated		
121	*Crack*	1
122	*Heroin_Diamorphine*	1
Tactics activated		
133	*Covert_Purchase*	1

14.3.2 Barnet

Turning now to the borough of Barnet, the answers to the queries posed to the CS network are very similar, and there is little difference between the two prototypes identified (Table 14.19).

Table 14.19 Profile of the variables in Barnet: first table refers to the borough as "place of residence"; second table refers to the borough as "place of arrest"

Query: Barnet		
Variables activated		**Values**
1	Sex_Male	1
5	Barnet	1
17	Harrow	1
47	UK	1
52	(EA4)_Asia	1
61	ConvictionsNumber	1
62	OffensesNumber	1
65	Off_Drug	1
66	Off_TheftKindred	1
67	Off_AgainstPerson	1
68	Off_OffensiveWeapons	1
70	Off_RelatedToPolice	1
71	Off_Fraud	1
72	Off_Total	1
73	Off_AgainstProperty	1
74	NumOfArrests	1
77	AR_OFF_Drug_trafficking_Offenses	1
78	AR_OFF_Drug_Possession_Offenses	1
80	AR_OFF_Offensive_Weapon_Offenses	1
81	AR_OFF_Firearms_Offenses	1
83	AR_OFF_Other_violent_Offenses	1
88	AR_PL_Brent	1
98	AR_PL_Harrow	1
113	AR_PL_Tower_Hamlets	1
118	NumOfDrugSeizures	1
124	NumOfCashSeizures	1
126	NumOfTactics	1
129	Police	1
139	ArrMode_NA	1
140	ArrMode_Direct	1
144	OnBailAtTimeOfOffence	1
Drugs activated		
119	*Cannabis*	0.47
Tactics activated		
131	*Search_of_Person*	1
132	*Search_of_Premises*	1
135	*Other_Generic_Tactic*	0.89

(continued)

Table 14.19 (continued)

Query: AR_PL_Barnet		
Variables activated		**Values**
1	Sex_Male	1
17	Harrow	1
47	UK	1
52	(EA4)_Asia	1
57	Age(21–25)	0.01
61	ConvictionsNumber	1
62	OffensesNumber	1
65	Off_Drug	1
66	Off_TheftKindred	1
67	Off_AgainstPerson	1
68	Off_OffensiveWeapons	1
70	Off_RelatedToPolice	1
71	Off_Fraud	1
72	Off_Total	1
73	Off_AgainstProperty	1
74	NumOfArrests	1
77	AR_OFF_Drug_trafficking_Offenses	1
78	AR_OFF_Drug_Possession_Offenses	1
80	AR_OFF_Offensive_Weapon_Offenses	1
81	AR_OFF_Firearms_Offenses	1
83	AR_OFF_Other_violent_Offenses	1
86	AR_PL_Barnet	1
88	AR_PL_Brent	1
98	AR_PL_Harrow	1
113	AR_PL_Tower_Hamlets	1
118	NumOfDrugSeizures	1
124	NumOfCashSeizures	1
126	NumOfTactics	1
129	Police	1
139	ArrMode_NA	1
140	ArrMode_Direct	1
144	OnBailAtTimeOfOffence	1
Drugs activated		
119	*Cannabis*	0.04
Tactics activated		
131	*Search_of_Person*	1
132	*Search_of_Premises*	1
135	*Other_Generic_Tactic*	1

The first table lists the characteristics of the prototypical arrestee as follows:

- Male.
- British.
- His ethnic appearance code is EA 4 – Asian.
- He has a criminal record comprising a wide range of offenses.

- At the time of arrest, the majority had cannabis in their possession.
- The boroughs of Brent, Harrow, and Tower Hamlets are activated as places of arrest, and in terms of generating arrests and seizures, the police used three of the six generic tactics.

The answers pertaining to the second query are practically identical save for the following: the age of the arrestees (21–25) and the activation value of *Cannabis*.

14.3.3 Camden

By way of contrast, the answers to the two questions posed in relation to the borough of Camden have produced very different prototypes. This difference is readily apparent; in Table 14.20, we see that only five variables were activated when the borough is "place of residence," while 38 variables were activated when the borough is "place of arrest."

14.3.4 Sutton

Finally, for the borough of Sutton, the prototypical arrestees identified in Table 14.21 are rather different. Those arrestees residing in the borough are described in first table as:

- Male.
- Aged between 35 and 45.
- British.
- Of *White-European* stock.
- Very similar to subjects living in *Kingston upon Thames* and *Merton.*
- In addition, they have criminal records for a wide range of offenses, and they were arrested for *drug possession*, *offensive weapon offenses*, and *firearm offenses*.
- The boroughs of Merton and Sutton are activated as *place of arrest.*
- No specific drugs or generic tactics are activated by the query.

Prototypical arrestees identified in second table are described as:

- Male.
- Aged between 35 and 45.
- British.
- Of *Afro-Caribbean* stock.
- Resident in *Kensington and Chelsea.*
- Their criminal records comprise a wide range of offenses, and they were arrested for drug trafficking offenses.

Table 14.20 Profile of the variables in Camden: first table refers to the borough as "place of residence"; second table refers to the borough as "place of arrest"

Query: Camden		
Variables activated		**Values**
1	Sex_Male	1
9	Camden	1
39	EASTEU	1
49	(EA1)_White_European	1
58	Age(25–35)	0.99
Query: AR_PL_Camden		
Variables activated		**Values**
1	Sex_Male	1
9	Camden	0.49
42	JAM	1
51	(EA3)_Afro-Caribbean	1
59	Age(35–45)	1
61	ConvictionsNumber	1
62	OffensesNumber	1
65	Off_Drug	1
66	Off_TheftKindred	1
67	Off_AgainstPerson	1
68	Off_OffensiveWeapons	1
69	Off_Sexual	1
70	Off_RelatedToPolice	1
71	Off_Fraud	1
72	Off_Total	1
73	Off_AgainstProperty	1
74	NumOfArrests	1
77	AR_OFF_Drug_trafficking_Offenses	1
88	AR_PL_Brent	1
89	AR_PL_Bromley	1
90	AR_PL_Camden	1
97	AR_PL_Haringey	1
103	AR_PL_Kensington_and_Chelsea	1
106	AR_PL_Lewisham	0.98
118	NumOfDrugSeizures	1
126	NumOfTactics	1
129	Police	1
136	NumOfTacticSequences	1
137	InOperation	1
138	ViolentOnArrest	1
140	ArrMode_Direct	1
141	ArrMode_Result_of_Enquiries	1
143	ArrMode_Other	1
144	OnBailAtTimeOfOffence	1
Drugs activated		
121	*Crack*	1
122	*Heroin_Diamorphine*	1
Tactics activated		
133	*Covert_Purchase*	1
135	*Other_Generic_Tactic*	1

Table 14.21 Profile of the variables in Sutton: first table refers to the borough as "place of residence"; second table refers to the borough as "place of arrest"

Query: Sutton		
Variables activated		**Values**
1	Sex_Male	1
23	Kingston_upon_Thames	1
26	Merton	1
31	Sutton	1
47	UK	1
49	(EA1)_White_European	1
59	Age(35–45)	1
61	ConvictionsNumber	1
62	OffensesNumber	1
65	Off_Drug	1
66	Off_TheftKindred	1
67	Off_AgainstPerson	1
68	Off_OffensiveWeapons	1
69	Off_Sexual	1
70	Off_RelatedToPolice	1
71	Off_Fraud	1
72	Off_Total	1
73	Off_AgainstProperty	1
74	NumOfArrests	1
78	AR_OFF_Drug_Possession_Offenses	1
80	AR_OFF_Offensive_Weapon_Offenses	1
81	AR_OFF_Firearms_Offenses	1
107	AR_PL_Merton	1
112	AR_PL_Sutton	1
139	ArrMode_NA	1
Query: AR_PL_Sutton		
Variables activated		**Values**
1	Sex_Male	1
22	Kensington_and_Chelsea	1
47	UK	1
51	(EA3)_Afro-Caribbean	1
59	Age(35–45)	1
61	ConvictionsNumber	1
62	OffensesNumber	1
65	Off_Drug	1
66	Off_TheftKindred	1
67	Off_AgainstPerson	1
68	Off_OffensiveWeapons	1
69	Off_Sexual	1
70	Off_RelatedToPolice	1
71	Off_Fraud	1
72	Off_Total	1
73	Off_AgainstProperty	1
74	NumOfArrests	1
77	AR_OFF_Drug_trafficking_Offenses	1

(continued)

Table 14.21 (continued)

88	AR_PL_Brent	1
89	AR_PL_Bromley	1
90	AR_PL_Camden	1
97	AR_PL_Haringey	1
103	AR_PL_Kensington_and_Chelsea	1
112	AR_PL_Sutton	1
118	NumOfDrugSeizures	1
126	NumOfTactics	1
129	Police	1
136	NumOfTacticSequences	1
137	InOperation	1
140	ArrMode_Direct	1
141	ArrMode_Result_of_Enquiries	1
143	ArrMode_Other	1
144	OnBailAtTimeOfOffence	1
Drugs activated		
121	*Crack*	1
122	*Heroin_Diamorphine*	1
Tactics activated		
133	*Covert_Purchase*	1

- The boroughs of Brent, Bromley, Camden, Haringey, and Kensington and Chelsea are activated as *place of arrest*, while crack and heroin are the most activated drugs.
- Covert purchase is the predominant generic tactic.

14.4 Conclusion

A complete analysis of the London Metropolitan Police Drug Database comprised of 144 variables encompassing some 1,120 cases that represented arrests throughout the 32 boroughs of London for the time ending in June 2006 was undertaken and analyzed using the constraint satisfaction artificial neural network (CS ANN) described in the previous chapter. The resulting investigatory tool was the minimal spanning tree which was thoroughly described and associated with each dataset used in its analysis. The input database for each analysis was binary and required a substantial effort to assemble the data in an acceptable format. The analysis also showed which variables in a given data subset were closest in relationship to each other. Those associations were used to explain the results of the MSTs.

Using this method, law enforcement agencies can transform their traditional, SQL-based databases into dynamic resources for which heretofore unknown connections of individuals that associate with certain other individuals can be identified and accurate profiles can be generated that could greatly aid the police in focusing their limited manpower resources onto groups and areas identified as being most probably for illegal drug activity.

A.1 Appendix

Next, we show the tables relating to each borough when is consider "place of residence" and "place of arrest" and the activated variables with respective values.

Query: Bexley		
Variables activated		**Values**
1	Sex_Male	1
6	Bexley	1
29	Richmond_upon_Thames	1
47	UK	1
49	(EA1)_White_European	1
56	Age(18–21)	1
75	AR_OFF_Theft_and_Kindred_Offenses	1
80	AR_OFF_Offensive_Weapon_Offenses	1
84	AR_OFF_Other_Offenses	1
87	AR_PL_Bexley	1
110	AR_PL_Richmond_upon_Thames	1
Drugs activated		
119	*Cannabis*	1
Tactics activated		
132	*Search_of_Premises*	1
Query: AR_PL_ Bexley		
Variables activated		**Values**
1	Sex_Male	1
6	Bexley	1
17	Harrow	1
47	UK	1
52	(EA4)_Asia	1
56	Age (18–21)	1
80	AR_OFF_Offensive_Weapon_Offenses	1
87	AR_PL_Bexley	1
98	AR_PL_Harrow	1
124	NumOfCashSeizures	1
Drug activated		
119	*Cannabis*	1
Tactics activated		
131	*Search_of_Person*	1
132	*Search_of_Premises*	1

(continued)

(continued)

Query: Brent		
Variables activated		**Values**
1	Sex_Male	1
7	Brent	1
22	Kensington_and_Chelsea	0.32
47	UK	1
51	(EA3)_Afro-Caribbean	1
59	Age(35–45)	1
61	ConvictionsNumber	1
62	OffensesNumber	1
65	Off_Drug	1
66	Off_TheftKindred	1
67	Off_AgainstPerson	1
68	Off_OffensiveWeapons	1
69	Off_Sexual	1
70	Off_RelatedToPolice	1
71	Off_Fraud	1
72	Off_Total	1
73	Off_AgainstProperty	1
74	NumOfArrests	1
77	AR_OFF_Drug_trafficking_Offenses	1
88	AR_PL_Brent	1
89	AR_PL_Bromley	1
90	AR_PL_Camden	1
97	AR_PL_Haringey	1
103	AR_PL_Kensington_and_Chelsea	1
118	NumOfDrugSeizures	1
126	NumOfTactics	1
129	Police	1
136	NumOfTacticSequences	1
137	InOperation	1
138	ViolentOnArrest	0.51
140	ArrMode_Direct	1
141	ArrMode_Result_of_Enquiries	1
143	ArrMode_Other	1
144	OnBailAtTimeOfOffence	1
Drugs activated		
121	*Crack*	1
122	*Heroin_Diamorphine*	1
Tactics activated		
133	*Covert_Purchase*	1
135	*Other_Generic_Tactic*	1
Query: AR_PL_ Brent		
Variables activated		**Values**
1	Sex_Male	1
17	Harrow	1
47	UK	1
51	(EA3)_Afro-Caribbean	1

(continued)

(continued)

61	ConvictionsNumber	1
62	OffensesNumber	1
65	Off_Drug	1
66	Off_TheftKindred	1
67	Off_AgainstPerson	1
68	Off_OffensiveWeapons	1
70	Off_RelatedToPolice	1
71	Off_Fraud	1
72	Off_Total	1
73	Off_AgainstProperty	1
74	NumOfArrests	1
76	AR_OFF_Offenses_the_Person_Offenses	1
77	AR_OFF_Drug_trafficking_Offenses	1
80	AR_OFF_Offensive_Weapon_Offenses	1
81	AR_OFF_Firearms_Offenses	1
83	AR_OFF_Other_violent_Offenses	1
88	AR_PL_Brent	1
98	AR_PL_Harrow	1
113	AR_PL_Tower_Hamlets	1
124	NumOfCashSeizures	1
126	NumOfTactics	1
129	Police	1
138	ViolentOnArrest	1
139	ArrMode_NA	1
140	ArrMode_Direct	1
144	OnBailAtTimeOfOffence	1
Tactics activated		
131	*Search_of_Person*	1
135	*Other_Generic_Tactic*	1
Query: Bromley		
Variables activated		**Values**
1	Sex_Male	1
8	Bromley	1
22	Kensington_and_Chelsea	1
47	UK	1
51	(EA3)_Afro-Caribbean	1
59	Age(35–45)	1
61	ConvictionsNumber	1
62	OffensesNumber	1
65	Off_Drug	1
66	Off_TheftKindred	1
67	Off_AgainstPerson	1
68	Off_OffensiveWeapons	1
69	Off_Sexual	1
70	Off_RelatedToPolice	1
71	Off_Fraud	1
72	Off_Total	1
73	Off_AgainstProperty	1
74	NumOfArrests	1

(continued)

(continued)

77	AR_OFF_Drug_trafficking_Offenses	1
88	AR_PL_Brent	1
89	AR_PL_Bromley	1
97	AR_PL_Haringey	1
103	AR_PL_Kensington_and_Chelsea	1
118	NumOfDrugSeizures	1
126	NumOfTactics	1
129	Police	1
136	NumOfTacticSequences	1
137	InOperation	1
140	ArrMode_Direct	1
141	ArrMode_Result_of_Enquiries	1
143	ArrMode_Other	1
144	OnBailAtTimeOfOffence	1
Drugs activated		
121	*Crack*	1
122	*Heroin_Diamorphine*	1
Tactics activated		
133	*Covert_Purchase*	1
Query: AR_PL_ Bromley		
Variables activated		**Values**
1	Sex_Male	1
22	Kensington_and_Chelsea	1
47	UK	1
51	(EA3)_Afro-Caribbean	1
59	Age(35–45)	1
61	ConvictionsNumber	1
62	OffensesNumber	1
65	Off_Drug	1
66	Off_TheftKindred	1
67	Off_AgainstPerson	1
68	Off_OffensiveWeapons	1
69	Off_Sexual	1
70	Off_RelatedToPolice	1
71	Off_Fraud	1
72	Off_Total	1
73	Off_AgainstProperty	1
74	NumOfArrests	1
77	AR_OFF_Drug_trafficking_Offenses	1
88	AR_PL_Brent	1
89	AR_PL_Bromley	1
90	AR_PL_Camden	1
97	AR_PL_Haringey	1
103	AR_PL_Kensington_and_Chelsea	1
118	NumOfDrugSeizures	1
126	NumOfTactics	1
129	Police	1
136	NumOfTacticSequences	1

(continued)

(continued)

137	InOperation	1
140	ArrMode_Direct	1
141	ArrMode_Result_of_Enquiries	1
143	ArrMode_Other	1
144	OnBailAtTimeOfOffence	1
Drugs activated		
121	*Crack*	1
122	*Heroin_Diamorphine*	1
Tactics activated		
133	*Covert_Purchase*	1
Query: Croydon		
Variables activated		**Values**
1	Sex_Male	1
10	Croydon	1
22	Kensington_and_Chelsea	1
47	UK	1
51	(EA3)_Afro-Caribbean	1
59	Age(35–45)	1
61	ConvictionsNumber	1
62	OffensesNumber	1
65	Off_Drug	1
66	Off_TheftKindred	1
67	Off_AgainstPerson	1
68	Off_OffensiveWeapons	1
69	Off_Sexual	1
70	Off_RelatedToPolice	1
71	Off_Fraud	1
72	Off_Total	1
73	Off_AgainstProperty	1
74	NumOfArrests	1
77	AR_OFF_Drug_trafficking_Offenses	1
88	AR_PL_Brent	1
89	AR_PL_Bromley	1
90	AR_PL_Camden	1
97	AR_PL_Haringey	1
103	AR_PL_Kensington_and_Chelsea	1
118	NumOfDrugSeizures	1
126	NumOfTactics	1
129	Police	1
136	NumOfTacticSequences	1
137	InOperation	1
140	ArrMode_Direct	1
141	ArrMode_Result_of_Enquiries	1
143	ArrMode_Other	1
144	OnBailAtTimeOfOffence	1
Drugs activated		
121	*Crack*	1
122	*Heroin_Diamorphine*	1

(continued)

(continued)

Tactics activated		
133	*Covert_Purchase*	1
Query: AR_PL_Croydon		
Variables activated		**Values**
1	Sex_Male	1
42	JAM	1
51	(EA3)_Afro-Caribbean	1
59	Age(35–45)	0.62
61	ConvictionsNumber	1
62	OffensesNumber	1
65	Off_Drug	1
66	Off_TheftKindred	1
67	Off_AgainstPerson	1
68	Off_OffensiveWeapons	1
69	Off_Sexual	1
70	Off_RelatedToPolice	1
71	Off_Fraud	0.9
72	Off_Total	1
73	Off_AgainstProperty	1
74	NumOfArrests	1
77	AR_OFF_Drug_trafficking_Offenses	1
88	AR_PL_Brent	1
89	AR_PL_Bromley	1
90	AR_PL_Camden	1
91	AR_PL_Croydon	1
97	AR_PL_Haringey	1
103	AR_PL_Kensington_and_Chelsea	1
106	AR_PL_Lewisham	1
118	NumOfDrugSeizures	1
126	NumOfTactics	1
129	Police	1
136	NumOfTacticSequences	1
137	InOperation	1
138	ViolentOnArrest	0.97
140	ArrMode_Direct	1
141	ArrMode_Result_of_Enquiries	1
143	ArrMode_Other	1
144	OnBailAtTimeOfOffence	1
Drugs activated		
121	*Crack*	1
122	*Heroin_Diamorphine*	1
Tactics activated		
133	Covert_Purchase	1
135	Other_Generic_Tactic	0.65

(continued)

(continued)

Query: Ealing		
Variables activated		**Values**
1	Sex_Male	1
11	Ealing	1
17	Harrow	1
47	UK	1
51	(EA3)_Afro-Caribbean	1
57	Age(21–25)	0.01
61	ConvictionsNumber	1
62	OffensesNumber	1
65	Off_Drug	1
66	Off_TheftKindred	1
67	Off_AgainstPerson	1
68	Off_OffensiveWeapons	1
70	Off_RelatedToPolice	1
71	Off_Fraud	1
72	Off_Total	1
73	Off_AgainstProperty	1
74	NumOfArrests	1
77	AR_OFF_Drug_trafficking_Offenses	0.83
78	AR_OFF_Drug_Possession_Offenses	1
80	AR_OFF_Offensive_Weapon_Offenses	1
81	AR_OFF_Firearms_Offenses	1
83	AR_OFF_Other_violent_Offenses	1
98	AR_PL_Harrow	1
113	AR_PL_Tower_Hamlets	0.94
124	NumOfCashSeizures	1
126	NumOfTactics	1
129	Police	1
138	ViolentOnArrest	1
139	ArrMode_NA	1
140	ArrMode_Direct	1
144	OnBailAtTimeOfOffence	1
Drug activated		
119	*Cannabis*	0.01
Tactics activated		
131	*Search_of_Person*	1
132	*Search_of_Premises*	0.62
135	*Other_Generic_Tactic*	0.99
Query: AR_PL_ Ealing		
Variables activated		**Values**
1	Sex_Male	1
17	Harrow	1
47	UK	1
52	(EA4)_Asia	1
57	Age(21–25)	1

(continued)

(continued)

65	Off_Drug	0.18
68	Off_OffensiveWeapons	0.01
74	NumOfArrests	1
78	AR_OFF_Drug_Possession_Offenses	1
80	AR_OFF_Offensive_Weapon_Offenses	1
81	AR_OFF_Firearms_Offenses	1
82	AR_OFF_Kidnapping_and_Abduction_Offenses	1
83	AR_OFF_Other_violent_Offenses	1
86	AR_PL_Barnet	1
92	AR_PL_Ealing	1
98	AR_PL_Harrow	1
113	AR_PL_Tower_Hamlets	0.2
114	AR_PL_Waltham_Forest	0.99
124	NumOfCashSeizures	1
126	NumOfTactics	1
129	Police	1
140	ArrMode_Direct	1
144	OnBailAtTimeOfOffence	1
Drugs activated		
119	*Cannabis*	1
Tactics activated		
131	*Search_of_Person*	1
132	*Search_of_Premises*	1
Query: Enfield		
Variables activated		**Values**
1	Sex_Male	1
12	Enfield	1
42	JAM	1
51	(EA3)_Afro-Caribbean	1
58	Age(25–35)	1
61	ConvictionsNumber	1
62	OffensesNumber	1
65	Off_Drug	1
66	Off_TheftKindred	1
67	Off_AgainstPerson	1
68	Off_OffensiveWeapons	1
69	Off_Sexual	1
70	Off_RelatedToPolice	1
72	Off_Total	1
73	Off_AgainstProperty	1
74	NumOfArrests	1
77	AR_OFF_Drug_trafficking_Offenses	1
88	AR_PL_Brent	1
89	AR_PL_Bromley	1
90	AR_PL_Camden	1
97	AR_PL_Haringey	1
103	AR_PL_Kensington_and_Chelsea	1
118	NumOfDrugSeizures	1
126	NumOfTactics	1

(continued)

(continued)

129	Police	1
136	NumOfTacticSequences	1
137	InOperation	1
138	ViolentOnArrest	1
140	ArrMode_Direct	1
141	ArrMode_Result_of_Enquiries	1
143	ArrMode_Other	1
144	OnBailAtTimeOfOffence	1
Drugs activated		
121	*Crack*	1
122	*Heroin_Diamorphine*	1
Tactics activated		
133	*Covert_Purchase*	1
135	*Other_Generic_Tactic*	1
Query: AR_PL_Enfield		
Variables activated		**Values**
1	Sex_Male	1
17	Harrow	1
47	UK	1
51	(EA3)_Afro-Caribbean	1
57	Age(21–25)	0.02
61	ConvictionsNumber	1
62	OffensesNumber	1
65	Off_Drug	1
66	Off_TheftKindred	1
67	Off_AgainstPerson	1
68	Off_OffensiveWeapons	1
70	Off_RelatedToPolice	1
71	Off_Fraud	1
72	Off_Total	1
73	Off_AgainstProperty	1
74	NumOfArrests	1
77	AR_OFF_Drug_trafficking_Offenses	0.94
78	AR_OFF_Drug_Possession_Offenses	0.95
80	AR_OFF_Offensive_Weapon_Offenses	1
81	AR_OFF_Firearms_Offenses	1
83	AR_OFF_Other_violent_Offenses	1
93	AR_PL_Enfield	1
98	AR_PL_Harrow	1
113	AR_PL_Tower_Hamlets	1
124	NumOfCashSeizures	1
126	NumOfTactics	1
129	Police	1
138	ViolentOnArrest	1
139	ArrMode_NA	1
140	ArrMode_Direct	1
144	OnBailAtTimeOfOffence	1

(continued)

(continued)

Tactics activated		
131	*Search_of_Person*	1
132	*Search_of_Premises*	0.73
135	*Other_Generic_Tactic*	0.79
Query: Greenwich		
Variables activated		**Values**
1	Sex_Male	1
6	Bexley	1
13	Greenwich	1
29	Richmond_upon_Thames	1
47	UK	1
49	(EA1)_White_European	1
56	Age(18–21)	1
75	AR_OFF_Theft_and_Kindred_Offenses	1
84	AR_OFF_Other_Offenses	1
87	AR_PL_Bexley	1
110	AR_PL_Richmond_upon_Thames	1
Drugs activated		
119	*Cannabis*	1
Tactics activated		
132	*Search_of_Premises*	1
Query: AR_PL_Greenwich		
Variables activated		**Values**
1	Sex_Male	1
6	Bexley	1
29	Richmond_upon_Thames	1
47	UK	1
49	(EA1)_White_European	1
56	Age(18–21)	1
75	AR_OFF_Theft_and_Kindred_Offenses	1
84	AR_OFF_Other_Offenses	1
87	AR_PL_Bexley	1
94	AR_PL_Greenwich	1
110	AR_PL_Richmond_upon_Thames	1
139	ArrMode_NA	1
Drugs activated		
119	*Cannabis*	1
Tactics activated		
132	*Search_of_Premises*	1
Query: Hackney		
Variables activated		**Values**
1	Sex_Male	1
14	Hackney	1
42	JAM	1
51	(EA3)_Afro-Caribbean	1
59	Age(35–45)	1

(continued)

(continued)

61	ConvictionsNumber	1
62	OffensesNumber	1
65	Off_Drug	1
66	Off_TheftKindred	1
67	Off_AgainstPerson	1
68	Off_OffensiveWeapons	1
69	Off_Sexual	1
70	Off_RelatedToPolice	1
71	Off_Fraud	0.42
72	Off_Total	1
73	Off_AgainstProperty	1
74	NumOfArrests	1
77	AR_OFF_Drug_trafficking_Offenses	1
88	AR_PL_Brent	1
89	AR_PL_Bromley	1
90	AR_PL_Camden	1
97	AR_PL_Haringey	1
103	AR_PL_Kensington_and_Chelsea	1
118	NumOfDrugSeizures	1
126	NumOfTactics	1
129	Police	1
136	NumOfTacticSequences	1
137	InOperation	1
138	ViolentOnArrest	0.88
140	ArrMode_Direct	1
141	ArrMode_Result_of_Enquiries	1
143	ArrMode_Other	1
144	OnBailAtTimeOfOffence	1
Drugs activated		
121	*Crack*	1
122	*Heroin_Diamorphine*	1
Tactics activated		
133	*Covert_Purchase*	1
135	*Other_Generic_Tactic*	1
Query: AR_PL_Hackney		
Variables activated		**Values**
1	Sex_Male	1
42	JAM	1
51	(EA3)_Afro-Caribbean	1
58	Age(25–35)	1
61	ConvictionsNumber	1
62	OffensesNumber	1
65	Off_Drug	1
66	Off_TheftKindred	1
67	Off_AgainstPerson	1
68	Off_OffensiveWeapons	1
69	Off_Sexual	1
70	Off_RelatedToPolice	1

(continued)

(continued)

71	Off_Fraud	0.42
72	Off_Total	1
73	Off_AgainstProperty	1
74	NumOfArrests	1
77	AR_OFF_Drug_trafficking_Offenses	1
88	AR_PL_Brent	1
89	AR_PL_Bromley	1
90	AR_PL_Camden	1
95	AR_PL_Hackney	1
97	AR_PL_Haringey	1
103	AR_PL_Kensington_and_Chelsea	1
118	NumOfDrugSeizures	1
126	NumOfTactics	1
129	Police	1
136	NumOfTacticSequences	1
137	InOperation	1
138	ViolentOnArrest	1
140	ArrMode_Direct	1
141	ArrMode_Result_of_Enquiries	1
143	ArrMode_Other	1
144	OnBailAtTimeOfOffence	1
Drugs activated		
121	*Crack*	1
122	*Heroin_Diamorphine*	1
Tactics activated		
133	*Covert_Purchase*	1
135	*Other_Generic_Tactic*	1
Query: Hammersmith_and_Fulham		
Variables activated		**Values**
1	Sex_Male	1
15	Hammersmith_and_Fulham	1
22	Kensington_and_Chelsea	1
47	UK	1
51	(EA3)_Afro-Caribbean	1
59	Age(35–45)	1
61	ConvictionsNumber	1
62	OffensesNumber	1
65	Off_Drug	1
66	Off_TheftKindred	1
67	Off_AgainstPerson	1
68	Off_OffensiveWeapons	1
69	Off_Sexual	1
70	Off_RelatedToPolice	1
71	Off_Fraud	1
72	Off_Total	1
73	Off_AgainstProperty	1
74	NumOfArrests	1
77	AR_OFF_Drug_trafficking_Offenses	1

(continued)

(continued)

88	AR_PL_Brent	1
89	AR_PL_Bromley	1
90	AR_PL_Camden	1
97	AR_PL_Haringey	1
103	AR_PL_Kensington_and_Chelsea	1
118	NumOfDrugSeizures	1
126	NumOfTactics	1
129	Police	1
136	NumOfTacticSequences	1
137	InOperation	1
140	ArrMode_Direct	1
141	ArrMode_Result_of_Enquiries	1
143	ArrMode_Other	1
144	OnBailAtTimeOfOffence	1
Drugs activated		
121	*Crack*	1
122	*Heroin_Diamorphine*	1
Tactics activated		
133	*Covert_Purchase*	1
Query: AR_PL_Hammersmith_and_Fulham		
Variables activated		**Values**
1	Sex_Male	1
17	Harrow	1
47	UK	1
51	(EA3)_Afro-Caribbean	1
61	ConvictionsNumber	1
62	OffensesNumber	1
65	Off_Drug	1
66	Off_TheftKindred	1
67	Off_AgainstPerson	1
68	Off_OffensiveWeapons	1
70	Off_RelatedToPolice	1
71	Off_Fraud	1
72	Off_Total	1
73	Off_AgainstProperty	1
74	NumOfArrests	1
76	AR_OFF_Offenses_the_Person_Offenses	1
77	AR_OFF_Drug_trafficking_Offenses	1
80	AR_OFF_Offensive_Weapon_Offenses	1
81	AR_OFF_Firearms_Offenses	1
83	AR_OFF_Other_violent_Offenses	1
90	AR_PL_Camden	0.01
96	AR_PL_Hammersmith_and_Fulham	1
98	AR_PL_Harrow	1
113	AR_PL_Tower_Hamlets	0.01
124	NumOfCashSeizures	0.99

(continued)

(continued)

126	NumOfTactics	1
129	Police	1
138	ViolentOnArrest	1
139	ArrMode_NA	1
140	ArrMode_Direct	1
144	OnBailAtTimeOfOffence	1
Tactics activated		
131	*Search_of_Person*	1
135	*Other_Generic_Tactic*	1
Query: Haringey		
Variables activated		**Values**
1	Sex_Male	1
16	Haringey	1
42	JAM	1
51	(EA3)_Afro-Caribbean	1
58	Age(25–35)	1
61	ConvictionsNumber	1
62	OffensesNumber	1
65	Off_Drug	1
66	Off_TheftKindred	1
67	Off_AgainstPerson	1
68	Off_OffensiveWeapons	1
69	Off_Sexual	1
70	Off_RelatedToPolice	1
71	Off_Fraud	0.03
72	Off_Total	1
73	Off_AgainstProperty	1
74	NumOfArrests	1
77	AR_OFF_Drug_trafficking_Offenses	1
88	AR_PL_Brent	1
89	AR_PL_Bromley	1
90	AR_PL_Camden	1
97	AR_PL_Haringey	1
103	AR_PL_Kensington_and_Chelsea	1
106	AR_PL_Lewisham	0.04
118	NumOfDrugSeizures	1
126	NumOfTactics	1
129	Police	1
136	NumOfTacticSequences	1
137	InOperation	1
140	ArrMode_Direct	1
141	ArrMode_Result_of_Enquiries	1
143	ArrMode_Other	1
144	OnBailAtTimeOfOffence	1
Drugs activated		
121	*Crack*	1
122	*Heroin_Diamorphine*	1

(continued)

(continued)

Tactics activated		
133	*Covert_Purchase*	1
135	*Other_Generic_Tactic*	1
Query: AR_PL_Haringey		
Variables activated		**Values**
1	Sex_Male	1
9	Camden	1
42	JAM	1
51	(EA3)_Afro-Caribbean	1
59	Age(35–45)	1
61	ConvictionsNumber	1
62	OffensesNumber	1
65	Off_Drug	1
66	Off_TheftKindred	1
67	Off_AgainstPerson	1
68	Off_OffensiveWeapons	1
69	Off_Sexual	1
70	Off_RelatedToPolice	1
71	Off_Fraud	1
72	Off_Total	1
73	Off_AgainstProperty	1
74	NumOfArrests	1
77	AR_OFF_Drug_trafficking_Offenses	1
88	AR_PL_Brent	1
89	AR_PL_Bromley	1
90	AR_PL_Camden	1
97	AR_PL_Haringey	1
103	AR_PL_Kensington_and_Chelsea	1
118	NumOfDrugSeizures	1
126	NumOfTactics	1
129	Police	1
136	NumOfTacticSequences	1
137	InOperation	1
140	ArrMode_Direct	1
141	ArrMode_Result_of_Enquiries	1
143	ArrMode_Other	1
144	OnBailAtTimeOfOffence	1
Drugs activated		
121	*Crack*	1
122	*Heroin_Diamorphine*	1
Tactics activated		
133	*Covert_Purchase*	1
135	*Other_Generic_Tactic*	1

(continued)

(continued)

Query: Harrow		
Variables activated		**Values**
1	Sex_Male	1
17	Harrow	1
47	UK	1
51	(EA3)_Afro-Caribbean	1
59	Age(35–45)	1
61	ConvictionsNumber	1
62	OffensesNumber	1
65	Off_Drug	1
66	Off_TheftKindred	1
67	Off_AgainstPerson	1
68	Off_OffensiveWeapons	1
69	Off_Sexual	1
70	Off_RelatedToPolice	1
71	Off_Fraud	1
72	Off_Total	1
73	Off_AgainstProperty	1
74	NumOfArrests	1
77	AR_OFF_Drug_trafficking_Offenses	1
88	AR_PL_Brent	1
89	AR_PL_Bromley	1
90	AR_PL_Camden	1
97	AR_PL_Haringey	1
98	AR_PL_Harrow	1
103	AR_PL_Kensington_and_Chelsea	1
118	NumOfDrugSeizures	1
126	NumOfTactics	1
129	Police	1
136	NumOfTacticSequences	1
137	InOperation	1
138	ViolentOnArrest	1
140	ArrMode_Direct	1
141	ArrMode_Result_of_Enquiries	1
143	ArrMode_Other	1
144	OnBailAtTimeOfOffence	1
Drugs activated		
121	*Crack*	1
122	*Heroin_Diamorphine*	1
Tactics activated		
133	*Covert_Purchase*	1
135	*Other_Generic_Tactic*	1
Query: AR_PL_Harrow		
Variables activated		**Values**
1	Sex_Male	1
9	Camden	1
47	UK	1

(continued)

(continued)

51	(EA3)_Afro-Caribbean	1
59	Age(35–45)	1
61	ConvictionsNumber	1
62	OffensesNumber	1
65	Off_Drug	1
66	Off_TheftKindred	1
67	Off_AgainstPerson	1
68	Off_OffensiveWeapons	1
69	Off_Sexual	1
70	Off_RelatedToPolice	1
71	Off_Fraud	1
72	Off_Total	1
73	Off_AgainstProperty	1
74	NumOfArrests	1
77	AR_OFF_Drug_trafficking_Offenses	1
88	AR_PL_Brent	1
89	AR_PL_Bromley	1
90	AR_PL_Camden	1
97	AR_PL_Haringey	1
98	AR_PL_Harrow	1
103	AR_PL_Kensington_and_Chelsea	1
118	NumOfDrugSeizures	1
126	NumOfTactics	1
129	Police	1
136	NumOfTacticSequences	1
137	InOperation	1
139	ArrMode_NA	1
140	ArrMode_Direct	1
141	ArrMode_Result_of_Enquiries	1
143	ArrMode_Other	1
144	OnBailAtTimeOfOffence	1
Drugs activated		
121	*Crack*	1
122	*Heroin_Diamorphine*	1
Tactics activated		
133	*Covert_Purchase*	1
135	*Other_Generic_Tactic*	1
Query: Havering		
Variables activated		**Values**
1	Sex_Male	1
18	Havering	1
22	Kensington_and_Chelsea	1
47	UK	1
51	(EA3)_Afro-Caribbean	1
59	Age(35–45)	1
61	ConvictionsNumber	1

(continued)

(continued)

62	OffensesNumber	1
65	Off_Drug	1
66	Off_TheftKindred	1
67	Off_AgainstPerson	1
68	Off_OffensiveWeapons	1
69	Off_Sexual	1
70	Off_RelatedToPolice	1
71	Off_Fraud	1
72	Off_Total	1
73	Off_AgainstProperty	1
74	NumOfArrests	1
77	AR_OFF_Drug_trafficking_Offenses	1
88	AR_PL_Brent	1
89	AR_PL_Bromley	1
97	AR_PL_Haringey	1
103	AR_PL_Kensington_and_Chelsea	1
118	NumOfDrugSeizures	1
126	NumOfTactics	1
129	Police	1
136	NumOfTacticSequences	1
137	InOperation	1
140	ArrMode_Direct	1
141	ArrMode_Result_of_Enquiries	1
143	ArrMode_Other	1
144	OnBailAtTimeOfOffence	1
Drugs activated		
121	*Crack*	1
122	*Heroin_Diamorphine*	1
Tactics activated		
133	*Covert_Purchase*	1
Query: AR_PL_Havering		
Variables activated		**Values**
1	Sex_Male	1
22	Kensington_and_Chelsea	1
47	UK	1
51	(EA3)_Afro-Caribbean	1
60	Age(>45)	1
61	ConvictionsNumber	1
62	OffensesNumber	1
65	Off_Drug	1
66	Off_TheftKindred	1
67	Off_AgainstPerson	1
68	Off_OffensiveWeapons	1
69	Off_Sexual	1
70	Off_RelatedToPolice	1
71	Off_Fraud	1

(continued)

(continued)

72	Off_Total	1
73	Off_AgainstProperty	1
74	NumOfArrests	1
77	AR_OFF_Drug_trafficking_Offenses	1
88	AR_PL_Brent	1
89	AR_PL_Bromley	1
97	AR_PL_Haringey	1
99	AR_PL_Havering	1
103	AR_PL_Kensington_and_Chelsea	1
118	NumOfDrugSeizures	1
126	NumOfTactics	1
129	Police	1
136	NumOfTacticSequences	1
137	InOperation	1
140	ArrMode_Direct	1
141	ArrMode_Result_of_Enquiries	1
143	ArrMode_Other	1
144	OnBailAtTimeOfOffence	1
Drugs activated		
121	*Crack*	1
122	*Heroin_Diamorphine*	1
Tactics activated		
133	*Covert_Purchase*	1
Query: Hillingdon		
Variables activated		**Values**
1	Sex_Male	1
19	Hillingdon	1
29	Richmond_upon_Thames	1
47	UK	1
49	(EA1)_White_European	1
56	Age(18–21)	1
100	AR_PL_Hillingdon	1
110	AR_PL_Richmond_upon_Thames	1
124	NumOfCashSeizures	1
125	Pounds	1
139	ArrMode_NA	1
Tactics activated		
132	*Search_of_Premises*	1
Query: AR_PL_Hillingdon		
Variables activated		**Values**
1	Sex_Male	1
19	Hillingdon	1
47	UK	1
52	(EA4)_Asia	1
56	Age (18–21)	1
100	AR_PL_Hillingdon	1

(continued)

(continued)

Query: Hounslow		
Variables activated		**Values**
1	Sex_Male	1
17	Harrow	1
20	Hounslow	1
47	UK	1
52	(EA4)_Asia	1
57	Age(21–25)	0.06
61	ConvictionsNumber	1
62	OffensesNumber	1
65	Off_Drug	1
66	Off_TheftKindred	1
67	Off_AgainstPerson	1
68	Off_OffensiveWeapons	1
70	Off_RelatedToPolice	1
71	Off_Fraud	1
72	Off_Total	1
73	Off_AgainstProperty	1
74	NumOfArrests	1
77	AR_OFF_Drug_trafficking_Offenses	1
78	AR_OFF_Drug_Possession_Offenses	1
80	AR_OFF_Offensive_Weapon_Offenses	1
81	AR_OFF_Firearms_Offenses	1
83	AR_OFF_Other_violent_Offenses	1
88	AR_PL_Brent	0.99
98	AR_PL_Harrow	1
113	AR_PL_Tower_Hamlets	1
118	NumOfDrugSeizures	0.98
124	NumOfCashSeizures	1
126	NumOfTactics	1
129	Police	1
138	ViolentOnArrest	1
139	ArrMode_NA	1
140	ArrMode_Direct	1
144	OnBailAtTimeOfOffence	1
Tactics activated		
131	*Search_of_Person*	1
132	*Search_of_Premises*	1
135	*Other_Generic_Tactic*	1
Query: AR_PL_Hounslow		
Variables activated		**Values**
1	Sex_Male	1
17	Harrow	0.34
32	Tower_Hamlets	1
47	UK	1
52	(EA4)_Asia	1
57	Age(21–25)	1

(continued)

(continued)

61	ConvictionsNumber	1
62	OffensesNumber	1
65	Off_Drug	1
66	Off_TheftKindred	1
67	Off_AgainstPerson	1
68	Off_OffensiveWeapons	1
70	Off_RelatedToPolice	1
71	Off_Fraud	1
72	Off_Total	1
73	Off_AgainstProperty	1
74	NumOfArrests	1
77	AR_OFF_Drug_trafficking_Offenses	0.33
78	AR_OFF_Drug_Possession_Offenses	1
80	AR_OFF_Offensive_Weapon_Offenses	1
81	AR_OFF_Firearms_Offenses	1
83	AR_OFF_Other_violent_Offenses	1
98	AR_PL_Harrow	1
101	AR_PL_Hounslow	1
113	AR_PL_Tower_Hamlets	1
124	NumOfCashSeizures	1
126	NumOfTactics	1
129	Police	1
138	ViolentOnArrest	1
139	ArrMode_NA	1
140	ArrMode_Direct	1
144	OnBailAtTimeOfOffence	1
Drugs activated		
119	Cannabis	0.01
Tactics activated		
131	*Search_of_Person*	1
132	*Search_of_Premises*	1
135	*Other_Generic_Tactic*	1
Query: Islington		
Variables activated		**Values**
1	Sex_Male	1
21	Islington	1
22	Kensington_and_Chelsea	1
47	UK	1
51	(EA3)_Afro-Caribbean	1
59	Age(35–45)	1
61	ConvictionsNumber	1
62	OffensesNumber	1
65	Off_Drug	1
66	Off_TheftKindred	1
67	Off_AgainstPerson	1
68	Off_OffensiveWeapons	1
69	Off_Sexual	1

(continued)

(continued)

70	Off_RelatedToPolice	1
71	Off_Fraud	1
72	Off_Total	1
73	Off_AgainstProperty	1
74	NumOfArrests	1
77	AR_OFF_Drug_trafficking_Offenses	1
88	AR_PL_Brent	1
89	AR_PL_Bromley	1
90	AR_PL_Camden	1
97	AR_PL_Haringey	1
103	AR_PL_Kensington_and_Chelsea	1
118	NumOfDrugSeizures	1
126	NumOfTactics	1
129	Police	1
136	NumOfTacticSequences	1
137	InOperation	1
140	ArrMode_Direct	1
141	ArrMode_Result_of_Enquiries	1
143	ArrMode_Other	1
144	OnBailAtTimeOfOffence	1
Drugs activated		
121	*Crack*	1
122	*Heroin_Diamorphine*	1
Tactics activated		
133	*Covert_Purchase*	1
Query: AR_PL_Islington		
Variables activated		**Values**
1	Sex_Male	1
9	Camden	1
42	JAM	1
51	(EA3)_Afro-Caribbean	1
59	Age(35–45)	1
61	ConvictionsNumber	1
62	OffensesNumber	1
65	Off_Drug	1
66	Off_TheftKindred	1
67	Off_AgainstPerson	1
68	Off_OffensiveWeapons	1
69	Off_Sexual	1
70	Off_RelatedToPolice	1
71	Off_Fraud	1
72	Off_Total	1
73	Off_AgainstProperty	1
74	NumOfArrests	1
77	AR_OFF_Drug_trafficking_Offenses	1
88	AR_PL_Brent	1
89	AR_PL_Bromley	1

(continued)

(continued)

90	AR_PL_Camden	1
97	AR_PL_Haringey	1
102	AR_PL_Islington	1
103	AR_PL_Kensington_and_Chelsea	1
118	NumOfDrugSeizures	1
126	NumOfTactics	1
129	Police	1
136	NumOfTacticSequences	1
137	InOperation	1
140	ArrMode_Direct	1
141	ArrMode_Result_of_Enquiries	1
143	ArrMode_Other	1
144	OnBailAtTimeOfOffence	1
Drugs activated		
121	*Crack*	1
122	*Heroin_Diamorphine*	1
Tactics activated		
133	*Covert_Purchase*	1
135	*Other_Generic_Tactic*	1
Query: Kensington & Chelsea		
Variables activated		**Values**
1	Sex_Male	1
22	Kensington & Chelsea	1
47	UK	1
51	(EA3)_Afro-Caribbean	1
59	Age(35–45)	1
61	ConvictionsNumber	1
62	OffensesNumber	1
65	Off_Drug	1
66	Off_TheftKindred	1
67	Off_AgainstPerson	1
68	Off_OffensiveWeapons	1
69	Off_Sexual	1
70	Off_RelatedToPolice	1
71	Off_Fraud	1
72	Off_Total	1
73	Off_AgainstProperty	1
74	NumOfArrests	1
77	AR_OFF_Drug_trafficking_Offenses	1
88	AR_PL_Brent	1
89	AR_PL_Bromley	1
90	AR_PL_Camden	1
97	AR_PL_Haringey	1
103	AR_PL_Kensington_and_Chelsea	1
118	NumOfDrugSeizures	1
126	NumOfTactics	1

(continued)

(continued)

129	Police	1
136	NumOfTacticSequences	1
137	InOperation	1
140	ArrMode_Direct	1
141	ArrMode_Result_of_Enquiries	1
143	ArrMode_Other	1
144	OnBailAtTimeOfOffence	1
Drugs activated		
121	*Crack*	1
122	*Heroin_Diamorphine*	1
Tactics activated		
133	*Covert_Purchase*	1
Query: AR_PL_Kensington & Chelsea		
Variables activated		**Values**
1	Sex_Male	1
22	Kensington & Chelsea	1
47	UK	1
51	(EA3)_Afro-Caribbean	1
59	Age(35–45)	1
61	ConvictionsNumber	1
62	OffensesNumber	1
65	Off_Drug	1
66	Off_TheftKindred	1
67	Off_AgainstPerson	1
68	Off_OffensiveWeapons	1
69	Off_Sexual	1
70	Off_RelatedToPolice	1
71	Off_Fraud	1
72	Off_Total	1
73	Off_AgainstProperty	1
74	NumOfArrests	1
77	AR_OFF_Drug_trafficking_Offenses	1
88	AR_PL_Brent	1
89	AR_PL_Bromley	1
90	AR_PL_Camden	1
97	AR_PL_Haringey	1
103	AR_PL_Kensington_and_Chelsea	1
118	NumOfDrugSeizures	1
126	NumOfTactics	1
129	Police	1
136	NumOfTacticSequences	1
137	InOperation	1
140	ArrMode_Direct	1
141	ArrMode_Result_of_Enquiries	1
143	ArrMode_Other	1
144	OnBailAtTimeOfOffence	1

(continued)

(continued)

Drugs activated		
121	*Crack*	1
122	*Heroin_Diamorphine*	1
Tactics activated		
133	*Covert_Purchase*	1
Query: Kingston_upon_Thames		
Variables activated		**Values**
3	Sex_notknown	1
23	Kingston_upon_Thames	1
35	Westminster	1
47	UK	1
49	(EA1)_White_European	1
59	Age(35–45)	1
61	ConvictionsNumber	1
62	OffensesNumber	1
65	Off_Drug	1
66	Off_TheftKindred	1
67	Off_AgainstPerson	1
68	Off_OffensiveWeapons	1
69	Off_Sexual	1
70	Off_RelatedToPolice	1
71	Off_Fraud	1
72	Off_Total	1
73	Off_AgainstProperty	1
78	AR_OFF_Drug_Possession_Offenses	1
104	AR_PL_Kingston_upon_Thames	1
116	AR_PL_Westiminster	1
139	ArrMode_NA	1
142	ArrMode_Given_into_custody	1
Drugs activated		
123	*MDMA*	1
Query: AR_PL_Kingston_upon_Thames		
Variables activated		**Values**
3	Sex_notknown	1
23	Kingston_upon_Thames	1
35	Westminster	1
47	UK	1
49	(EA1)_White_European	1
59	Age(35–45)	1
61	ConvictionsNumber	1
62	OffensesNumber	1
65	Off_Drug	1
66	Off_TheftKindred	1
67	Off_AgainstPerson	1
68	Off_OffensiveWeapons	1
69	Off_Sexual	1

(continued)

(continued)

70	Off_RelatedToPolice	1
71	Off_Fraud	1
72	Off_Total	1
73	Off_AgainstProperty	1
78	AR_OFF_Drug_Possession_Offenses	1
104	AR_PL_Kingston_upon_Thames	1
116	AR_PL_Westiminster	1
139	ArrMode_NA	1
142	ArrMode_Given_into_custody	1
Drugs activated		
123	*MDMA*	1
Query: Lambeth		
Variables activated		**Values**
1	Sex_Male	1
24	Lambeth	1
42	JAM	1
51	(EA3)_Afro-Caribbean	1
59	Age(35–45)	1
61	ConvictionsNumber	1
62	OffensesNumber	1
65	Off_Drug	1
66	Off_TheftKindred	1
67	Off_AgainstPerson	1
68	Off_OffensiveWeapons	1
69	Off_Sexual	1
70	Off_RelatedToPolice	1
71	Off_Fraud	1
72	Off_Total	1
73	Off_AgainstProperty	1
74	NumOfArrests	1
77	AR_OFF_Drug_trafficking_Offenses	1
88	AR_PL_Brent	1
89	AR_PL_Bromley	1
90	AR_PL_Camden	1
97	AR_PL_Haringey	1
103	AR_PL_Kensington_and_Chelsea	1
106	AR_PL_Lewisham	1
118	NumOfDrugSeizures	1
126	NumOfTactics	1
129	Police	1
136	NumOfTacticSequences	1
137	InOperation	1
138	ViolentOnArrest	1
140	ArrMode_Direct	1
141	ArrMode_Result_of_Enquiries	1
143	ArrMode_Other	1
144	OnBailAtTimeOfOffence	1

(continued)

(continued)

Drugs activated		
121	*Crack*	1
122	*Heroin_Diamorphine*	1
Tactics activated		
133	*Covert_Purchase*	1
135	*Other_Generic_Tactic*	1
Query: AR_PL_Lambeth		
Variables activated		**Values**
1	Sex_Male	1
42	JAM	1
51	(EA3)_Afro-Caribbean	1
58	Age(25–35)	0.97
61	ConvictionsNumber	1
62	OffensesNumber	1
65	Off_Drug	1
66	Off_TheftKindred	1
67	Off_AgainstPerson	1
68	Off_OffensiveWeapons	1
69	Off_Sexual	1
70	Off_RelatedToPolice	1
71	Off_Fraud	0.39
72	Off_Total	1
73	Off_AgainstProperty	1
74	NumOfArrests	1
77	AR_OFF_Drug_trafficking_Offenses	1
88	AR_PL_Brent	1
89	AR_PL_Bromley	1
90	AR_PL_Camden	1
97	AR_PL_Haringey	1
103	AR_PL_Kensington_and_Chelsea	1
105	AR_PL_Lambeth	1
106	AR_PL_Lewisham	0.98
118	NumOfDrugSeizures	1
126	NumOfTactics	1
129	Police	1
136	NumOfTacticSequences	1
137	InOperation	1
138	ViolentOnArrest	1
140	ArrMode_Direct	1
141	ArrMode_Result_of_Enquiries	1
143	ArrMode_Other	1
144	OnBailAtTimeOfOffence	1
Drugs activated		
121	*Crack*	1
122	*Heroin_Diamorphine*	1

(continued)

(continued)

Tactics activated		
133	*Covert_Purchase*	1
135	*Other_Generic_Tactic*	1
Query: Lewisham		
Variables activated		**Values**
1	Sex_Male	1
25	Lewisham	1
42	JAM	1
51	(EA3)_Afro-Caribbean	1
59	Age(35–45)	1
61	ConvictionsNumber	1
62	OffensesNumber	1
65	Off_Drug	1
66	Off_TheftKindred	1
67	Off_AgainstPerson	1
68	Off_OffensiveWeapons	1
69	Off_Sexual	1
70	Off_RelatedToPolice	1
71	Off_Fraud	1
72	Off_Total	1
73	Off_AgainstProperty	1
74	NumOfArrests	1
77	AR_OFF_Drug_trafficking_Offenses	1
88	AR_PL_Brent	1
89	AR_PL_Bromley	1
90	AR_PL_Camden	1
97	AR_PL_Haringey	1
103	AR_PL_Kensington_and_Chelsea	1
106	AR_PL_Lewisham	1
118	NumOfDrugSeizures	1
126	NumOfTactics	1
129	Police	1
136	NumOfTacticSequences	1
137	InOperation	1
138	ViolentOnArrest	1
140	ArrMode_Direct	1
141	ArrMode_Result_of_Enquiries	1
143	ArrMode_Other	1
144	OnBailAtTimeOfOffence	1
Drugs activated		
121	*Crack*	1
122	*Heroin_Diamorphine*	1
Tactics activated		
133	*Covert_Purchase*	1
135	*Other_Generic_Tactic*	1

(continued)

(continued)

Query: AR_PL_Lewisham		
Variables activated		**Values**
1	Sex_Male	1
9	Camden	0.9
42	JAM	1
51	(EA3)_Afro-Caribbean	1
59	Age(35–45)	1
61	ConvictionsNumber	1
62	OffensesNumber	1
65	Off_Drug	1
66	Off_TheftKindred	1
67	Off_AgainstPerson	1
68	Off_OffensiveWeapons	1
69	Off_Sexual	1
70	Off_RelatedToPolice	1
71	Off_Fraud	1
72	Off_Total	1
73	Off_AgainstProperty	1
74	NumOfArrests	1
77	AR_OFF_Drug_trafficking_Offenses	1
88	AR_PL_Brent	1
89	AR_PL_Bromley	1
90	AR_PL_Camden	1
97	AR_PL_Haringey	1
103	AR_PL_Kensington_and_Chelsea	1
106	AR_PL_Lewisham	1
118	NumOfDrugSeizures	1
126	NumOfTactics	1
129	Police	1
136	NumOfTacticSequences	1
137	InOperation	1
138	ViolentOnArrest	0.94
140	ArrMode_Direct	1
141	ArrMode_Result_of_Enquiries	1
143	ArrMode_Other	1
144	OnBailAtTimeOfOffence	1
Drugs activated		
121	*Crack*	1
122	*Heroin_Diamorphine*	1
Tactics activated		
133	*Covert_Purchase*	1
135	*Other_Generic_Tactic*	1
Query: Merton		
Variables activated		**Values**
1	Sex_Male	1
23	Kingston_upon_Thames	0.12
26	Merton	1

(continued)

(continued)

31	Sutton	0.36
47	UK	1
49	(EA1)_White_European	1
59	Age(35–45)	1
61	ConvictionsNumber	1
62	OffensesNumber	1
65	Off_Drug	1
66	Off_TheftKindred	1
67	Off_AgainstPerson	1
68	Off_OffensiveWeapons	1
69	Off_Sexual	1
70	Off_RelatedToPolice	1
71	Off_Fraud	1
72	Off_Total	1
73	Off_AgainstProperty	1
74	NumOfArrests	1
78	AR_OFF_Drug_Possession_Offenses	1
80	AR_OFF_Offensive_Weapon_Offenses	1
81	AR_OFF_Firearms_Offenses	1
107	AR_PL_Merton	1
112	AR_PL_Sutton	1
124	NumOfCashSeizures	0.05
126	NumOfTactics	0.03
129	Police	0.23
139	ArrMode_NA	1
143	ArrMode_Other	0.21
Tactics activated		
131	*Search_of_Person*	0.93
132	*Search_of_Premises*	0.16
Query: AR_PL_Merton		
Variables activated		**Values**
1	Sex_Male	1
22	Kensington_and_Chelsea	1
47	UK	1
51	(EA3)_Afro-Caribbean	1
59	Age(35–45)	1
61	ConvictionsNumber	1
62	OffensesNumber	1
65	Off_Drug	1
66	Off_TheftKindred	1
67	Off_AgainstPerson	1
68	Off_OffensiveWeapons	1
69	Off_Sexual	1
70	Off_RelatedToPolice	1
71	Off_Fraud	1
72	Off_Total	1
73	Off_AgainstProperty	1

(continued)

(continued)

74	NumOfArrests	1
77	AR_OFF_Drug_trafficking_Offenses	1
88	AR_PL_Brent	1
89	AR_PL_Bromley	1
97	AR_PL_Haringey	1
103	AR_PL_Kensington_and_Chelsea	1
107	AR_PL_Merton	1
118	NumOfDrugSeizures	1
126	NumOfTactics	1
129	Police	1
136	NumOfTacticSequences	1
137	InOperation	1
140	ArrMode_Direct	1
141	ArrMode_Result_of_Enquiries	1
143	ArrMode_Other	1
144	OnBailAtTimeOfOffence	1
Drugs activated		
121	*Crack*	1
122	*Heroin_Diamorphine*	1
Tactics activated		
133	*Covert_Purchase*	1
Query: Newham		
Variables activated		**Values**
1	Sex_Male	1
17	Harrow	1
27	Newham	1
47	UK	1
52	(EA4)_Asia	1
57	Age(21–25)	0.2
61	ConvictionsNumber	1
62	OffensesNumber	1
65	Off_Drug	1
66	Off_TheftKindred	1
67	Off_AgainstPerson	1
68	Off_OffensiveWeapons	1
70	Off_RelatedToPolice	1
71	Off_Fraud	1
72	Off_Total	1
73	Off_AgainstProperty	1
74	NumOfArrests	1
77	AR_OFF_Drug_trafficking_Offenses	0.71
78	AR_OFF_Drug_Possession_Offenses	1
80	AR_OFF_Offensive_Weapon_Offenses	1
81	AR_OFF_Firearms_Offenses	1
83	AR_OFF_Other_violent_Offenses	1
98	AR_PL_Harrow	1

(continued)

(continued)

113	AR_PL_Tower_Hamlets	1
124	NumOfCashSeizures	1
126	NumOfTactics	1
129	Police	1
138	ViolentOnArrest	1
139	ArrMode_NA	1
140	ArrMode_Direct	1
144	OnBailAtTimeOfOffence	1
Tactics activated		
131	*Search_of_Person*	1
132	*Search_of_Premises*	1
135	*Other_Generic_Tactic*	1
Query: AR_PL_Newham		
Variables activated		**Values**
1	Sex_Male	1
32	Tower_Hamlets	1
47	UK	1
52	(EA4)_Asia	1
57	Age(21–25)	1
61	ConvictionsNumber	1
62	OffensesNumber	1
65	Off_Drug	1
66	Off_TheftKindred	1
67	Off_AgainstPerson	1
68	Off_OffensiveWeapons	1
70	Off_RelatedToPolice	1
71	Off_Fraud	1
72	Off_Total	1
73	Off_AgainstProperty	1
74	NumOfArrests	1
78	AR_OFF_Drug_Possession_Offenses	1
80	AR_OFF_Offensive_Weapon_Offenses	1
81	AR_OFF_Firearms_Offenses	1
83	AR_OFF_Other_violent_Offenses	1
98	AR_PL_Harrow	1
108	AR_PL_Newham	1
113	AR_PL_Tower_Hamlets	1
124	NumOfCashSeizures	1
126	NumOfTactics	1
129	Police	1
139	ArrMode_NA	1
140	ArrMode_Direct	1
144	OnBailAtTimeOfOffence	1
Tactics activated		
131	*Search_of_Person*	1
132	*Search_of_Premises*	1
135	*Other_Generic_Tactic*	1

(continued)

(continued)

Query: Redbridge		
Variables activated		**Values**
1	Sex_Male	1
28	Redbridge	1
31	Sutton	0.9
47	UK	1
49	(EA1)_White_European	1
60	Age(>45)	0.99
61	ConvictionsNumber	1
62	OffensesNumber	1
65	Off_Drug	0.99
66	Off_TheftKindred	1
67	Off_AgainstPerson	1
68	Off_OffensiveWeapons	1
70	Off_RelatedToPolice	0.9
71	Off_Fraud	1
72	Off_Total	1
73	Off_AgainstProperty	1
74	NumOfArrests	0.74
75	AR_OFF_Theft_and_Kindred_Offenses	0.92
80	AR_OFF_Offensive_Weapon_Offenses	1
81	AR_OFF_Firearms_Offenses	1
109	AR_PL_Redbridge	0.27
112	AR_PL_Sutton	1
124	NumOfCashSeizures	0.98
139	ArrMode_NA	1
Drugs activated		
119	*Cannabis*	0.6
Tactics activated		
132	*Search_of_Premises*	1
135	*Other_Generic_Tactic*	0.64
Query: AR_PL_Redbridge		
Variables activated		**Values**
1	Sex_Male	1
22	Kensington_and_Chelsea	1
47	UK	1
51	(EA3)_Afro-Caribbean	1
59	Age(35–45)	1
61	ConvictionsNumber	1
62	OffensesNumber	1
65	Off_Drug	1
66	Off_TheftKindred	1
67	Off_AgainstPerson	1
68	Off_OffensiveWeapons	1
69	Off_Sexual	1
70	Off_RelatedToPolice	1
71	Off_Fraud	1
72	Off_Total	1

(continued)

(continued)

73	Off_AgainstProperty	1
74	NumOfArrests	1
77	AR_OFF_Drug_trafficking_Offenses	1
88	AR_PL_Brent	1
89	AR_PL_Bromley	1
90	AR_PL_Camden	1
97	AR_PL_Haringey	1
103	AR_PL_Kensington_and_Chelsea	1
109	AR_PL_Redbridge	1
118	NumOfDrugSeizures	1
126	NumOfTactics	1
129	Police	1
136	NumOfTacticSequences	1
137	InOperation	1
139	ArrMode_NA	1
140	ArrMode_Direct	1
141	ArrMode_Result_of_Enquiries	1
143	ArrMode_Other	1
144	OnBailAtTimeOfOffence	1
Drugs activated		
121	*Crack*	1
122	*Heroin_Diamorphine*	1
Tactics activated		
133	*Covert_Purchase*	1
Query: Richmond_upon_Thames		
Variables activated		**Values**
1	Sex_Male	1
6	Bexley	1
29	Richmond_upon_Thames	1
47	UK	1
49	(EA1)_White_European	1
56	Age(18–21)	1
75	AR_OFF_Theft_and_Kindred_Offenses	1
87	AR_PL_Bexley	1
110	AR_PL_Richmond_upon_Thames	1
139	ArrMode_NA	1
Drugs activated		
119	*Cannabis*	1
Tactics activated		
132	*Search_of_Premises*	1
Query: AR_PL_Richmond_upon_Thames		
Variables activated		**Values**
1	Sex_Male	1
6	Bexley	1
29	Richmond_upon_Thames	1
47	UK	1
49	(EA1)_White_European	1

(continued)

(continued)

56	Age(18–21)	1
75	AR_OFF_Theft_and_Kindred_Offenses	1
87	AR_PL_Bexley	1
110	AR_PL_Richmond_upon_Thames	1
139	ArrMode_NA	1
Drugs activated		
119	*Cannabis*	1
Tactics activated		
132	*Search_of_Premises*	1
Query: Southwark		
Variables activated		**Values**
1	Sex_Male	1
30	Southwark	1
42	JAM	1
51	(EA3)_Afro-Caribbean	1
58	Age(25–35)	1
61	ConvictionsNumber	1
62	OffensesNumber	1
65	Off_Drug	1
66	Off_TheftKindred	1
67	Off_AgainstPerson	1
68	Off_OffensiveWeapons	1
69	Off_Sexual	1
70	Off_RelatedToPolice	1
72	Off_Total	1
73	Off_AgainstProperty	1
74	NumOfArrests	1
77	AR_OFF_Drug_trafficking_Offenses	1
88	AR_PL_Brent	1
89	AR_PL_Bromley	1
90	AR_PL_Camden	1
97	AR_PL_Haringey	1
103	AR_PL_Kensington_and_Chelsea	1
106	AR_PL_Lewisham	1
118	NumOfDrugSeizures	1
126	NumOfTactics	1
129	Police	1
136	NumOfTacticSequences	1
137	InOperation	1
138	ViolentOnArrest	1
140	ArrMode_Direct	1
141	ArrMode_Result_of_Enquiries	1
143	ArrMode_Other	1
144	OnBailAtTimeOfOffence	1

(continued)

(continued)

Drugs activated		
121	*Crack*	1
122	*Heroin_Diamorphine*	1
Tactics activated		
133	*Covert_Purchase*	1
135	*Other_Generic_Tactic*	1
Query: AR_PL_Southwark		
Variables activated		**Values**
1	Sex_Male	1
42	JAM	1
51	(EA3)_Afro-Caribbean	1
59	Age(35–45)	1
61	ConvictionsNumber	1
62	OffensesNumber	1
65	Off_Drug	1
66	Off_TheftKindred	1
67	Off_AgainstPerson	1
68	Off_OffensiveWeapons	1
69	Off_Sexual	1
70	Off_RelatedToPolice	1
71	Off_Fraud	1
72	Off_Total	1
73	Off_AgainstProperty	1
74	NumOfArrests	1
77	AR_OFF_Drug_trafficking_Offenses	1
88	AR_PL_Brent	1
89	AR_PL_Bromley	1
90	AR_PL_Camden	1
97	AR_PL_Haringey	1
103	AR_PL_Kensington_and_Chelsea	1
111	AR_PL_Southwark	1
118	NumOfDrugSeizures	1
126	NumOfTactics	1
129	Police	1
136	NumOfTacticSequences	1
137	InOperation	1
138	ViolentOnArrest	1
140	ArrMode_Direct	1
141	ArrMode_Result_of_Enquiries	1
143	ArrMode_Other	1
144	OnBailAtTimeOfOffence	1
Drugs activated		
121	*Crack*	1
122	*Heroin_Diamorphine*	1

(continued)

(continued)

Tactics activated		
133	*Covert_Purchase*	1
135	*Other_Generic_Tactic*	1
Query: Tower_Hamlets		
Variables activated		**Values**
1	Sex_Male	1
17	Harrow	1
32	Tower_Hamlets	1
47	UK	1
52	(EA4)_Asia	1
57	Age(21–25)	0.47
61	ConvictionsNumber	1
62	OffensesNumber	1
65	Off_Drug	1
66	Off_TheftKindred	1
67	Off_AgainstPerson	1
68	Off_OffensiveWeapons	1
70	Off_RelatedToPolice	1
71	Off_Fraud	1
72	Off_Total	1
73	Off_AgainstProperty	1
74	NumOfArrests	1
77	AR_OFF_Drug_trafficking_Offenses	0.74
78	AR_OFF_Drug_Possession_Offenses	1
80	AR_OFF_Offensive_Weapon_Offenses	1
81	AR_OFF_Firearms_Offenses	1
83	AR_OFF_Other_violent_Offenses	1
98	AR_PL_Harrow	1
113	AR_PL_Tower_Hamlets	1
124	NumOfCashSeizures	1
126	NumOfTactics	1
129	Police	1
138	ViolentOnArrest	1
139	ArrMode_NA	1
140	ArrMode_Direct	1
144	OnBailAtTimeOfOffence	1
Tactics activated		
131	*Search_of_Person*	1
132	*Search_of_Premises*	1
135	*Other_Generic_Tactic*	1
Query: AR_PL_Tower_Hamlets		
Variables activated		**Values**
1	Sex_Male	1
17	Harrow	1
32	Tower_Hamlets	1
47	UK	1
52	(EA4)_Asia	1

(continued)

(continued)

57	Age(21–25)	0.51
61	ConvictionsNumber	1
62	OffensesNumber	1
65	Off_Drug	1
66	Off_TheftKindred	1
67	Off_AgainstPerson	1
68	Off_OffensiveWeapons	1
70	Off_RelatedToPolice	1
71	Off_Fraud	1
72	Off_Total	1
73	Off_AgainstProperty	1
74	NumOfArrests	1
77	AR_OFF_Drug_trafficking_Offenses	0.72
78	AR_OFF_Drug_Possession_Offenses	1
80	AR_OFF_Offensive_Weapon_Offenses	1
81	AR_OFF_Firearms_Offenses	1
83	AR_OFF_Other_violent_Offenses	1
98	AR_PL_Harrow	1
113	AR_PL_Tower_Hamlets	1
124	NumOfCashSeizures	1
126	NumOfTactics	1
129	Police	1
138	ViolentOnArrest	1
139	ArrMode_NA	1
140	ArrMode_Direct	1
144	OnBailAtTimeOfOffence	1
Tactics activated		
131	*Search_of_Person*	1
132	*Search_of_Premises*	1
135	*Other_Generic_Tactic*	1
Query: Waltham_Forest		
Variables activated		**Values**
1	Sex_Male	1
17	Harrow	0.83
33	Waltham_Forest	1
47	UK	1
51	(EA3)_Afro-Caribbean	1
57	Age(21–25)	0.81
61	ConvictionsNumber	1
62	OffensesNumber	1
65	Off_Drug	1
66	Off_TheftKindred	1
67	Off_AgainstPerson	0.99
68	Off_OffensiveWeapons	1
70	Off_RelatedToPolice	0.98

(continued)

(continued)

71	Off_Fraud	0.98
72	Off_Total	1
73	Off_AgainstProperty	0.99
74	NumOfArrests	1
77	AR_OFF_Drug_trafficking_Offenses	0.83
78	AR_OFF_Drug_Possession_Offenses	1
80	AR_OFF_Offensive_Weapon_Offenses	1
81	AR_OFF_Firearms_Offenses	1
82	AR_OFF_Kidnapping_and_Abduction_off	1
83	AR_OFF_Other_violent_Offenses	1
86	AR_PL_Barnet	0.01
98	AR_PL_Harrow	1
113	AR_PL_Tower_Hamlets	0.62
114	AR_PL_Waltham_Forest	1
124	NumOfCashSeizures	1
126	NumOfTactics	1
129	Police	1
138	ViolentOnArrest	0.93
139	ArrMode_NA	1
140	ArrMode_Direct	1
144	OnBailAtTimeOfOffence	1
Drugs activated		
119	*Cannabis*	0.47
Tactics activated		
131	*Search_of_Person*	1
132	*Search_of_Premises*	1
135	Other_Generic_Tactic	0.83
Query: AR_PL_Waltham_Forest		
Variables activated		**Values**
1	Sex_Male	1
17	Harrow	1
47	UK	1
51	(EA3)_Afro-Caribbean	1
57	Age(21–25)	0.03
61	ConvictionsNumber	1
62	OffensesNumber	1
65	Off_Drug	1
66	Off_TheftKindred	1
67	Off_AgainstPerson	1
68	Off_OffensiveWeapons	1
70	Off_RelatedToPolice	1
71	Off_Fraud	1
72	Off_Total	1
73	Off_AgainstProperty	1
74	NumOfArrests	1
77	AR_OFF_Drug_trafficking_Offenses	0.99
78	AR_OFF_Drug_Possession_Offenses	1

(continued)

(continued)

80	AR_OFF_Offensive_Weapon_Offenses	1
81	AR_OFF_Firearms_Offenses	1
83	AR_OFF_Other_violent_Offenses	1
88	AR_PL_Brent	0.14
98	AR_PL_Harrow	1
113	AR_PL_Tower_Hamlets	1
114	AR_PL_Waltham_Forest	1
124	NumOfCashSeizures	1
126	NumOfTactics	1
129	Police	1
138	ViolentOnArrest	1
139	ArrMode_NA	1
140	ArrMode_Direct	1
144	OnBailAtTimeOfOffence	1
Tactics activated		
131	*Search_of_Person*	1
132	*Search_of_Premises*	0.98
135	Other_Generic_Tactic	1
Query: Wandsworth		
Variables activated		**Values**
1	Sex_Male	1
34	Wandsworth	1
42	JAM	1
51	(EA3)_Afro-Caribbean	1
59	Age(35–45)	1
61	ConvictionsNumber	1
62	OffensesNumber	1
65	Off_Drug	1
66	Off_TheftKindred	1
67	Off_AgainstPerson	1
68	Off_OffensiveWeapons	1
69	Off_Sexual	1
70	Off_RelatedToPolice	1
71	Off_Fraud	1
72	Off_Total	1
73	Off_AgainstProperty	1
74	NumOfArrests	1
77	AR_OFF_Drug_trafficking_Offenses	1
88	AR_PL_Brent	1
89	AR_PL_Bromley	1
90	AR_PL_Camden	1
97	AR_PL_Haringey	1
103	AR_PL_Kensington_and_Chelsea	1
118	NumOfDrugSeizures	1
126	NumOfTactics	1

(continued)

(continued)

129	Police	1
136	NumOfTacticSequences	1
137	InOperation	1
140	ArrMode_Direct	1
141	ArrMode_Result_of_Enquiries	1
143	ArrMode_Other	1
144	OnBailAtTimeOfOffence	1
Drugs activated		
121	*Crack*	1
122	*Heroin_Diamorphine*	1
Tactics activated		
133	*Covert_Purchase*	1
135	*Other_Generic_Tactic*	1
Query: AR_PL_Wandsworth		
Variables activated		**Values**
1	Sex_Male	1
22	Kensington_and_Chelsea	1
47	UK	1
51	(EA3)_Afro-Caribbean	1
59	Age(35–45)	1
61	ConvictionsNumber	1
62	OffensesNumber	1
65	Off_Drug	1
66	Off_TheftKindred	1
67	Off_AgainstPerson	1
68	Off_OffensiveWeapons	1
69	Off_Sexual	1
70	Off_RelatedToPolice	1
71	Off_Fraud	1
72	Off_Total	1
73	Off_AgainstProperty	1
74	NumOfArrests	1
77	AR_OFF_Drug_trafficking_Offenses	1
88	AR_PL_Brent	1
89	AR_PL_Bromley	1
90	AR_PL_Camden	0.02
97	AR_PL_Haringey	1
103	AR_PL_Kensington_and_Chelsea	1
115	AR_PL_Wandsworth	1
118	NumOfDrugSeizures	1
126	NumOfTactics	1
129	Police	1
136	NumOfTacticSequences	1
137	InOperation	1
140	ArrMode_Direct	1
141	ArrMode_Result_of_Enquiries	1
143	ArrMode_Other	1
144	OnBailAtTimeOfOffence	1

(continued)

(continued)

Drugs activated		
121	*Crack*	1
122	*Heroin_Diamorphine*	1
Tactics activated		
133	*Covert_Purchase*	1
Query: Westminster		
Variables activated		**Values**
1	Sex_Male	1
9	Camden	1
35	Westminster	1
47	UK	1
51	(EA3)_Afro-Caribbean	1
59	Age(35–45)	1
61	ConvictionsNumber	1
62	OffensesNumber	1
65	Off_Drug	1
66	Off_TheftKindred	1
67	Off_AgainstPerson	1
68	Off_OffensiveWeapons	1
69	Off_Sexual	1
70	Off_RelatedToPolice	1
71	Off_Fraud	1
72	Off_Total	1
73	Off_AgainstProperty	1
74	NumOfArrests	1
77	AR_OFF_Drug_trafficking_Offenses	1
88	AR_PL_Brent	1
89	AR_PL_Bromley	1
90	AR_PL_Camden	1
97	AR_PL_Haringey	1
103	AR_PL_Kensington_and_Chelsea	1
118	NumOfDrugSeizures	1
126	NumOfTactics	1
129	Police	1
136	NumOfTacticSequences	1
137	InOperation	1
139	ArrMode_NA	0.29
140	ArrMode_Direct	1
141	ArrMode_Result_of_Enquiries	1
143	ArrMode_Other	1
144	OnBailAtTimeOfOffence	1
Drugs activated		
121	*Crack*	1
122	*Heroin_Diamorphine*	1
Tactics activated		
133	*Covert_Purchase*	1
135	*Other_Generic_Tactic*	0.34

(continued)

(continued)

Query: AR_PL_Westminster		
Variables activated		**Values**
1	Sex_Male	1
9	Camden	1
47	UK	1
51	(EA3)_Afro-Caribbean	1
59	Age(35–45)	1
61	ConvictionsNumber	1
62	OffensesNumber	1
65	Off_Drug	1
66	Off_TheftKindred	1
67	Off_AgainstPerson	1
68	Off_OffensiveWeapons	1
69	Off_Sexual	1
70	Off_RelatedToPolice	1
71	Off_Fraud	1
72	Off_Total	1
73	Off_AgainstProperty	1
74	NumOfArrests	1
77	AR_OFF_Drug_trafficking_Offenses	1
88	AR_PL_Brent	1
89	AR_PL_Bromley	1
90	AR_PL_Camden	1
97	AR_PL_Haringey	1
103	AR_PL_Kensington_and_Chelsea	1
116	AR_PL_Westminster	1
118	NumOfDrugSeizures	1
126	NumOfTactics	1
129	Police	1
136	NumOfTacticSequences	1
137	InOperation	1
140	ArrMode_Direct	1
141	ArrMode_Result_of_Enquiries	1
143	ArrMode_Other	1
144	OnBailAtTimeOfOffence	1
Drugs activated		
121	*Crack*	1
122	*Heroin_Diamorphine*	1
Tactics activated		
133	*Covert_Purchase*	1
135	*Other_Generic_Tactic*	1

References

Buscema, M. (1998). Constraint satisfaction neural networks. *Substance Use & Misuse, 33*(2), 389–408.

Buscema, M. (2004, May). Genetic doping algorithm (GenD). Theory and applications. *Expert Systems, 21*(2), 63–79.

Buscema, M., & Sacco, P. L. (2010). Auto-contractive maps, the H function, and the maximally regular graph (MRG): A new methodology for data mining. In V. Capecchi, et al. (Eds.), *Applications of mathematics in models, artificial neural networks and arts*, Chapter 11. DOI 10.1007/978-90-481-8581-8_11. London: Springer.

Kruskal, J. B. (1956). On the shortest spanning subtree of a graph and the traveling salesman problem. *Proceedings of the American Mathematical Society, 7*(1), 48–50.

Rumelhart, D. E., Smolensky, P., McClelland, J. L., & Hinton, G. E. (1986). Schemata and sequential thought processes in PDP models. In J. L. McClelland & D. E. Rumelhart (Eds.), *PDP, exploration in the microstructure of cognition* (Vol. II). Cambridge, MA: The MIT Press.

Software

Buscema, M. (2007a). *Contractive maps. Software for programming auto contractive maps* (Semeion Software #15, v. 2), Rome.

Buscema, M. (2007b). *Constraints satisfaction networks. Software for programming non linear auto-associative networks* (Semeion Software #14, v. 10), Rome.

Buscema, M. (2008a). *MST. Software for programming Trees from artificial networks weights matrix* (Semeion Software #38, v 5), Rome.

Buscema, M. (2008b). *Pst cluster. Software for clustering based on Gend algorithm* (Semeion Software #34, v5.2).

Massini, G. (2007). *Tree visualizer. Software to draw and manipulate tree graph* (Semeion Software #40, v. 3), Rome.

Chapter 15
Auto-Contractive Maps, H Function, and the Maximally Regular Graph: A New Methodology for Data Mining

Massimo Buscema

15.1 Learning Equations

The auto-contractive map (AutoCM) consists of a three-layer architecture: an input layer in which the signal is captured from the environment, a hidden layer where the signal is modulated inside the CM, and an output layer by which the CM influences the environment according to the stimuli previously received (Fig. 15.1).

Each layer is composed of N units. Then the whole CM is composed by $3N$ units. The connections between the input layer and the hidden layer are monodedicated, whereas the ones between the hidden layer and the output layer are at maximum gradient. Therefore, with respect to the number of units, the corresponding number of the connections Nc is given by $Nc = N\,(N + 1)$.

All the connections of CM may be initialized both by equal values and by values at random. The best practice is to initialize all the connections with the same positive value, close to zero.

The auto-contractive map learning algorithm may be characterized by four orderly steps (see the AutoCM theory in Buscema and Sacco 2010; Buscema 2007a, b and effective applications in medical field of AutoCM from Buscema 2007a, b; Buscema and Grossi 2008, 2009; Buscema et al. 2008a, b; Licastro et al. 2010a, b; Grossi et al. 2011; Eller-Vainicher et al. 2011):

1. Signal transfer from the input into the hidden layer
2. Adaptation of the connections value between the input layer and the hidden layer*

M. Buscema (✉)
Semeion Research Centre, Rome, Italy
e-mail: m.buscema@semeion.it

M. Buscema and W.J. Tastle (eds.), *Intelligent Data Mining in Law Enforcement Analytics: New Neural Networks Applied to Real Problems*, DOI 10.1007/978-94-007-4914-6_15,

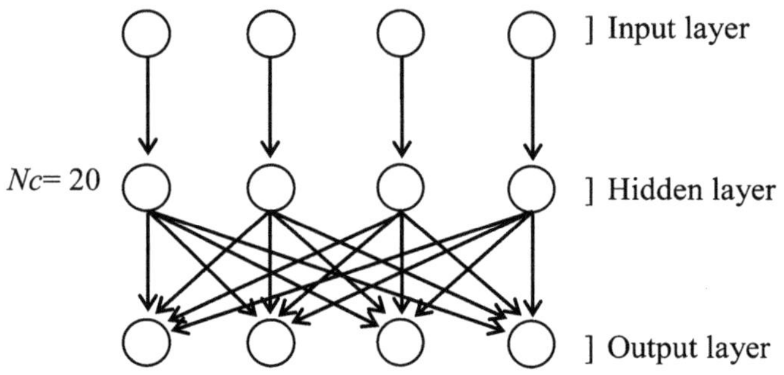

Fig. 15.1 The figure gives an example of an AutoCM with $N = 4$

3. Signal transfer from the hidden layer into the output layer*
4. Adaptation of the connections value between the hidden layer and the output layer

(*): steps 2 and 3 may take place in parallel.

We define as $m^{[s]}$ the units of the input layer (*s*ensors), scaled between 0 and 1, as $m^{[h]}$ the ones of the *h*idden layer, and as $m^{[t]}$ the ones of the output layer (system *t*arget). We define **v** as the vector of monodedicated connections, **w** the matrix of the connections between the hidden layer and output layer, and n indicates a moment in discrete time in which the weights in **w** are calculated, or rather, n is the number of cycles of elaboration that, beginning at zero, increases one unit at each successive cycle: $n \in N$.

There are signal forward transfer and learning equations:

(a) Signal transfer from the input to the hidden layer:

$$m_{i_{(n)}}^{[h]} = m_i^{[s]} \left(1 - \frac{v_{i_{(n)}}}{C}\right) \tag{15.1}$$

where $C =$ positive real number not less than 1, named the contractive factor, and where the (n) subscript has been omitted from the input layer units for simplicity given that they are constant at every elaboration cycle.

(b) Adaptation of the connections $v_{i_{(n)}}$ through the $\Delta v_{i_{(n)}}$ trapping the energy difference generated by Eq. (15.1):

$$\Delta v_{i_{(n)}} = \left(m_i^{[s]} - m_{i_{(n)}}^{[h]}\right) \cdot \left(1 - \frac{v_{i_{(n)}}}{C}\right); \tag{15.2}$$

$$v_{i_{(n+1)}} = v_{i_{(n)}} + \Delta v_{i_{(n)}}. \tag{15.3}$$

(c) Signal transfer from the hidden layer to the output layer:

$$\text{Net}_{i_{(n)}} = \sum_{j}^{N} m_{j_{(n)}}^{[h]} \cdot \left(1 - \frac{w_{i,j_{(n)}}}{C}\right); \tag{15.4}$$

$$m_{i_{(n)}}^{[t]} = m_{i_{(n)}}^{[h]} \left(1 - \frac{\text{Net}_{i_{(n)}}}{C}\right). \tag{15.5}$$

(d) Adaptation of the connections $w_{i,j_{(n)}}$ through the $\Delta w_{i,j_{(n)}}$ trapping the energy differences generated by Eq. (15.5):

$$\Delta w_{i,j_{(n)}} = \left(m_{i_{(n)}}^{[h]} - m_{i_{(n)}}^{[t]}\right) \cdot \left(1 - \frac{w_{i,j_{(n)}}}{C}\right) \cdot m_{j_{(n)}}^{[h]}; \tag{15.6}$$

$$w_{i,j_{(n+1)}} = w_{i,j_{(n)}} + \Delta w_{i,j_{(n)}}. \tag{15.7}$$

The value $m_{j_{(n)}}^{[h]}$ of (15.6) is used for proportioning the change of the connection $w_{i,j_{(n)}}$ to the quantity of energy liberated by node $m_{j_{(n)}}^{[h]}$ in favor of node $m_{i_{(n)}}^{[t]}$.

In auto-contractive mapping, the learning process, conceived as an adjustment in the connections with respect to the minimization of energy, corresponds to the continuous acceleration and deceleration of velocities of the learning signals (corrections $\Delta w_{i,j_{(n)}}$ and $\Delta v_{i_{(n)}}$) inside the artificial neural network (ANN) connection matrix.

The initial step is to make the precedent sentence evident by showing the CM convergence equation:

$$\lim_{n \to \infty} v_{i_{(n)}} = C. \tag{15.8}$$

In fact, when $v_{i_{(n)}} = C$, then $\Delta v_{i_{(n)}} = 0$ (Eq. 15.2) and $m_{i_{(n)}}^{[h]} = 0$ (Eq. 15.1) and, consequently, $\Delta w_{i,j_{(n)}} = 0$ (Eq. 15.6).

During this mathematic analysis, we will introduce four new variables that we consider the key points of the AutoCM learning process:

1. $\varepsilon_{i_{(n)}}$ is the contractive factor of the first layer of AutoCM weights:

$$\varepsilon_{i_{(n)}} = 1 - \frac{v_{i_{(n)}}}{C}.$$

Clearly, the choice of C and the initialization of the connection weights are done so that this factor is a number always included in the inclusive [0, 1] range and it decreases at every nth elaboration cycle.

We will observe that it is infinitesimal as n tends to ∞.

2. $\eta_{i,j_{(n)}}$ is the contractive factor of the second layer of AutoCM weights for which the initializations are nonnegative, negative, and less than 1:

$$\eta_{i,j_{(n)}} = 1 - \frac{w_{i,j_{(n)}}}{C}.$$

3. $\phi_{i_{(n)}}$ is the difference between the hidden nodes and the input nodes:

$$\phi_{i_{(n)}} = m_i^{[s]} - m_{i\,(n)}^{[h]}.$$

   This is an n real function, positive for values lower than 1.

4. $\lambda_{i_{(n)}}$ is the difference between the output nodes and the hidden nodes:

$$\lambda_{i_{(n)}} = m_{i_{(n)}}^{[h]} - m_{i\,(n)}^{[t]}.$$

   It, too, is a real function with positive values decreasing with n.

The second step during the CM learning phase is to demonstrate how $\Delta v_{i_{(n)}}$ increases and decreases according to an always positive parabola arc law.

At this point, we can rewrite Eq. (15.2) as:

$$\Delta v_{i_{(n)}} = \left(m_i^{[s]} - m_i^{[s]}\left(1 - \frac{v_{i_{(n)}}}{C}\right)\right) \cdot \left(1 - \frac{v_{i_{(n)}}}{C}\right) = m_i^{[s]} \cdot \frac{v_{i_{(n)}}}{C}\left(1 - \frac{v_{i_{(n)}}}{C}\right). \quad (15.2a)$$

According to the $\varepsilon_{i_{(n)}}$ definition, we can write $\frac{v_{i_{(n)}}}{C} = 1 - \varepsilon_{i\,(n)}$ and then rewrite the (15.2a) function of $\varepsilon_{i_{(n)}}$:

$$\Delta v_{i_{(n)}} = m_i^{[s]}\left(1 - \varepsilon_{i\,(n)}\right) \cdot \left(1 - \left(1 - \varepsilon_{i_{(n)}}\right)\right) = m_i^{[s]}\left(1 - \varepsilon_{i_{(n)}}\right) \cdot \varepsilon_{i_{(n)}}. \quad (15.2b)$$

Following the $\varepsilon_{i_{(n)}}$ definition and having decreased, between 0 and 1, the values of the input layer units, it can be seen that the $\Delta v_{i_{(n)}}$ parabola arc described in the equation will verify the condition:

$$0 < \Delta v_{i_{(n)}} < \varepsilon_{i_{(n)}} \leq C \cdot \varepsilon_{i_{(n)}}. \quad (15.2c)$$

Equation (15.2c) means that the increment of $\Delta v_{i_{(n)}}$ will always be smaller than the quantity that $v_{i_{(n)}}$ needs to reach up C.

In fact, $v_{i_{(n+1)}} = v_{i_{(n)}} + \Delta v_{i_{(n)}} = C - C \cdot \varepsilon_{i_{(n)}} + \Delta v_{i_{(n)}}$, but from (15.2c) it is known that $\Delta v_{i_{(n)}} - C \cdot \varepsilon_{i_{(n)}} \leq 0$.

Or rather, what algebraically sums to C is always a nonpositive quantity whose absolute value decreases with n increasing. It follows that $v_{i_{(n)}}$ will never exceed C: $\lim_{n\to\infty} v_{i_{(n)}} = C$, and from this also:

$$\lim_{n\to\infty} \varepsilon_{i_{(n)}} = 0, \quad \lim_{n\to\infty} m^{[h]}_{i_{(n)}} = 0, \quad \lim_{n\to\infty} m^{[t]}_{i_{(n)}} = 0, \quad \lim_{n\to\infty} \phi_{i_{(n)}} = m^{[s]}_i,$$
$$\text{and } \lim_{n\to\infty} \lambda_{i_{(n)}} = 0. \tag{15.8}$$

Further, the contractive factor in Eqs. (15.1) and (15.5) makes this relation evident:

$$m^{[t]}_{i\,(n)} \le m^{[h]}_{i(n)} \le m^{[s]}_i. \tag{15.1–15.5}$$

In fact:

$$m^{[h]}_{i\,(n)} = m^{[s]}_i \cdot \varepsilon_{i(n)} \tag{15.1a}$$

and

$$m^{[t]}_{i\,(n)} = m^{[s]}_i \cdot \varepsilon_{i(n)} \cdot \left(1 - \frac{\text{Net}_{i(n)}}{C}\right) \tag{15.5a}$$

Now it is possible to clarify the relationship between $\Delta v_{i_{(n)}}$and$\Delta w_{i,j(n)}$. From Eqs. (15.11, 15.5), we can assume:

$$m^{[h]}_{i\,(n)} = m^{[s]}_i - \phi_{i_{(n)}}, \tag{15.1b}$$

where $\varphi_{i_{(n)}}$ is a small positive real number close to 1,

$$m^{[t]}_{i(n)} = m^{[h]}_{i(n)} - \lambda_{i_{(n)}}, \tag{15.5b}$$

where $\lambda_{i_{(n)}}$ is a small positive real number close to 0; it has real positive values which become close to 0 with increasing n, and

$$m^{[t]}_{i(n)} = m^{[s]}_i - (\phi_{i_{(n)}} + \lambda_{i_{(n)}}). \tag{15.5c}$$

The introduction of the functions defined in key points 1 and 3 transforms (15.2) in

$$\Delta v_{i_{(n)}} = \left(m^{[s]}_i - m^{[h]}_{i_{(n)}}\right) \cdot \left(1 - \frac{v_{i_{(n)}}}{C}\right) = \varphi_{i_{(n)}} \cdot \varepsilon_{i_{(n)}}, \tag{15.2d}$$

and for key points 2 and 4 and Eq. (15.1a), then (15.6) becomes:

$$\Delta w_{i,j_{(n)}} = \left(m^{[h]}_{i_{(n)}} - m^{[t]}_{i_{(n)}}\right) \cdot \left(1 - \frac{w_{i,j(n)}}{C}\right) \cdot m^{[h]}_{j_{(n)}} = \lambda_{i_{(n)}} \cdot \eta_{i,j_{(n)}} \cdot m^{[s]}_j \cdot \varepsilon_{i_{(n)}} \tag{15.6b}$$

and

$$\lim_{\varepsilon \to 0} \Delta w_{i,j(n)} = 0\,. \tag{15.6e}$$

Now we can consider Eq. (15.7) and rewrite it as:

$$w_{i,j(n+1)} = C \cdot \left(1 - \eta_{i,j(n)}\right) + \lambda_{i(n)} \cdot \eta_{i(n)} \cdot m_j^{[s]} \cdot \varepsilon_{i(n)}\,. \tag{15.7a}$$

From (15.7a) and the previously derived Eq. (15.8) results, we have:

$$\lim_{n \to \infty} w_{i,j(n)} = C \cdot \left(1 - \eta_{i,j(n)}\right)\,. \tag{15.7b}$$

Thus, at the beginning of the training of the input and hidden layers, their units will be very similar (Eq. 15.1) and, consequently, $\Delta v_{i(n)}$ will be very small (Eq. 15.2d); for the same reason, $\lambda_{i(n)}$, from definition 4, will initially be very big and $\Delta w_{i,j(n)}$ will be bigger than $\Delta v_{i(n)}$ (Eqs. 15.2d and 15.6b).

During the training, while $v_{i(n)}$ rapidly increases according to the increase of n, $m_{i(n)}^{[h]}$ decreases, so $\lambda_{i(n)}$ and $\varepsilon_{i(n)}$ and, consequently, $\Delta w_{i,j(n)}$ monotonically continue to decrease, while $\Delta v_{i(n)}$ increases slowly. When $\lambda_{i(n)}$ becomes close to zero, $m_{i(n)}^{[h]}$ will only be a bit bigger than $m_{i(n)}^{[t]}$ (see Eq. 15.5b). At this point, $\Delta v_{i(n)}$ is on the global maximum of the equation $\Delta v_{i(n)} = m_i^{[s]}\left(1 - \varepsilon_{i(n)}\right) \cdot \varepsilon_{i(n)}$, so after this critical point, $\Delta v_{i(n)}$ will decrease toward zero.

This can be illustrated on a numerical simulation of a toy dataset.

We consider a three-bit dataset:

3 BITS	Var 1	Var 2	Var 3
Rec 1	0	0	0
Rec 2	0	0	1
Rec 3	0	1	0
Rec 4	0	1	1
Rec 5	1	0	0
Rec 6	1	0	1
Rec 7	1	1	0
Rec 8	1	1	1

After 48 epochs, the auto-contractive mapping artificial neural network, with $C = 1$, completely learns this dataset (RMSE = 0.0000).

If we name **v** the three weights of the first layer, at the end of the training, we have this situation:

v(1)	1.00
v(2)	1.00
v(3)	1.00

The weights of the second layer, instead, will be (with the main diagonal not trained):

w(1,1)	**0.00**
w(1,2)	**0.89**
w(1,3)	**0.89**
w(2,2)	**0.00**
w(2,3)	**0.89**
w(3,3)	**0.00**

In the next table, we show the dynamics along the training of the weights of first layer and of only three weights in the second layer, connecting the different nodes (**w(1,2) – w(1,3) – w(2,3)**):

3bits	**v(1)**	**v(2)**	**v(3)**	**w(1,2)**	**w(1,3)**	**w(2,3)**
Epoch1	0.000161	0.000161	0.000161	0.370855	0.370856	0.370857
Epoch2	0.000259	0.000259	0.000259	0.533956	0.533957	0.533959
Epoch3	0.000418	0.000418	0.000418	0.627853	0.627855	0.627857
Epoch4	0.000672	0.000672	0.000672	0.689412	0.689414	0.689416
Epoch5	0.001083	0.001083	0.001083	0.733061	0.733064	0.733066
Epoch6	0.001742	0.001742	0.001742	0.765688	0.765692	0.765695
Epoch7	0.002803	0.002803	0.002803	0.791018	0.791022	0.791026
Epoch8	0.004508	0.004508	0.004508	0.811242	0.811248	0.811253
Epoch9	0.007242	0.007242	0.007242	0.827736	0.827743	0.827750
Epoch10	0.011617	0.011617	0.011617	0.841397	0.841407	0.841416
Epoch11	0.018589	0.018589	0.018589	0.852828	0.852841	0.852854
Epoch12	0.029631	0.029631	0.029631	0.862436	0.862454	0.862471
Epoch13	0.046947	0.046947	0.046947	0.870492	0.870515	0.870539
Epoch14	0.073678	0.073678	0.073678	0.877163	0.877195	0.877227
Epoch15	0.113961	0.113961	0.113961	0.882545	0.882589	0.882633
Epoch16	0.172487	0.172487	0.172487	0.886696	0.886753	0.886811
Epoch17	0.253100	0.253100	0.253100	0.889672	0.889746	0.889820
Epoch18	0.356198	0.356198	0.356198	0.891592	0.891682	0.891773
Epoch19	0.475946	0.475946	0.475946	0.892666	0.892770	0.892875
Epoch20	0.599992	0.599992	0.599992	0.893172	0.893285	0.893400
Epoch21	0.713680	0.713680	0.713680	0.893370	0.893487	0.893608
Epoch22	0.806393	0.806393	0.806393	0.893435	0.893554	0.893677
Epoch23	0.874853	0.874853	0.874853	0.893453	0.893573	0.893697
Epoch24	0.921696	0.921696	0.921696	0.893458	0.893578	0.893702
Epoch25	0.952067	0.952067	0.952067	0.893459	0.893579	0.893703
Epoch26	0.971068	0.971068	0.971068	0.893459	0.893579	0.893703
Epoch27	0.982688	0.982688	0.982688	0.893459	0.893579	0.893703
Epoch28	0.989697	0.989697	0.989697	0.893459	0.893579	0.893703
Epoch29	0.993887	0.993887	0.993887	0.893459	0.893579	0.893703
Epoch30	0.996380	0.996380	0.996380	0.893459	0.893579	0.893703
Epoch31	0.997859	0.997859	0.997859	0.893459	0.893579	0.893703

(continued)

(continued)

3bits	v(1)	v(2)	v(3)	w(1,2)	w(1,3)	w(2,3)
Epoch32	0.998735	0.998735	0.998735	0.893459	0.893579	0.893703
Epoch33	0.999252	0.999252	0.999252	0.893459	0.893579	0.893703
Epoch34	0.999558	0.999558	0.999558	0.893459	0.893579	0.893703
Epoch35	0.999739	0.999739	0.999739	0.893459	0.893579	0.893703
Epoch36	0.999846	0.999846	0.999846	0.893459	0.893579	0.893703
Epoch37	0.999909	0.999909	0.999909	0.893459	0.893579	0.893703
Epoch38	0.999946	0.999946	0.999946	0.893459	0.893579	0.893703
Epoch39	0.999968	0.999968	0.999968	0.893459	0.893579	0.893703
Epoch40	0.999981	0.999981	0.999981	0.893459	0.893579	0.893703
Epoch41	0.999989	0.999989	0.999989	0.893459	0.893579	0.893703
Epoch42	0.999993	0.999993	0.999993	0.893459	0.893579	0.893703
Epoch43	0.999996	0.999996	0.999996	0.893459	0.893579	0.893703
Epoch44	0.999998	0.999998	0.999998	0.893459	0.893579	0.893703
Epoch45	0.999999	0.999999	0.999999	0.893459	0.893579	0.893703
Epoch46	0.999999	0.999999	0.999999	0.893459	0.893579	0.893703
Epoch47	0.999999	0.999999	0.999999	0.893459	0.893579	0.893703
Epoch48	1.000000	1.000000	1.000000	0.893459	0.893579	0.893703

As a further simplification, we show also the graph of the weight **v(1,1)** and **w(1,2)** dynamics:

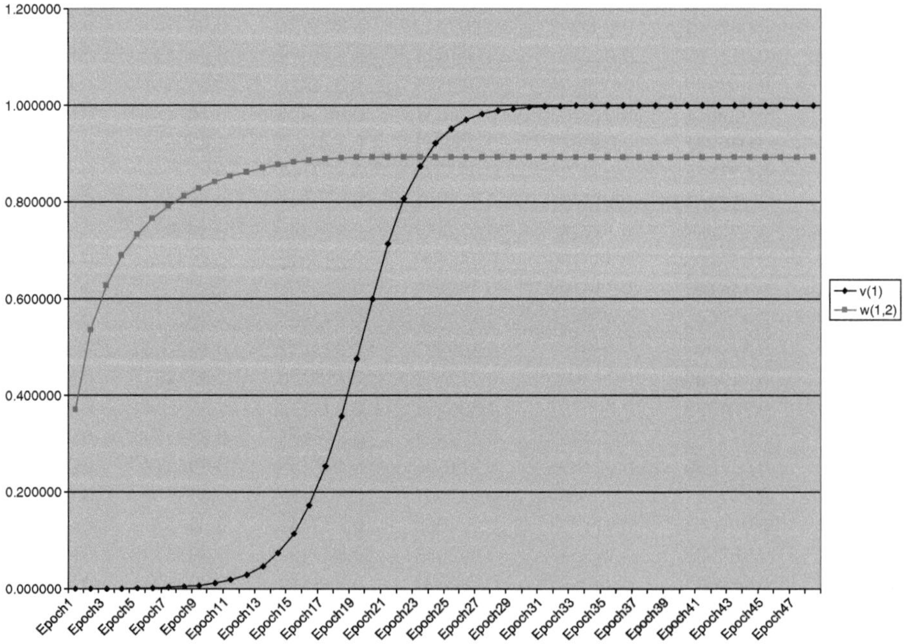

We can compare this graph with the graph showing the updating of the same weights during the training:

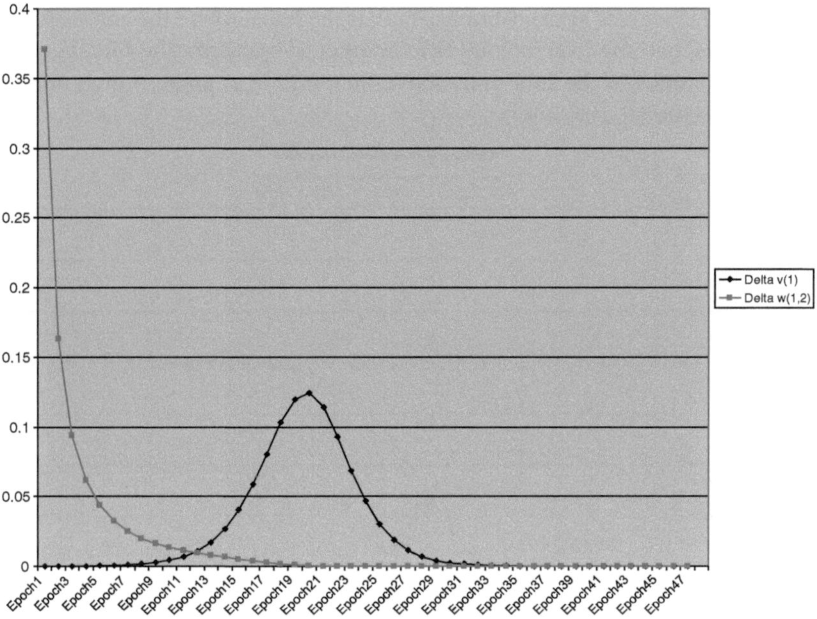

This graph shows in an exemplar way the dynamics described by Eq. (15.7b).

The same phenomenon happens when we show the graph of the dynamics of the first hidden node and the first output node:

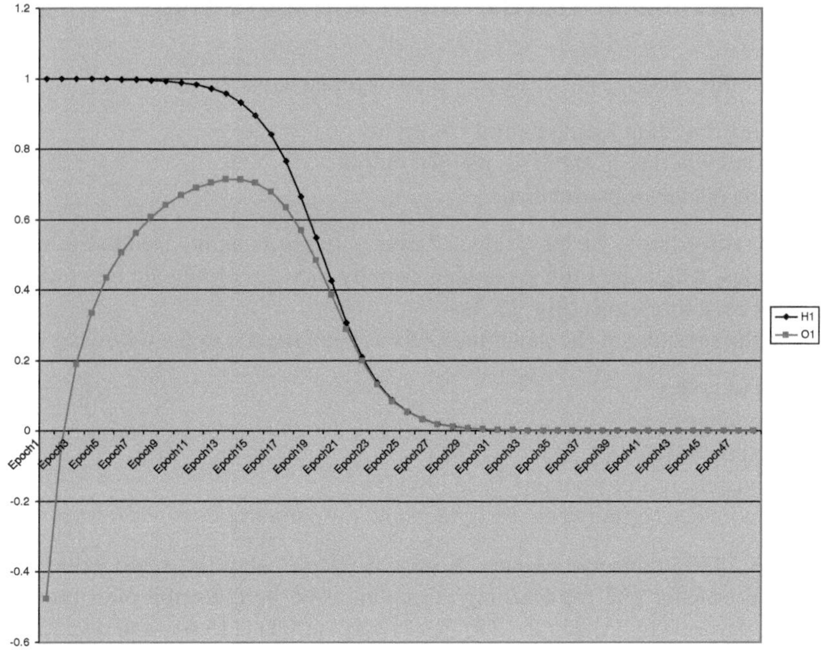

The second layer of CM connections, then, is the place where the energy liberated from the nodes, passing from one layer to another, is *trapped*. The following figure shows the dynamics of the four contractive factors $\varepsilon, \eta, \varphi,$ andλ, which mirror the weights and the units' evolution:

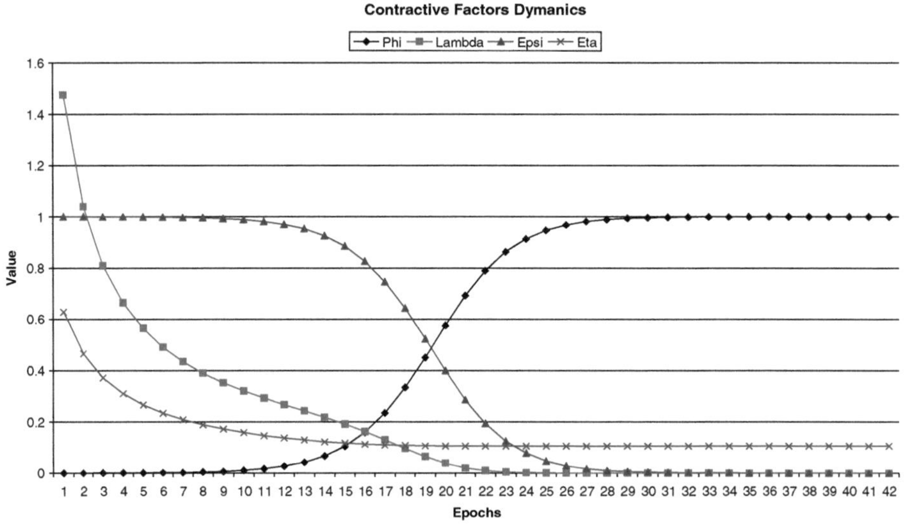

15.2 Experiments with the Auto-Contractive Map

The aim of this section is to evaluate the performance of CM:

(a) How it behaves in facing a group of inputs
(b) How and whether it stabilizes its own output
(c) How its connections stabilize

For this illustration, we have selected nine patterns as input, each one composed of 121 nodes, which are little more than sketchy pictures of nine human faces with nine different expressions (Fig. 15.2):

Given the structure of the input, the CM has been shaped in the following way:

- 121 input nodes
- 121 hidden nodes
- 121 output nodes
- 121 connections between inputs and hidden
- 14641 connections between the hidden and output units

All the 14762 connections have been initialized using the same value (0.01). The signal transfer and the learning equations used here are the ones previously described [from (15.1), (15.2), (15.3), (15.4), (15.5), (15.6), and (15.7)]. The

Fig. 15.2 Each face is pictured inside a matrix X with value one and zero. It must be noted that the CM lacks any knowledge on the space ordering

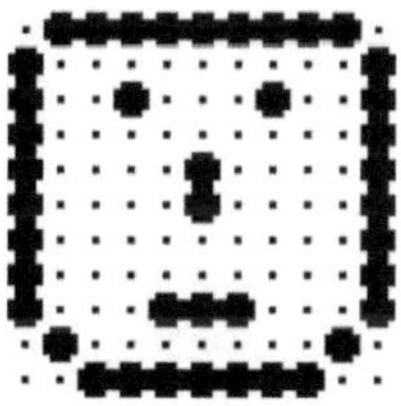

Fig. 15.3 Output shared to all the nine patterns after two epochs

Fig. 15.4 Union among all patterns after around ten epochs

learning has been performed presenting randomly the nine patterns. We have used the notion of *epoch* in its traditional meaning: 1 *epoch = a complete presentation to the ANNs of all the training patterns.*

The performance of the ANN during training may be divided into five phases:

(a) In the initial training, the output of all the nine patterns tends to assume the value 1 for all the input nodes belonging to the subset shared by all the patterns and the value 0 for the remainder of the output nodes. An example of such an output is the following (Fig. 15.3):
(b) In a second phase, each input vector generates the union of all the patterns as output. See Fig. 15.4.
(c) In the third phase, each input pattern is reproduced exactly in the output vector.
(d) In the fourth phase, the output vector presents only the differences that each pattern has with respect to the global intersection of all patterns. See Fig. 15.5.
(e) In the last phase, every input produces a null (zero) output.

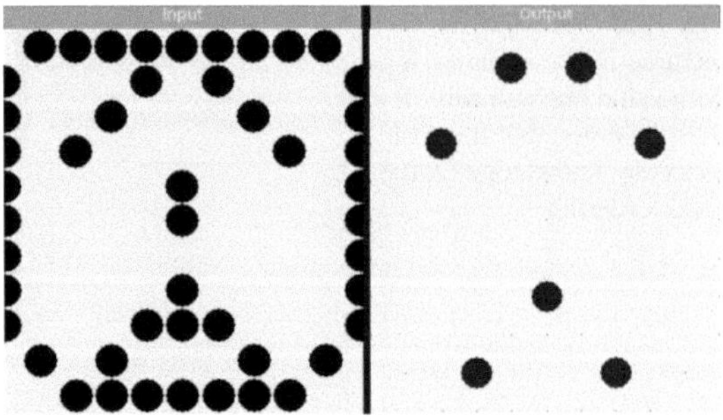

Fig. 15.5 Each pattern generates in output only its specific features, after around 100 epochs

But the most noteworthy characteristic can be found out by analyzing the structure of the 14641 connections stabilized between the hidden and the output layer of the ANN.

Such connections, really, represent a single-step *fractal projection* of the N-dimensional space of the input in an N^2-dimensional space represented by the weights matrix of the CM (Fig. 15.6).

We can easily note some characteristics of this matrix:

(a) The component "faces" draws a "face" which is the *union*, projected on a square scale, of all the patterns of training.
(b) The components "faces" are not all equal; that is, there are all the "expressions of the face" and all the "expressions of the eyes" occurring in the nine patterns of training.
(c) Each component "face" is not equal to any of the nine patterns of training: if the "expression of the mouth" is equal to the expression of the mouth of some patterns of training, then the "expression of the eyes" is the *union* of all the expressions of the eyes of the patterns of training. Similarly, if the "expression of the eyes" is equal to the one of some patterns of training, then the expression of the mouth is the *union* of all the expressions of the mouths of its patterns of training.
(d) Finally, CM is able to find out from a set of patterns the global statistics of their associations among variables.

Such a performance of the matrix of the hidden-output connections of the CM has been noted in hundreds of tests made with input classes different from each other in relation to quantity and orthogonally.

15.3 AutoCM: Theoretical Consideration

Auto-contractive maps do not behave as a regular ANN:

(a) They begin learning with all connections set up with the same value, so they do not suffer the symmetric connection problem.

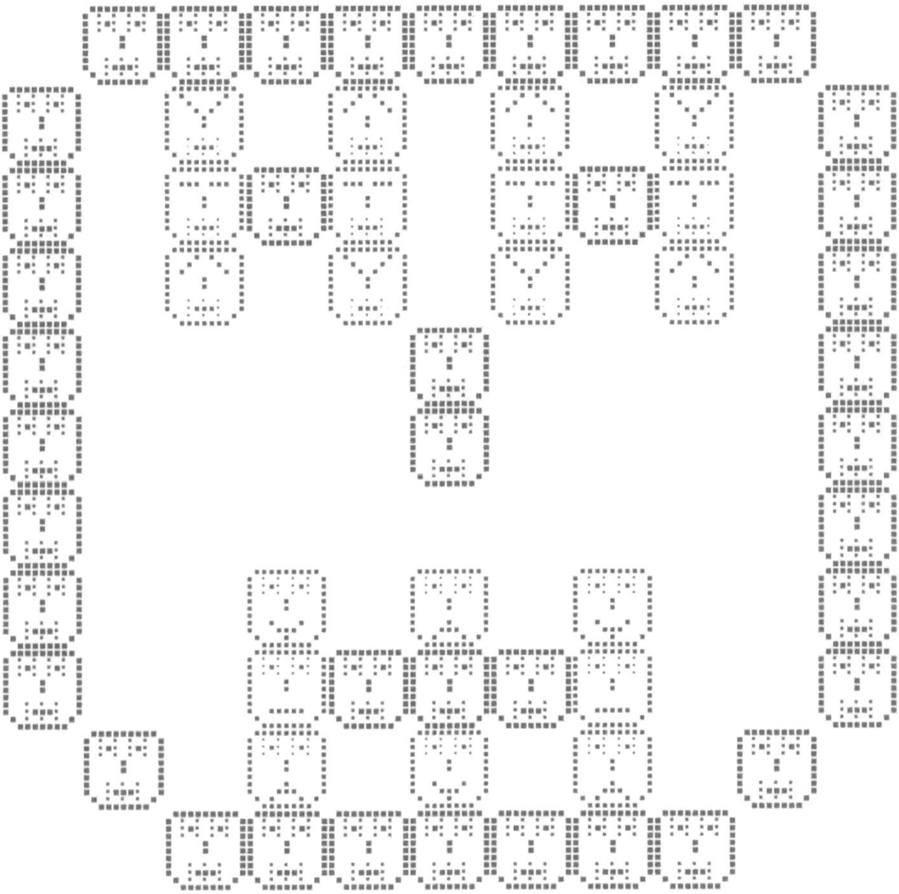

Fig. 15.6 Matrix of the value of the hidden-output connections according to the *B* operator (for *B* operator)

(b) During training, they develop only positive values for each connection. Therefore, AutoCM does not present inhibitory relations among nodes, but only different strengths of excitatory connections.

(c) AutoCM can also learn under harsh conditions, that is, when the connections of the main diagonal of the second connection matrix are removed. When the learning process is organized in this way, AutoCM seems to find a specific relationship between each variable and any other. Consequently, from an experimental point of view, it seems that the ranking of its connection matrix is equal to the ranking of the joint probability between each variable and the others.

(d) After the learning process, any input vector belonging to the training set will generate a null output vector. Thus, the energy minimization of the training

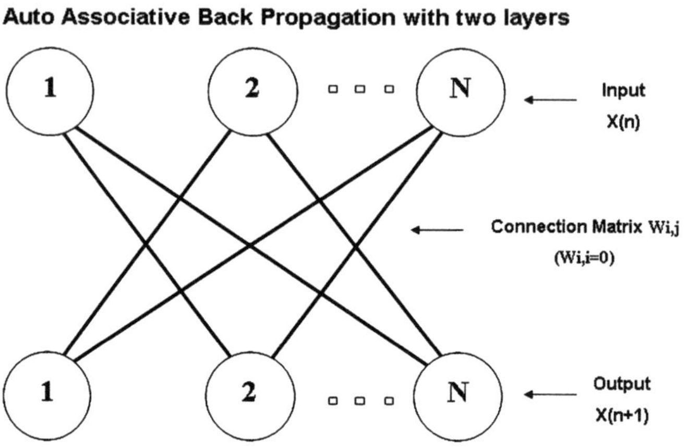

Fig. 15.7 A two-layer BP with $w_{i,i} = 0$

vectors is represented by a function through which the trained connections absorb completely the input training vectors. AutoCM seems to learn to transform itself in a dark body.

(e) At the end of the training phase ($\Delta w_{i,j} = 0$), all the components of the weights vector **v** attain the same value:

$$\lim_{n\to\infty} v_{i(n)} = C. \tag{15.8}$$

The matrix **w**, then, represents the CM knowledge about all the dataset.

It is possible to transform the **w** matrix also in probabilistic joint association among the variables m:

$$p_{i,j} = \frac{w_{i,j}}{\sum_{j=1}^{N} w_{i,j}}; \tag{15.9}$$

$$P\left(m_j^{[s]}\right) = \sum_{i}^{N} p_{i,j} = 1. \tag{15.10}$$

The new matrix **p** can be read as the probability of transition from any state variable to anyone else:

$$P\left(m_i^{[t]} \middle| m_j^{[s]}\right) = p_{i,j}. \tag{15.11}$$

(f) At the same time, the matrix **w** may be transformed into a non-Euclidean distance metric (semimetric), when we train the CM with the main diagonal of the **w** matrix fixed at value N.

Now, if we consider N as a limit value for all the weights of the **w** matrix, we can write:

$$d_{i,j} = N - w_{i,j}. \tag{15.12}$$

The new matrix **d** is also a squared symmetric matrix where the main diagonal represents the zero distance between each variable from itself.

15.4 The Contractive Factor

There is another way to interpret the squared weights matrix of the AutoCM system. We have to assume each variable of the dataset as a vector composed of the all its values. At this point, the dynamic value of each connection between two variables represents the local velocity of their mutual attraction caused by their mutual vectors similarity: the greater the similarity of the vectors, the greater their attraction speed. When two variables are attracted by each other, they proportionally contract the original Euclidean space between them. The limit case occurs when two variables are identical; the space contraction should be infinitive and the two variables should collapse in the same point.

We can extract from each weight of a trained AutoCM this specific contractive factor:

$$F_{i,j} = \left(1 - \frac{w_{i,j}}{C}\right)^{-1}; \quad 1 \leq F_{i,j} \leq \infty. \tag{15.9a}$$

This equation is interesting for three reasons:

1. It is the inverse of the equation used as the contractive factor during the AutoCM training.
2. With respect to Eq. (15.3b), each monoconnection v_i at the end of the training will reach the value C. In this case, the contractive factor will be infinitive because the two variables connected by the weight are really the same variable.
3. Otherwise, in Eq. (15.7b), each weight $w_{i,j}$ at the end of the training will always be smaller than C. This means that the contractive factor for each weight of the matrix that we are considering will be always noninfinitive. In fact, in the case of the weight $w_{i,i}$, the variable is connected with itself, but the same variable has also received the influences of the other variables (recall that the matrix **w** is a squared matrix where each variable is linked to the other). Consequently, this variable cannot be exactly the same.

At this point, we are able to calculate the contractive distance between each variable and the other, modifying the original Euclidean distance with a specific contractive factor.

The Euclidean distance among the variables in the dataset is given by the following equation:

$$d_{i,j}^{[\text{Eucliden}]} = \sqrt{\sum_{k}^{R} \left(x_{i,k} - x_{j,k}\right)^2},$$

where :

$R = \text{thenumberoftherecordsoftheassigneddataset}$

$x_{i,k}\text{and}x_{j,k} = \text{the}i\text{–thvalueandthe}j\text{–thvalueoftwovariablesinthe}k\text{–threcord}$ (15.10a)

And, consequently, the AutoCM distance matrix among the same variables is:

$$d_{i,j}^{[\text{AutoCM}]} = \frac{d_{i,j}^{[\text{Euclidean}]}}{F_{i,j}}. \quad (15.11a)$$

15.5 AutoCM and the Minimum Spanning Tree

Equation (15.12) transforms the squared weights matrix of AutoCM into a squared matrix of distances among nodes. Each distance between a pair of node becomes, consequently, the weighted edge between these pair of nodes. At this point, the matrix **d** may be analyzed through the graph theory.

A graph is a mathematical abstraction that is useful for solving many kinds of problems. Fundamentally, a graph consists of a set of vertices and a set of edges where an edge is something that connects two vertices in the graph. More precisely, a graph is a pair (*V*,*E*), where *V* is a finite set and *E* is a binary relation on *V*, on which it is possible to attribute a scalar value (in this case the weight is the distance $d_{i,j}$).

V is called a vertex set whose elements are called vertices. *E* is a collection of edges where an edge is a pair (*u*,*v*) with *u*,*v* in *V*. In a directed graph, edges are ordered pairs connecting a source vertex to a target vertex. In an undirected graph, edges are unordered pairs and connect the two vertices in both directions; hence, in an undirected graph, (*u*,*v*) and (*v*,*u*) are two ways of writing the same edge.

It does not say what a vertex or edge represents. They could be cities with connecting roads, or web pages with hyperlinks. These details are left out of the definition of a graph for an important reason; they are not a necessary part of the graph abstraction.

Table 15.1 Adjacency matrix of a distance matrix

	A	B	C	D	...	E
A	0	1	1	1	1	1
B	1	0	1	1	1	1
C	1	1	0	1	1	1
D	1	1	1	0	1	1
...	1	1	1	1	0	1
E	1	1	1	1	1	0

An *adjacency-matrix* representation of a graph is a two-dimensional $V \times V$ array, where rows represent the list of vertices and the columns represent the edges among the vertices. Each element in the array is stored with a Boolean value saying whether the edge (u,v) is in the graph.

A *distance matrix* among V vertices represents an undirected graph, where each vertex is linked with all other vertices except for itself (Table 15.1).

At this point, it is useful to introduce the concept of *minimum spanning tree* (MST).

The minimum spanning tree problem is defined as follows: find an acyclic subset T of E that connects all of the vertices in the graph and whose total weight is minimized, where the total weight is given by:

$$d(\mathrm{T}) = \sum_{i=0}^{N-1} \sum_{j=i+1}^{N} d_{i,j}, \forall d_{i,j}. \tag{15.13}$$

T is called the spanning tree, and MST is the T with the minimum sum of its edges weighted:

$$\mathrm{MST} = \mathrm{Min}\left\{d(T_k)\right\}. \tag{15.14}$$

Given an undirected graph G, representing a **d** matrix of distances with V vertices in which all vertices are completely linked to each other, the total number of edges (E) is:

$$E = \frac{V \cdot (V-1)}{2}, \tag{15.15}$$

and the number of possible trees is:

$$T = V^{V-2}. \tag{15.16}$$

Kruskal (1956) found an algorithm (Zsuzsanna 2001) able to determine the MST of any undirected graph in a quadratic number of steps, that being the worst case. Obviously, the Kruskal algorithm generates one tree of the total number of possible MSTs. In fact, in a weighted graph, more than one MST is possible.

From the conceptual perspective, the MST represents the *energy minimization* state of a structure. In fact, if we consider the atomic elements of a structure as vertices of a graph and the strength among them as the weight of each edge linking a pair of vertices, the MST represents the minimum of energy needed because all the elements of the structure continue to stay together.

In a closed system, all the components tend to minimize the overall energy. So the MST, in specific situations, can represent the most probable state to which a system can tend.

To define the MST of an undirected graph, each edge of the graph has to be weighted. Equation (15.12) shows a way to weight each edge whose nodes are the variables of a dataset and whose weights of a trained AutoCM provides the metrics.

Obviously, it is possible to use any kind of auto-associative ANN or any kind of linear auto-associator to generate a weight matrix among the variables of an assigned dataset. But it is hard to train a two-layer auto-associative backpropagation with the weights on the main diagonal fixed (to avoid autocorrelation). In most of the cases, the root mean square error (RMSE) ceases to decrease after few epochs, especially when the orthogonality of the records increases. This is usual when it is necessary to weight the distance among the records of the assigned dataset. In this case, in fact, it is necessary to train the transposed matrix of the assigned dataset.

By the way, if a linear auto-associator is used, all the nonlinear associations among variables will be lost.

So, actually, AutoCM seems to be the best choice to compute a complete and nonlinear matrix of weights among variables or among records of any assigned dataset.

15.6 Other Algorithms for MST

Theoretically, as previously stated, it is possible to use any algorithm to weight the graph edges. It will be useful to provide some of the more commonly used ones.

15.6.1 Linear Correlation

First, it is necessary to calculate the linear correlation between each pair of variables in the assigned dataset:

$$R_{i,j} = \frac{\sum_{k=1}^{N} (x_{i,k} - \bar{x}_i) \cdot (x_{j,k} - \bar{x}_j)}{\sqrt{\sum_{k=1}^{N} (x_{i,k} - \bar{x}_i)^2 \cdot \sum_{k=1}^{N} (x_{j,k} - \bar{x}_j)^2}};$$

$$-1 \leq R_{i,j} \leq 1; \quad i, j \in [1, 2, ..., M], \tag{15.17}$$

where :

$R_{i,j}$ = linear correlation between any couple of variables x_i and x_j of the assigned dataset

$\bar{x}_i$ = mean value of any variable x_i

N = number of records of the assigned dataset

M = number of variables of the assigned dataset

Equation (15.17) will generate a symmetric squared matrix with null diagonal, defining the linear correlation between each variable and any other.

Equation (15.18) will transform the matrix of correlation into a matrix of linear distances among the variables:

$$d_{i,j}^{[R]} = \sqrt{2 \cdot \left(1 - R_{i,j}\right)}. \tag{15.18}$$

At this point, the assigned dataset is transformed in an undirected weighted graph, where MST is applicable.

15.6.2 Prior Probability

First, it is necessary to calculate the prior probability of co-occurrence between any couple of variables in the assigned dataset:

$$A_{i,j} = -\ln \frac{\frac{1}{N^2} \cdot \sum_{k=1}^{N} x_{i,k} \cdot \left(1 - x_{j,k}\right) \cdot \sum_{k=1}^{N} \left(1 - x_{i,k}\right) \cdot x_{j,k}}{\frac{1}{N^2} \cdot \sum_{k=1}^{N} x_{i,k} \cdot x_{j,k} \cdot \sum_{k=1}^{N} \left(1 - x_{i,k}\right) \cdot \left(1 - x_{j,k}\right)};$$
$$-\infty \leq A_{i,j} \leq +\infty; x \in [0,1]; i,j \in [1,2,...,M], \tag{15.19}$$

where :

$A_{i,j}$ = association strength between any couple of variables x_i and x_j of the assigned dataset

x_i = value of any variable scaled between 0 and 1

N = number of records of the assigned dataset

M = number of variables of the assigned dataset

At this point, it is possible to transform the matrix of the variables' association into a nonlinear distance matrix:

$$d_{i,j}^{[A]} = \text{Max}\,A - A_{i,j}, \quad \text{whereMax}\,A = \text{maximum}A\text{matrixvalue}. \tag{15.20}$$

15.6.3 Euclidean Distance

The Euclidean distance among variables is easy to generate. It is necessary, first, to scale the value of the variables to between 0 and 1 and then to transpose the matrix of the assigned dataset:

$$d_{i,j}^{[E]} = \sqrt{\sum_{k=1}^{M} \left(x_{i,k} - x_{j,k}\right)^2 ; i,j \in [1,2,\ldots N] ; x \in [0,1]},$$

where :

$d_{i,j}^{[E]}$ = Euclidean distance among any couple of variables

x_i = value of any record scaled between 0 and 1

N = number of variables of the assigned dataset

M = number of records of the assigned dataset (15.21)

All these algorithms shown above, in any case, have the advantage of being computationally very fast, but they are limited to defining the distance among variables or records, grouping all the records or all the variables in isolated couples. That means that each weight explains the association between two variables or two records, but it does not take into account the influence of the other variables or the other records on it.

This situation is quite similar to the case of ten children playing all together in a swimming pool. It is not realistic to explain their global behavior, making the statistics of their interaction, by grouping them into all possible pairs.

15.6.4 Auto-Associative BP

A backpropagation without a hidden unit layer and without connections on the main diagonal can also be used to compute a metric among variables.

This is an ANN featuring an extremely simple learning algorithm:

$$\text{Output}_i = f\left(\sum_j^N \text{Input}_j \cdot W_{i,j} + \text{Bias}_i\right) = \frac{1}{1 + e^{-\left(\sum_j^N \text{Input}_j \cdot W_{i,j} + \text{Bias}_i\right)}} \qquad W_{i,i} = 0. \tag{15.22}$$

$$\begin{aligned} \delta_i &= (\text{Input}_i - \text{Output}_i) \cdot f'(\text{Output}_i) \\ &= (\text{Input}_i - \text{Output}_i) \cdot \text{Output}_i \cdot (1 - \text{Output}_i)\,. \end{aligned} \tag{15.23}$$

$$\Delta W_{i,j} = \text{LCoef} \cdot \delta_i \cdot \text{Input}_j; \quad \text{LCoef} \in [0, 1]\,. \tag{15.24}$$

$$\Delta \text{Bias}_i = \text{LCoef} \cdot \delta_i\,. \tag{15.25}$$

$$\begin{aligned} W_{i,j}^{[n+1]} &= W_{i,j}^{[n]} + \frac{1}{2} \cdot \left(\Delta W_{i,j} + \Delta W_{j,i}\right); \\ \text{Bias}_i^{[n+1]} &= \text{Bias}_i^{[n]} + \Delta \text{Bias}_i\,. \end{aligned} \tag{15.26}$$

AutoBP is an ANN featuring $N^2 - N$ internode connections and N bias inside every exit node, for a total of N^2 adaptive weights. It is an algorithm that works similarly to logistic regression and can be used to establish the dependency of every variable from each other.

The advantage of AutoBP is its learning speed, which is due to the small number of connections and to the simplicity of its topology and algorithm. Moreover, at the end of the learning phase, the connections between variables, because they are direct, have a clear conceptual meaning. Every connection indicates a relationship of faded excitement, inhibition, or indifference between every pair of variables or records.

The disadvantage of AutoBP is its limited convergence capacity, due to that same topological simplicity. That is to say, complex relationships between variables may be approximated or ignored.

At the end of the training phase, it is necessary to convert each connection in a nonlinear symmetric distance (semimetric):

$$\begin{aligned} V_{i,j} &= V_{j,i} = \frac{1}{2} \cdot \left(W_{i,j} + \text{Bias}_i + W_{j,i} + \text{Bias}_j\right); \\ d_{i,j}^{[Bp]} &= \text{Max}\,V - V_{i,j}; \quad \text{where Max}\,V = \text{Max}\left\{V_{i,j}\right\}. \end{aligned} \tag{15.27}$$

15.7 Some Qualitative Features of MST

Once we have a distance matrix among nodes, $d_{i,j}^{[...]}$with$i,j \in [1,2,...,N]$, the MST of the implicit graph is easy to define using Kruskal algorithm. The MST adjacency matrix, then, must undergo analysis.

For the adjacency matrix, the easiest criterion to study is to rank the number of links of each node; this algorithm defines the *connectivity* of each node:

$$C_i = \sum_j^N l_{i,j},$$

$$\text{where :}$$

$$\text{If} l_{i,j} \in \text{Mst, then} l_{i,j} = 1$$

$$\text{If} l_{i,j} \notin \text{Mst, then} l_{i,j} = 0$$

$$l_{i,j} = \text{possible direct connection between Nod e}_i \text{and Nod e}_j \tag{15.28}$$

(a) Nodes with only one link are named *leaves*. Leaves define the boundaries of the MST graph.
(b) Nodes with two links are named *connectors*.
(c) Nodes with more than two connections are named *hubs*. Each hub has a specificity degree of connectivity:

$$\text{HubDegree}_i = C_i - 2. \tag{15.29}$$

A second indicator qualifying a MST graph is the *clustering strength* of each of its node.

The clustering strength of each node is proportional to the number of its links and to the number of links of the nodes directly connected to it:

$$S_i = \frac{C_i^2}{\sum_{j=1}^{C_i} C_j}. \tag{15.30}$$

A third indicator is the *degree of protection* of each node in any adjacency-directed graph. This indicator defines the rank of centrality of each node within the graph, when an iterative pruning algorithm is applied to the graph. This algorithm was created by Massini at Semeion Research Center in 2006 (software: Massini 2007) and applied to a problem for the first time as a global indicator for a graph complexity.

```
Pruning Algorithm

Rank = 0;

Rank = 0;

{

   Rank + +;

   Consider_All_Nodes_with_The_Minimum_Number_of_Links();

   Delete_These_Links();

   Assign_a_Rank_To_All_Nodes_Without_Link (Rank) ;

   Update_The_New_Graph();

   Check_Number_of_Links();

}whileat_least_a_link_is_present;                                    (15.31)
```

The higher the rank of a node, the greater is the centrality of its position within the graph. The latest nodes to be pruned are also the kernel nodes of the graph.

Similarly, the pruning algorithm was applied by Massini (software: Massini 2007) to measure the complexity of any tree. Here it is generalized to measure the global complexity of any kind of graph.

15.8 Graph Complexity: The H Function

The pruning algorithm can also be used to define the *quantity of graph complexity* of any graph.

In fact, if we assume μ as the mean number of nodes without any link in each iteration, then during the pruning algorithm, we can write the hubness index, H_0, of a graph with N nodes as:

$$H_0 = \frac{\mu \cdot \varphi - 1}{A}; \quad 0 < H_0 < 2, \tag{15.32}$$

where $\mu = 1/M \sum_i^M Nd_i = A/M$,$\phi = 1/P \sum_j^P S_{TG\,j}$,$A$ = number of links of the graph ($N-1$ for tree graphs), M = number of iterations of pruning algorithm, P = number of types of pruning, Nd_i = number of nodes without link at the jth iteration, and $S_{TG\,j}$ = series of pruning gradient types.

Using H_0, as global indicator, it is possible to define how much a graph is *hub-oriented*.

This simple example can show three possible cases when $N = 6$ ($N =$ number of nodes):

Case 1: $H_0 = 0.2$, tree is for 1 /5 hub-oriented:

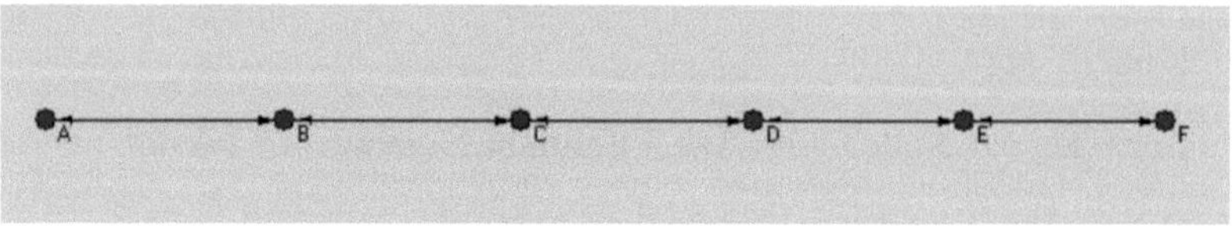

Case 2: $H_0 = 1$, tree is completely hub-oriented:

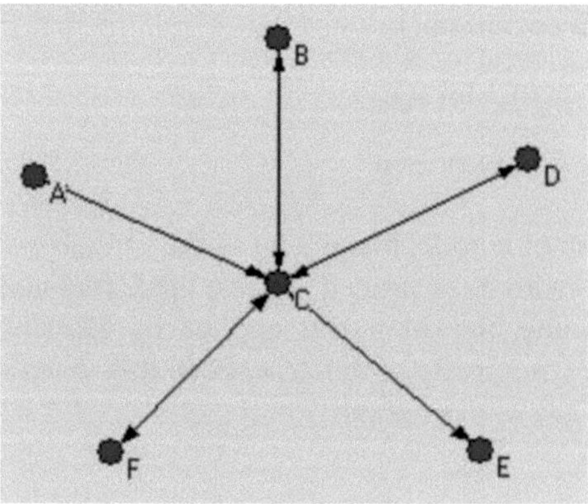

Case 3: $H_0 = 0.4$, tree is for 2 /5 hub-oriented:

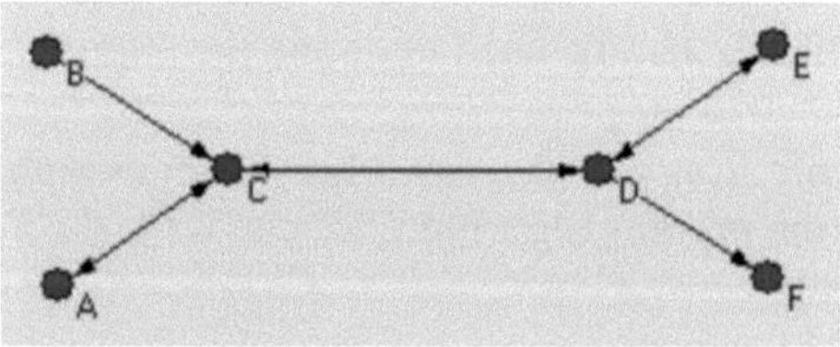

This simple Eq. (15.32) is also shown to be correct in the limit case of a tree with only three nodes. In this case, $H_0 = 1$ is correct because this type of tree shows the limit where a hub collapses into a chain.

This case limit has relevance when the number of nodes x is odd and their topology is a chain.

In fact,

if:

- $S =$ progressive index for pruning steps
- $G =$ gradient of the erased nodes at step j

- $L =$ number of link erased at step j
- $N^* =$ number of erased nodes at step j

then,

$$\left|\begin{array}{cccc} S & G & L & N* \\ 1 & 1 & 2 & 2 \\ 2 & 1 & 2 & 2 \\ \vdots & \vdots & \vdots & \vdots \\ \frac{(x-1)}{2} & 1 & 2 & 3 \end{array}\right| \tag{15.33}$$

$$\phi^{[C]} = 1 \quad \mu^{[C]} = \frac{2x}{x-1}$$

$$N^{[C]} = x$$

$$H_0^{[C]} = \frac{\mu^{[C]} \cdot \phi^{[C]} - 1}{N^{[C]} - 1} = \frac{x+1}{(x-1)^2} = \frac{1}{x-1} \cdot \frac{x+1}{x-1}. \tag{15.34}$$

In other words,

$$\lim_{x \to \infty} H = 0. \tag{15.35}$$

So, in a case of a "chain tree" composed of an odd number of nodes, the last pruning interaction has to delete three nodes representing the limit case where "hub tree" and "chain tree" collapse into each other. In this condition, a "chain tree" will present a H_0 value always a little bigger than 0. Increasing the number of the odd nodes in the "chain tree," this value squared decreases asymptotically to zero.

The H index, in any case, finds a structural difference between trees composed of an even number of nodes and trees composed of an odd number of nodes (Fig. 15.8).

15.9 Graph and MST Complexity

The H indicator (Eq. 15.32) represents the *global hubness* of a graph. When $H = 0$, the tree is a one-dimensional line and its complexity is minimal. When $H = 1$, the tree presents only one hub, and its complexity is maximum for a tree. The complexity of a graph, in fact, is connected to its entropy. The quantity of information in a graph is linked to the graph diameter and number of vertices connected: with the same number of vertices, the shorter the diameter of the graph, the greater is its entropy.

Starting from the classical concept of entropy, we can write:

$$E = -K \cdot \sum_{i}^{N} p_i \cdot \ln(p_i). \tag{15.36}$$

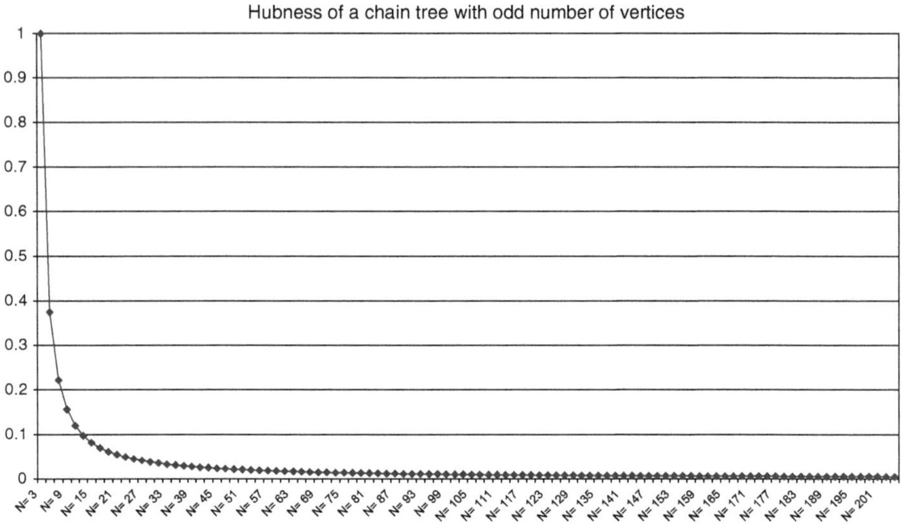

Fig. 15.8 Evaluation of Eq. (15.34)

If we name $E(G)$ the topological entropy of a generic tree graph, we can write:

$$E(G) = -\frac{A}{M} \cdot \sum_{i}^{N} \frac{C_i}{A} \cdot \ln\left(\frac{C_i}{A}\right); \quad 0 < E(G) < \infty, \tag{15.37}$$

where $A =$ number of graph arcs ($N-1$, when the graph is a tree), $N =$ number of the graph vertices, $M =$ number of pruning cycles necessary to completely disconnect the graph, and $C_i =$ degree of connectivity of each vertex.

The equation $C_i\,/A$ measures the probability that a generic node C_j, where $j \neq i$ has to be linked directly to node C_i. This means that the entropy of a graph, $E(G)$, will increase when the number of vertices with a large number of links increases. In the same way, the probability of arranging the links of N vertices into a linear chain, using a random process, is the lowest. Consequently, when the number of pruning cycles, M, needed for a graph is greatest, its graph entropy is least.

Equation (15.37) shows clearly that a "hub tree" has more entropy than a "chain tree." Consequently, as the H index of a tree increases, so does its redundancy increase. At this point, it is necessary to give some example of the H function and the relative topological entropy as applied to any generic adjacency-directed graph. According to the H function, the complexity of any graph is ruled by Eq. (15.32):

$$0 < H_0 < 2.$$

More specifically, $0 < H_0 < 1\,/2$ for any kind of tree, except for the "star tree" in which $H_0 = 1$. For a regular graph, the H function has the interval:

$$1.\overline{6} \leq H_0 < 2.$$

For any other kind of graph (with the tree graph being the exception), the H function can take any value of the interval (Eq. 15.32) according to its symmetry: the greater the symmetry, the greater the H value. It is useful, now, to provide a practical example: let us show how to compute the H function and the topological entropy of a generic graph.

First of all, we introduce the concept of a *pruning table* as useful tool in which to organize the graph analysis:

$$\begin{vmatrix} M & G & L & N \\ 1 & g_1 & l_1 & n_1 \\ \ldots & \ldots & \ldots & \ldots \\ k & g_k & l_k & n_k \end{vmatrix}$$

Legend: $M =$ *j*th pruning cycle, $G =$ gradient of *j*th pruning cycle, $L =$ number of deleted links at *j*th pruning cycle, and $N =$ number of deleted nodes at *j*th pruning cycle.

Given a generic graph:

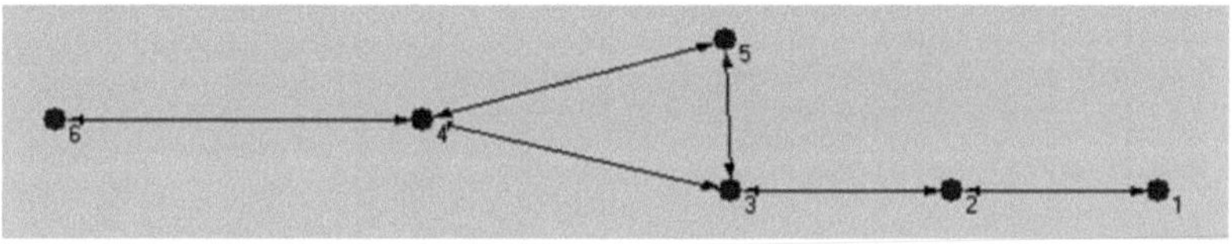

its pruning table will be:

$$\begin{vmatrix} M & G & L & N \\ 1 & 1 & 2 & 2 \\ 2 & 1 & 1 & 1 \\ 3 & 2 & 3 & 3 \end{vmatrix}$$

At this point, by applying Eqs. (15.32) and (15.37), it is possible to compute the H function and the topological entropy of the graph:

$$\varphi = \frac{1}{P}\sum_{j}^{P} S_{TG\,j} = \frac{1}{2}\sum_{j}^{2} S_{TG\,j} = \frac{1}{2}(1+2) = \frac{3}{2} = 1.5;$$

$$\mu = \frac{A}{M} = \frac{6}{3} = 2;$$

$$H_0 = \frac{\mu \cdot \varphi - 1}{A} = \frac{2 \cdot 1.5 - 1}{6} = \frac{1}{3} = 0.33.$$

$$E(G) = -\frac{A}{M} \cdot \sum_{i}^{N} \frac{c_i}{A} \ln\left(\frac{c_i}{A}\right) = 4.04561706.$$

Now, as an example, we present the different H functions and topological entropy of a different graph with only six nodes:

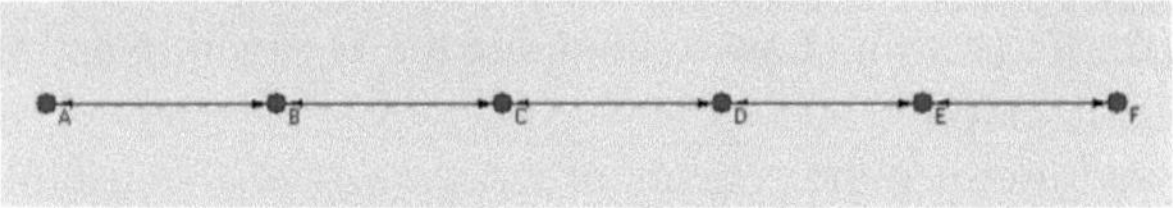

Chain: $H = 0.2 \quad E(G) = 3.5116$

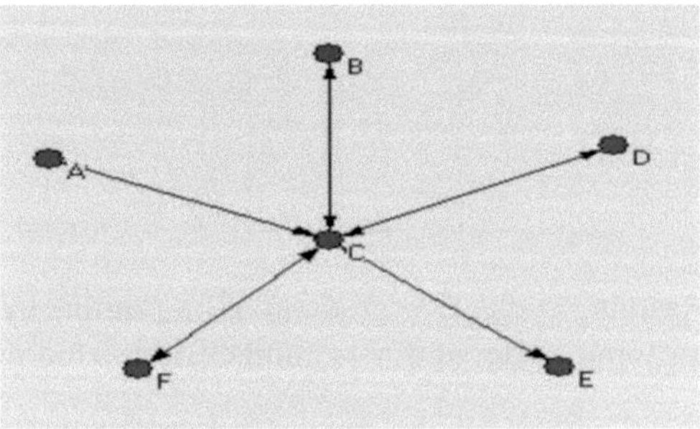

Star: $H = 1 \quad E(G) = 8.0471$

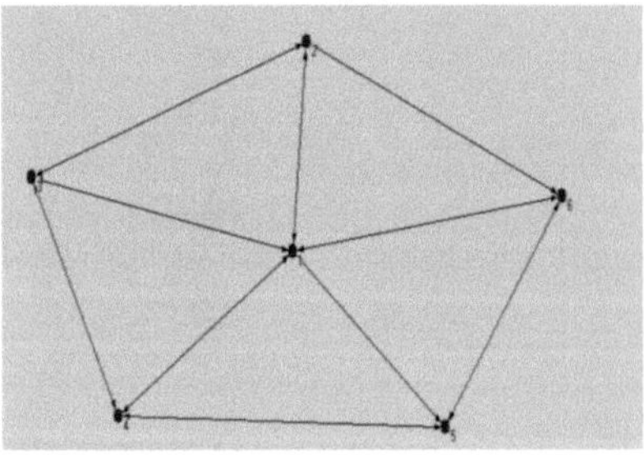

Closed star: $H = 1.7 \quad E(G) = 21.5253$

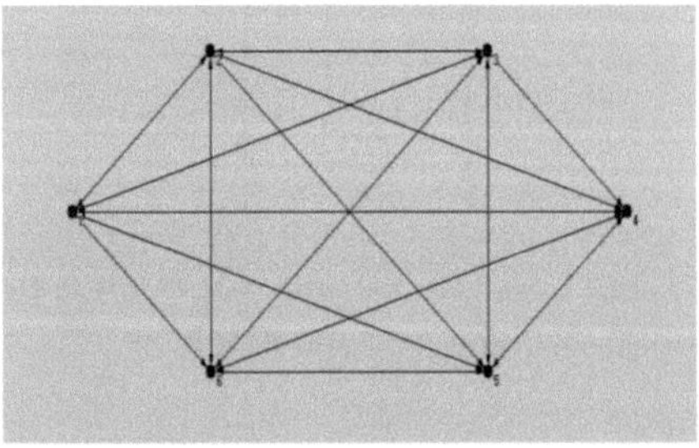

Complete graph: $H = 1.93 \quad E(G) = 32.9583$

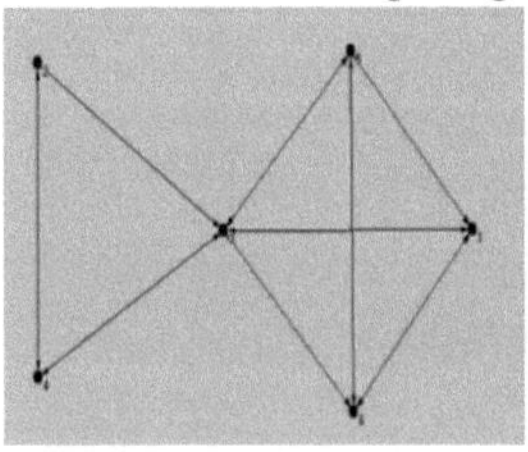

R-graph: $H = 0.72$ $E(G) = 9.4213$

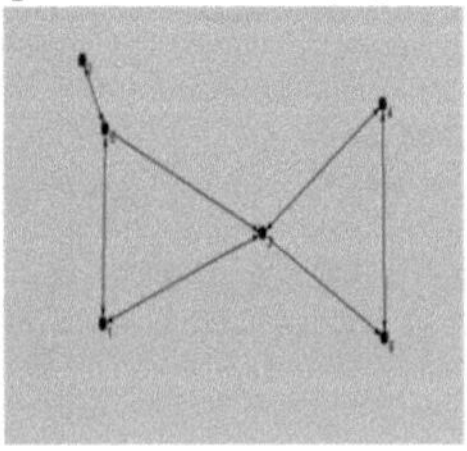

R-graph: $H = 0.5$ $E(G) = 7.1214$

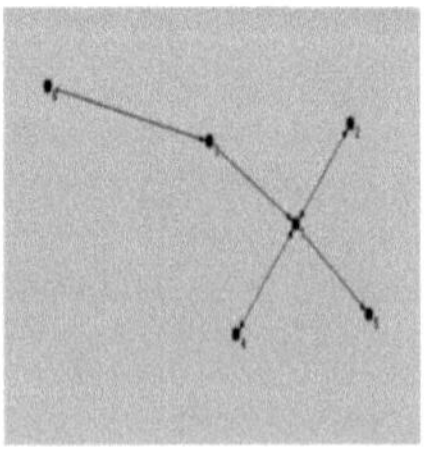

R-tree: $H = 0.4$ $E(G) = 4.5814$

15.10 The Delta H Function

Now, we should consider every graph, and specially the MST, from a dynamical point of view. When one or more vertices are deleted from a graph, the other vertices rearrange their links according to their specific metrics and constraints to connect to each other once again.

We can define an H index for an N number of MST, each one generated from the original distance matrix, by deleting one vertex at any one calculation:

$$H_i = \frac{\mu_i \cdot \phi_i - 1}{A - 1}; \quad 0 \leq H_i < 2, \tag{15.38}$$

where $\mu_i = 1/M \sum_j^M Nd_j = N/M$,$\phi = 1/P \sum_k^P S_{TG\,k}$,$A =$ number of links of the graph ($N - 1$ for tree graphs),

$M =$ number of iterations of pruning algorithm, $P =$ number of types of pruning, $Nd_j =$ number of nodes without link at the jth iteration, and $S_{TG\,k} =$ series of pruning gradient types.

Each H_i represents the tree complexity of the same distance matrix when the $i$thvertex is deleted. Consequently, the difference between the complexity (i.e., H_0) of the whole MST and the complexity of any MST generated without one of the graph vertices (H_i) is the measure of the contribution of each vertex of the graph to the global complexity:

$$\delta H_i = H_0 - H_i. \tag{15.39}$$

This new index states how much each vertex of a graph contributes either to increase ($\delta H_i < 0$) or to decrease ($\delta H_i > 0$) the redundancy of the assigned graph. We have named this function the *delta H function*, and it can be applied to any kind of graph.

15.11 AutoCM, MST, and Delta H Function: An Illustration

The Gang dataset (with apologies to the musical West Side Story) is a small dataset composed of 27 records and 5 variables (Table 15.2):

The structure of the dataset is:

- Gang = {Jets, Sharks}
- Age = {20s, 30s, 40s}
- Education = {Junior School, High School, College}
- Status = {Married, Single, Divorced}
- Profession = {Pusher, Bookie, Burglar}

Table 15.2 Gang dataset

Name	Gang	Age	Education	Status	Profession
ART	Jets	40 age	Junior School	Single	Pusher
AL	Jets	30 age	Junior School	Married	Burglar
SAM	Jets	20 age	College	Single	Bookie
CLYDE	Jets	40 age	Junior School	Single	Bookie
MIKE	Jets	30 age	Junior School	Single	Bookie
JIM	Jets	20 age	Junior School	Divorced	Burglar
GREG	Jets	20 age	High School	Married	Pusher
JOHN	Jets	20 age	Junior School	Married	Burglar
DOUG	Jets	30 age	High School	Single	Bookie
LANCE	Jets	20 age	Junior School	Married	Burglar
GEORGE	Jets	20 age	Junior School	Divorced	Burglar
PETE	Jets	20 age	High School	Single	Bookie
FRED	Jets	20 age	High School	Single	Pusher
GENE	Jets	20 age	College	Single	Pusher
RALPH	Jets	30 age	Junior School	Single	Pusher
PHIL	Sharks	30 age	College	Married	Pusher
IKE	Sharks	30 age	Junior School	Single	Bookie
NICK	Sharks	30 age	High School	Single	Pusher
DON	Sharks	30 age	College	Married	Burglar
NED	Sharks	30 age	College	Married	Bookie
KARL	Sharks	40 age	High School	Married	Bookie
KEN	Sharks	20 age	High School	Single	Burglar
EARL	Sharks	40 age	High School	Married	Burglar
RICK	Sharks	30 age	High School	Divorced	Burglar
OL	Sharks	30 age	College	Married	Pusher
NEAL	Sharks	30 age	High School	Single	Bookie
DAVE	Sharks	30 age	High School	Divorced	Pusher

First of all, it is necessary to transform each string variable in a Boolean expression (Table 15.3):

The new dataset is now composed of 14 binary variables, the most of which orthogonal.

Because we want to use an AutoCM ANN to process the records, we must transpose this matrix (Table 15.4):

AutoCM ANN will learn this dataset using the variables as hyperpoints and the records as the coordinates of the hyperpoints.

After about 30 epochs, the AutoCM, with a contractive factor of 6.19615221, is completely trained (RMSE = 0.00000000), and the weights matrix is ready (Table 15.5):

Through Eq. (15.12), we transform the weights matrix into a distance matrix (Table 15.6):

At this point, the MST of the dataset is the following (Fig. 15.9):

Table 15.3 The binary Gang Dataset. Notice the absence of empty cells

Gang 14 × 27	Jet	Sharks	20'	30'	40'	JS	COL	HS	Single	Married	Divorced	Pusher	Bookie	Burglar
ART	1	0	0	0	1	1	0	0	1	0	0	1	0	0
AL	1	0	0	1	0	1	0	0	0	1	0	0	0	1
SAM	1	0	1	0	0	0	1	0	1	0	0	0	1	0
CLYDE	1	0	0	0	1	1	0	0	1	0	0	0	1	0
MIKE	1	0	0	1	0	1	0	0	1	0	0	0	1	0
JIM	1	0	1	0	0	1	0	0	0	0	1	0	0	1
GREG	1	0	1	0	0	0	0	1	0	1	0	1	0	0
JOHN	1	0	1	0	0	1	0	0	0	1	0	0	0	1
DOUG	1	0	0	1	0	0	0	1	1	0	0	0	1	0
LANCE	1	0	1	0	0	1	0	0	0	1	0	0	0	1
GEORGE	1	0	1	0	0	1	0	0	0	0	1	0	0	1
PETE	1	0	1	0	0	0	0	1	1	0	0	0	1	0
FRED	1	0	1	0	0	0	0	1	1	0	0	1	0	0
GENE	1	0	1	0	0	0	1	0	1	0	0	1	0	0
RALPH	1	0	0	1	0	1	0	0	1	0	0	1	0	0
PHIL	0	1	0	1	0	0	1	0	0	1	0	1	0	0
IKE	0	1	0	1	0	1	0	0	1	0	0	0	1	0
NICK	0	1	0	1	0	0	0	1	1	0	0	1	0	0
DON	0	1	0	1	0	0	1	0	0	1	0	0	0	1
NED	0	1	0	1	0	0	1	0	0	1	0	0	1	0
KARL	0	1	0	0	1	0	0	1	0	1	0	0	1	0
KEN	0	1	1	0	0	0	0	1	1	0	0	0	0	1
EARL	0	1	0	0	1	0	0	1	0	1	0	0	0	1
RICK	0	1	0	1	0	0	0	1	0	0	1	0	0	1
OL	0	1	0	1	0	0	1	0	0	1	0	1	0	0
NEAL	0	1	0	1	0	0	0	1	1	0	0	0	1	0
DAVE	0	1	0	1	0	0	0	1	0	0	1	1	0	0

Table 15.4 Gang Dataset transposed

Gang 27 × 14	ART	AL	SAM	CLYDE	MIKE	JIM	GREG	JOHN	DOUG	LANCE	GEORGE	PETE	FRED	GENE	RALPH	PHIL
Jet	1	1	1	1	1	1	1	1	1	1	1	1	1	1	1	0
Sharks	0	0	0	0	0	0	0	0	0	0	0	0	0	0	0	1
20’	0	0	1	0	0	1	1	1	0	1	1	1	1	1	0	0
30’	0	1	0	0	1	0	0	0	1	0	0	0	0	0	1	1
40’	1	0	0	1	0	0	0	0	0	0	0	0	0	0	0	0
JS	1	1	0	1	1	1	0	1	0	1	1	0	0	0	1	0
COL	0	0	1	0	0	0	0	0	0	0	0	0	0	1	0	1
HS	0	0	0	0	0	0	1	0	1	0	0	1	1	0	0	0
Single	1	0	1	1	1	0	0	0	1	0	0	1	1	1	1	0
Married	0	1	0	0	0	0	1	1	0	1	0	0	0	0	0	1
Divorced	0	0	0	0	0	1	0	0	0	0	1	0	0	0	0	0
Pusher	1	0	0	0	0	0	1	0	0	0	0	0	1	1	1	1
Bookie	0	0	1	1	1	0	0	0	1	0	0	1	0	0	0	0
Burglar	0	1	0	0	0	1	0	1	0	1	1	0	0	0	0	0

(continued)

Table 15.4 (continued)

Gang 27 × 14	IKE	NICK	DON	NED	KARL	KEN	EARL	RICK	OL	NEAL	DAVE
Jet	0	0	0	0	0	0	0	0	0	0	0
Sharks	1	1	1	1	1	1	1	1	1	1	1
20’	0	0	0	0	0	1	0	0	0	0	0
30’	1	1	1	1	0	0	0	1	1	1	1
40’	0	0	0	0	1	0	1	0	0	0	0
JS	1	0	0	0	0	0	0	0	0	0	0
COL	0	0	1	1	0	0	0	0	1	0	0
HS	0	1	0	0	1	1	1	1	0	1	1
Single	1	1	0	0	0	1	0	0	0	1	0
Married	0	0	1	1	1	0	1	0	1	0	0
Divorced	0	0	0	0	0	0	0	1	0	0	1
Pusher	0	1	0	0	0	0	0	0	1	0	1
Bookie	1	0	0	1	1	0	0	0	0	1	0
Burglar	0	0	1	0	0	1	1	1	0	0	0

Table 15.5 AutoCM weights matrix

Weights	ART	AL	SAM	CLYDE	MIKE	JIM	GREG	JOHN	DOUG	LANCE	GEORGE	PETE	FRED	GENE	RALPH	PHIL
ART	6.1962	5.2850	5.4981	5.9948	5.8317	5.3123	5.3767	5.2930	5.4659	5.2930	5.3123	5.4435	5.8808	5.9064	6.0280	3.6752
AL	5.2850	6.1962	4.3788	5.2894	5.9166	5.8041	6.5038	6.1319	5.6655	6.0319	5.8041	4.3625	4.3556	4.3721	5.9061	5.3856
SAM	5.4981	4.3788	6.1962	5.8730	5.8248	5.3103	5.3025	5.2865	5.8171	5.2865	5.3103	6.0214	5.8339	6.0422	5.4381	2.9256
CLYDE	5.9948	5.2894	5.8730	6.1962	6.0135	5.3167	4.3493	5.2974	5.8203	5.2974	5.3167	5.8386	5.4637	5.5178	5.8291	0.0001
MIKE	5.8317	5.9166	5.8248	6.0135	6.1962	5.2798	4.3265	5.2605	6.0657	5.2605	5.2798	5.7903	5.4067	5.4608	6.0735	4.2319
JIM	5.3123	5.8041	5.3103	5.3167	5.2798	6.1962	5.3253	6.0110	4.4193	6.0110	6.0871	5.2763	5.2654	5.2946	5.2715	0.0001
GREG	5.3767	5.5038	5.3025	4.3493	4.3265	5.3253	6.1962	5.8737	5.5393	5.8737	5.3253	5.8674	6.0418	5.7770	5.2577	5.1466
JOHN	5.2930	6.0319	5.2865	5.2974	5.2605	6.0110	5.8737	6.1962	4.3875	6.1257	6.0110	5.2525	5.2416	5.2708	5.2522	3.8849
DOUG	5.4659	5.6655	5.8171	5.8203	6.0657	4.4193	5.5393	4.3875	6.1962	4.3875	4.4193	6.0384	5.9002	5.4753	5.9471	4.1552
LANCE	5.2930	6.0319	5.2865	5.2974	5.2605	6.0110	5.8737	6.1257	4.3875	6.1962	6.0110	5.2525	5.2416	5.2708	5.2522	3.8849
GEORGE	5.3123	5.8041	5.3103	5.3167	5.2798	6.0871	5.3253	6.0110	4.4193	6.0110	6.1962	5.2763	5.2654	5.2946	5.2715	0.0001
PETE	5.4435	4.3625	6.0214	5.8386	5.7903	5.2763	5.8674	5.2525	6.0384	5.2525	5.2763	6.1962	6.0521	5.8318	5.3835	0.0001
FRED	5.8808	4.3556	5.8339	5.4637	5.4067	5.2654	6.0418	5.2416	5.9002	5.2416	5.2654	6.0521	6.1962	6.0342	5.8126	3.5168
GENE	5.9064	4.3721	6.0422	5.5178	5.4608	5.2946	5.7770	5.2708	5.4753	5.2708	5.2946	5.8318	6.0342	6.1962	5.8382	4.7576
RALPH	6.0280	5.9061	5.4381	5.8291	6.0735	5.2715	5.2577	5.2522	5.9471	5.2522	5.2715	5.3835	5.8126	5.8382	6.1962	5.2719
PHIL	3.6752	5.3856	2.9256	0.0001	4.2319	0.0001	5.1466	3.8849	4.1552	3.8849	0.0001	0.0001	3.5168	4.7576	5.2719	6.1962
IKE	5.3281	5.2891	5.1172	5.7535	6.0217	3.6919	0.0001	3.6763	5.7314	3.6763	3.6919	5.0173	3.9287	4.0800	5.8611	5.3551
NICK	5.2430	4.0979	4.0358	4.0623	5.4345	0.0001	4.9881	0.0001	5.8464	0.0001	0.0001	5.2777	5.7640	5.1931	5.8022	5.7766
DON	0.0001	5.8303	3.0786	0.0001	4.2896	3.6271	3.6827	5.1309	4.2129	5.1309	3.6271	0.0001	0.0001	2.9942	4.2407	6.0007
NED	0.0001	5.3848	4.8009	3.5016	5.2802	0.0001	3.8553	3.9285	5.2011	3.9285	0.0001	3.4447	0.0001	2.8590	4.1425	6.0006
KARL	2.3591	3.7673	3.5629	4.4991	3.3814	0.0001	5.3836	4.0077	5.0541	4.0077	0.0001	5.2104	3.9247	0.0001	0.0001	5.3013
KEN	4.2137	3.4048	5.2858	4.1760	4.0223	5.0342	5.0804	4.9435	5.2297	4.9435	5.0342	5.7954	5.8026	5.3038	4.0523	3.9815
EARL	2.5185	5.0313	0.0001	2.3812	0.0001	3.6180	5.3490	5.1529	3.8881	5.1529	3.6180	4.0174	4.0084	0.0001	0.0001	5.2474
RICK	0.0001	5.3627	0.0001	0.0001	4.2792	4.4973	3.9957	3.5176	5.4355	3.5176	4.4973	3.9606	3.9516	0.0001	4.2303	5.3638
OL	3.6752	5.3856	2.9256	0.0001	4.2319	0.0001	5.1466	3.8849	4.1552	3.8849	0.0001	0.0001	3.5168	4.7576	5.2719	6.1162
NEAL	4.0836	4.1174	5.1090	5.1164	5.7922	0.0001	3.7810	0.0001	6.0064	0.0001	0.0001	5.7288	5.3049	4.0871	5.4243	5.3000
DAVE	3.6474	4.2389	0.0001	0.0001	4.2475	2.3525	5.1893	0.0001	5.4023	0.0001	2.3525	3.8954	5.1880	3.4981	5.2690	5.8502

(continued)

Table 15.5 (continued)

Weights	IKE	NICK	DON	NED	KARL	KEN	EARL	RICK	OL	NEAL	DAVE
ART	5.3281	5.2430	0.0001	0.0001	2.3591	4.2137	2.5185	0.0001	3.6752	4.0836	3.6474
AL	5.2891	4.0979	5.8303	5.3848	3.7673	3.4048	5.0313	5.3627	5.3856	4.1174	4.2389
SAM	5.1172	4.0358	3.0786	4.8009	3.5629	5.2858	0.0001	0.0001	2.9256	5.1090	0.0001
CLYDE	5.7535	4.0623	0.0001	3.5016	4.4991	4.1760	2.3812	0.0001	0.0001	5.1164	0.0001
MIKE	6.0217	5.4345	4.2896	5.2802	3.3814	4.0223	0.0001	4.2792	4.2319	5.7922	4.2475
JIM	3.6919	0.0001	3.6271	0.0001	0.0001	5.0342	3.6180	4.4973	0.0001	0.0001	2.3525
GREG	0.0001	4.9881	3.6827	3.8553	5.3836	5.0804	5.3490	3.9957	5.1466	3.7810	5.1893
JOHN	3.6763	0.001	5.1309	3.9285	4.0077	4.9435	5.1529	3.5176	3.8849	0.0001	0.0001
DOUG	5.7314	5.8464	4.2129	5.2011	5.0541	5.2297	3.8881	5.4355	4.1552	6.0064	5.4023
LANCE	3.6763	0.0001	5.1309	3.9285	4.0077	4.9435	5.1529	3.5176	3.8849	0.0001	0.0001
GEORGE	3.6919	0.0001	3.6271	0.0001	0.0001	5.0342	3.6180	4.4973	0.0001	0.0001	2.3525
PETE	5.0173	5.2777	0.0001	3.4447	5.2104	5.7954	4.0174	3.9606	0.0001	5.7288	3.8954
FRED	3.9287	5.7640	0.0001	0.0001	3.9247	5.8026	4.0084	3.9516	3.5168	5.3049	5.1880
GENE	4.0800	5.1931	2.9942	2.8590	0.0001	5.3038	0.0001	0.0001	4.7576	4.0871	3.4981
RALPH	5.8611	5.8022	4.2407	4.1425	0.0001	4.0523	0.0001	4.2303	5.2719	5.4243	5.2690
PHIL	5.3551	5.7766	6.0007	6.0006	5.3013	3.9815	5.2474	5.3638	6.1162	5.3000	5.8502
IKE	6.1962	5.8856	5.3577	5.7929	5.1612	5.5253	3.9717	5.3406	5.3551	6.0298	5.3400
NICK	5.8856	6.1962	5.2953	5.2697	5.1870	5.8699	5.2320	5.8235	5.7766	6.0526	6.0164
DON	5.3577	5.2953	6.1962	5.9998	5.2239	5.2634	5.7808	5.8687	6.0007	5.3026	5.3659
NED	5.7929	5.2697	5.9998	6.1962	5.8447	3.9611	5.2565	5.3408	6.0006	5.7581	5.3403
KARL	5.1612	5.1870	5.2239	5.8447	6.1962	5.3089	6.0012	5.2920	5.3013	5.7242	5.2792
KEN	5.5253	5.8699	5.2634	3.9611	5.3089	6.1962	5.8487	5.8227	3.9815	5.8720	5.2537
EARL	3.9717	5.2320	5.7808	5.2565	6.0012	5.8487	6.1962	5.8565	5.2474	5.2474	5.3243
RICK	5.3406	5.8235	5.8687	5.3408	5.2920	5.8227	5.8565	6.1962	5.3638	5.8328	5.9915
OL	5.3551	5.7766	6.0007	6.0006	5.3013	3.9815	5.2474	5.3638	6.1962	5.3000	5.8502
NEAL	6.0298	6.0526	5.3026	5.7581	5.7242	5.8720	5.2474	5.8328	5.3000	6.1962	5.8237
DAVE	5.3400	6.0164	5.3659	5.3403	5.2792	5.2537	5.3243	5.9915	5.8502	5.8237	6.1962

Table 15.6 AutoCM distance matrix

Distances	ART	AL	SAM	CLYDE	MIKE	JIM	GREG	JOHN	DOUG	LANCE	GEORGE	PETE	FRED	GENE	RALPH	PHIL
ART	0.0000	0.9112	0.6981	0.2014	0.3645	0.8839	0.8195	0.9032	0.7303	0.9032	0.8839	0.7527	0.3154	0.2898	0.1682	2.5210
AL	0.9112	0.0000	1.8174	0.9068	0.2796	0.3921	0.6924	0.1643	0.5307	0.1643	0.3921	1.8337	1.8406	1.8241	0.2901	0.8106
SAM	0.6981	1.8174	0.0000	0.3232	0.3714	0.8859	0.8937	0.9097	0.3791	0.9097	0.8859	0.1748	0.3623	0.1540	0.7581	3.2706
CLYDE	0.2014	0.9068	0.3232	0.0000	0.1827	0.8795	1.8469	0.8988	0.3759	0.8988	0.8795	0.3576	0.7325	0.6784	0.3671	6.1961
MIKE	0.3645	0.2796	0.3714	0.1827	0.0000	0.9164	1.8697	0.9357	0.1305	0.9357	0.9164	0.4059	0.7895	0.7354	0.1227	1.9643
JIM	0.8839	0.3921	0.8859	0.8795	0.9164	0.0000	0.8709	0.1852	1.7769	0.1852	0.1091	0.9199	0.9308	0.9016	0.9247	6.1961
GREG	0.8195	0.6924	0.8937	1.8469	1.8697	0.8709	0.0000	0.3225	0.6569	0.3225	0.8709	0.3288	0.1544	0.4192	0.9385	1.0496
JOHN	0.9032	0.1643	0.9097	0.8988	0.9357	0.1852	0.3225	0.0000	1.8087	0.0705	0.1852	0.9437	0.9546	0.9254	0.9440	2.3113
DOUG	0.7303	0.5307	0.3791	0.3759	0.1305	1.7769	0.6569	1.8087	0.0000	1.8087	1.7769	0.1578	0.2960	0.7209	0.2491	2.0410
LANCE	0.9032	0.1643	0.9097	0.8988	0.9357	0.1852	0.3225	0.0705	1.8087	0.0000	0.1852	0.9437	0.9546	0.9254	0.9440	2.3113
GEORGE	0.8839	0.3921	0.8859	0.8795	0.9164	0.1091	0.8709	0.1852	1.7769	0.1852	0.0000	0.9199	0.9308	0.9016	0.9247	6.1961
PETE	0.7527	1.8337	0.1748	0.3576	0.4059	0.9199	0.3288	0.9437	0.1578	0.9437	0.9199	0.0000	0.1441	0.3644	0.8127	6.1961
FRED	0.3154	1.8406	0.3623	0.7325	0.7895	0.9308	0.1544	0.9546	0.2960	0.9546	0.9308	0.1441	0.0000	0.1620	0.3836	2.6794
GENE	0.2898	1.8241	0.1540	0.6784	0.7354	0.9016	0.4192	0.9254	0.7209	0.9254	0.9016	0.3644	0.1620	0.0000	0.3580	1.4386
RALPH	0.1682	0.2901	0.7581	0.3671	0.1227	0.9247	0.9385	0.9440	0.2491	0.9440	0.9247	0.8127	0.3836	0.3580	0.0000	0.9343
PHIL	2.5210	0.8106	3.2706	6.1961	1.9643	6.1961	1.0496	2.3113	2.0410	2.3113	6.1961	6.1961	2.6794	1.4386	0.9243	0.0000
IKE	0.8681	0.9071	1.0790	0.4427	0.1745	2.5043	6.1961	2.5199	0.4648	2.5199	2.5043	1.1789	2.2675	2.1162	0.3351	0.8411
NICK	0.9532	2.0983	2.1604	2.1339	0.7617	6.1961	1.2081	6.1961	0.3498	6.1961	6.1961	0.9185	0.4322	1.0031	0.3940	0.4196
DON	6.1961	0.3659	3.1176	6.1961	1.9066	2.5691	2.5135	1.0653	1.9833	1.0653	2.5691	6.1961	6.1961	3.2020	1.9555	0.1955
NED	6.1961	0.8114	1.3953	2.6946	0.9160	6.1961	2.6409	2.2677	0.9951	2.2677	6.1961	2.7515	6.1961	3.3372	2.0537	0.1956
KARL	3.8371	2.4289	2.6333	1.6971	2.8148	6.1961	0.8126	2.1885	1.1421	2.1885	6.1961	0.9858	2.2715	6.1961	6.1961	0.8949
KEN	1.9825	2.7914	0.9104	2.0202	2.1739	1.1620	1.1158	1.2527	0.9665	1.2527	1.1620	0.4008	0.3936	0.8924	02.1439	2.2147
EARL	3.6777	1.1649	6.1961	3.8150	6.1961	2.5782	0.8472	1.0433	2.3081	1.0433	2.5782	2.1788	2.1878	6.1961	6.1961	0.9488
RICK	6.1961	0.8335	6.1961	6.1961	1.9170	1.6989	2.2005	2.6786	0.7607	2.6786	1.6989	2.2356	2.2446	6.1961	1.9659	0.8324
OL	2.5210	0.8106	3.2706	6.1961	1.9643	6.1961	1.0496	2.3113	2.0410	2.3113	6.1961	6.1961	2.6794	1.4386	0.9243	0.0800
NEAL	2.1126	2.0788	1.0872	1.0798	0.4040	6.1961	2.4252	6.1961	0.1898	6.1961	6.1961	0.4674	0.8913	2.1091	0.7719	0.8962
DAVE	2.5488	1.9573	6.1961	6.1961	1.9487	3.8437	1.0069	6.1961	0.7939	6.1961	3.8437	2.3008	1.0082	2.6981	0.9272	0.3460

(continued)

Table 15.6 (continued)

Distances	IKE	NICK	DON	NED	KARL	KEN	EARL	RICK	OL	NEAL	DAVE
ART	0.8681	0.9532	6.1961	6.1961	3.8371	1.9825	3.6777	6.1961	2.5210	2.1126	2.5488
AL	0.9071	2.0983	0.3659	0.8114	2.4289	2.7914	1.1649	0.8335	0.8106	2.0788	1.9573
SAM	1.0790	2.1604	3.1176	1.3953	2.6333	0.9104	6.1961	6.1961	3.2706	1.0872	6.1961
CLYDE	0.4427	2.1339	6.1961	2.6946	1.6971	2.0202	3.8150	6.1961	6.1961	1.0798	6.1961
MIKE	0.1745	0.7617	1.9066	0.9160	2.8148	2.1739	6.1961	1.9170	1.9643	0.4040	1.9487
JIM	2.5043	6.1961	2.5691	6.1961	6.1961	1.1620	2.5782	1.6989	6.1961	6.1961	3.8437
GREG	6.1961	1.2081	2.5135	2.3409	0.8126	1.1158	0.8472	2.2005	1.0496	2.4152	1.0069
JOHN	2.5199	6.1961	1.0653	2.2677	2.1885	1.2527	1.0433	2.6786	2.3113	6.1961	6.1961
DOUG	0.4648	0.3498	1.9833	0.9951	1.1421	0.9665	2.3081	0.7607	2.0410	0.1898	0.7939
LANCE	2.5199	6.1961	1.0653	2.2677	2.1885	1.2527	1.0433	2.6786	2.3113	6.1961	6.1961
GEORGE	2.5043	6.1961	2.5691	6.1961	6.1961	1.1620	2.5782	1.6989	6.1961	6.1961	3.8437
PETE	1.1789	0.9185	6.1961	2.7515	0.9858	0.4008	2.1788	2.2356	6.1961	0.4674	2.3008
FRED	2.2675	0.4322	6.1961	6.1961	2.2715	0.3936	2.1878	2.2446	2.6794	0.8913	1.0082
GENE	2.1162	1.0031	3.2020	3.3372	6.1961	0.8924	6.1961	6.1961	1.4386	2.1091	2.6981
RALPH	0.3351	0.3940	1.9555	2.0537	6.1961	2.1439	6.1961	1.9659	0.9243	0.7719	0.9272
PHIL	0.8411	0.4196	0.1955	0.1956	0.8949	2.2147	0.9488	0.8324	0.0800	0.8962	0.3460
IKE	0.0000	0.3106	0.8385	0.4033	1.0350	0.6709	2.2245	0.8556	0.8411	0.1664	0.8562
NICK	0.3106	0.0000	0.9009	0.9265	1.0092	0.3263	0.9642	0.3727	0.4196	0.1436	0.1798
DON	0.8385	0.9009	0.0000	0.1964	0.9723	0.9328	0.4154	0.3275	0.1955	0.8936	0.8303
NED	0.4033	0.9265	0.1964	0.0000	0.3515	2.2351	0.9397	0.8554	0.1956	0.4381	0.8559
KARL	1.0350	1.0092	0.9723	0.3515	0.0000	0.8873	0.1950	0.9042	0.8949	0.4720	0.9170
KEN	0.6709	0.3263	0.9328	2.2351	0.8873	0.0000	0.3475	0.3735	2.2147	0.3242	0.9425
EARL	2.2245	0.9642	0.4154	0.9397	0.1950	0.3475	0.0000	0.3397	0.9488	0.9488	0.8719
RICK	0.8556	0.3727	0.3275	0.8554	0.9042	0.3735	0.3397	0.0000	0.8324	0.3634	0.2047
OL	0.8411	0.4196	0.1955	0.1956	0.8949	2.2147	0.9488	0.8324	0.0000	0.8962	0.3460
NEAL	0.1664	0.1436	0.8936	0.4381	0.4720	0.3242	0.9488	0.3634	0.8962	0.0000	0.3725
DAVE	0.8562	0.1798	0.8303	0.8559	0.9170	0.9425	0.8719	0.2047	0.3460	0.3725	0.0000

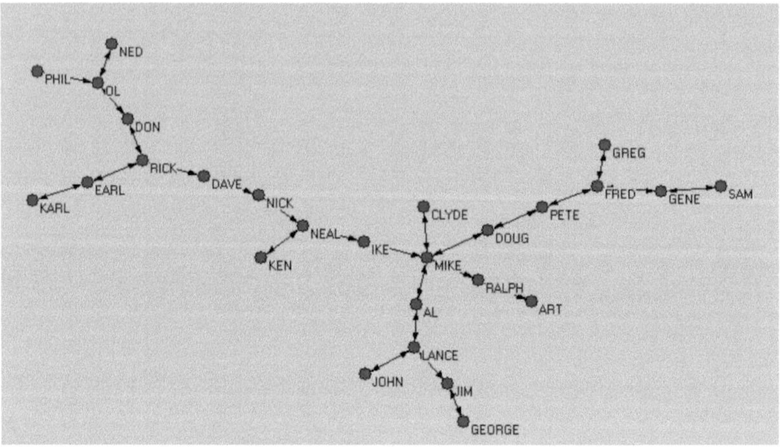

Fig. 15.9 The MST of the global networks ($H_0 = 0.10989$)

The local indexes for this tree are the following:

Vertex degree		Clustering strength		Pruning rank	
Equation (15.28)		Equation (15.30)		Equation (15.31)	
ART	1	CYLDE	0.2	ART	1
SAM	1	GREG	0.333333	SAM	1
CLYDE	1	JOHN	0.333333	CYLDE	1
GREG	1	PHIL	0.333333	GREG	1
JOHN	1	NED	0.333333	JOHN	1
GEORGE	1	KEN	0.333333	GEORGE	1
PHIL	1	GEORGE	0.5	PHIL	1
NED	1	ART	0.5	NED	1
KARL	1	IKE	0.5	KARL	1
KEN	1	AL	0.5	KEN	1
GENE	2	KARL	0.5	GENE	2
RALPH	2	SAM	0.5	RALPH	2
JIM	2	DOUG	0.571429	JIM	2
IKE	2	DON	0.666667	EARL	2
NICK	2	RALPH	0.666667	OL	2
DON	2	PETE	0.8	DON	3
DOUG	2	NICK	0.8	FRED	3
AL	2	DAVE	0.8	LANCE	3
PETE	2	GENE	1	AL	4
EARL	2	EARL	1	RICK	4
DAVE	2	JIM	1	PETE	4
FRED	3	RICK	1.5	DOUG	5
RICK	3	LANCE	1.8	DAVE	5
OL	3	NEAL	1.8	MIKE	6
NEAL	3	FRED	1.8	NICK	6
LANCE	3	OL	2.25	NEAL	7
MIKE	5	MIKE	2.777778	IKE	7

The global indexes for global hubness (H_0, H_i) and for the graph entropy (E_0, E_i) are:

Global hubness Equations (15.32), (15.33), (15.34), (15.35), (15.36), (15.37), and (15.38)		Graph entropy Equation (15.37)	
***H*(0)=**	**0.10989**	***E*(0)=**	**18.52077**
H(AL)=	0.09	E(AL)=	15.52418
H(ART)=	0.108571	E(ART)=	17.50778
H(SAM)=	0.108571	E(JIM)=	17.50778
H(CLYDE)=	0.108571	E(NICK)=	17.50778
H(JIM)=	0.108571	E(KARL)=	17.50778
H(GREG)=	0.108571	E(JOHN)=	17.58254
H(JOHN)=	0.108571	E(LANCE)=	17.58254
H(DOUG)=	0.108571	E(GEORGE)=	17.58254
H(LANCE)=	0.108571	E(PETE)=	17.58254
H(GEORGE)=	0.108571	E(FRED)=	17.58254
H(PETE)=	0.108571	E(GENE)=	17.58254
H(FRED)=	0.108571	E(PHIL)=	17.58254
H(GENE)=	0.108571	E(SAM)=	17.58254
H(RALPH)=	0.108571	E(NED)=	17.58254
H(PHIL)=	0.108571	E(GREG)=	17.58254
H(IKE)=	0.108571	E(KEN)=	17.58254
H(NICK)=	0.108571	E(EARL)=	17.58254
H(NED)=	0.108571	E(OL)=	17.58254
H(KARL)=	0.108571	E(DAVE)=	17.58254
H(KEN)=	0.108571	E(DOUG)=	17.66717
H(EARL)=	0.108571	E(IKE)=	17.66717
H(OL)=	0.108571	E(CLYDE)=	17.66717
H(DAVE)=	0.108571	E(RALPH)=	17.74192
H(RICK)=	0.133333	E(NEAL)=	20.42575
H(MIKE)=	0.133333	E(RICK)=	20.51296
H(NEAL)=	0.133333	E(DON)=	20.51296
H(DON)=	0.133333	E(MIKE)=	20.52449

From both the hubness point of view and the entropy point of view, if we remove Rick from the graph, or Mike, or Neal, or Don, the complexity of the graph, and consequently its entropy, increases; on the other hand, if we remove Al from the global graph, the complexity of the graph, and consequently its entropy, decreases.

That is not absolutely evident if we analyze the same graph comparing the local indexes. From a naïve point of view, one could think exactly the opposite: because Mike is a big hub (five links), if he was to be removed, then the global network must become simpler. But from a global viewpoint, the rearrangement of the networks without some of its vertices works in a completely different way (Fig. 15.10):

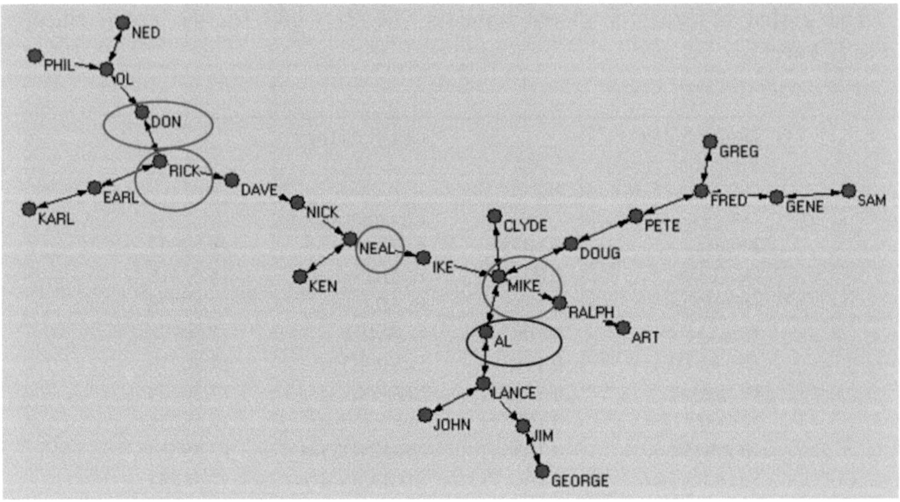

Fig. 15.10 The MST of the global networks marked ($H_0 = 0.10989$)

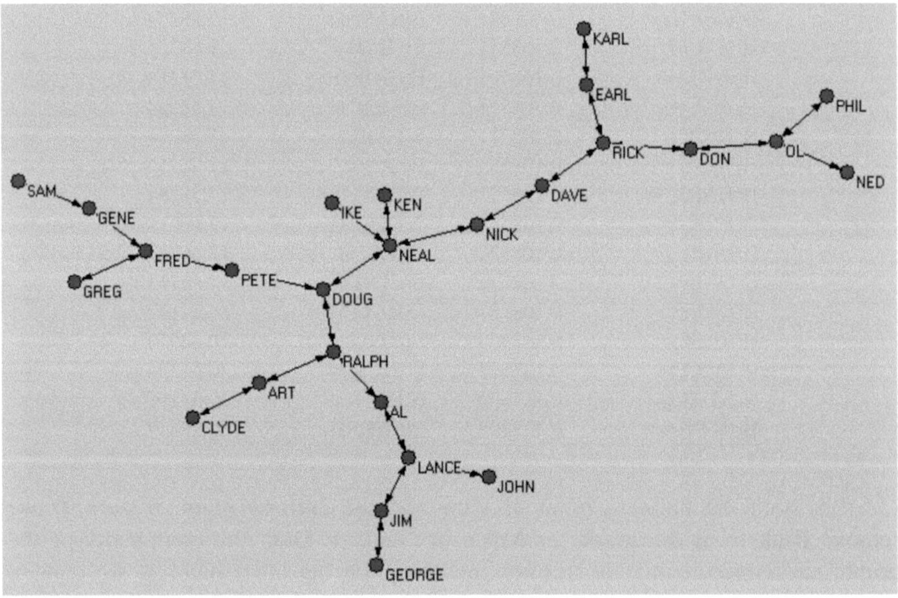

Fig. 15.11 New MST without Mike ($H_0 = 0.133333$)

If we remove the vertices within the red circles from Fig. 15.2, the new MST will show a more complex structure, while if we remove the vertex within the blue circle, the new MST will be simpler (Figs. 15.11 and 15.12).

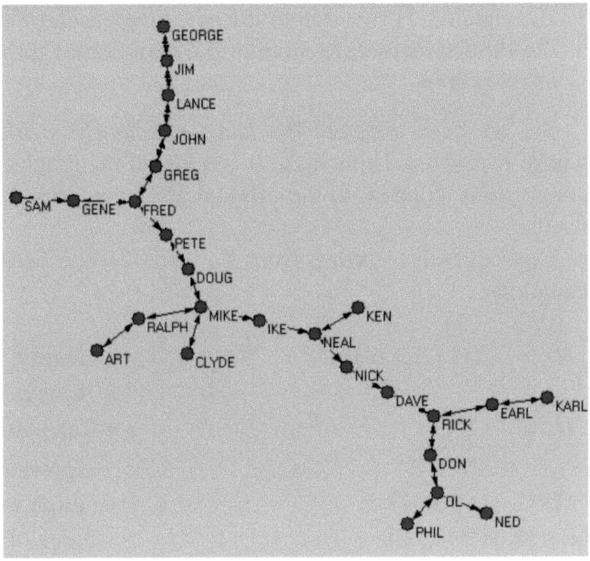

Fig. 15.12 New MST without Al ($H_0 = 0.09$)

15.12 AutoCM and Maximally Regular Graph (MRG)

The MST represents the nervous system of any dataset. In fact, the summation of the strength of the connection among all the variables represents the total energy of that system. The MST selects only the connections *that minimize this energy*. Consequently, all the links shown by the MST are fundamental, but not every fundamental link of the dataset is shown by MST.

Such a limit is intrinsic to the nature of MST itself: every link able to generate a cycle within the graph is eliminated, whatever its strength. To avoid this limit and to better explain the intrinsic complexity of a dataset, it is necessary to add more links to the graph according to two criteria:

1. The new links must be *relevant* from a quantitative point of view.
2. The new links must be able to generate new *cyclic regular microstructures*, from a qualitative point of view.

Consequently, the MST tree graph is transformed into an undirected graph with cycles. Because of the cycles, the new graph is a dynamic system, involving the *time* dimension in its structure.

This is the reason why this new graph should provide information not only about the structure but also about the *functions* of the variables of the dataset.

To build this new graph, we need to proceed in the following way:

1. Assume the MST structure as a starting point of the new graph.
2. Consider the sorted list of the connections skipped during the MST generation.

3. Estimate the H function of the new graph each time we add a new connection to the MST structure, to monitor the variation of the complexity of the new graph at every step.

So, we have assigned the name *maximally regular graph* (MRG) to a graph whose *H function* is the highest, among all the graphs generated by adding the new connections skipped, to the original MST but prior to the completion of the MST itself.

Consequently, starting from Eq. (15.32), the MRG is given by the following equations:

$$
\begin{array}{ll}
H_i = f\left(G\left(A_p, N\right)\right); & /*\text{ Generic Function on a graph with } A_p \text{ arcs and } N \text{ Nodes } */ \\
H_i = \frac{\mu_p \cdot \varphi_p - 1}{A_p}; & /*\text{ Calculation of H Function, where } H_0 \text{ represents MST complexity } */ \\
\text{MRG} = \text{Max}\left\{H_i\right\}. & /*\text{ Graph with highest H } */ \\
i \in [0, 1, 2, ..., R]; & /*\text{ Index of H Function } */ \\
p \in [N-1, N, N+1, ..., N-1+R]. & /*\text{ index for the number of graph arcs } */ \\
R \in \left[0, 1, .., \frac{(N-1)\cdot(N-2)}{2}\right]; & /*\text{ Number of the skipped arcs during the M.S.T. generation } */
\end{array}
\qquad (15.40)
$$

The "*R*" variable is a key variable during the MRG generation. "*R*," in fact, could be null when the generation of MST implies no connections to be skipped. In this case, there is no MRG for that dataset. Furthermore, the "*R*" variable makes sure that the last, and consequently the weakest, connection added to generate the MRG is always more relevant than the weakest connection of the MST. The MRG, finally, generates, starting from the MST, the graph representing the *highest number of regular microstructures using the most important connections* of the dataset.

Moreover, the greater the H function selected to generate the MRG, the more meaningful the microstructures shown in the MRG.

15.12.1 Maximally Regular Graph: An Example

Let us consider again the "Gang" dataset (Table 15.2) to generate the MRG of this dataset (Fig. 15.13):

In this example, the H function has its peak when the system adds the seventh connection (start counting from 0) skipped during the MST generation. So the MRG needs seven new connections to be added to the MST, and, consequently,

		MST	
H(0)=	0.1099		
H(1)=	0.1130	AL	JOHN
H(2)=	0.1250	SAM	PETE
H(3)=	0.0897	LANCE	GEORGE
H(4)=	0.0758	JIM	JOHN
H(5)=	0.1261	JOHN	GEORGE
H(6)=	0.0953	DOUG	NEAL
H(7)=	**0.1538**	**PHIL**	**DON**
H(8)=	0.1493	PHIL	NED
H(9)=	0.1450	DON	NED
H(10)=	0.1410	ART	CLYDE
H(11)=	0.1293	DOUG	RALPH
H(12)=	0.1090	ART	GENE
H(13)=	0.1128	AL	RALPH
H(14)=	0.0907	DOUG	FRED
H(15)=	0.0941	IKE	NICK
H(16)=	0.1190	ART	FRED
H(17)=	0.1163	GREG	JOHN
H(18)=	0.1000	GREG	LANCE
H(19)=	0.0978	SAM	CLYDE
H(20)=	0.0957	NICK	KEN
H(21)=	0.1367	GREG	PETE
H(22)=	0.1167	RALPH	IKE
H(23)=	0.1008	PHIL	DAVE
H(24)=	0.1343	OL	DAVE
H(25)=	0.1317	KEN	EARL
H(26)=	0.1440	NED	KARL
H(27)=	0.1212	DOUG	NICK

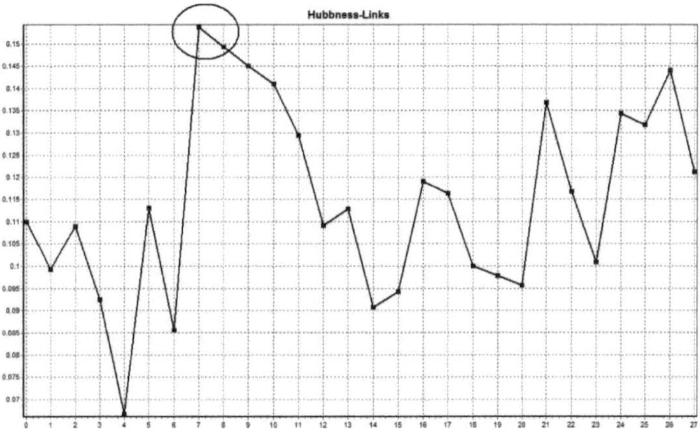

Fig. 15.13 Calculation of MRG hubness of "Gang" dataset

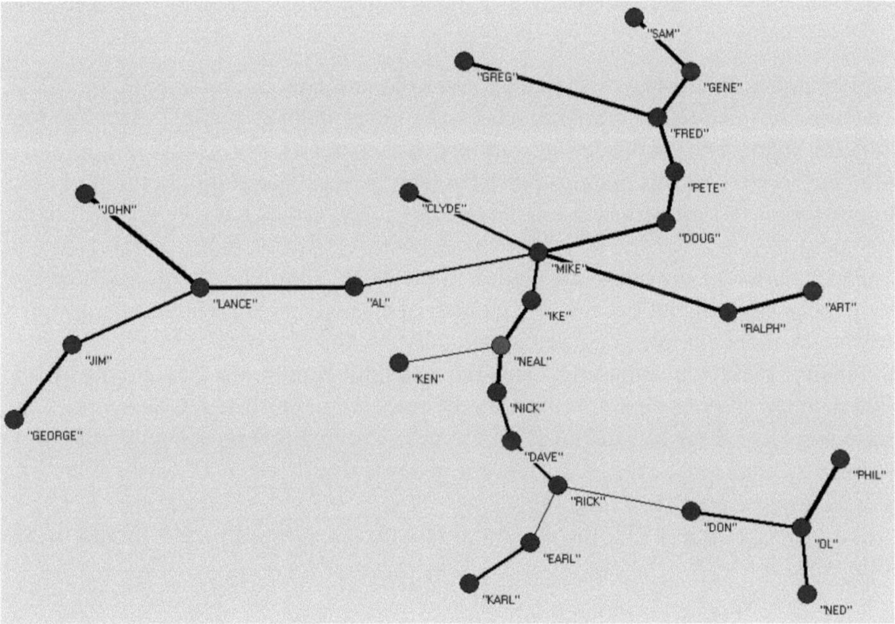

Fig. 15.14 MST of "Gang" dataset

the H function increases beyond 50 % with respect to the original MST H function: ($H(0) = 10.99$, $H(7) = 15.38$). Obviously, the quality of the two graphs must be quite different (Figs. 15.14 and 15.15):

The MRG increases the information contained in the MST. The boundary between the Jets and Sharks members is represented by a cycle of four subjects:

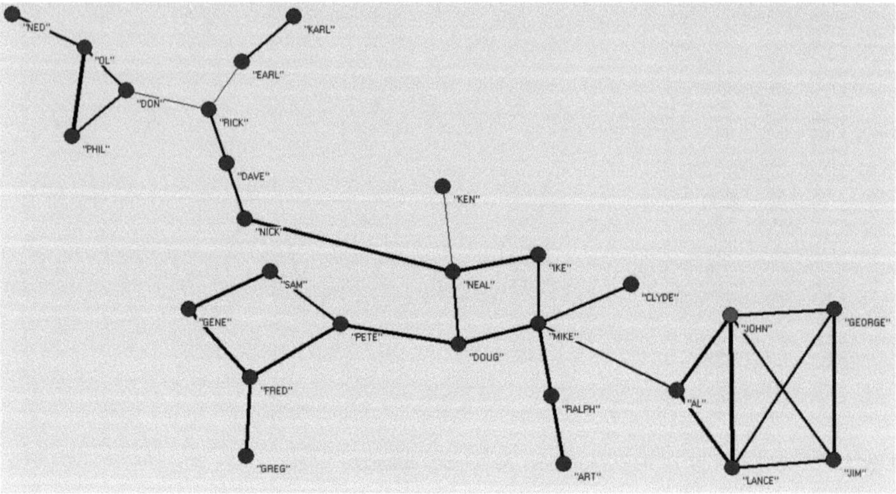

Fig. 15.15 MRG of "Gang" dataset

Neal and Ken are Sharks, while Doug and Mike are Jets. So, looking at the MRG, the edges between Jets and Sharks seem to be fuzzy and negotiable. In fact, the four subjects laying on this border are outliers in their respective gangs. Furthermore, Al, a member of the Jets gang, is placed at the top of an autonomous circuit of links among four Jets members, as he is the head of a new virtual gang hidden within the Jets gang.

Examination of the new information in the MRG about the structure of the two gangs becomes evident: because the number of cycles is greater, the Jets gang shows itself to be more complex and articulated than the Sharks gang.

Finally, the cycle including Don, Ol, and Phil represents a prototype of the Sharks member whose features are very different from the other Jets subjects. In the same way, Jets show two different prototypes. The first is represented by the cycle including Gene, Sam, Fred, and Pete; the second is outlined by the cycle including John, George, Lance, and Jim.

According to the MRG, the structural features of each prototype should be the following:

1. Prototype of gang hybridization:

$$\left[\frac{\text{Jets}}{\text{Sharks}}\right] + 30' + \left[\frac{\text{JuniorSchool}}{\text{HighSchool}}\right] + \text{Single} + \text{Bookie}.$$

2. Prototype of Sharks member:

$$30' + \text{College} + \text{Married} + \left[\frac{\text{Pusher}}{\text{Burglar}}\right].$$

3. First prototype of Jets member:

$$20' + \left[\frac{\text{College}}{\text{HighSchool}}\right] + \text{Single} + \left[\frac{\text{Bookie}}{\text{Pusher}}\right].$$

4. Second prototype of Jets member:

$$20' + \text{JuniorSchool} + \left[\frac{\text{Married}}{\text{Divorced}}\right] + \text{Burglar}.$$

15.13 Conclusions

This chapter presents new theoretical hypotheses, new mathematical algorithms, and new criteria to measure the complexity of the networks:

1. The math, topology, and algorithm of a new ANN, named the *auto-contractive map* (AutoCM), have been presented. The AutoCM system reshapes the distances among variables or records of any dataset, considering their global vectorial similarities and consequently defines the specific warped space in which variables or records can work.
2. A filter known as the *MST* can be used to cluster a distance matrix, generated from a dataset, in a very useful way.
3. A new index, named the *H function*, was shown to size the topological complexity of any kind of graph, and its mathematical consistency and an application has been shown.
4. This chapter further created, from the H function, a new index to measure the relevance and contribution of any node within a semantic graph (a graph generated by a dataset). We have named this new index the *delta H function*.
5. Finally, a new type of semantic graph has been defined using the H function, called the *maximally regular graph* (MRG). From an MST, generated from any metric, the MRG reshapes the links among the nodes in order to maximize the fundamental and the most regular structures implicated in any dataset.

A.1 Appendices

A.1.1 Appendix A: AutoCM Convergence

Giovanni Pieri

The first step is to show the CM convergence equation. Convergence means that in the long term (for growing n), the connections do not vary any more, that is,

$$\Delta v_{i_{(n)}} = 0; \Delta w_{i,j_{(n)}} = 0.$$

A sufficient condition for convergence is:

$$\lim_{n \to \infty} v_{i_{(n)}} = C. \tag{15.8}$$

In fact, when $v_{i_{(n)}} = C$, then $\Delta v_{i_{(n)}} = 0$ (Eq. 15.2) and $m_{i_{(n)}}^{[h]} = 0$ (Eq. 15.1) and, consequently, $\Delta w_{i,j_{(n)}} = 0$ (Eq. 15.6).

The second step is to demonstrate that the sufficient condition holds. At this point, we can rewrite Eqs. (15.2) and (15.3) in this way:

$$\Delta v_{i_{(n)}} = \left(m_{i_{(n)}}^{[s]} - m_{i_{(n)}}^{[s]} \cdot \left(1 - \frac{v_{i_{(n)}}}{C}\right)\right) \cdot \left(1 - \frac{v_{i_{(n)}}}{C}\right) = m_{i_{(n)}}^{[s]} \cdot \frac{v_{i_{(n)}}}{C} \cdot \left(1 - \frac{v_{i_{(n)}}}{C}\right); \tag{15.2a}$$

$$v_{i_{(n+1)}} = v_{i_{(n)}} + m_{i_{(n)}}^{[s]} \cdot \frac{v_{i_{(n)}}}{C} \cdot \left(1 - \frac{v_{i_{(n)}}}{C}\right). \tag{15.3a}$$

For the sake of clarity, we pose:

$$v_{i_{(n+1)}} = v_{n+1};$$

$$\frac{v_{i_{(n)}}}{C} = y;$$

$$m_{i_{(n)}}^{[s]} = m.$$

Obtaining a simplified version of Eq. (15.3a):

$$v_{n+1} = Cy + m \cdot y \cdot (1 - y) = y\,(C + m) - my^2. \tag{15.3b}$$

It has to be noted that in Eq. (15.3b), while C is a true constant, remaining unchanged during training, m is a variable, which is bounded both superiorly and inferiorly; in fact, the following inequalities hold: $0 \leq m \leq 1$. This property will be exploited to demonstrate Eq. (15.8).

A graphical representation of Eq. (15.3b) is helpful to make clear its properties. The general form of Eq. (15.3b) is parabolic, passing for two fixed points not dependent on m: the origin where both v_{n+1} and y are null and the point of coordinates $y = 1$; $v_{n+1} = C$. Between the two points, the function may have a maximum. This happens if $C < 1$ and m is close to 1. The lower C is and the higher m is, the more pronounced is the maximum. Otherwise ($C > 1$) the maximum is outside the interval and it is found for $y > 1$.

Three cases of Eq. (15.3b) are represented in Figs. A.1, A.2, A.3, which are obtained respectively for $C = 0.8$, $C = 1$, and $C = 1.5$. In each case, various values of m give origin to different curves.

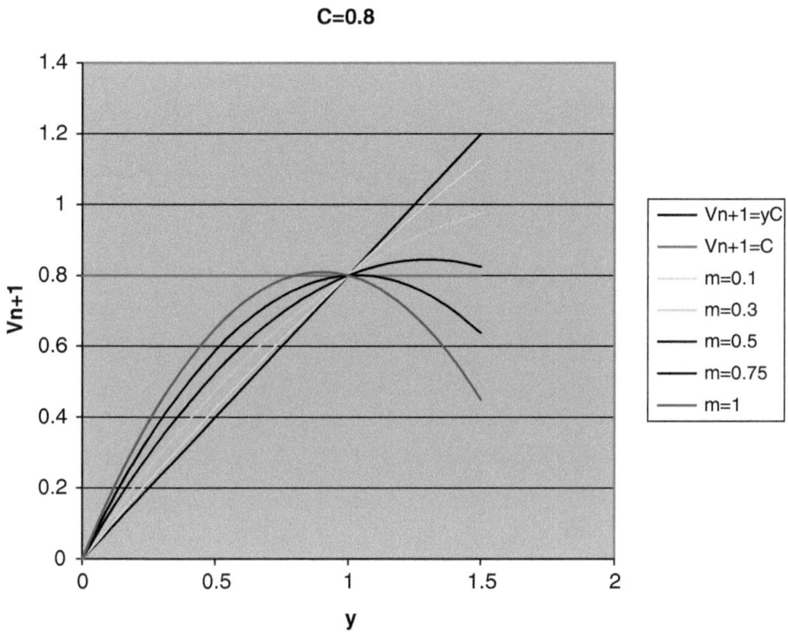

Fig. A.1 Equation (15.3b) in the case of $C = 0.8$

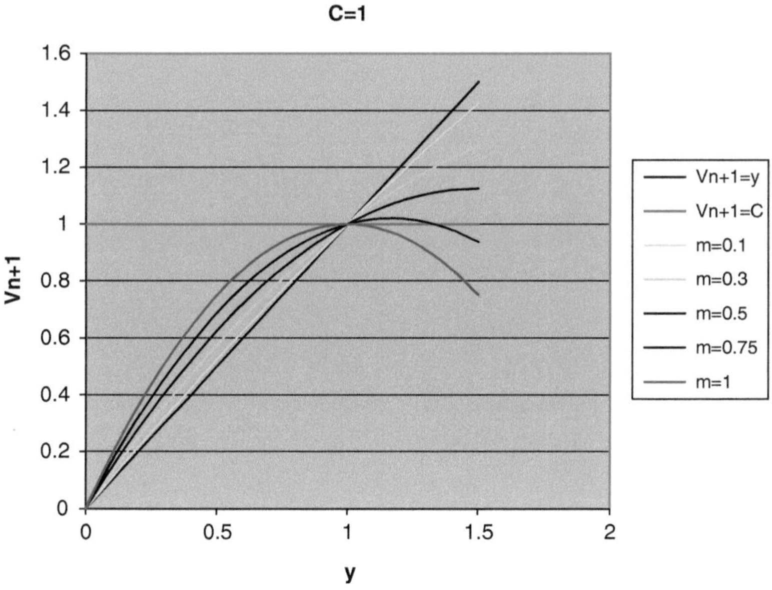

Fig. A.2 Equation (15.3b) in the case of $C = 1$

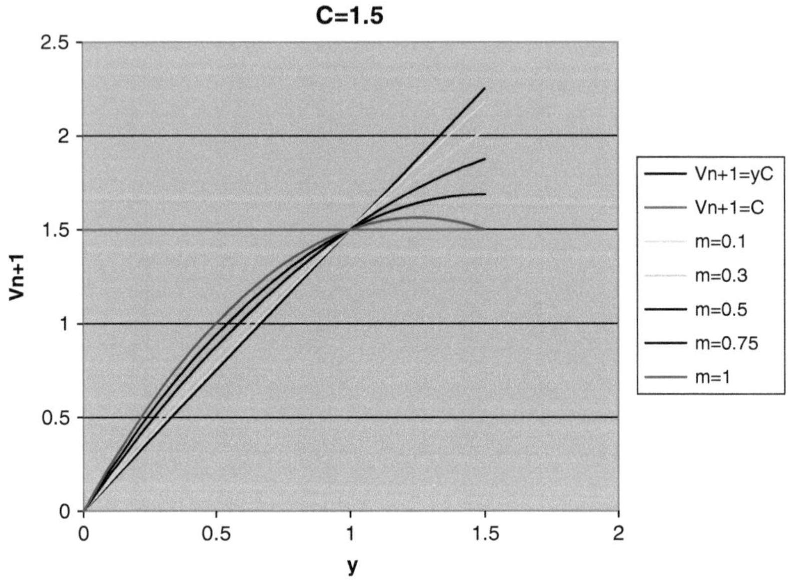

Fig. A.3 Equation (15.3b) in the case of $C = 1.5$

To elucidate what the use of the above diagrams is, let us assume for a moment a constant value for m: the curve corresponding to that value represents all the possible values for $v_{\text{n}+1}$. If $C \geq 1$ and $y < 1$, the following properties are readily seen:

(a) $v_{\text{n}+1}$ is always less than C.
(b) $v_{\text{n}+1}$ is always larger than v_{n}.
(c) When n grows, even y grows and so does $v_{\text{n}+1}$ indefinitely.

It is also readily seen that these properties hold even for a variable m, that is, m function of n.

If it is assumed that the property (c) is equivalent to say that for any positive ε exists at least one v_{n} which is:

$$v_{\text{n}} > C - \text{e},$$

it can be easily demonstrated that the sufficient condition (15.8) for convergence holds (if the above equation is satisfied, also the definition of limit is satisfied, and therefore, $\lim_{n \to \infty} v_{i(n)} = C$).

The equivalence cannot be assumed as unconditionally true, but only as being a reasonable conjecture, and, therefore, any demonstration based on it must be considered not completely sound.

The same holds for the condition $v_{i(n)}/C = 1 - \varepsilon$ which is in turn another form of the equivalent of the property (c) discussed above.

In particular, it is clear that the sufficient condition $\lim_{n\to\infty} v_{i_{(n)}} = C$ does not hold either where the parabolic curve shows a maximum at $y < 1$ and when the initial value v_0 is larger than C.

A.1.2 Appendix B: Operator "B"

Riccardo Petritoli[1]

Let $\bar{W}$ be the space of the square matrices whose elements satisfy these relations:

$$W \in \bar{W}, \quad n_{\dim} = \dim(W), \quad n_{\dim} \in \left(x | \exists\, y, z \in N^{+}, x = y \cdot z\right).$$

We define:

(a) The set D of the ordered pairs (n_R, n_C) with $n_R, n_C \in N^{+}, n_R \cdot n_C = n_{\dim}$
(b) The operator $B(n_R, n_C)$: $W \to W'$

$$w_{i,j} = w'_{k,l} \quad i, j, k, l \in \{1, ..., n_{\dim}\}$$

with:

$$k = [(i-1)/n_R] \cdot n_R + 1 + (j-1)/n_C$$

$$l = [(i-1) \bmod n_R] \cdot n_C + 1 + (j-1) \bmod n_C$$

where:
"/" is the division defined in N
"mod" is the modulo operation (also defined in N)

Notes

1. From definition (b) follows that the B operator makes a simple change of element positions in the matrix (a sort of "block transpose").
Consider this example:
Let $\bar{W}$ be a matrix space with $n_{\dim} = 12$ and the B operator with $n_R = 3$, $n_C = 4$.
The matrix W is:

$$W = \begin{pmatrix} w_{1,1} & w_{1,2} & \cdots & w_{1,12} \\ w_{2,1} & w_{2,2} & \cdots & w_{2,12} \\ \vdots & \vdots & \ddots & \vdots \\ w_{12,1} & w_{12,2} & \cdots & w_{12,12} \end{pmatrix}.$$

[1] The "B" operator was invented and implemented by M Buscema in 1998 at Semeion Research Center. The "B" operator is presented in this chapter for the first time.

Using the operator $B(3,\ 4)$, the result will be:

$$W' = \begin{pmatrix}
\overbrace{\begin{matrix} w_{1,1} & w_{1,2} & w_{1,3} & w_{1,4} \\ w_{1,5} & w_{1,6} & w_{1,7} & w_{1,8} \\ w_{1,9} & w_{1,10} & w_{1,11} & w_{1,12} \end{matrix}}^{1} & \overbrace{\begin{matrix} w_{2,1} & w_{2,2} & w_{2,3} & w_{2,4} \\ w_{2,5} & w_{2,6} & w_{2,7} & w_{2,8} \\ w_{2,9} & w_{2,10} & w_{2,11} & w_{2,12} \end{matrix}}^{2} & \overbrace{\begin{matrix} w_{3,1} & w_{3,2} & w_{3,3} & w_{3,4} \\ w_{3,5} & w_{3,6} & w_{3,7} & w_{3,8} \\ w_{3,9} & w_{3,10} & w_{3,11} & w_{3,12} \end{matrix}}^{3} \\
\overbrace{\begin{matrix} w_{4,1} & w_{4,2} & w_{4,3} & w_{4,4} \\ w_{4,5} & w_{4,6} & w_{4,7} & w_{4,8} \\ w_{4,9} & w_{4,10} & w_{4,11} & w_{4,12} \end{matrix}}^{4} & \overbrace{\begin{matrix} w_{5,1} & w_{5,2} & w_{5,3} & w_{5,4} \\ w_{5,5} & w_{5,6} & w_{5,7} & w_{5,8} \\ w_{5,9} & w_{5,10} & w_{5,11} & w_{5,12} \end{matrix}}^{5} & \overbrace{\begin{matrix} w_{6,1} & w_{6,2} & w_{6,3} & w_{6,4} \\ w_{6,5} & w_{6,6} & w_{6,7} & w_{6,8} \\ w_{6,9} & w_{6,10} & w_{6,11} & w_{6,12} \end{matrix}}^{6} \\
\overbrace{\begin{matrix} w_{7,1} & w_{7,2} & w_{7,3} & w_{7,4} \\ w_{7,5} & w_{7,6} & w_{7,7} & w_{7,8} \\ w_{7,9} & w_{7,10} & w_{7,11} & w_{7,12} \end{matrix}}^{7} & \overbrace{\begin{matrix} w_{8,1} & w_{8,2} & w_{8,3} & w_{8,4} \\ w_{8,5} & w_{8,6} & w_{8,7} & w_{8,8} \\ w_{8,9} & w_{8,10} & w_{8,11} & w_{8,12} \end{matrix}}^{8} & \overbrace{\begin{matrix} w_{9,1} & w_{9,2} & w_{9,3} & w_{9,4} \\ w_{9,5} & w_{9,6} & w_{9,7} & w_{9,8} \\ w_{9,9} & w_{9,10} & w_{9,11} & w_{9,12} \end{matrix}}^{9} \\
\overbrace{\begin{matrix} w_{10,1} & w_{10,2} & w_{10,3} & w_{10,4} \\ w_{10,5} & w_{10,6} & w_{10,7} & w_{10,8} \\ w_{10,9} & w_{10,10} & w_{10,11} & w_{10,12} \end{matrix}}^{10} & \overbrace{\begin{matrix} w_{1,1} & w_{1,2} & w_{1,3} & w_{1,4} \\ w_{1,5} & w_{1,6} & w_{1,7} & w_{1,8} \\ w_{11,9} & w_{11,10} & w_{11,11} & w_{11,12} \end{matrix}}^{11} & \overbrace{\begin{matrix} w_{1,1} & w_{1,2} & w_{1,3} & w_{1,4} \\ w_{1,5} & w_{1,6} & w_{1,7} & w_{1,8} \\ w_{12,9} & w_{12,10} & w_{12,11} & w_{12,12} \end{matrix}}^{12}
\end{pmatrix}.$$

From the previous example, we can extract a simple algorithm for the B operator.

Let the starting matrix be divided in 3×4 blocks:

$$W = \begin{pmatrix} A_1 & B_1 & C_1 \\ A_2 & B_2 & C_2 \\ A_3 & B_3 & C_3 \\ A_4 & B_4 & C_4 \end{pmatrix}.$$

From every ordered set (A_n, B_n, C_n), we can obtain a 3×12 block D_n; the resulting 4 blocks are going to be the final matrix:

$$W' = \begin{pmatrix} D_1 \\ D_2 \\ D_3 \\ D_4 \end{pmatrix}.$$

This is the procedure:

Step 1 Consider block A:

$$A = \begin{pmatrix} a_{1,1} & a_{1,2} & a_{1,3} & a_{1,4} \\ a_{2,1} & a_{2,2} & a_{2,3} & a_{2,4} \\ a_{3,1} & a_{3,2} & a_{3,3} & a_{3,4} \end{pmatrix}.$$

Let the block be "vectorized" obtaining V_A:

$$V_A = \begin{pmatrix} a_{1,1} & a_{1,2} & a_{1,3} & a_{1,4} & a_{2,1} & a_{2,2} & a_{2,3} & a_{2,4} & a_{3,1} & a_{3,2} & a_{3,3} & a_{3,4} \end{pmatrix}.$$

Step 2 Repeat step 1 for blocks B and C obtaining vectors V_B and V_C.
Step 3 Put the row vectors V_A, V_B, V_C in a column obtaining D:

$$D = \begin{pmatrix} V_A \\ V_B \\ V_C \end{pmatrix}.$$

The operator B is a bijective function: $B(n_R, n_C)$: $W \leftrightarrow W'$.
In fact, the inverse operator $B^{-1}(n_R, n_C)$ exists and is equal to $B(n_R, n_C)$: $W \overset{B}{\rightarrow} W' \overset{B}{\rightarrow} W$.
(Note that the transpose operator also has this feature.)

3. The B operator is linear. In fact, $B\,(a \cdot W' + b \cdot W'') = a \cdot B\,(W') + b \cdot B\,(W'')$.
4. It follows from definition (a) that, for every n_{dim}, the set D always contains at least two elements:

$$(1, n_{\text{dim}})\text{ e }(n_{\text{dim}}, 1).$$

In these cases, we have:

$$B(1, n_{\text{dim}}) \equiv I \quad \text{(Identityoperator)}$$

$$B(n_{\text{dim}}, 1) \equiv T \quad \text{(Transposeoperator)}$$

A.1.3 Appendix C: The Concept of Hubness

R. Petritoli and M. Buscema

A.1.3.1 C.1 Definition of Hubness

Pruning Algorithm

Consider a graph with N nodes and A links. We use the following algorithm (*pruning algorithm*):

1. Detect in the graph all the nodes with the *minimum gradient*, that is, all the nodes with the smallest number of links.
2. "Set free" all the detected nodes erasing their links.
3. Apply steps 1 and 2 until all the nodes of the graph are free (*complete disconnection of the graph*).

We define *pruning cycle number* as the number of iteration to disconnect completely the graph; we indicate this value with M.

Example 1 Consider the following graph ($N = 8$ and $A = 11$):

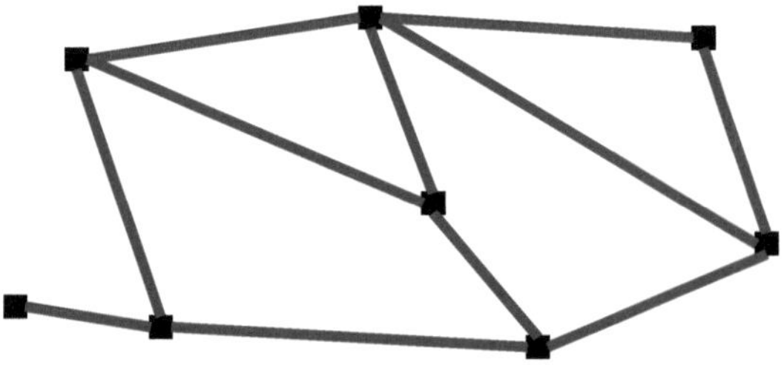

Let us apply the pruning algorithm:

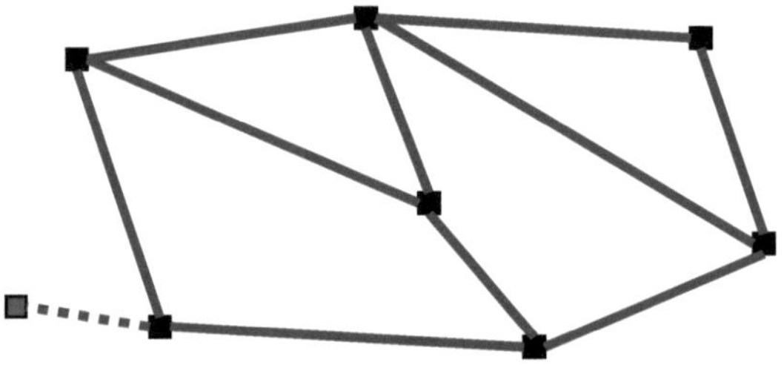

1st pruning cycle:
Minimum gradient = 1
Erased links = 1
Released nodes = 1

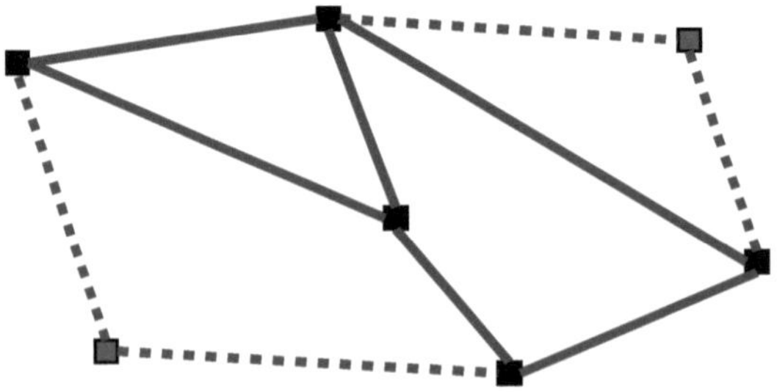

2nd pruning cycle:
Minimum gradient = 2
Erased links = 4
Released nodes = 2

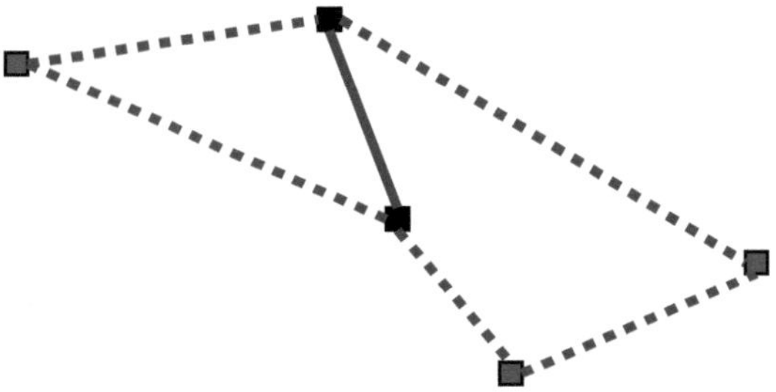

3rd pruning cycle:
Minimum gradient = 2
Erased links = 5
Released nodes = 3

4th pruning cycle:
Minimum gradient = 1
Erased links = 1
Released nodes = 2

The resulting number of pruning cycles M is 4.

Pruning Table

In order to take note of evolution of all the variables during the pruning process, we introduce the *pruning table*:

K	G	L	N_d
1	g_1	l_1	n_1
...	...	...	...
M	g_M	l_M	n_M

Every row is matched with a single pruning cycle; a row is composed by the following variables:

- K: progressive number which identify the jth pruning cycle
- G: pruning gradient of the jth cycle
- L: number of links erased in the jth cycle
- N_d: number of nodes released in the jth cycle

Example 2 Considering the graph of the previous example, we have the following pruning table:

K	G	L	N_d
1	1	1	1
2	2	4	2
3	2	5	3
4	1	1	2

We also define two variables that come from the pruning table: P e S_{TG}. To do that, we need a preliminary operation: *the partition of gradients*.

We have the sequence of gradients from the pruning:

$G = g_1, g_2, \ldots, g_M$.

Let us split the gradients in classes using the following rules:

- A class contains at least one element of the sequence.
- Two adjoining elements in the sequence with equal values are in the same class.

We name P the number of emerging classes and S_{TG} the common value of a class (i.e., the value of each element belonging to the class).

Example 3 Let us have a sequence of gradients with $M = 10$:

$G = 2, 1, 1, 1, 2, 4, 3, 2, 6, 6$.

The partition will be:

$C1 = \{2\}$, $C2 = \{1,1,1\}$, $C3 = \{2\}$, $C4 = \{4\}$, $C5 = \{3\}$, $C6 = \{2\}$, $C7 = \{6,6\}$.

The resulting number of classes will be equal to 7; the sequence S_{TG} will be 2, 1, 2, 4, 3, 2, 6.

Example 4 Consider the graph of Examples 1 and 2; we have:

$G = 1, 2, 2, 1$.

The partition will be:

$C1 = \{1\}$, $C2 = \{2,2\}$, $C3 = \{1\}$.

The resulting number of classes will be equal to 3; the sequence S_{TG} will be 1, 2, 1.

The μ and φ Parameters

We introduce two variables that will be used for the definition of hubness:

$$\mu = \frac{1}{M}\sum_{i}^{M} Nd_i = \frac{N}{M};$$

$$\phi = \frac{1}{P}\sum_{j}^{P} S_{TG\,j}.$$

Hubness of a Graph

Definition of hubness:

$$H_0 = \frac{\mu \cdot \phi - 1}{A}.$$

Example 5 Considering the graph of Examples 1 and 2, we have:

$$\mu = \frac{1}{M}\sum_{i}^{M} Nd_i = \frac{N}{M} = \frac{8}{4} = 2,$$

$$\phi = \frac{1}{P}\sum_{j}^{P} S_{TG\,j} = \frac{1}{3}(1+2+1) = \frac{4}{3},$$

$$H_0 = \frac{\mu \cdot \phi - 1}{A} = \frac{2 \cdot \frac{4}{3} - 1}{11} = \frac{\frac{8-3}{3}}{11} = \frac{5}{33} = 0.\overline{15}.$$

A.1.3.2 C.2 Remarkable Cases

C.2.1 Case No 1: The Chain

- Case 1.1: Chain with x nodes
  If x is even:

$$\left|\begin{matrix} M & G & L & N \\ 1 & 1 & 2 & 2 \\ 2 & 1 & 2 & 2 \\ \vdots & \vdots & \vdots & \vdots \\ \frac{x}{2} & 1 & 1 & 2 \end{matrix}\right|$$

$$\phi^{[C]} = 1 \quad \mu^{[C]} = 2$$

$$N^{[C]} = x$$

$$H_0^{[C]} = \frac{\mu^{[C]} \cdot \phi^{[C]} - 1}{N^{[C]} - 1} = \frac{2-1}{x-1} = \frac{1}{x-1}.$$

If x is odd:

$$\left|\begin{matrix} M & G & L & N \\ 1 & 1 & 2 & 2 \\ 2 & 1 & 2 & 2 \\ \vdots & \vdots & \vdots & \vdots \\ \frac{(x-1)}{2} & 1 & 2 & 3 \end{matrix}\right|$$

$$\phi^{[C]} = 1 \quad \mu^{[C]} = \frac{2x}{x-1}$$

$$N^{[C]} = x$$

$$H_0^{[C]} = \frac{\mu^{[C]} \cdot \phi^{[C]} - 1}{N^{[C]} - 1} = \frac{x+1}{(x-1)^2} = \frac{1}{x-1} \cdot \frac{x+1}{x-1}.$$

The value of the hubness depends on the level of connectivity of the graph, that is, the possibility of reaching any node starting from any other node using the shortest path (the smallest number of links). In that sense, the presence of hubs (nodes with a high number of links) increases the global connectivity of the graph.

In the case of chain, the connectivity of the graph is very low: to reach one edge from the other one, we need to use all the links of the chain. The longer is the chain, the more the compactness of the graph decreases. The consequence is that the hubness decreases as $1/x$ when the number of nodes x increases.

C.2.2 Case No 2: The Star

- Case 2.1: Star with x nodes:

$$\begin{vmatrix} M & G & L & N \\ 1 & 1 & x-1 & x \end{vmatrix}$$

$$\phi^{[S]} = 1 \quad \mu^{[S]} = x$$

$$N^{[S]} = x$$

$$H_0^{[S]} = \frac{\mu^{[S]} \cdot \phi^{[S]} - 1}{N^{[S]} - 1} = \frac{x-1}{x-1} = 1.$$

Comparing with the case of the chain, the star is the opposite: each node can reach any other node with at most 2 links (i.e., crossing only one node). Such level of connectivity holds steady as the number of nodes of the star increases. So the hubness is equal to 1 regardless of the number x of nodes.

- Case 2.2: Star with x nodes and 1 tail with x nodes:

$$\begin{vmatrix} M & G & L & N \\ 1 & 1 & 1 & 1 \\ \vdots & \vdots & \vdots & \vdots \\ x & 1 & 1 & 1 \\ x+1 & 1 & x-1 & x \end{vmatrix}$$

$$\phi' = 1 \quad \mu' = \frac{2x}{x+1}$$

$$N' = 2x$$

$$A' = 2x - 1$$

$$H_0' = \frac{\mu' \cdot \phi' - 1}{2x - 1} = \frac{x-1}{2x^2 + x - 1}$$

$$H_0'(3) = \frac{3-1}{18+3-1} = \frac{1}{10}$$

$$\lim_{x\to\infty} H_0' = \lim_{x\to\infty} \frac{x-1}{2x^2+x-1} = 0.$$

The tail may collapse dramatically the compactness of the whole graph; increasing the number of branches of the star and the number of nodes of the chain at the same time, the hubness goes to 0: the connectivity of the graph is lost.

The following cases show further this feature:

- Case 2.3: Star with x nodes and 2 tails with x nodes:

$$\left|\begin{array}{cccc} M & G & L & N \\ 1 & 1 & 2 & 2 \\ \vdots & \vdots & \vdots & \vdots \\ x & 1 & 2 & 2 \\ x+1 & 1 & x-1 & x \end{array}\right|$$

$$\phi'' = 1 \quad \mu'' = \frac{2x+x}{x+1} = \frac{3x}{x+1}$$

$$N'' = 3x$$

$$A'' = 3x-1$$

$$H_0'' = \frac{\mu''\cdot\phi''-1}{3x-1} = \frac{2x-1}{3x^2+2x-1}$$

$$H_0''(3) = \frac{6-1}{27+6-1} = \frac{5}{321}$$

$$\lim_{x\to\infty} H_0'' = \lim_{x\to\infty} \frac{2x-1}{3x^2+2x-1} = 0.$$

- Case 2.4: Star with x nodes and x tails with x nodes:

$$\left|\begin{array}{cccc} M & G & L & N \\ 1 & 1 & x & x \\ \vdots & \vdots & \vdots & \vdots \\ x & 1 & x & x \\ x+1 & 1 & x-1 & x \end{array}\right|$$

$$\phi^* = 1 \quad \mu^* = \frac{x^2 + x}{x + 1} = x$$

$$N^* = x^2 + x$$

$$A^* = x^2 + x - 1$$

$$H_0^* = \frac{\mu^* \cdot \phi^* - 1}{x^2 + x - 1} = \frac{x - 1}{x^2 + x - 1}$$

$$H_0^*(3) = \frac{3 - 1}{9 + 3 - 1} = \frac{2}{11}$$

$$\lim_{x \to \infty} H_0^* = \lim_{x \to \infty} \frac{x - 1}{x^2 + x - 1} = 0.$$

C.2.3 Case No 3: The Tree

- Case 3.1: Tree with x nodes and y pruning steps ($x \geq 2$; $y \leq$ x):

$$H_0^{[A]} = \frac{\mu^{[A]} \cdot \phi^{[A]} - 1}{N^{[A]} - 1} = \frac{\frac{x}{y} - 1}{x - 1}.$$

If $y = 1$ (star case):

$$H_0^{[A1]} = \frac{\frac{x}{y} - 1}{x - 1} = \frac{x - 1}{x - 1} = 1.$$

If $y = 2$:
$H_0^{[A2]} = \frac{\frac{x}{y}-1}{x-1} = \frac{\frac{x}{2}-1}{x-1}$; therefore, for $x = 2, 3, 4, \ldots$:
$H_0^{[A2]} = 0,\ \frac{1}{4},\ \frac{1}{3},\ \frac{3}{8},\ \cdots \to\ \frac{1}{2}$, that is, $\forall x,\ H_0^{[A2]} < \frac{1}{2}$
(note: $x = 2$ and $y = 2$ is impossible).
If $y = x - 1$:
$H_0^{[A(x-1)]} = \frac{\frac{x}{x-1}-1}{x-1} = \frac{x-x+1}{(x-1)^2} = \frac{1}{(x-1)^2}$, then with $x = 2, 3, 4, \ldots$:

$$H_0^{[A(x-1)]} = 1, \frac{1}{4}, \frac{1}{9}, \frac{1}{16}, \cdots \to 0$$

(note: $x = 2$ and $y = 1$ is the case of the star with two tails [case 7]).
If $y = x$ (impossible):

$$H_0^{[Ax]} = \frac{\frac{x}{x} - 1}{x - 1} = \frac{1 - 1}{x - 1} = 0.$$

Since for $x > 2$ and $y > 1$:

$$\frac{x}{y} - 1 \leq \frac{x}{2} - 1, \quad \text{that is} \quad \frac{\frac{x}{y} - 1}{x - 1} \leq \frac{\frac{x}{2} - 1}{x - 1}$$

We have for $x \geq 2$, $1 < y < x$:

$$H_0^{[A]} \leq H_0^{[A2]} < \frac{1}{2}.$$

This result highlights that the hubness of a tree usually is very small (<½; the star $[H = 1]$ is an exception). In fact, the lack of close loops decreases the level of connectivity of the graph: there is only a path between two nodes (there are no "shortcuts"!). Increasing the number of nodes, the compactness of the graph decreases and the hubness goes to 0.

C.2.4 Case No 4: The Complete Regular Graph

- Case 4.1: Complete regular graph with x nodes:

$$\begin{vmatrix} M & G & L & N \\ 1 & x-1 & \frac{x^2-x}{2} & x \end{vmatrix}$$

$$\phi^{[\text{GRC}]} = x - 1 \quad \mu^{[\text{GRC}]} = x$$

$$N^{[\text{GRC}]} = x$$

$$A^{[\text{GRC}]} = \frac{x^2 - x}{2}$$

$$\begin{aligned} H_0^{[\text{GRC}]} &= \frac{\mu^{[\text{GRC}]} \cdot \phi^{[\text{GRC}]} - 1}{\frac{x^2-x}{2}} \\ &= 2 \cdot \frac{x \cdot (x-1) - 1}{x^2 - x} = 2 \cdot \frac{x^2 - x - 1}{x^2 - x} = 2 - \frac{2}{x^2 - x}. \end{aligned}$$

In a complete regular graph, each node is directly linked to any other node; the compactness is the maximum possible and the hubness has a value >1.5, which goes to 2 as the number of nodes increases. Note that the hubness is an extensive variable, and therefore, it depends not only by the connectivity but also by the dimensions (number of nodes): between two complete regular graphs (maximum connectivity) the one with more nodes will have the highest hubness.

- Case 4.2: Complete regular graph with x nodes and 1 tail with x nodes:

$$\left|\begin{array}{cccc} M & G & L & N \\ 1 & 1 & 1 & 1 \\ \vdots & \vdots & \vdots & \vdots \\ x & 1 & 1 & 1 \\ x+1 & x-1 & \frac{x^2-x}{2} & x \end{array}\right|$$

$$\phi' = \frac{x}{2} \quad \mu' = \frac{2x}{x+1}$$

$$N' = 2x$$

$$A' = \frac{x^2-x}{2} + x = \frac{x^2+x}{2}$$

$$H_0' = \frac{\mu' \cdot \phi' - 1}{\frac{x^2+x}{2}} = \frac{2}{x} \cdot \frac{x^2-x+1}{x^2+2x+1}$$

$$\lim_{x\to\infty} H_0' = \lim_{x\to\infty} \frac{2}{x} \cdot \frac{x^2-x+1}{x^2+2x+1} = 0.$$

Like Case 2.2, the presence of nodes with gradient lower than the maximum one makes the hubness collapse under the unity. This "hypersensibility" is examined in the next case.

- Case 4.3: Complete regular graph with x nodes and 2 tails with x nodes:

$$\left|\begin{array}{cccc} M & G & L & N \\ 1 & 1 & 2 & 2 \\ \vdots & \vdots & \vdots & \vdots \\ x & 1 & 2 & 2 \\ x+1 & x-1 & \frac{x^2-x}{2} & x \end{array}\right|$$

$$\phi'' = \frac{x}{2} \quad \mu'' = \frac{2x+x}{x+1} = \frac{3x}{x+1}$$

$$N'' = 3x$$

$$A'' = \frac{x^2-x}{2} + 2x = \frac{x^2+3x}{2}$$

$$H_0'' = \frac{\mu'' \cdot \phi'' - 1}{\frac{x^2+3x}{2}} = \frac{1}{x} \cdot \frac{3x^2 - 2x - 2}{x^2 + 4x + 3}$$

$$\lim_{x\to\infty} H_0'' = \lim_{x\to\infty} \frac{1}{x} \cdot \frac{3x^2 - 2x - 2}{x^2 + 4x + 3} = 0.$$

- Case 4.4: Complete regular graph with x nodes and x tails with x nodes:

$$\left|\begin{matrix} M & G & L & N \\ 1 & 1 & x & x \\ \vdots & \vdots & \vdots & \vdots \\ x & 1 & x & x \\ x+1 & x-1 & \frac{x^2-x}{2} & x \end{matrix}\right|$$

$$\phi^* = \frac{x}{2} \quad \mu^* = \frac{x^2 + x}{x + 1} = x$$

$$N^* = x^2 + x$$

$$A^* = \frac{x^2 - x}{2} + x^2 = \frac{3x^2 - x}{2}$$

$$H_0^* = \frac{\mu^* \cdot \phi^* - 1}{\frac{3x^2-x}{2}} = \frac{x^2 - 2}{3x^2 - x}$$

$$\lim_{x\to\infty} H_0^* = \lim_{x\to\infty} \frac{x^2 - 2}{3x^2 - x} = \frac{1}{3}.$$

- Case 4.5: Complete regular graph with x nodes and y tails with x nodes $(1 < y \leq x)$:

$$\left|\begin{matrix} M & G & L & N \\ 1 & 1 & y & y \\ \vdots & \vdots & \vdots & \vdots \\ x & 1 & y & y \\ x+1 & x-1 & \frac{x^2-x}{2} & x \end{matrix}\right|$$

$$\phi''' = \frac{x}{2}$$

$$\mu''' = \frac{xy + x}{x + 1} = \frac{x(y + 1)}{x + 1}$$

$$N''' = x + xy = x(y+1)$$

$$A''' = \frac{x^2 - x}{2} + xy = \frac{x^2 + 2xy - x}{2}$$

$$H_0''' = \frac{\frac{x(y+1)}{x+1} \cdot \frac{x}{2} - 1}{\frac{x^2+2xy-x}{2}} = \frac{1}{2} \cdot \frac{1}{x+1}\left(x - \frac{x^3 - 3x^2 + 4x + 4}{x^2 + 2xy - x}\right)$$

$$= \frac{1}{2} \cdot \frac{x}{x+1}\left(1 - \frac{x^3 - 3x^2 + 4x + 4}{x^3 + 2x^2 y - x^2}\right).$$

If $y \neq x$:

$$\lim_{x\to\infty} H_0''' = \lim_{x\to\infty} \frac{1}{2} \cdot \frac{x}{x+1}\left(1 - \frac{x^3 - 3x^2 + 4x + 4}{x^3 + 2x^2 y - x^2}\right) = 0.$$

If $y = x$:

$$\lim_{x\to\infty} H'''_0 = \lim_{x\to\infty} \frac{1}{2} \cdot \frac{x}{x+1}\left(1 - \frac{x^3 - 3x^2 + 4x + 4}{x^3 + 2x^2 y - x^2}\right)$$

$$= \lim_{x\to\infty} \frac{1}{2} \cdot \frac{x}{x+1}\left(1 - \frac{x^3 - 3x^2 + 4x + 4}{x^3 + 2x^3 - x^2}\right)$$

$$= \lim_{x\to\infty} \frac{1}{2} \cdot \frac{x}{x+1}\left(1 - \frac{x^3 - 3x^2 + 4x + 4}{3x^3 - x^2}\right)$$

$$= \frac{1}{2} \cdot \left(1 - \frac{1}{3}\right) = \frac{1}{2} \cdot \left(\frac{2}{3}\right) = \frac{1}{3}.$$

C.2.5 Case No 5: The Closed Star

- Case 5.1: Closed star with x nodes:

$$\begin{vmatrix} M & G & L & N \\ 1 & 3 & 2x-2 & x \end{vmatrix}$$

$$\phi^{[SC]} = 3 \quad \mu^{[SC]} = x$$

$$N^{[SC]} = x$$

$$A^{[SC]} = 2(x-1)$$

$$H_0^{[SC]} = \frac{\mu^{[SC]} \cdot \phi^{[SC]} - 1}{2(x-1)} = \frac{3x-1}{2(x-1)}$$

$$H_0^{[SC]}(5) = \frac{15-1}{2(5-1)} = \frac{7}{4}$$

$$\lim_{x\to\infty} H_0^{[SC]} = \lim_{x\to\infty} \frac{3x-1}{2x-2} = \frac{3}{2}.$$

The closed star represents the case of a not regular graph with hubness greater than 1. In fact the star, already with a very high connectivity, is upgraded by the connections between the "spokes" of the wheel, increasing the space of possible paths (and possible "shortcuts"). This increased compactness explains the above-mentioned high levels of hubness.

The following cases once more explain how easily the value of hubness decreases adding low connected nodes.

- Case 5.2: Closed star with x nodes and 1 tail with x nodes:

M	G	L	N
1	1	1	1
$\vdots$	$\vdots$	$\vdots$	$\vdots$
x	1	1	1
$x+1$	3	$2x-2$	x

$$\phi' = 2 \quad \mu' = \frac{2x}{x+1}$$

$$N' = 2x$$

$$A' = 3x - 2$$

$$H_0' = \frac{\mu' \cdot \phi' - 1}{3x-2} = \frac{3x-1}{3x^2+x-2}$$

$$H_0'(5) = \frac{15-1}{75+5-2} = \frac{7}{39}$$

$$\lim_{x\to\infty} H_0' = \lim_{x\to\infty} \frac{3x-1}{3x^2+x-2} = 0.$$

- Case 5.3: Closed star with x nodes and 2 tails with x nodes:

$$\begin{vmatrix} M & G & L & N \\ 1 & 1 & 2 & 2 \\ \vdots & \vdots & \vdots & \vdots \\ x & 1 & 2 & 2 \\ x+1 & 3 & 2x-2 & x \end{vmatrix}$$

$$\phi'' = 2 \quad \mu'' = \frac{2x + x}{x+1} = \frac{3x}{x+1}$$

$$N'' = 3x$$

$$A'' = 4x - 2$$

$$H_0'' = \frac{\mu'' \cdot \phi'' - 1}{2(x-1)} = \frac{5x - 1}{4x^2 + 2x - 2}$$

$$H_0''(5) = \frac{25 - 1}{100 + 10 - 2} = \frac{2}{9}$$

$$\lim_{x\to\infty} H_0'' = \lim_{x\to\infty} \frac{5x - 1}{4x^2 + 2x - 2} = 0.$$

- Case 5.4: Closed star with x nodes and x tails with x nodes:

$$\begin{vmatrix} M & G & L & N \\ 1 & 1 & x & x \\ \vdots & \vdots & \vdots & \vdots \\ x & 1 & x & x \\ x+1 & 3 & 2x-2 & x \end{vmatrix}$$

$$\phi^* = 2 \quad \mu^* = \frac{x^2 + x}{x+1} = x$$

$$N^* = x^2 + x$$

$$A^* = x^2 + 2x - 2$$

$$H_0^* = \frac{\mu^* \cdot \phi^* - 1}{x^2 + 2x - 2} = \frac{2x - 1}{x^2 + 2x - 2}$$

$$H_0^*(5) = \frac{10-1}{25+10-2} = \frac{3}{11}$$

$$\lim_{x\to\infty} H_0^* = \lim_{x\to\infty} \frac{2x-1}{x^2+2x-2} = 0.$$

- Case 5.5: Closed star with x nodes and maximum gradient
  - Construction procedure of the graph with gradient $x-2$ (only with x odd):
    - Take a complete regular graph with dimension $x-1$.
    - Erase one link in each couple of nodes ($x-1$ will be even).
    - Add a new node and link it to the others.

      We have:

$$\mu^{[SC]} = x$$

$$\phi^{[SC]} = x-2$$

$$N^{[SC]} = x$$

$$A^{[SC]} = \frac{(x-1)^2-(x-1)}{2} - \frac{x-1}{2} + (x-1)$$
$$= \frac{(x-1)^2-2(x-1)+2(x-1)}{2} = \frac{(x-1)^2}{2}$$

$$H_0^{[SC]} = \frac{\mu^{[SC]}\cdot\phi^{[SC]}-1}{\frac{(x-1)^2}{2}}$$
$$= 2\cdot\frac{x(x-2)-1}{(x-1)^2} = 2\cdot\frac{x^2-2x-1}{(x-1)^2} = 2\cdot\frac{x^2-2x-1}{x^2-2x+1}$$

$$\lim_{x\to\infty} H_0^{[SC]} = \lim_{x\to\infty} 2\cdot\frac{x^2-2x-1}{x^2-2x+1} = 2.$$

- Construction procedure of the graph with gradient $x-3$:
  - Take a complete regular graph with dimension $x-1$.
  - Erase two links in each node (we can do that with x even as well as x odd).
  - Add a new node and link it to the others.

    We have:

$$\mu^{[SC]} = x$$

$$\phi^{[SC]} = x - 3$$

$$N^{[SC]} = x$$

$$\begin{aligned} A^{[SC]} &= \frac{(x-1)^2 - (x-1)}{2} - (x-1) + (x-1) \\ &= \frac{(x-1)^2 - (x-1)}{2} = \frac{(x-1)(x-2)}{2} \end{aligned}$$

$$H_0^{[SC]} = \frac{\mu^{[SC]} \cdot \phi^{[SC]} - 1}{\frac{(x-1)^2}{2}} = 2 \cdot \frac{x(x-3) - 1}{(x-1)(x-2)} = 2 \cdot \frac{x^2 - 3x - 1}{x^2 - 3x + 2}$$

$$\lim_{x \to \infty} H_0^{[SC]} = \lim_{x \to \infty} 2 \cdot \frac{x^2 - 3x - 1}{x^2 - 3x + 2} = 2.$$

References

Buscema, M. (2007a). *A novel adapting mapping method for emergent properties discovery in data bases: Experience in medical field.* In 2007 IEEE international conference on systems, man and cybernetics (SMC 2007). Montreal, Canada.

Buscema, M. (Ed.). (2007b). *Squashing theory and contractive map network* (Semeion Technical Paper #32). Rome.

Buscema, M., & Grossi, E. (2008). The semantic connectivity map: An adapting self-organizing knowledge discovery method in data bases. Experience in gastro-oesophageal reflux disease. *International Journal of Data Mining and Bioinformatics, 2*(4), 362–404.

Buscema, M., & Grossi, E. (Eds.). (2009). *Artificial adaptive systems in medicine* (pp. 25–47). Saif Zone: Bentham e-books.

Buscema, M., & Sacco, P. L. (2010). Auto-contractive maps, the H function, and the Maximally Regular Graph (MRG): A new methodology for data mining, Chapter 11. In V. Capecchi et al. (Eds.), *Applications of mathematics in models, artificial neural networks and arts*. New York/London: Springer. doi:10.1007/978-90-481-8581-8_11.

Buscema, M., Grossi, E., Snowdon, D., & Antuono, P. (2008a). Auto-contractive maps: An artificial adaptive system for data mining. An application to Alzheimer disease. *Current Alzheimer Research, 5*, 481–498.

Buscema, M., Helgason, C., & Grossi, E. (2008b). *Auto contractive maps, H function and maximally regular graph: Theory and applications*. In Special session on "Artificial adaptive systems in medicine: Applications in the real world, NAFIPS 2008 (IEEE)", New York.

Eller-Vainicher, C., Zhukouskaya, V. V., Tolkachev, Y. V., Koritko, S. S., Cairoli, E., Grossi, E., Beck-Peccoz, P., Chiodini, I., & Shepelkevich, A. P. (2011). Low bone mineral density and its predictors in type 1 diabetic patients evaluated by the classic statistics and artificial neural network analysis. *Diabetes Care, 34*, 2186–2191.

Grossi, E., Blessi, G., Sacco, P. L., & Buscema, M. (2011). The interaction between culture, health and psychological well-being: Data mining from the Italian culture and well-being project. *Journal of Happiness Studies, 13*, 129–148.

Kruskal, J. B. (1956). On the shortest spanning subtree of a graph and the traveling salesman problem. *Proceedings of the American Mathematical Society, 7*(1), 48–50.

Licastro, F., Porcellini, E., Chiappelli, M., Forti, P., & Buscema, M. (2010a). Multivariable network associated with cognitive decline and dementia. *International Neurobiology of Aging, 1*(2), 257–269.

Licastro, F., Porcellini, E., Forti, P., Buscema, M., Carbone, I., Ravaglia, G., & Grossi, E. (2010b). Multi factorial interactions in the pathogenesis pathway of Alzheimer's disease: a new risk charts for prevention of dementia. *Immunity & Ageing, 7*(Suppl 1), S4.

Zsuzsanna, A. R. (2001). *Statistical mechanics of complex networks.* Dissertation, Department of Physics, Notre Dame University, Indiana.

Software

Massini, G. (2007) Tree visualizer. Software to draw and manipulate tree graph, Semeion Software #40, v. 3, Rome.

Chapter 16
Analysis of a Complex Dataset Using the Combined MST and Auto-Contractive Map

Giovanni Pieri

16.1 Introduction

An auto-contractive map (Auto CM) learns the relationships among data elements in a database with a new and extremely efficient process in which the *output* of the Auto CM is discarded in the final phase of analysis. At this point, *all* the information contained in the database as represented by the variables and its instances is transferred into the Auto CM, including the relationships between variables and dependences between records. All the information lies in a special structure called the W matrix; it is this *weights* matrix that contains the information concerning the connections of the nodes which are present within the artificial network represented as the *hidden* layer and on the output layer (Buscema and Sacco 2010).

Although all the information is actually contained in this matrix, access to it is nevertheless difficult, since, in general, the W matrix is not easier to read than the initial database though in some cases in which a graphic representation of the W matrix can be given, a direct reading is possible though not necessarily easy.

In order to make the information readable to the non-scientist, it is necessary to understand its elements as 'distances'. Through a mathematical device, it is possible to operate on the data such that the diagonal elements of the W matrix are all the same and equal to a prefixed number d greater than all the other elements of the matrix, which are positive by definition. The expression

$$\text{dist}_{ij} = d - w_{ij}$$

is then interpreted as the distance between the variable i and the variable j. The dist_{ij} elements cannot be less than zero by definition. Therefore, the W matrix can be

G. Pieri (✉)
Research Associate at Semeion, Research Center of Sciences of Communication, Rome, Italy
e-mail: semeion@semeion.it

M. Buscema and W.J. Tastle (eds.), *Intelligent Data Mining in Law Enforcement Analytics: New Neural Networks Applied to Real Problems*, DOI 10.1007/978-94-007-4914-6_16,

represented by a non-directional graph whose nodes are the variables and whose arcs are the distances given by the preceding expression.

The bigger an element of the *W* matrix, the greater the link between the variables connected by the element. In other words, the elements of the *W* matrix represent the intensity of the connections between the variables. It is easy to note that the smaller the distances between variables (defined as above), the stronger will be the link between them and therefore the greater is the importance that must be assigned to them in the analysis of the results concerning the Auto CM.

The *minimum spanning tree* (MST) is a mathematical tree diagram that is extracted from a non-directional graph and has two characteristics (Kruskal 1956):

1. It joins all nodes.
2. The sum of the lengths of all the arcs is minimal.

The MST therefore represents only a part of the graph, i.e. it exposes only the arcs minimising the total length, and therefore generally speaking they are some of the shortest of the graph, though not necessarily all the shortest arcs. Therefore, the MST represents the links between the strongest variables, disregarding the others. The MST does not remove the other arcs; it just omits showing them, but the relationships represented by the arcs, which do not form part of the MST (in the graph each node is connected to all the others), continue to exist and are not necessarily weak. On the contrary, some of them can have intensity comparable to those represented.

It is always advisable to take this aspect into account when analysing the results of an Auto CM. One should take the MST graph from an operative point of view, that is to say, through a calculation of sensitivity aimed to see how the MST changes as one of the nodes is removed (which is equivalent to disregarding the variable in the analysis). Generally speaking, we can say (and we shall see this later) that sometimes the removal of a node changes little in the MST. It is therefore confirmed that the links which are not represented in the original MST may still be very strong and may effectively replace those removed by elimination of one of the nodes. The interpretations of the relationships between variables that are revealed on the map are based on the characteristics revealed by the MST, and one should always take into account that these links between nodes are prevalent relationships, which are never exclusive.

16.2 Interpretation of the Global Graph

The Global Graph (the MST resulted from the Auto CM that has learned the database of the drug arrests in London) may be described schematically as consisting of a central trunk interconnecting the drugs seized, onto which fairly complex side branches are grafted (Fig. 16.1). The node representing *Crack* is directly connected to the node representing *Heroin-Diamorphine*, whilst it is connected to the

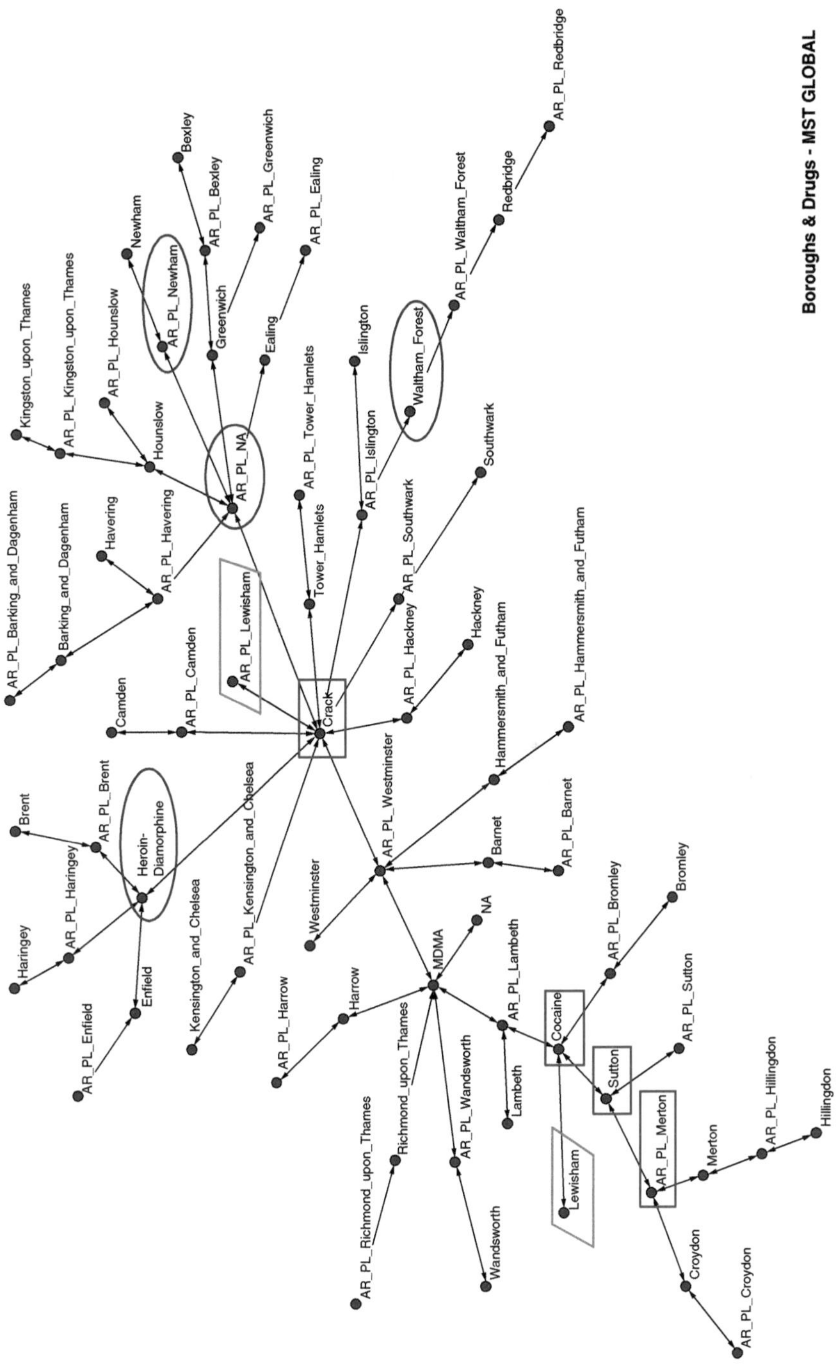

Fig. 16.1 Global MST graph

one representing *MDMA* [(3,4-methylenedioxymethamphetamine) is a synthetic, psychoactive drug that is chemically similar to the stimulant methamphetamine and the hallucinogen mescaline] through a *hub* representing the arrests carried out in the borough of *Westminster*. *MDMA* is connected in turn to the *hub Cocaine* through the node representing the borough of *Lambeth*. Leaving aside for the moment the significance of the direct or indirect connections between substances, we analyse the significance of the branches.

The branches leaving from the central trunk consist exclusively of nodes representing borough variables and they all end with leaves having similar structures, with one exception that we shall discuss later. The simplest branches have two possible structures formed as follows:

Drug ⟷ Residents in Borough ⟷ Arrests in the same Borough

Drug ⟷ Arrests in Borough ⟷ Residents in the same Borough

Examples of these structures can be found in the leaves connected to *Heroin-Diamorphine*:

Heroin-Diamorphine ⟷ Enfield ⟷ AR_PL_Enfield

Heroin-Diamorphine ⟷ AR_PL_Haringey ⟷ Haringey

Although similar, both structures do not mean exactly the same thing. In the first case, the residents of *Enfield* are more linked to the drug than their place of residence. This means that they deal in the borough where they live, but they partly go over its boundaries with the result that they may also be arrested elsewhere. In the second case, in *Haringey* the opposite occurs: without any doubt, the residents deal in *Heroin-Diamorphine*, but they are not the only ones on their territory. In other words, the residents of *Enfield* have a more active part in dealing in *Heroin-Diamorphine* than the residents of *Haringey*.

In each case, the link between the place of residence and the place of arrest is very strong and this means that the drug distribution system has strongly territorial characteristics, which are certainly at the level of the individual borough and also at a wider level, as we shall see. Smaller territorial structures may exist in the different boroughs, but our analysis tool, by its very nature, does not reveal them. If in the future there was an interest also in investigating on smaller scales, the database would have to be segmented into smaller territorial units contained in the boroughs.

This difference enables one to create a hierarchy of the boroughs according to their degree of activity. For example, among the three boroughs linked to *Heroin-Diamorphine, Enfield* has a driving role, and its residents cover the market of their own territory well and partly spill over into the territory of the other two boroughs (Haringey, Brent), collaborating moreover with the local residents, who nevertheless maintain the pre-eminence on their territory. This picture reveals a structure of a

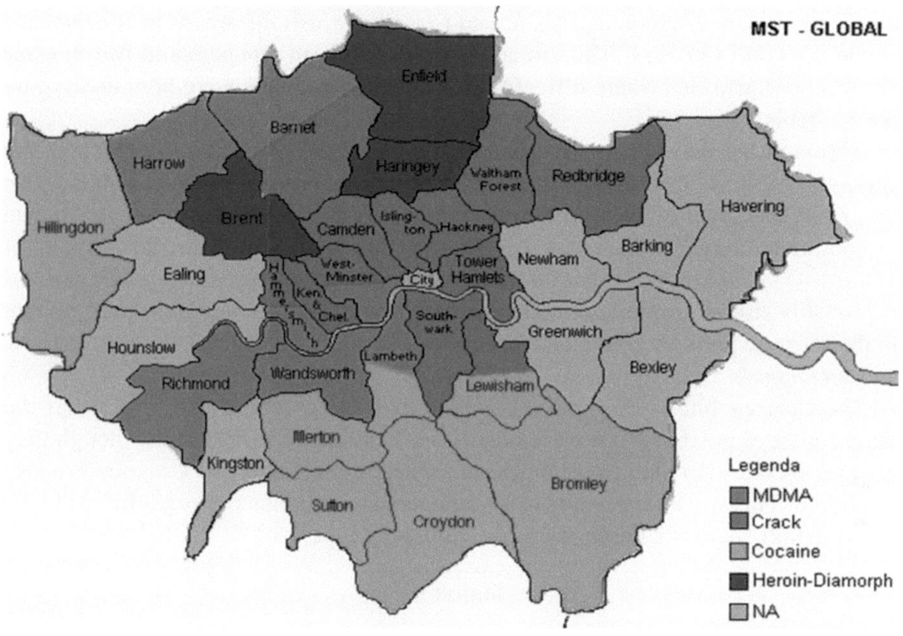

Fig. 16.2 Map of London with the boroughs coloured according to their connection with the type of drug

higher level than the one concerning the individual boroughs. It can be identified with the territory which has a predominant diffusion of a given drug. So, even the drugs show a certain degree of territoriality, and it can be represented as in Fig. 16.2. In Fig. 16.2, the boroughs connecting two drugs in the graph are represented with a split colour.

The territorial structure of dealing can be likened to a feudal structure in which a feudal vassal of a higher level controls a territory, part of it directly and part of it through lower-ranking vassals subordinate to him. This may suggest that in *Enfield* there is a higher power controlling all the *Heroin-Diamorphine* trafficking in the three boroughs (and also elsewhere where this trafficking is not typifying) and the sources of supply, running the storage and distribution centres in his territory. In this situation, territorial adjacency counts but is not essential. Even in the Middle Ages, when the communications were much more difficult than today, one feudal vassal could control territories not adjacent to one another that had entered his orbit as a result of historical and political accidents.

The effect of the lack of territorial adjacency can be clearly seen in the case of *MDMA,* where the boroughs with a more active profile (represented by a branch of the first type) are two in number: *Harrow* and *Richmond.* The latter is adjacent to *Wandsworth,* which has a less active profile and therefore should be considered dependent on *Richmond,* with probable joint control of *Harrow.* The other boroughs just gravitate on *MDMA* indirectly through nodes that connect two different drugs:

the arrests in *Lambeth* connect *MDMA* to *Cocaine*, whilst the arrests in *Westminster* connect *MDMA* to *Crack*. This situation shows that both *Lambeth* and *Westminster* are frontier territories where different drugs coexist and they were both dealt in by the residents.

Although the database is not dynamic (it does not contain time variables), the situation suggests that in the frontier territories a process of substitution is in progress. This seems even more evident if it is considered that the present territorial discontinuity may be the result of the penetration of *Crack* in the domain of *MDMA* which was previously continuous from *Barnet* to *Westminster*. The process is probably still in progress in the frontier areas and started some time back, so that in the separate parts two independent centres of control came into being (in *Harrow* and *Richmond*).

The *Cocaine* hub is even more complex than the previous ones, and it has the main centre of power in *Sutton*: residents predominate in local arrests, though they are also linked with the arrests in Merton where the residents of Croydon are also important. This situation permits us to deduce that within the boroughs linked to the Cocaine hub, there is a hierarchy of the following type:

Sutton
↓
Bromley
↓
Croydon
↓
Merton
↓
Hillingdon

There is no physical continuity between the latter two boroughs, but there is a relationship of dependence where the residents of *Merton* deal in Cocaine in *Hillingdon* together with the residents. It is not possible to say exactly whether the present discontinuity is the remnant of adjacency in past times, or if it is the result of an expansive process of the *Cocaine* area. In each case, given the links between the two parts of the separate territory, it involves recent events.

The territory of *Lambeth* as a frontier territory seems to enjoy a degree of independence, since arrests concern only the residents, and either *Cocaine* or *MDMA* is seized. There is no significant infiltration of residents of other territories involved in *Cocaine* or *MDMA* dealing. In other words, it is a buffer area having good relationships with both the macro-areas with which it borders.

The territory of *Lewisham* is an exception, where it is the only territory whose residents are not linked to the arrests in that territory, but they are directly linked with *Cocaine* dealing, whilst the arrests in *Lewisham* are linked with *Crack*. This anomaly can be explained by a recent change from *Cocaine* to *Crack* in the territory of *Lewisham* that has forced the residents remained linked with *Cocaine* to go and deal elsewhere. The lack of that close correspondence between residents and dealing, found in all the other territories, makes Lewisham a territory in transition: in the long term, it should return to equilibrium with the residents abandoning the link with *Cocaine* and returning to their own territory dealing in *Crack*.

By far, the most complex hub is *Crack*. The borough of *Tower Hamlets* is the only one to be connected to the *Crack* hub through its own residents and not by a node representing arrests, which makes it the dominant territory around which the other numerous *Crack* territories aggregate. That is to say, the mastermind and the supply, storage and distribution centre should be found there. A precise hierarchy between the territories is not possible because of the presence of the node AR_PL_NA (place of arrest not available) whose meaning is not completely clear and therefore prevents a precise classification of the numerous territories depending on that hub.

The *Crack* hub also has other specific characteristics worthwhile discussing which concerns the relationship with the other three drugs present on the territory:

- The relationship with *Cocaine* is mediated only by the residents of the territory of *Lewisham* which represent a recent acquisition concerning *Crack* and which is still awaiting a definitive set-up.
- The relationship with *MDMA* is mediated by a rather complex hub involving three boroughs: *Westminster*, *Hammersmith* and *Barnet*. There seems to be a conflict mainly in the territory of *Westminster*, involving those people of the other two boroughs who, however, maintain the control of their territories as well as the local residents. The aim of the conflict could be to get the *Crack* on the left bank of the Thames where there is still significant presence of *MDMA*.
- On the other hand, the relationship with *Heroin-Diamorphine* is direct, without nodes representing bordering territories. This situation suggests that there is a lack of progress in this direction and that Crack is not trying to extend into the territory covered by *Heroin-Diamorphine*. This could be due to a formal or de facto alliance between the two centres of power of *Tower Hamlets* and *Enfield* or to a commercial type of synergism between the two types of drug.

16.3 Effect of Removing the AR_PL_NA Node

What is the information content of the variable *AR_PL_NA*? The answer depends entirely on its definition. If it is really lack of the datum as one can literally assume, it actually does not have any additional content, given the difficulty in imagining the significance of what is not available.

The alternative is that it involves arrests made outside the circle of the 33 London boroughs, but it seems very unlikely that data on operations carried out outside the city will end up in a database of police operations carried out in London. However, this definition would clearly explain the territoriality of this variable that is very difficult to explain by a random loss of information.

In any case, it is not a variable on which the researcher can act at will. For this reason, the removal of the node generates a new graph showing the real territorial connections that was partially obscured by the presence of a non-significant variable.

It is in this way that Fig. 16.3, in our opinion, in actual fact represents the basic graph in its more direct reality and Fig. 16.4 is its graphic representation. Excluding the variable *AR_PL_NA*, with respect to the map of the Global MST (Fig. 16.2), the variations are as follows (Fig. 16.3):

- Havering, Barking, Newham, Kingston, Ealing and Hounslow are connected to Crack.
- Bexley and Greenwich are connected to MDMA.

The conclusions drawn from an examination of the original graph (Fig. 16.2) remain valid; in fact, they come out of it reinforced:

- *Crack's* power centre remains *Tower Hamlets,* and the hierarchy relationships with the surrounding territories remain unchanged.
- The relationships with the other drugs remain unchanged: conflict and penetration into the *MDMA* territories in the bordering boroughs; penetration occurring in the *Lewisham* territory, recently grabbed from *Cocaine*, in equilibrium (through alliance or vassalage) with *Heroin-Diamorphine*.
- The *Crack* territory is very extensive and includes a band cutting London from the west to the east, predominantly in the north of the Thames and practically in contiguous territories (except for the territory of Kingston). The areas occupied by the other drugs seem fragmented and marginalised.
- The main direction of *Crack's* expansion is from the north to the south towards or across the Thames, and it is concentrated in the central boroughs. A secondary direction of penetration is from the south to the north, approximately from *Westminster* to *Barnet.*

It is now possible to establish a hierarchy between the various boroughs belonging to the *Crack* area, excluding for simplicity the bordering territories:

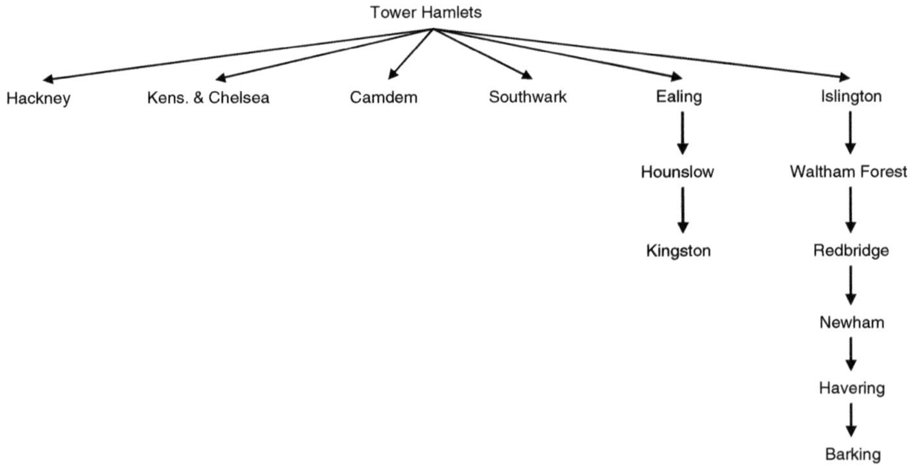

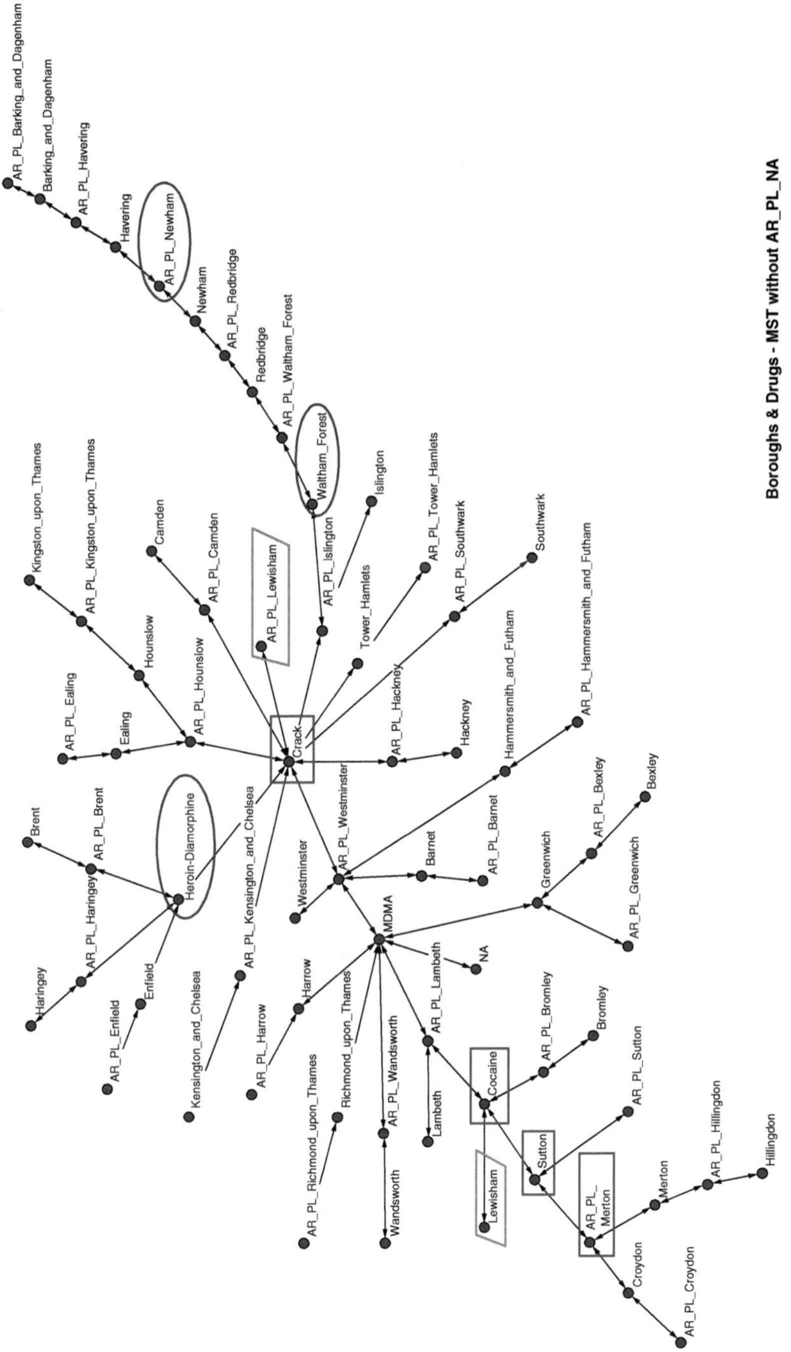

Fig. 16.3 Resultant MST graph after exclusion of the variable AR_PL_NA

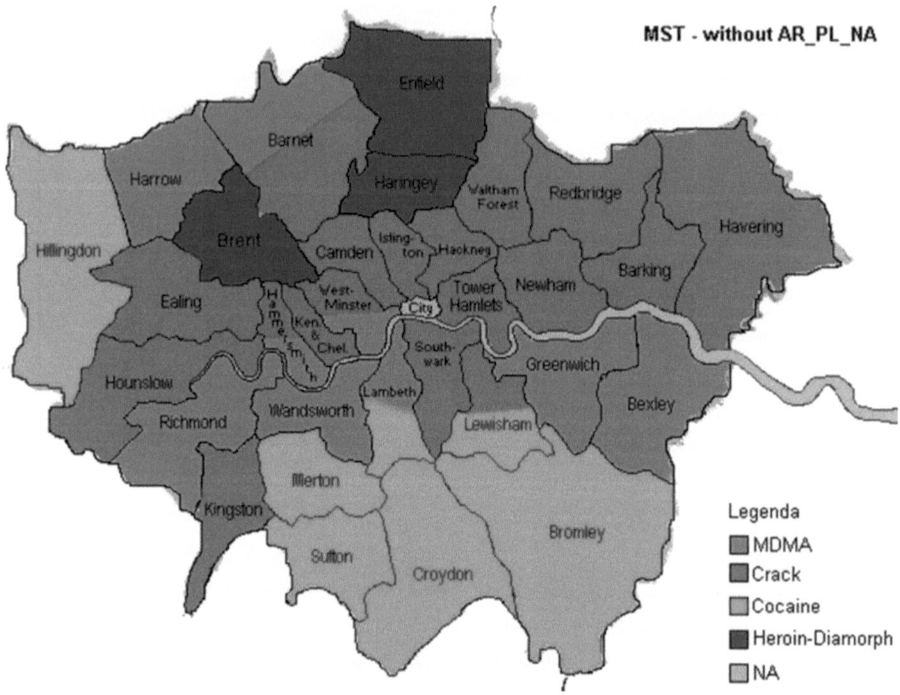

Fig. 16.4 Resultant MST graph after exclusion of the variable AR_PL_NA

Returning to the feudal model, it is clear that *Tower Hamlets'* power centre controls directly the territory in which it is established, and it is very close, approaching the territories surrounding it. The extremities of *Crack's* domain are controlled by a chain of vassalage relationships, as it is logical for distant territories. The chain is longer easterly than westerly. If this is significant, it may be due to the time factor: the western territories were acquired successively with respect to the eastern territories, believing that more time would have made it possible to set up longer chains of command in a safe way. The eastern extremity of the northern Thames is therefore the relatively quiet hinterland supporting Crack's expansion southwards and northwards.

The hierarchy of the *Cocaine* hub can also be better now, if it is explained as follows:

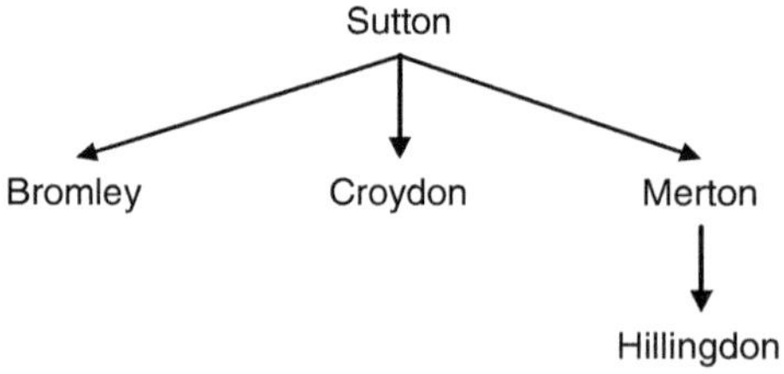

Although the structure of this hub has smaller sizes, it is similar to *Crack's,* lending the possibility to it being similarly considered.

The hierarchical structure of the MDMA hub is less coherent and may be represented as follows:

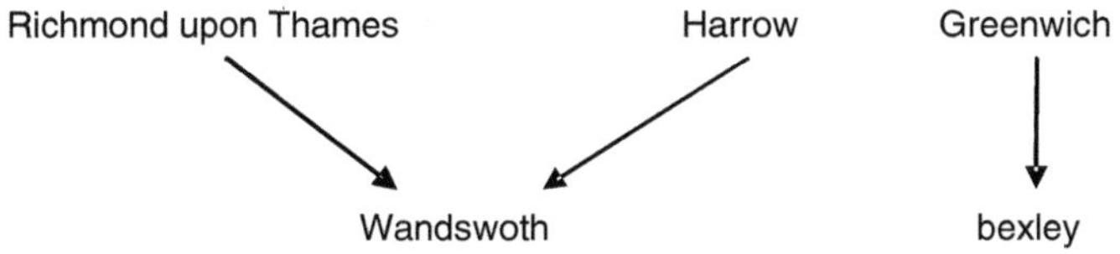

This situation is showing an apparent territorial discontinuity that can be interpreted as the remnant of a previously more ordered structure but actually undone by the attack of *Crack.*

The structure of Heroin-Diamorphine is ordered, but minimal, as follows:

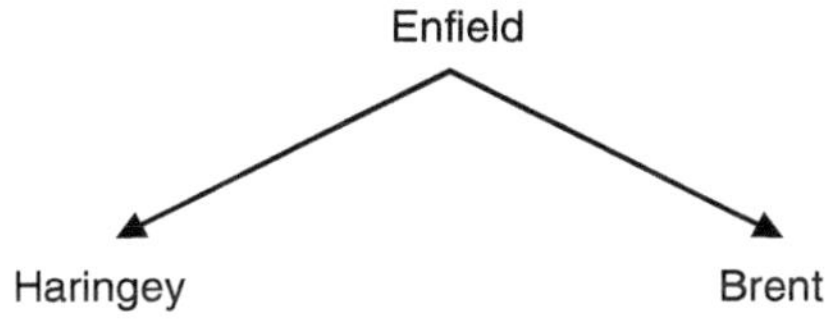

This situation may be interpreted as the result of the equilibrium reached with the stronger *Crack.*

16.4 Effect of the Removal of Nodes Relating to Drugs

Having exhausted the discussion concerning the removal of *AR_PL_NA*, we shall attempt to give a strategic significance to the removal of other nodes. Removing the node of a substance from a graph corresponds to the removal of the drug itself from the market. In practice, it means an opportunity to act directly on the trafficking of the substance itself, interrupting the sources of supply, seizing the stashes and intercepting deliveries between the storage and the dealing places.

Naturally, the removal of nodes marked with a red rectangle increases the complexity of the resultant network, thereby reinforcing it. From this point of view, the removal of red nodes seems unadvisable, whilst it is advisable to remove nodes surrounded by a blue ellipse, after which the complexity of the network, and therefore its robustness, decreases.

According to this interpretation, the only substance that should be removed is *Heroin-Diamorphine*: the territory now occupied by this substance will be occupied by *Crack* as can be seen in Figs. 16.5 and 16.6. The weakening of the network should be caused by the sudden excess of territory for Crack which is occupied but not well controlled.

The result of the removal both of *Heroin_Diamorphine* and of the other variable marked by a blue ellipse (*Waltham Forest)* should represent a considerable simplifi-

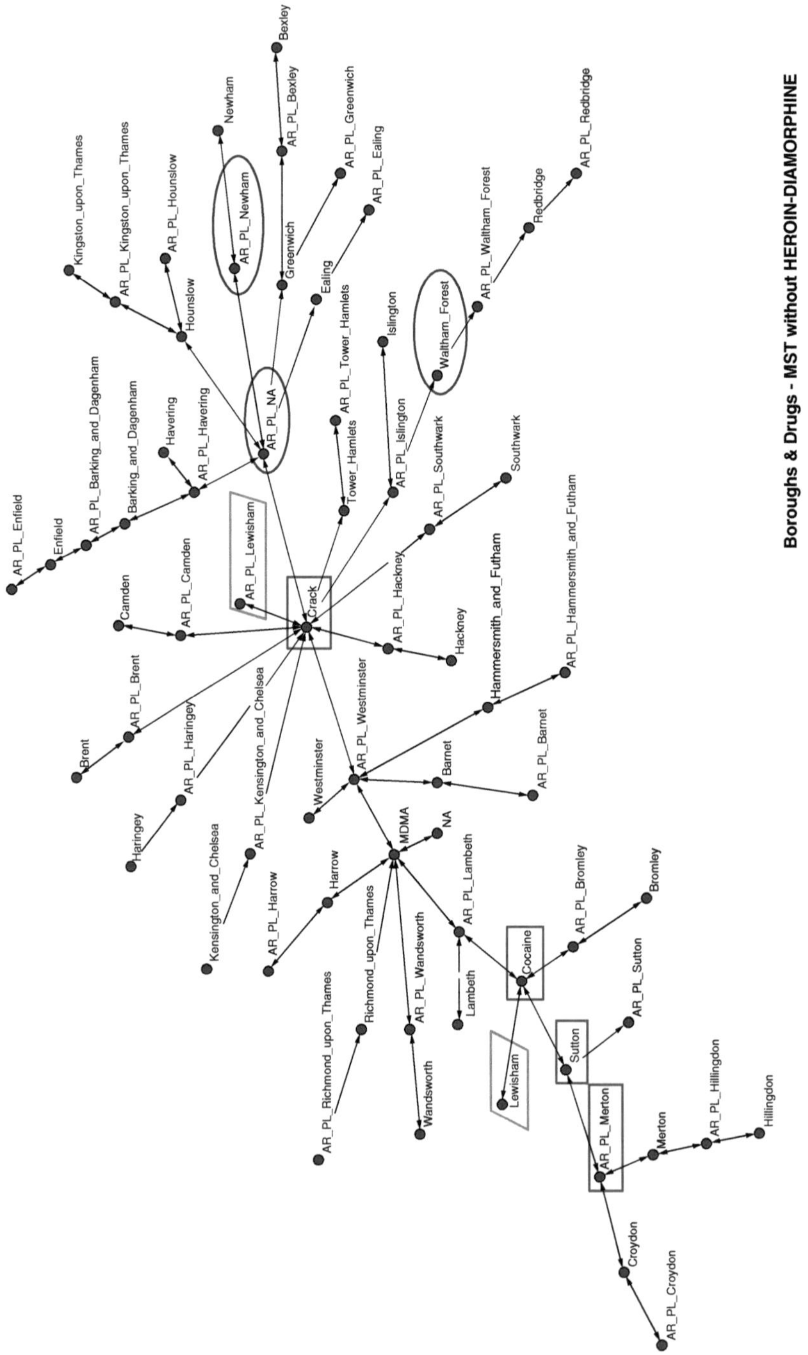

Fig. 16.5 Resultant MST graph after exclusion of the variable Heroin-Diamorphine

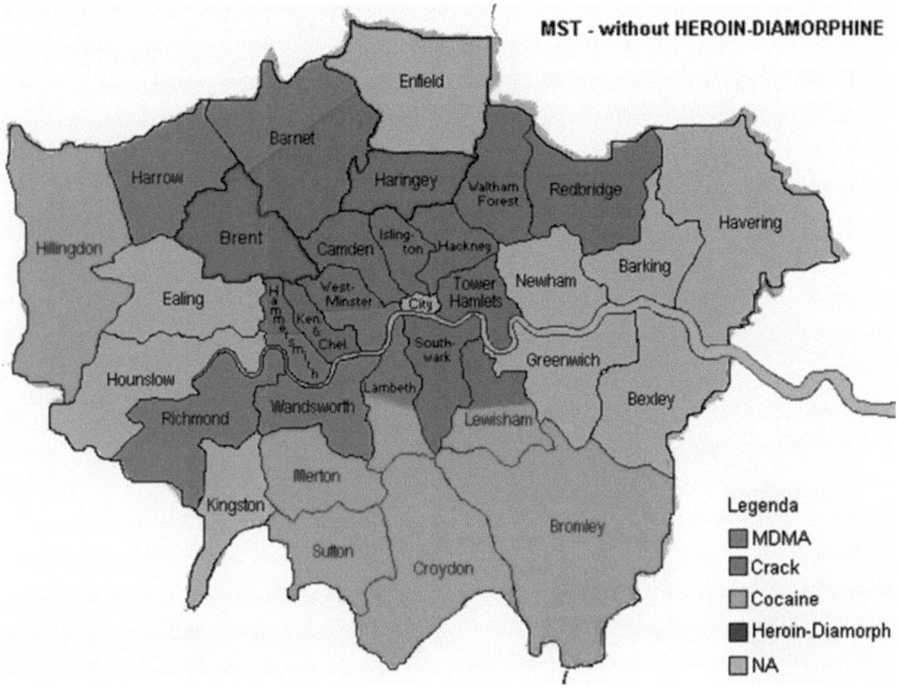

Fig. 16.6 Resultant map of London after exclusion of the variable Heroin-Diamorphine

cation of the graph, and therefore it is a substantial weakening of the entire system. A measurement of this simplification is given by the decrease of the graph's entropy measure before and after the removal.

The removal of *Waltham Forest* corresponds to interrupting the long chain of command that extends eastward into the area of *Crack*: even though secondary, the removal of a power centre weakens the overall structure, and therefore it would be advisable.

Although removing one of those nodes does not increase or decrease the network's complexity in an important way, it may be interesting to see the effect of excluding *Tower Hamlets* which should represent the place where *Crack's* maximum power is concentrated.

The effect can be seen in Fig. 16.7.

Even if the graph is shown upside down, the MST graph is practically the same as the graph of Fig. 16.3. Any other borough takes the place of *Tower Hamlets,* because no variable representing the residents of a borough is directly connected to the drug. This means that the dominant role of *Tower Hamlets* is not replaced and that *Crack's* area is to all intents and purposes not influenced by the removal. The connection with *Heroin-Diamorphine* remains unchanged and likewise is *Waltham Forest's* role in the chain of command. The system continues to function without a dominant centre. The effect of an attack on *Tower Hamlets* would be slight, at least if it is not accompanied by other measures.

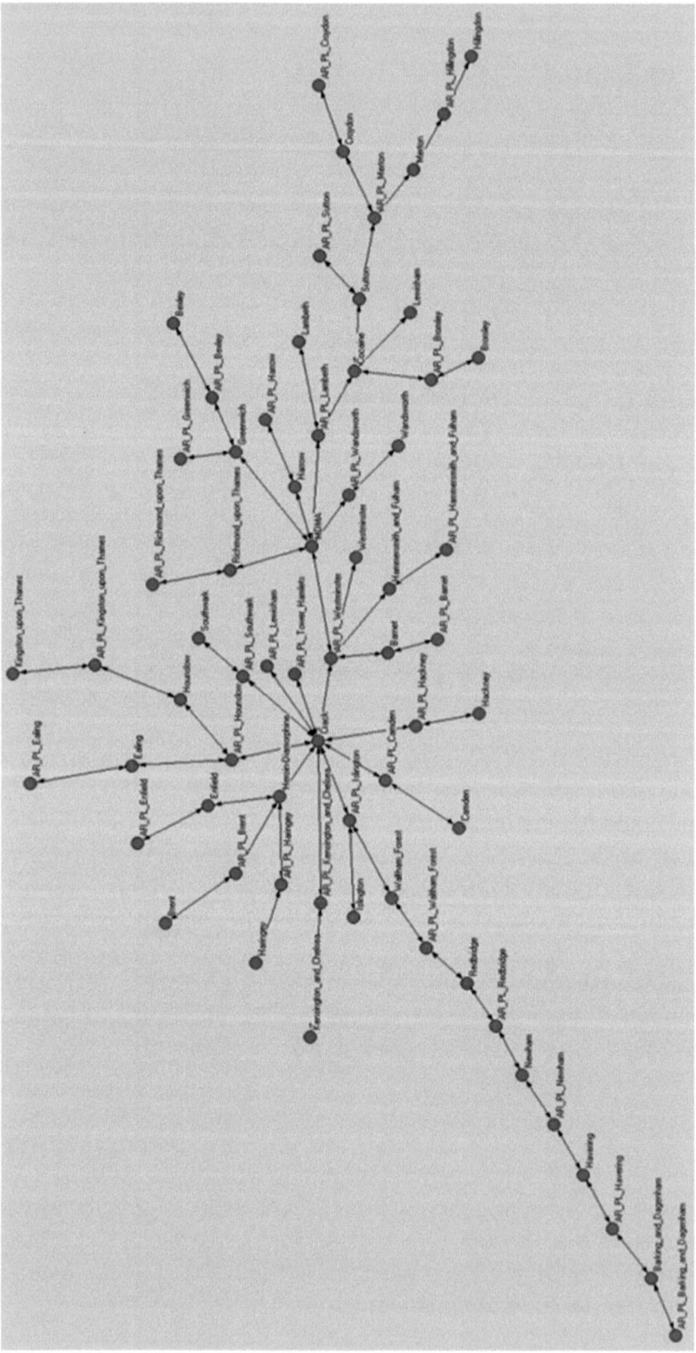

Fig. 16.7 Resultant MST graph after exclusion of the variables AR_PL_NA and Tower Hamlets

16.5 Conclusion

The power of the auto-connective map and its accompanying MST permits the extraction of non-linear information and a visual display of those results. Each database is different as is each subset of data derived from a database, and each dataset must be interpreted in meaningful ways. Using the dataset derived from a police database, certain information can be extracted that, using SQL as the means for analysis, would not ordinarily be available to the investigator. With this in mind, we now summarise the rules of interpretation we set out when commenting on the various graphs produced by the police dataset.

1. The leaf structures connecting the arrests in a borough and the residents belonging to that borough indicate the territoriality of dealing.
2. If (in case 5.1) the residents are connected to the drug, it means that their area of influence includes their territory, but it also extends elsewhere.
3. If (in case 5.1) the arrests are connected to the drug it means that the area of influence of the residents is in their territory, but others may also operate there.
4. If a drug is a hub joining different branches relating to boroughs, it also means that the drug is territorial, and its (super) territory is given by joining of the boroughs involved in the hub.
5. A hierarchy of dominance between the different boroughs can be drawn up within the territory of a drug: the one closest to the hub dominates over the others.
6. In case 5.5, when the distance from the hub is the same, the dominating borough is the one eventually connected to the market of the other.

There are some rules that can be used to formulate hypotheses on the dynamics of the dealing market. These are only hypotheses, because the database is not dynamic.

7. The connectors positioned between two drugs involve bordering areas where it is supposed a drug's active penetration into the territory of the other.
8. If two drugs are directly connected, it is supposed a situation of equilibrium without any attempts at reciprocal penetration. It cannot be supposed how stable the equilibrium will be.
9. If there is an exception to rule 1 (residents in a borough connected to a drug and arrests in that borough connected to another drug), it is supposed a recent substitution of the first drug by the second one.

References

Kruskal, J. B. (1956). On the shortest spanning subtree of a graph and the traveling salesman problem. *Proceedings of the American Mathematical Society, 7*(1), 48–50.

Buscema, M., & Sacco, P. L. (2010). Auto-contractive maps, the H function, and the maximally regular graph (MRG): a new methodology for data mining (chapter 11). In V. Capecchi et al. (Eds.), *Applications of mathematics in models, artificial neural networks and arts*. Dordrecht/London: Springer. doi:10.1007/978-90-481-8581-8_11.

Software

Buscema, M. (2007). *Contractive Maps. Software for programming Auto Contractive Maps.* (Semeion Software #15, v. 2), Rome.

Buscema, M. (2007). *Constraints Satisfaction Networks. Software for programming Non Linear Auto-Associative Networks* (Semeion Software #14, v. 10), Rome.

Buscema, M. (2008). *MST. Software for programming Trees from artificial networks weights matrix* (Semeion Software #38, v 5), Rome.

Massini, G. (2007). *Tree Visualizer. Software to draw and manipulate tree graph* (Semeion Software #40, v. 3), Rome.

Chapter 17
Auto-Contractive Maps and Minimal Spanning Tree: Organization of Complex Datasets on Criminal Behavior to Aid in the Deduction of Network Connectivity

Giulia Massini and Massimo Buscema

17.1 Introduction

The aim of this chapter is to show how it is possible to organize a database of substantial size so as to be able to effectively control the information contained within it and therefore be able to make valid deductions that identify possible connections between the subjects.

For example, let us assume we are analyzing a human population and have to deduce, on the basis of data concerning the individual subjects, which persons in actual fact associate with one another. The aim of this analysis is similar: having data on individuals whose criminal history has been archived, but concerning whom it is not known, or at any rate it is not recorded within this archive, if any relationship exists between them. In short, one must find some mechanism by which it can be determined those individuals for whom a relationship does exist. For example, to determine if they are part of the same "gang" or at any rate whether they are involved in the same drug trafficking "circle."

We referred to a database that is part of the Metropolitan Police Service Central Drug Trafficking Database (CDTD), which has already been analyzed elsewhere in this book, consisting of 1,120 subjects who had been arrested in connection with drugs in various boroughs of London each identified by 144 variables.[1]

For each person, they involve the following profiles:

- Sex (*male, female, not known*)
- Borough of residence of the subject arrested (32 boroughs + 1 NotAvailable)

---

[1]This dataset was extracted in June 2006 when the CDTD Database contained 1,590 tactic sequences, 1,667 persons and 1,190 accused persons (and also 70 incomplete cases).

G. Massini (✉) • M. Buscema
Semeion Research Center of Sciences of Commuication, via Sersale 117, Rome, Italy
e-mail: g.massini@semeion.i

M. Buscema and W.J. Tastle (eds.), *Intelligent Data Mining in Law Enforcement Analytics: New Neural Networks Applied to Real Problems*, DOI 10.1007/978-94-007-4914-6_17,

- Borough where each subject was arrested *Arrest Place* (32 boroughs + 1 NotAvailable)
- Nationality of the subject arrested (*Africa AFR, Asia ASIA, Eastern Europe EASTEU, Europe EU, Ireland IRE, Jamaica JAM, ME, NK, SAME, Turkey and Sylon TU-CY, United Kingdom UK, Vietnam VTN*)
- The somatic stock of the subject arrested (*White European EA1, Dark European EA2, Afro-Caribbean EA3, Asian EA4, Oriental EA5, Arab EA6*)
- Age (expressed in six classes: *<18, 18–21, 21–25, 25–35, 35–45, >45*)
- Number of previous convictions
- Number of previous offenses
- Details of the previous offenses (*FirstConvAge, LastConvAge, Drug, TheftKindred, AgainstPerson, OffensiveWeapons, Sexual, RelatedToPolice, Fraud, Total, AgainstProperty*)
- Number of previous arrests
- Details of previous arrests (*Theft and Kindred offenses, offenses the Person offenses, Drug Trafficking offenses, Drug Possession offenses, Other Drug offenses, Offensive Weapon offenses, Firearms offenses, Kidnapping and Abduction offenses, Other Violent offenses, Other offenses*)
- Number of drug seizures
- Type of drugs seized from the subject arrested (*Cannabis, Cocaine, Crack, Heroin/Diamorphine, MDMA*)
- Amount of money found in their possession at the time of the arrest (*NumOfCashSeizures, Pounds*)
- Number of tactics used by the police (*NumOfTactics*)
- Type of agent who made the arrest (*Police, Non-Law Enforcement Agent, Other Law Enforcement Agent)*
- Types of tactics used by the police to make the arrest (*Search of Object, Search of Person, Search of Premises, Covert Purchase, Controlled Delivery, Other Generic Tactic*)
- Number of tactic sequences
- Arrest made as part of a more widely organized operation (*InOperation*)
- Aims of the police operations *Arrest Mode* (*Direct, Result of Enquiries, Given into Custody, Other, not defined NA*)
- Behavior of the person at the time of the arrest, whether or not violent
- Whether at the time of the arrest the subject was on bail

Upon close inspection of the above attributes, it can be seen that there is no explicit information connecting the subjects with one another, but there does exist, for each subject, a summation of his illicit drug activities. It is with this connection, that of drug activity, to which each variable is assigned that characterizes the sum of all the crimes that the subject has committed rather than to an assignment of individual crimes to an individual. This distinction is subtle, but critical.

17.2 Method

The procedure used to attain the objective consists of three main steps (Kruskal 1956):

1. Processing of the data by an Auto-Contractive Map network that fixes the distance between the various records of the DB and each record with all the others (*CS SW, ver.5.0 Semeion© 2001–2006*)
2. Identification of the connections between the records, based on the distances identified by the Auto-Contractive Map, by the minimal spanning tree, MST (*MST Detector SW, ver.3.0 Semeion© 2006*)
3. Creation of the graph that visualizes the relationships between the records identified by the MST (*Tree Visualizer SW, ver.1.0 Semeion© 2006*)

In the MST "tree" graph, the "leaves" (also called nodes) are the individual points that represent the 1,120 records contained in the database and can therefore be associated with an individual physical person, while the relationships between the records are represented by the "branches" lines (also called edges).

In Fig. 17.1, we show a visualization of the overall graph.

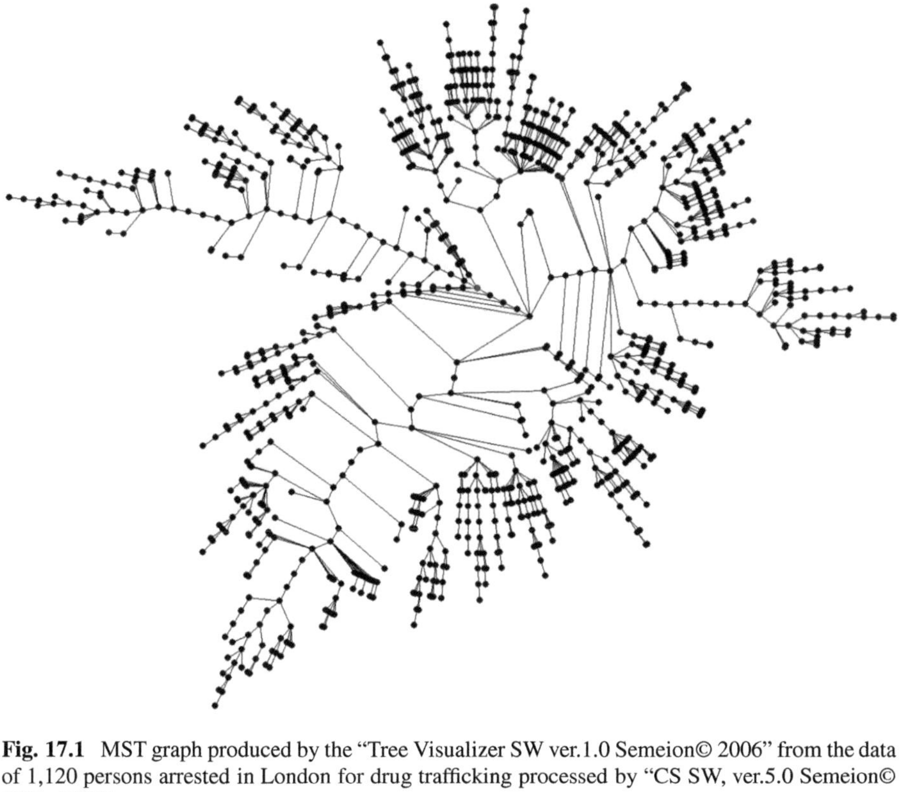

Fig. 17.1 MST graph produced by the "Tree Visualizer SW ver.1.0 Semeion© 2006" from the data of 1,120 persons arrested in London for drug trafficking processed by "CS SW, ver.5.0 Semeion© 2001–2006"

From an analysis of the graph, it was possible to establish, as we shall later see, that the complexity of the information contained in it, given by the high number of subjects and variables, has been organized effectively for classifying the subjects and making deductions concerning the possible meanings that can be attributed to the connections between them.

17.3 Organization of the Subjects of the DB on the MST with Respect to the Variables

Some variables are used in the system as guidelines for carrying out a complex classification of the DB, that is, *sex* (Fig. 17.2), *age* (Fig. 17.3), *nationality* (Fig. 17.4), and *ethnicity* (Fig. 17.5); with respect to these individual variables, the subjects have been placed in specific branchings.

The variable *sex* (*male/female/not known*) sees *female* subjects distributed together with the *not known* subjects on a branch that starts from an area close to the central point of the graph (in the graph, the central point is represented in red and the variable of interest is in green), while the *male* subjects are distributed throughout the remainder of the graph.

Also with respect to the variable *age* (Fig. 17.3), the system was able to combine the subjects into groups. One can observe that persons aged between *25 and 35* occupy the center of the graph insofar as men are concerned. However, the subjects belonging to the other age bands are positioned in groups that leave from the center and are arranged into areas that are more or less peripheral in relation to their nationality and/or ethnicity. This suggests that the age between 25 and 35 is the most active phase of the subject's criminal life in this area and a point at which the subject opens up new criminal contacts in the drugs network. It should also be noted that the age between *25 and 35* years is the age that contains the greatest number of

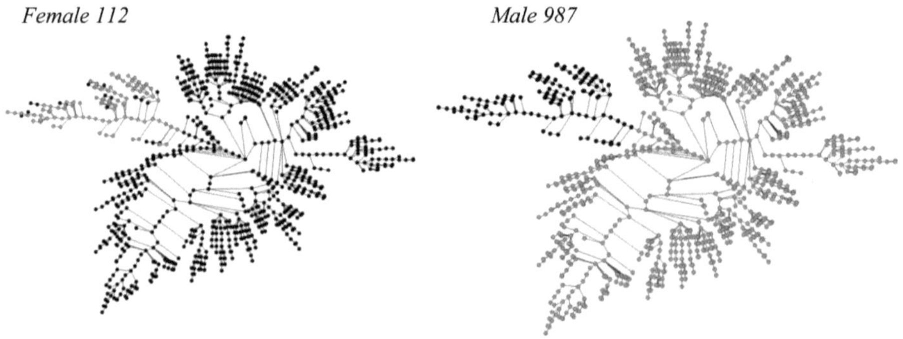

Fig. 17.2 The figure shows the distribution of the 1,120 subjects according to sex

subjects, as the following histogram shows, where the curve is at a maximum in this age band.

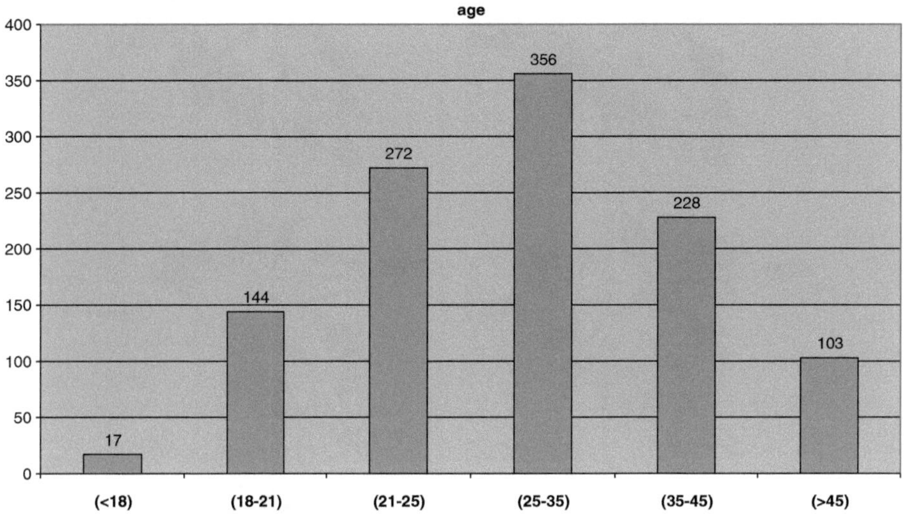

Histogram of the distribution of criminal subjects based on age

Histogram of the distribution of criminal subjects based on age

Regarding *nationality*, a macro-group is identified (Fig. 17.4) positioned centrally and consisting of *723* British subjects (*UK*). These subjects belong to various ethnic groups of which the most important at the strategic level for drug trafficking is, by this analysis, assigned to *Afro-Caribbeans* having British nationality and aged between *25 and 35* years (Fig. 17.5). With regard to ethnicity, it is seen that the two groups opposite in the graph are *Afro-Caribbean/White European*.

It was possible, therefore, to identify this central group in the MST graph by querying the system and placing the operator <AND> between the variables *sex*, *age*, *nationality*, and *ethnicity* (Fig. 17.6).

17.4 Deductions Concerning the Relationships Between the Subjects

To verify whether the system had actually connected the subjects of the present DB in real mode, a comparison was made with other data belonging to the same subjects but relating to the CRIS archive. The CRIS (Central Drug Trafficking Database) is the original source archive of the current DB, having more extensive information on the subjects. Some of this additional information concerns data that establishes the relationship between subjects. In the CRIS, it was possible to find information on subjects who had been arrested in the same police operation or who had a degree of relationship with others, etc.

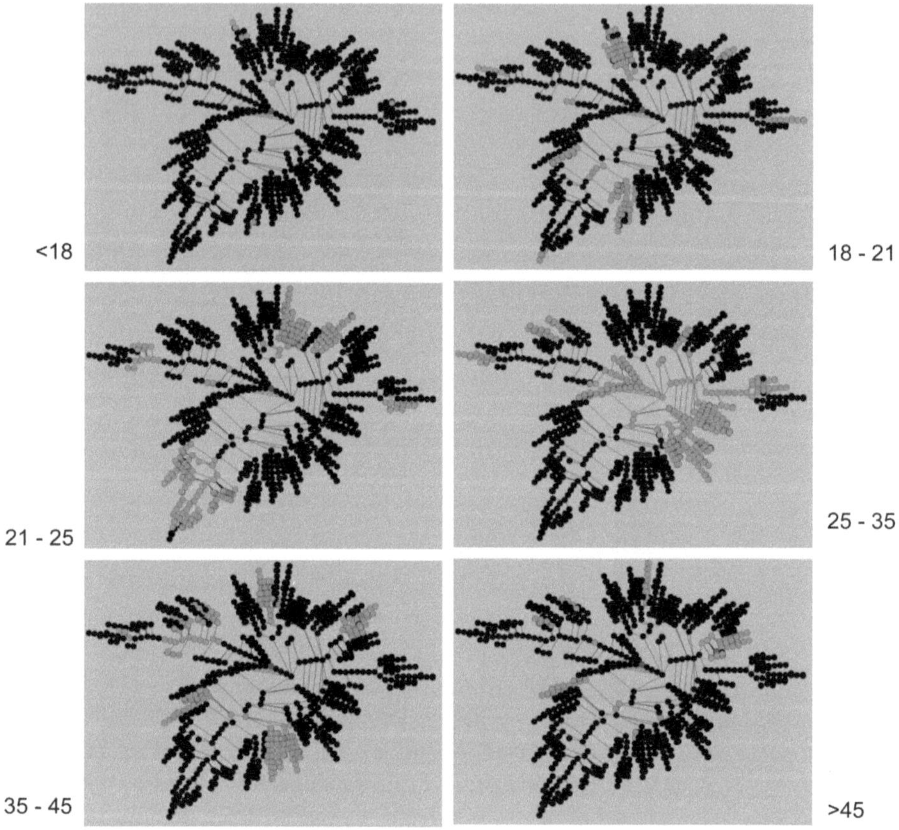

Fig. 17.3 The figure shows the distribution of the subjects according to their age band

It was therefore possible to establish that some of the subjects who were connected in the CRIS were in a position of proximity on the branchings of the MST graph.

Since the CRIS contains only information that proves the actual relationship between the subjects, it may be hoped that the MST tree might suggest *other possible relationships between the subjects, even if they have not emerged explicitly through the information already in the possession of the police.*

Next, we show some examples of the information present in the CRIS on the connection between the subjects and concerning which it was found that in the MST graph, these subjects were placed in a relationship of proximity in specific branchings. For each example, the following information is defined:

- Image of the overall graph and identification of the branch where the subjects who were connected in the CRIS are identified
- Zoom onto the branch selected and identification of the subjects

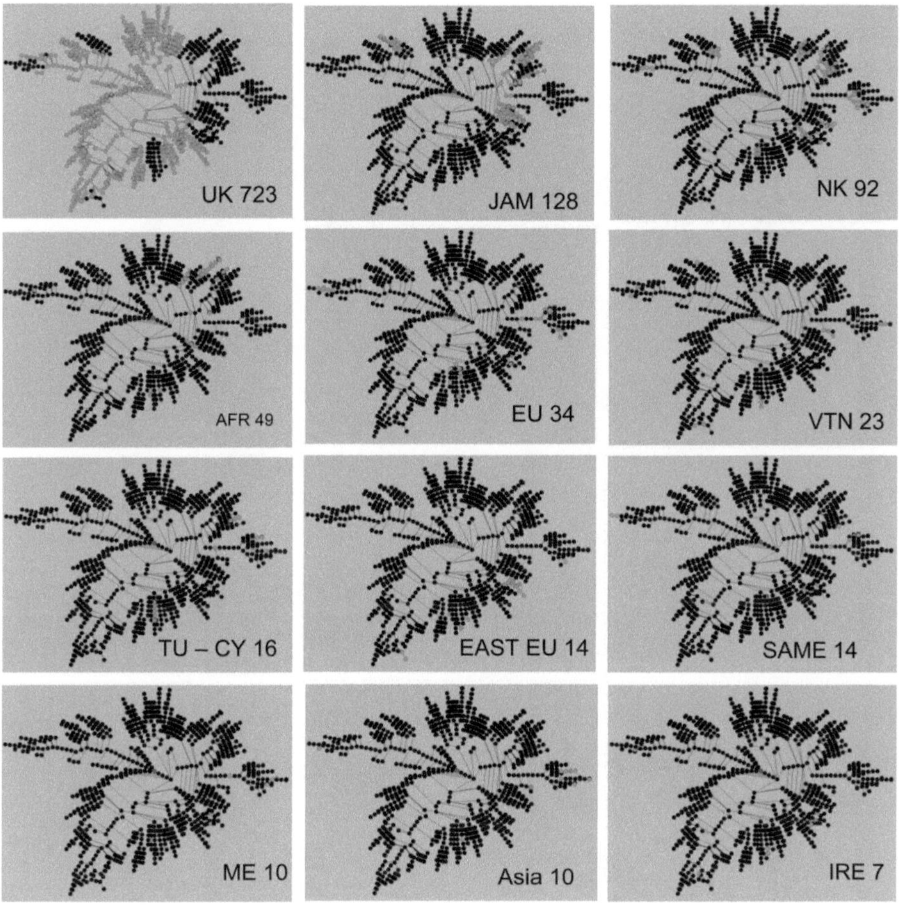

Fig. 17.4 The figure shows the distribution of the subjects according to their nationality

- List of the common variables that characterize the subjects in the current DB
- Comments on the information obtained by consulting the CRIS

In this example (Fig. 17.7), we show the positioning on the MST of some subjects who the CRIS had connected with one another and who in the MST are placed close to the base of common branchings. The information shared for these subjects is possession of Cannabis at the time of the arrest. However, other information can be found in the CRIS that provides a connection between pairs of individuals: 84 and 137, 137 and 463, 463 and 624, 137 and 624.

Subjects 84 and 137 were both in possession of weapons. Subject 137 was also linked to 463 because both were involved in growing Cannabis. Subject 463 was in turn also connected in the CRIS to 624 in a strange way in that they were both

Fig. 17.5 The figure shows the distribution of the subjects according to their ethnicity (somatic stock)

caught with Cannabis hidden in a bag of dirty laundry. Obviously, this information on the hiding place was not present in the DB on which the present MST is based.

Subject 137 was also connected to 624, both being found in possession of Cannabis during "execution of a warrant."

In this branch (Fig. 17.8), a connection was established between subjects 191 and 196 in that the CRIS had information concerning the fact that both had been located by CCTV (closed-circuit television) and were arrested in Coldharbour Lane, Lambeth. This information was also not present in the current DB.

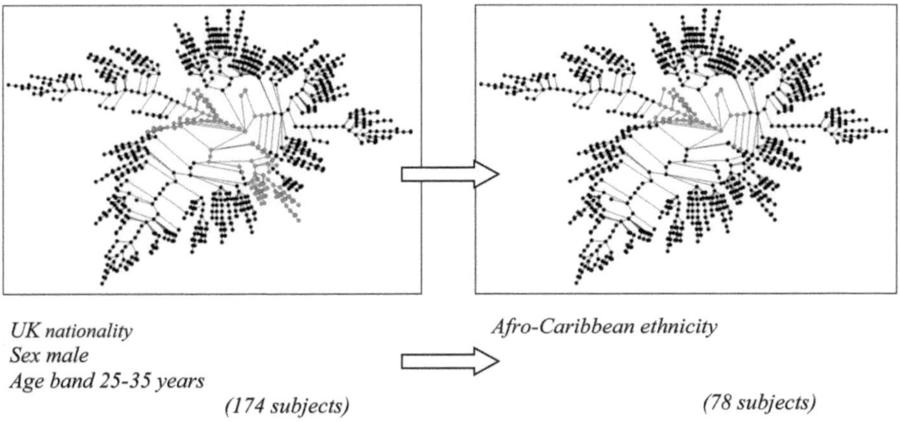

Fig. 17.6 The figure on the *left* shows the distribution of the subjects having certain characteristics (UK/male/25–35 years); the one on the *right* shows a selection made from the former when a further characteristic is activated: Afro-Caribbean ethnicity

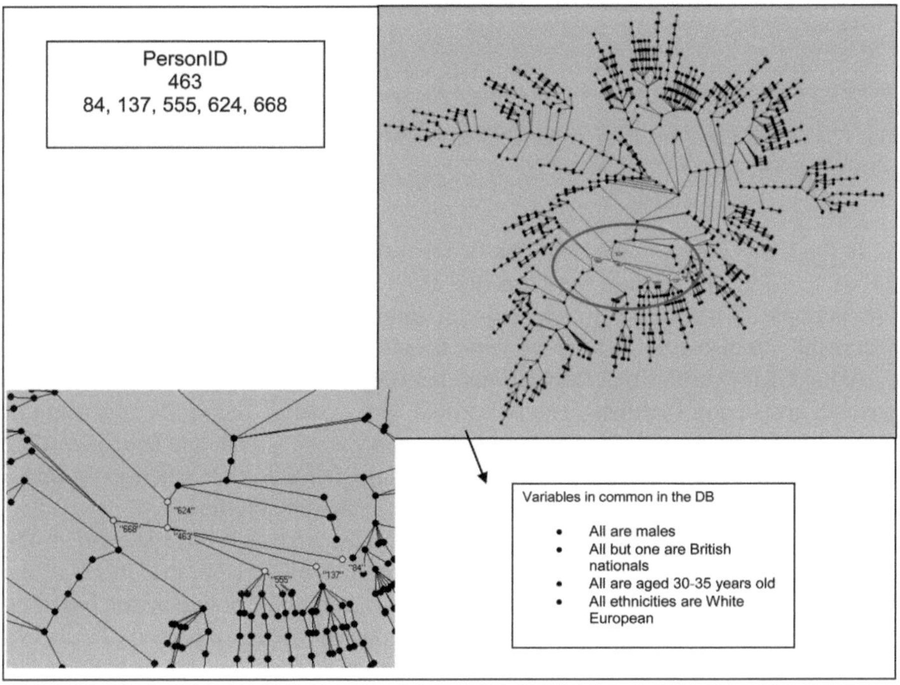

Fig. 17.7 The figure shows the positioning on the MST graph of subjects focused on in the comparison with the data from the CRIS

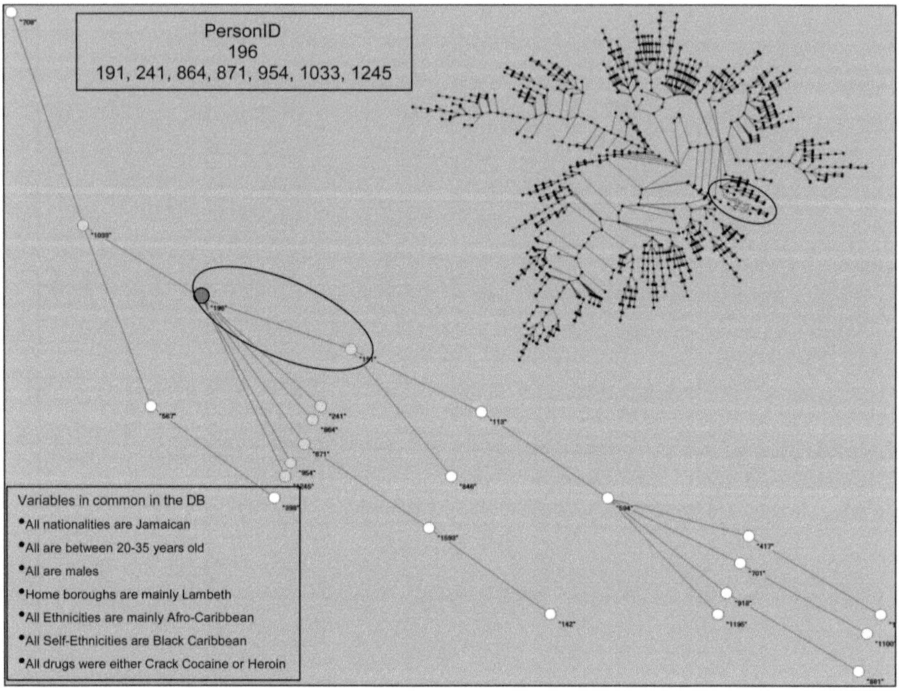

Fig. 17.8 The figure shows the positioning on the MST graph of subjects focused on in the comparison with the data from the CRIS

In the CRIS data, connections can be identified between subjects 164, 183, and 557 (Fig. 17.9) in that they were arrested as a result of special police surveillance, for example, with CCTV (closed-circuit television), which had targeted them specifically. It also emerged that all these episodes occurred in Camden.

In the CRIS, similarities can be found between subjects 140, 566, and 602: they were all arrested in Lambeth. The police had "got to them" indirectly as a result of other investigations, not having identified them previously. One had been identified as a result of enquiries at the home of a missing girl, the other when executing a warrant for another case, and lastly offering drugs in plain clothes.

From the CRIS reports for persons 200, 528, 561, 806, 1,292, and 1,438 (Fig. 17.10), it emerged that all shared a common characteristic; that is, of all the drug seizures it should be noted that none were the result of police action based on tests but rather the result of the execution of a warrant through a person of trust in a previous arrest who warned the police. All were found in possession of MDMA.

Furthermore, 50% of them were arrested in nightclubs, although not the same one, but in clubs in different areas.

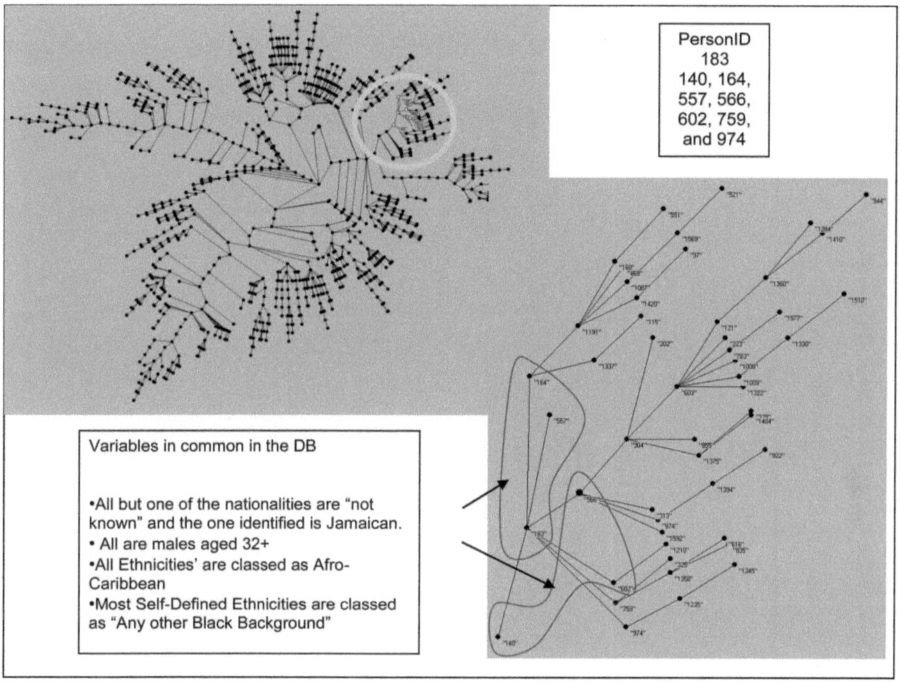

Fig. 17.9 The figure shows the positioning on the MST graph of subjects focused on in the comparison with the data from the CRIS

In the CRIS report, it is established that persons 779 and 781 are brothers (Fig. 17.11) who had been arrested together in the same "tactic sequence."

This information was not present in the DB that enabled the MST graph to be drawn up.

This example (Fig. 17.12) is very similar to the previous one. In fact, in the CRIS report, it is established that persons 350 and 352 are brothers. Clearly, in this case too, appropriate data was not present in the DB that enabled the MST graph to be drawn up.

In the CRIS reports, it emerges that subjects 475, 476, 477, and 1,317 (Fig. 17.13) are linked in that they are part of the same Middle Market Drug Project and were arrested together. Subject 1,312 is also linked in that, although not part of the same operation, he is nevertheless part of the Middle Market Drug Project. Subject 1,445 is linked to these in that he was arrested with a warrant in a National Crime Squad operation.

Absolutely none of this information was present in the DB that allowed the MST graph to be drawn up, yet their relationships are nonetheless defined in the graphs.

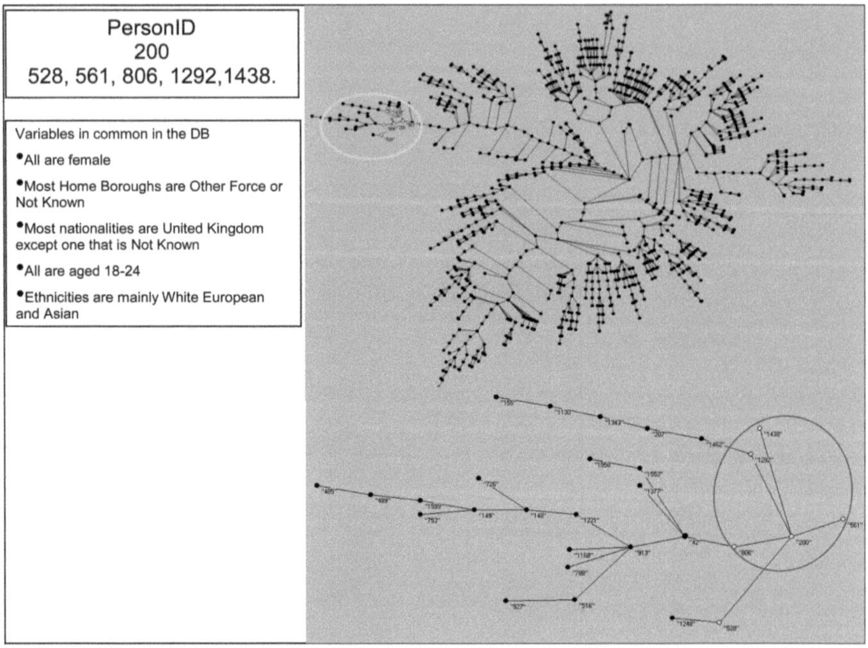

Fig. 17.10 The figure shows the positioning on the MST graph of subjects focused on in the comparison with the data from the CRIS

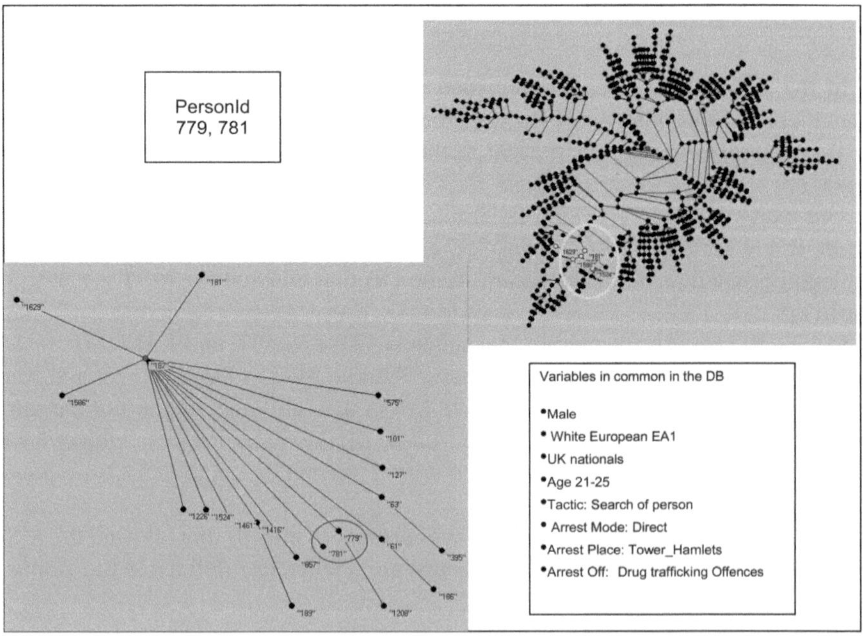

Fig. 17.11 The figure shows the positioning on the MST graph of subjects focused on in the comparison with the data from the CRIS

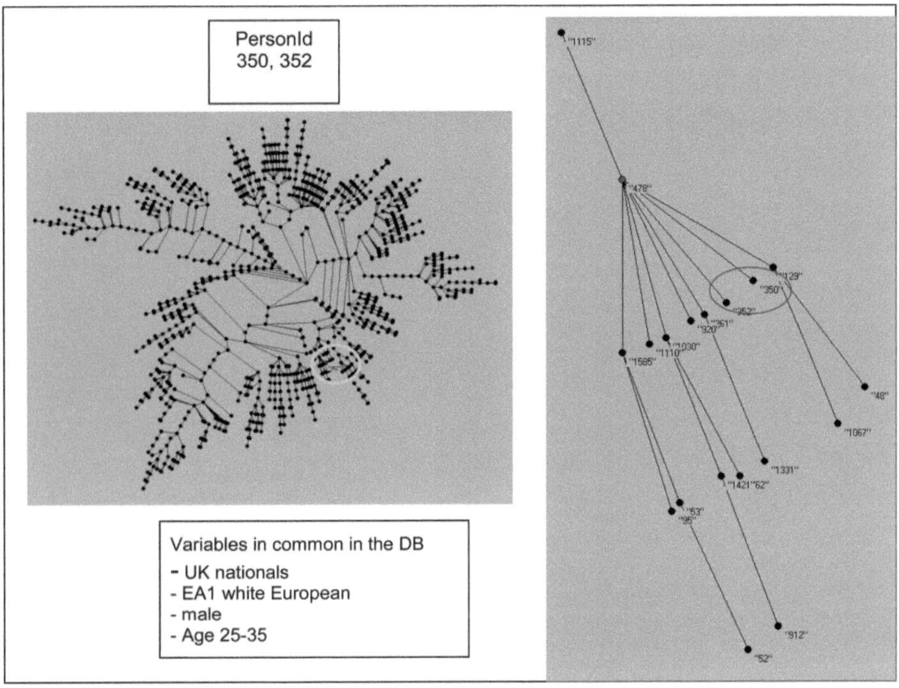

Fig. 17.12 The figure shows the positioning on the MST graph of subjects focused on in the comparison with the data from the CRIS

17.5 Conclusion

We conclude with this very significant example with some observations. The system that we used was capable of identifying the connections between records, even though the presence of an explicit connection was not available in the data. This gives us pause to consider the possibility that other identifiable connections are real even if they not are known at the moment and are not present in the CRIS archive. This leads us to a thought to ponder: how does one interpret all the connections present on the graph?

The conceptual system on which the MST is based is that the Auto CM networks carry out a global evaluation of the whole of the DB and assign a value to the connection between each record individually with respect to all the others. By this operation, the solution space has a dimensionality equal to the number of the records (in this case 1,120), and each record contracts and/or expands the space that separates it from the others. The resultant topology is stored on the connections matrix that links each record with all the others. Thus, at the end, a numerical value is assigned representing each relationship between pairs of records. This value is inversely proportional to the strength of relationship between these two records: smaller if the relationship between the records is very strong, greater if the relationship is weak. Therefore, the MST acts on all these values and selects one

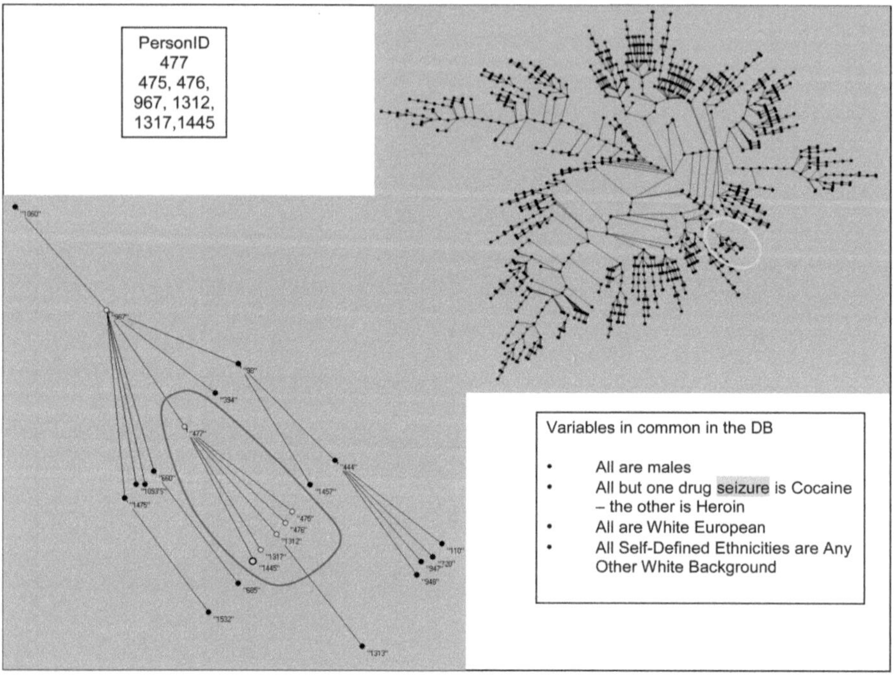

Fig. 17.13 The figure shows the positioning on the MST graph of subjects focused on in the comparison with the data from the CRIS

for each record that shows the strongest relationship contributed to the economy of the global graph whose length must be the smallest possible. The final graph must, by definition, be completely connected (i.e., to a single graph) and not have circuits between the branches.

Thus, the MST has as an obligatory link between each record representing the single choice made from among all those possible. However, in the graph, it is apparent that a node can be connected to many others, since several records can maintain the strongest relationship, that is, a tie, with the same node. It is likely that this system may be useful for suggesting alternative routes to investigate for a better understanding of the dynamic existing in drug trafficking.

References

Kruskal, J. B. (1956). On the shortest spanning subtree of a graph and the traveling salesman problem. *Proceedings of the American Mathematical Society, 7*(1), 48–50.

Massimo, B., & Sacco, P. L. (2010). Auto-contractive maps, the H function, and the maximally regular graph (MRG): a new methodology for data mining (chapter 11). In V. Capecchi et al. (Eds.), *Applications of mathematics in models, artificial neural networks and arts*. Dordrecht/London: Springer. doi:10.1007/978-90-481-8581-8_11.

Software

Buscema, M. (2007). *Contractive Maps. Software for programming Auto Contractive Maps* (Semeion Software #15, v. 2), Rome.

Buscema, M. (2007). Constraints Satisfaction Networks. Software for programming Non Linear Auto-Associative Networks, Semeion Software #14 (v. 10), Rome.

Buscema, M. (2008). *MST. Software for programming Trees from artificial networks weights matrix* (Semeion Software #38, v 5), Rome.

Massini, G. (2007). *Tree Visualizer. Software to draw and manipulate tree graph* (Semeion Software #40, v. 3), Rome.

Chapter 18
Data Mining Using Nonlinear Auto-Associative Artificial Neural Networks: The Arrestee Dataset

Massimo Buscema

18.1 Data Description

From 2004 to 2006, the London Metropolitan Police in partnership with Semeion Research Center of the Sciences of Communication (Rome, Italy) activated the Central Drug Trafficking Database (CDTD). The main purpose of this project was to organize all the data about drug trafficking in London into a relational database suitable for the application of a new powerful activity of information intelligence, that of using a new set of artificial intelligence algorithms, patented by Semeion over the last few years.

The results of this project were included in a special report of March 2006. The report was evaluated enthusiastically in May 2006 by independent British academics. Actually, the CDTD project is awaiting management and use from the new MET Intelligence Bureau (MIB) (Source: MPS Drugs Strategy 2007–2010 and Delivery Plan, Chaps. 6.1–6.13).

From the relational database of the CDTD, we were able to extract a dataset of 1,117 persons arrested in the previous 4 months in London during 2006. The characteristics of the database are described in the following variables and areas:

M. Buscema (✉)
Semeion Research Center of Sciences of Communication, Rome, Italy
e-mail: m.buscema@semeion.it

M. Buscema and W.J. Tastle (eds.), *Intelligent Data Mining in Law Enforcement Analytics: New Neural Networks Applied to Real Problems*, DOI 10.1007/978-94-007-4914-6_18,

Areas	Variables
Places	Home_Boroughs
	Borough_of_Arrest
Anagraphic_Data	Gender
	Nation_Group
	Ethnic_group
	Age
Past_Criminal_Curriculum	Convictions_Number
	Offenses_Number
	Age_at_first_Conviction
	Years_From_Last_Convictioin
	Drug_Offenses
	TheftKindered_Offenses
	Offenses_vs_persons
	Weapons_Offenses
	Sexual_Offenses
	Offenses_VS_Police
	Fraud_Offenses
	Offenses_VS_Property
	Drug_Trafficking
	Other_Offenses
	Total_Offenses
Findings_of_the_Arrest	Number_of_Arrests
	Number_of_Drug_Seizures
	Cash_Seizures
	Pounds_Seizures
	Type_of_Drugs_Seizures
Organization_and_Modalities_of_the_Arrest	Tactic
	Arrest_Mode

We have further articulated each variable into micro-variables:

18.1.1 Places

Home Boroughs:

Micro-variables	Cases	%	Micro-variables	Cases	%
Barking_and_dagenham	8	0.72	Hounslow	23	2.06
Barnet	18	1.61	Islington	54	4.83
Bexley	10	0.90	Kensington_and_Chelsea	37	3.31
Brent	45	4.03	Kingston_upon_Thames	3	0.27
Bromley	32	2.86	Lambeth	71	6.36
Camden	43	3.85	Lewisham	65	5.82

(continued)

(continued)

Micro-variables	Cases	%	Micro-variables	Cases	%
Croydon	40	3.58	Merton	20	1.79
Ealing	40	3.58	Newham	46	4.12
Enfield	27	2.42	Redbridge	13	1.16
Greenwich	18	1.61	Richmond_upon_Thames	5	0.45
Hackney	50	4.48	Southwark	73	6.54
Hammersmith_and_Fulham	22	1.97	Sutton	13	1.16
Haringey	40	3.58	Tower_Hamlets	39	3.49
Harrow	13	1.16	Waltham_Forest	34	3.04
Havering	12	1.07	Wandsworth	20	1.79
Hillingdon	19	1.70	Westminster	22	1.97
Total	975		Borough not available	142	12.71

Borough of Arrest:

Micro-variables	Cases	%	Micro-variables	Cases	%
Arr_Barking_and_dagenham	8	0.36	Arr_Hounslow	43	1.94
Arr_Barrnet	44	1.98	Arr_Islington	116	5.23
Arr_Bexley	23	1.04	Arr_kensington_and_Chelsea	180	8.12
Arr_Brent	76	3.43	Arr_Kingston_upon_Thames	10	0.45
Arr_Bromley	128	5.77	Arr_Lambeth	130	5.86
Arr_Camden	168	7.57	Arr_Lewisham	120	5.41
Arr_Croydon	32	1.44	Arr_Merton	39	1.76
Arr_Ealing	76	3.43	Arr_Newham	65	2.93
Arr_enfield	25	1.13	Arr_Redbridge	16	0.72
Arr_Greenwich	41	1.85	Arr_Richmond_upon_Thames	8	0.36
Arr_Hackney	162	7.30	Arr_Southwark	130	5.86
Arr_Hammersmith_and_Fulham	24	1.08	Arr_Sutton	39	1.76
Arr_Haringey	94	4.24	Arr_TowerHamlets	56	2.52
Arr_Harrow	16	0.72	Arr_WalthamForest	40	1.80
Arr_Havering	21	0.95	Arr_Wandsworth	54	2.43
Arr_Hillingdon	29	1.31	Arr_Westminster	114	5.14
Total[a]	2, 127		Borough not available	91	4.10

[a]Same persons can be arrested more times in the same borough

18.1.2 Graphic Data

Gender:

Micro-variables	Cases	%
Sex_Male	985	88.18
Sex_Female	111	9.94
Sex_notknown	21	1.88
Total	1,117	

Nation Group:

Micro-variables	Code	Cases %	%
Africans	AFR	49	4.39
Asiatic	ASIA	10	0.90
People from east of Europe	EASTEU	14	1.25
European	EU	34	3.04
Irish	IRE	7	0.63
Jamaicans	JAM	128	11.46
People from Middle-East	ME	10	0.90
Non UK citizens	NK	91	8.15
People from South America	SAME	14	1.25
Turkish and Cypriots	TU-CY	16	1.43
Uk citizens	UK	721	64.55
People from Vietnam	VTN	23	2.06
	Total	1,117	

Ethnic Group:

Micro-variables	Code	Cases	%
White-European	EA1	374	33.48
Dark-European	EA2	86	7.70
Afro-Caribbean	EA3	531	47.54
Asia	EA4	80	7.16
Oriental	EA5	32	2.86
Arab	EA6	1	1.16
Not available		1	0.09
total	1,117		

Age:

Micro-variables	Cases	%
Age(UpTo18)	17	1.52
Age(18–21)	144	12.89
Age(21–25)	272	24.35
Age(25–35)	355	31.78
Age(35–45)	226	20.23
Age(Over45)	103	9.22
Total	1,117	

18.1.3 Past Criminal Curriculum

Convictions Number:

Micro-variables	Cases	%
ConvictionsNumber(0)	177	15.85
ConvictionsNumber(1)	271	24.26
ConvictionsNumber(2)	120	10.74
ConvictionsNumber(3)	96	8.59
ConvictionsNumber(4)	78	6.98
ConvictionsNumber(5–10)	215	19.25
ConvictionsNumber(11–20)	107	9.58
ConvictionsNumber(over20)	53	4.74
Total	1,117	

Offenses Number:

Micro-variables	Cases	%
OffensesNumber(0)	175	15.67
OffensesNumber(1)	145	12.98
OffensesNumber(2)	114	10.21
OffensesNumber(3–5)	202	18.08
OffensesNumber(6–10)	184	16.47
OffensesNumber(11–20)	160	14.325
OffensesNumber(20–50)	106	9.49
OffensesNumber(over-50)	31	2.78
Total	1,117	

Age of the First Conviction:

Micro-variables	Cases	%
FirstConvAge(up-to-18)	415	37.15
FirstConvAge(19–21)	221	19.79
FirstConvAge(22–27)	217	19.43
FirstConvAge(28–33)	122	10.92
FirstConvAge(34–39)	75	6.71
FirstConvAge(40–45)	38	3.405
FirstConvAge(46–51)	20	1.79
FirstConvAge(over-51)	9	0.81
Total	1,117	

Years from the Last Conviction:

Micro-variables	Cases	%
Off_LastConvAge(0)	218	19.52
Off_LastConvAge(1)	466	41.72
Off_LastConvAge(2)	353	31.60
Off_LastConvAge(3–5)	45	4.03
Off_LastConvAge(6–10)	24	2.15
Off_LastConvAge(11–20)	6	0.54
Off_LastConvAge(over20)	5	0.45
Total	1,117	

18.1.4 *Type and Number of Offenses*

Drug Offenses:

Micro-variables	Cases	%
Drug_Possession_Offenses	194	14.76
AR_OFF_Other_Drug_Offenses	38	2.89
Off_Drug(0)	250	19.03
Off_Drug(1)	277	21.08
Off_Drug(2)	189	14.38
Off_Drug(3–5)	232	17.66
Off_Drug(6–10)	134	10.20
Off_Drug(Over-10)	35	2.66
Total[a]	1,314	

[a]One person can be present in more items

Theft and Kindred Offenses:

Micro-variables	Cases	%
Only Theft_and_Kindred_Offenses	86	7.15
Off_TheftKindred(0)	659	54.78
Off_TheftKindred(1–5)	301	25.02
Off_TheftKindred(6–10)	74	6.15
Off_TheftKindred(11–20)	48	3.99
Off_TheftKindred(over-20)	35	2.91
Total[a]	1,203	

[a]One person can be present in more items

Offenses Against Personnel:

Micro-variables	Cases	%
OffensesAgainstPerson(only)	9	0.80
Off_AgainstPerson(0)	927	82.33
Off_AgainstPerson(1)	91	8.08
Off_AgainstPerson(2)	54	4.80
Off_AgainstPerson(3–5)	37	3.29
Off_AgainstPerson(over-5)	8	0.715
Total[a]	1,126	

[a]one person can be present in more items

Weapons Offenses

Micro-variable	Cases	%
Offensive_Weapon_Offenses(Only)	41	3.47
Firearms_Offenses	23	1.95
Off_OffensiveWeapons(0)	895	75.78
Off_OffensiveWeapons(1)	127	10.75
Off_OffensiveWeapons(2)	52	4.40
Off_OffensiveWeapons(over-2)	43	3.64
Total[a]	1,181	

[a]One person can be present in more items

Sexual Offenses:

Micro-variables	Cases	%
Off_Sexual(0)	1,091	96.67
Off_Sexual(1)	12	1.07
Off_Sexual(2)	5	0.45
Off_Sexual(over-2)	9	0.81
Total	1,117	

Offenses Against Police:

Micro-variables	Cases	%
Off_RelatedToPolice(0)	812	72.69
Off_RelatedToPolice(1)	131	11.73
Off_RelatedToPolice(2–5)	128	11.46
Off_RelatedToPolice(over-5)	46	4.12
Total	1,117	

Fraud Offenses:

Micro-variables	Cases	%
Off_Fraud(0)	969	86.75
Off_Fraud(1)	76	6.80
Off_Fraud(2–5)	54	4.835
Off_Fraud(over-5)	18	1.615
Total	1,117	

Offenses Against Property:

Micro-variables	Cases	%
Off_AgainstProperty(0)	966	86.48
Off_AgainstProperty(1)	89	7.97
Off_AgainstProperty(2)	28	2.51
Off_AgainstProperty(more)	34	3.04
Total	1,117	

Drug Trafficking Offenses:

Micro-variables	Cases	%
Drug_trafficking_Offenses(0)	211	18.89
Drug_trafficking_Offenses(1)	725	64.91
Drug_trafficking_Offenses(2–5)	137	12.265
Drug_trafficking_Offenses(over-5)	44	3.94
Total	1,117	

Other Offenses:

Micro-variables	Cases	%
AR_OFF_Kidnapping_and_Abduction_offenses	8	5.84
Other_violent_offenses	74	54.01
Other_offenses	55	40.15
Total	137	

Total Number of Offenses:

Micro-variables	Cases	%
Off_Total(0)	178	15.94
Off_Total(1)	148	13.25
Off_Total(2)	111	9.94
Off_Total(3–5)	199	17.82
Off_Total(6–10)	185	16.56
Off_Total(11–20)	161	14.41
Off_Total(20–50)	104	9.31
Off_Total(over-50)	31	2.78
Total	1,117	

18.1.5 Findings of the Arrest

Total Number of Arrests:

Micro-variables	Cases	%
NumOfArrests(1)	600	53.72
NumOfArrests(2)	302	27.04
NumOfArrests(3)	113	10.12
NumOfArrests(over-3)	102	9.13
Total	1,117	

Number of Drug Seizures:

Micro-variables	Cases	%
NumOfDrugSeizures(0)	123	11.01
NumOfDrugSeizures(1)	360	32.23
NumOfDrugSeizures(2)	285	25.51
NumOfDrugSeizures(3–5)	241	21.58
NumOfDrugSeizures(over-5)	108	9.675
Total	1,117	

Number of Cash Seizures:

Micro-variables	Cases	%
NumOfCashSeizures(0)	807	72.25
NumOfCashSeizures(1)	260	23.28
NumOfCashSeizures(over-1)	50	4.48
Total	1,117	

Number of Pounds Seizures:

Micro-variables	Cases	%
Pounds(0)	840	75.20
Pounds(up100)	58	5.19
Pounds(up500)	118	10.56
Pounds(up1000)	47	4.21
Pounds(up5,000)	39	3.49
Pounds(over-5000)	15	1.34
Total	1,117	

Type of Drugs Found:

Micro-variables	Cases	%
Cannabis	1,019	36.43
Cocaine	471	16.84
Crack	591	21.13
Heroin	588	21.02
MDMA	128	4.58
Total	2,797	

18.1.6 Organization and Modalities of the Arrest

Number of Tactics, Type of Tactics, Number of Sequences, Type of Action: Arrest Mode:

Micro-variables	Cases	%
NumOfTactics(1)	409	6.87
NumOfTactics(2)	358	6.01
NumOfTactics(3)	190	3.19
NumOfTactics(over-3)	160	2.69
GenericTactic_Search_of_Object	61	1.02
GenericTactic_Search_of_Person	873	14.66
GenericTactic_Search_of_Premises	973	16.34
GenericTactic_Convert_Purchase	534	8.97
GenericTactic_Controlled_Delivery	11	0.18
GenericTactic_Other	153	2.57
NumOfTacticsSequences(1)	986	16.55
NumOfTacticsSequences(2)	54	0.91
NumOfTacticsSequences(3–5)	34	0.57
NumOfTacticsSequences(over-5)	43	0.72
InOperation(0)	935	15.70
InOperation(1)	82	1.38
InOperation(2–5)	57	0.96
InOperation(over-5)	43	0.72
Total	5,956	

18.2 Explorative Analysis Using Self-Organizing Maps (SOM)

The "Persons Arrested" dataset seems to be very rich in information. But to explore this information world, any bivariate analysis is useless, from cross-tabulation to the chi-squared test.

"Persons Arrested" dataset, in fact, is formed by 1,117 records and 28 variables, further articulated into 246 micro-variables. Any analysis of this dataset, to provide useful information, has to consider the interaction of all the 246 atomic variables together between each record and all the others. Any other reduction in variables, without a valid motivation, may lead to deep misunderstandings and invalid outcomes. Thus, we must approach this dataset using *multivariate* analysis.

We may not process this dataset simply using a *linear multivariate* technique because the assumption of linearity in analyzing human activity is completely arbitrary. The reduction of the number of the variables, for example, according to the explained variance criterion, is rough. No one can know a priori if marginal differences between a pair of variables represent noise or key information.

To make a first serious exploration of the "Persons Arrested" dataset, we need to process these data using a *non-highly linear multivariate* technique such as a Self-Organizing Map (SOM) (see Kohonen 1990, 1995a, b).

Micro-variables	Cases	%
Non-Law_Enforcement_Agent	49	0.86
Other-Law_Enforcement_Agent	25	0.44
ViolentOnArrest(0)	1,068	18.76
ViolentOnArrest(1)	33	0.58
ViolentOnArrest(more-1)	16	0.28
ArrMode_NA	6	0.11
ArrMode_Direct(0)	304	5.34
ArrMode_Direct(1)	517	9.08
ArrMode_Direct(2)	187	3.29
ArrMode_Direct(3–5)	78	1.37
ArrMode_Direct(over-5)	31	0.54
ArrMode_Result_of_Enquiries(0)	1,035	18.18
ArrMode_Result_of_Enquiries(1)	44	0.77
ArrMode_Result_of_Enquiries(2)	19	0.33
ArrMode_Result_of_Enquiries(over-2)	19	0.33
ArrMode_Given_into_custody	27	0.47
ArrMode_Other(0)	1,002	17.67
ArrMode_Other(1)	83	1.46
ArrMode_Other(2)	16	0.28
ArrMode_Other(over-2)	16	0.28
OnBailAtTimeOfOffence(0)	1,047	18.39
OnBailAtTimeOfOffence(1)	42	0.74
OnBailAtTimeOfOffence(2)	17	0.30
OnBailAtTimeOfOffence(over-2)	11	0.19
Total	5,692	

SOM, in fact, presents many suitable features:

- SOM is not sensitive to the cardinality of the variables.
- SOM processes all the records and all the variables simultaneously.
- SOM is an artificial neural network (ANN) and consequently provides its best organization of data, processing the same data many times along the time. This is a fundamental way to consider the nonlinear relationships among data themselves.
- SOM clusters the whole data according to their global similarities.
- SOM squashes variables and records of the dataset into two dimensions, and during this multidimensional scaling, it selects only the most important features of the dataset.
- SOM results are tables and maps very easy to understand.

In the experiments presented here, we have used two Semeion research software[1] products. We have experimented with varying SOM configurations containing different map formats: 15 × 15, 20 × 20, and 30 × 30.

Because the results demonstrated stability in every experiment, we select the SOM 30 × 30 map that produces the minimum Topographic Error:

SOM 30 × 30					
Date file	Codebook file	Topographic error	Quantization error	Map compactness error	Codebook error
Persons(246 × 1,117)	Persons30 × 30_Sigmoid	0.00%	9.61%	1.12%	0.13%

We have consider Topographic Error as relevant in the evaluation of SOM performance, only because this cost function, with the Quantization Error, is well known in the SOM literature. However, we have found that the *Codebook Error* and the *Map Compactness Error* are the more consistent cost functions, very suitable for defining the SOM codebook after the training phase.

We now define these concepts:

Topographic Error: after the SOM training, the two nearest codebooks for each record are inspected. If the two codebooks are adjacent, the projection is considered correct; if the two codebooks do not belong to the same cell neighborhood, a Topographic Error is counted (in a SOM squared grid, each cell has a neighborhood composed of the eight nearest cells).

Quantization Error: This index is not really an error but rather the mean of the variance that the SOM has collected inside each codebook. This index is zero when the SOM places every record in a different cell. But in this case, the SOM ANN has not executed its task, that is, to cluster the dataset.

So the quantization error works only as a clustering index of the SOM training.

Codebook Error: This index, developed by M Buscema in 2006 at Semeion, is very useful in measuring the compactness of each codebook after the SOM training phase. The Codebook Error traces, for each cell (codebook), the circle of minimal radius whose center is the codebook itself and includes all the records belonging to that cell. All the other records included in that circle are considered a compactness error associated with that codebook.

[1]M Buscema, MOS: Maps Organizing System, version 2.0, Semeion Software #26, Rome 2002–2007; G Massini, SOM, Shell for programming Self-Organizing Maps, Version 7.0, Semeion Software #19, Rome 2000–2007.

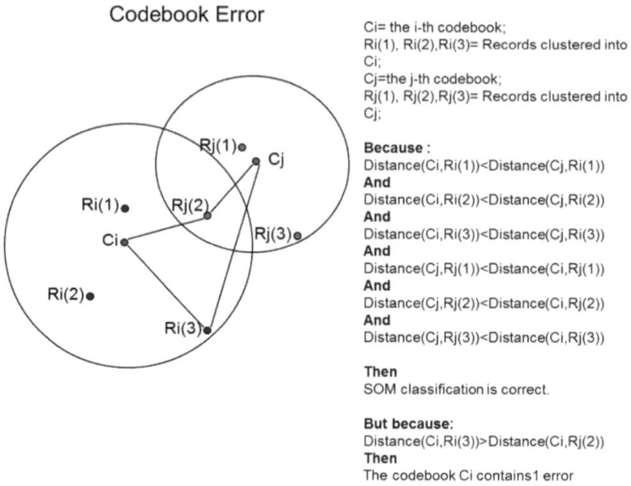

Codebook Error Equations

- Distance between the kth record clustered into the ith cell and the codebook of the same cell:

$$d_{i,i_k} = \sqrt{\sum_s \left(C_{i_s} - R_{i_{k_s}}\right)^2}.$$

- $R_{i_{\text{Max}}}$ is the record clustered into the ith cell, whose distance from its codebook is the highest:

$$d_{i,i_{\text{Max}}} = \operatorname*{Max}_k \{d_{i,i_k}\}.$$

- Distance between the kth record, clustered into the jth cell, and the codebook of the ith cell:

$$d_{i,j_k} = \sqrt{\sum_s \left(C_{i_s} - R_{j_{k_s}}\right)^2};$$

$$if\ \left(d_{i,j_k} < d_{i,i_{Max}}\right) \text{error}_{i,j} + +;$$

$$E_i = \frac{1}{T - N_i} \cdot \sum_{j \neq i}^{C} \text{error}_{i,j},$$

$$N_i = \text{Total Records in Cell } i\text{th}, T = \text{Total Records};$$

$$E^* = \frac{1}{C} \cdot \sum_i^C E_i, \quad C = \text{Total Number of Cells}.$$

Map Compactness Error: a more restrictive form of the codebook error. For each cell, the minimal circle includes all the records clustered in that cell and is centered each time on every record of the codebook. All the records clustered in other cells, including being present in another circle, are considered an error of map compactness.

Map Compactness Error Equations

R_{i_k} is kth record clustered into the ith cell.
Distance between the kth record and the zth record, clustered into the same cell:

$$d_{i,i_{k,z}} = \sqrt{\sum_s \left(R_{i_{k_s}} - R_{i_{z_s}}\right)^2}.$$

Max distance into the ith cell from the kth record:

$$d_{i,i_{k,\mathrm{Max}}} = \operatorname*{Max}_z \left\{d_{i,i_{k,z}}\right\}.$$

Distance between the kth record clustered into ith cell and zth record clustered into jth cell:

$$d_{i,j_{k,z}} = \sqrt{\sum_s \left(R_{i_{k_s}} - R_{j_{z_s}}\right)^2},$$

$$if\ \left(d_{i,j_{k,z}} < d_{i,i_{k,\mathrm{Max}}}\right) \mathrm{error}_{i,j} ++,$$

$$E_i = \frac{1}{N_i \cdot (T - N_i)} \sum_{j \neq i}^{C} \mathrm{error}_{i,j},$$

$$E^* = \frac{1}{C} \cdot \sum_i^C E_{i.}$$

N_i = Total records in cell ith; T = total records; C = total number of cells.

The SOM trained matrix was composed of 900 codebooks organized in a topology of 30 × 30 cells (Fig. 18.1).

Each codebook of the trained SOM is an ideal prototype for the similar records clustered in that cell. Consequently, each codebook is a vector of 246 features, where every vector component (variable) can be more or less active, according to the prototype that the SOM selects for that cell: close cells have similar prototypes; distant cells have different prototypes.

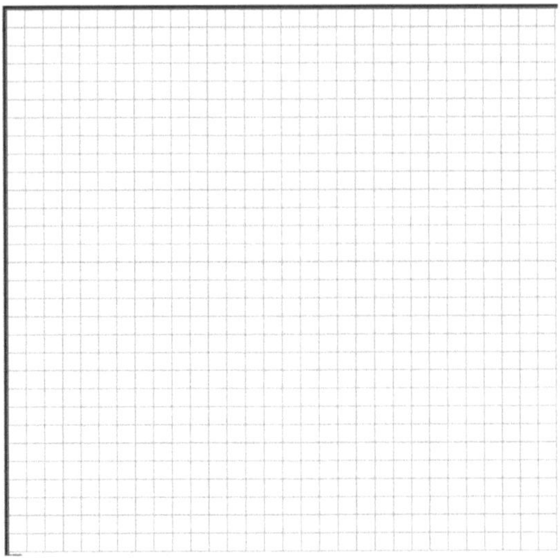

Fig. 18.1 The 30 × 30 SOM matrix grid

So, the double task of the SOM during the training phase is:

1. To distribute all the records of the sample in the same cell or in close cells or distant cells, according to their global similarities
2. Simultaneously, to generate dynamically for each cell its specific codebook (prototype of the records clustered together)

After 100 epochs of training, the SOM clustering ended.

Figure 18.2 shows that data were clearly clustered into the grid: many cells are empty and others have attracted many records. This is methodologically relevant: the number of the available cells is 900, and the number of the records to be projected onto the grid is 1,117. Consequently, a random distribution of records onto the cell should have been possible: quite a record for one cell. All the same, the SOM concentrates similar records in the closer cells and creates a large cell distance among different records. This is a qualitative feature showing the consistency of the SOM clustering.

The SOM also generates a different codebook for every cell: each cell of the trained SOM, in fact, presents a prototype of the all dataset variables in its specific locations. We must determine the 246 slices of the cube where each slice represents the distribution of the activation value of each variable in the 30 × 30 grid.

An effective example could be the slices generated by the SOM about the type of drugs found during the arrest (Fig. 18.3):

Comparing these maps, we can observe:

1. Cannabis distribution is extensive and quite spread out, with two large concentrations:

   (a) The first one in the center of the map
   (b) The second one in the east side of the map

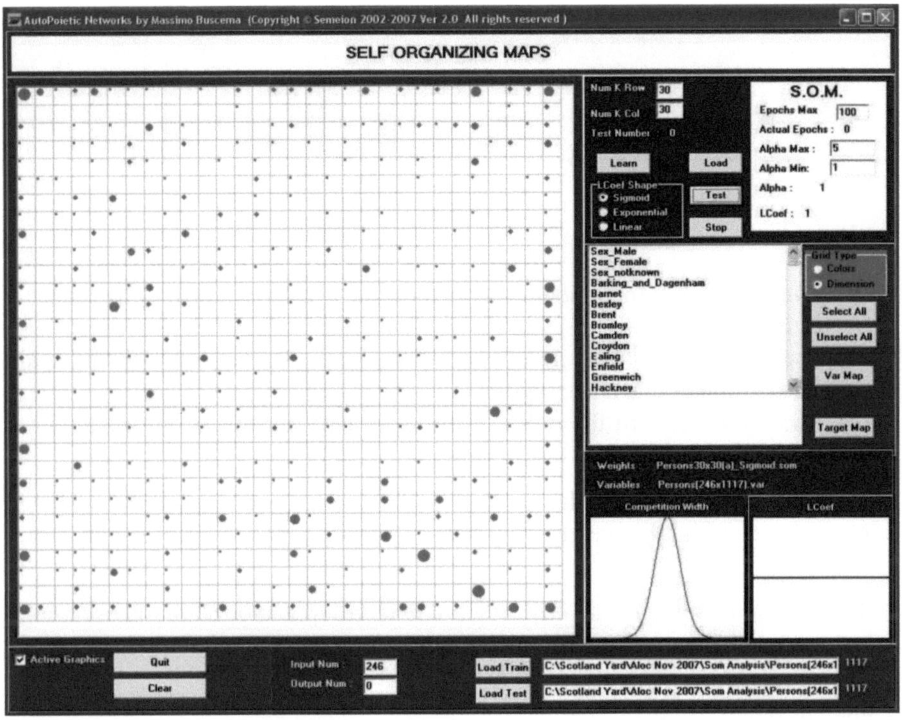

Fig. 18.2 One example of SOM software after the training phase: the *larger the circle*, the more records that are clustered into a cell

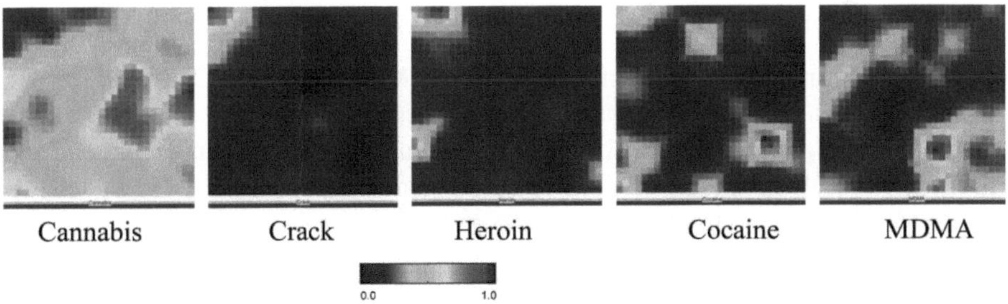

Fig. 18.3 Clusters according to the type of drugs

2. Crack and heroin have a similar and specific distribution, but heroin has also two other small clusters:
   (a) The first in the southwest of the map
   (b) The second, smaller, in the southeast of the map
3. Cocaine has a complex distribution, clustered into at least four groups:
   (a) The biggest, in the east-southeast of the map, close to the main cannabis distribution group.

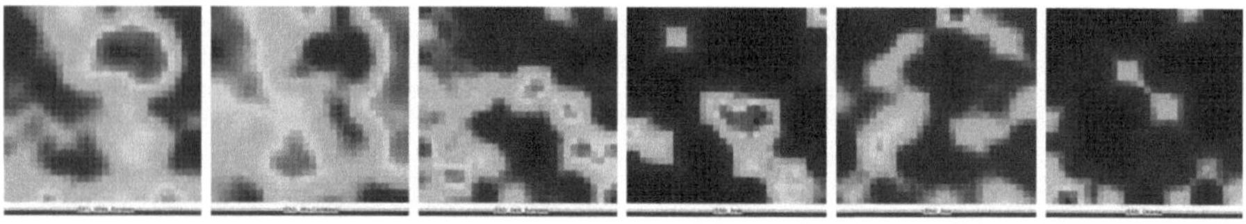

Fig. 18.4 Clusters according to the ethnic group

(b) The second one, in the southwest of the map, overlapped with some cannabis activity.
(c) The third one, in the northwest of the map, overlapped with the main distribution of crack and heroin.
(d) The fourth, toward the north of the map, is a free space, not particularly frequented by the other three drugs.

4. MDMA seems to be similar to that of the distribution of cocaine, with the exception that MDMA prefers to be distributed with cannabis, but not with heroin and crack.

The next problem is to identify which ethnic group and/or nation is eventually associated with this map of these various types of drugs.

The slices representing the ethnic groups' distribution can provide some answers to these questions (Fig. 18.4):

The White European group seems to be specialized in cocaine and partially in cannabis, while the Dark European group seems divided into two groups:

1. One group seems to manage most of the cocaine and MDMA trafficking.
2. The second group suggests activity in heroin and other drugs.

Arabs appear concentrated in cannabis and MDMA, while Orientals have a strong niche in a small and separate market of cocaine.

The Asiatic group seems to be weekly linked to cannabis, but another group (in the north of the map) appears to work as a "generic dealer."

Afro-Caribbean are divided into three big clusters:

1. The first one is completely dedicated to the crack and heroin trafficking (northeast of the map).
2. The second one in a more generic trafficking activities, working and/or competing also with Asiatic and Dark European (the east side of the map).
3. The third one, linked to the first group, more dedicated to the cannabis market.

The nation variables are also clustered by SOM in a very informative way (Fig. 18.5):

The Turkey and Cypriot groups are part of the dark European group who has built a monolithic niche for heroin. South Americans, instead, seem to be split into two groups: one specialized in cannabis and the other specialized as most important in cocaine trafficking.

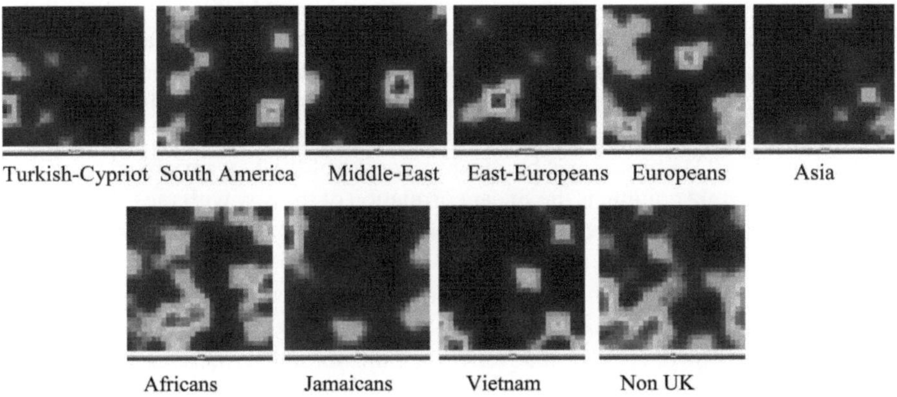

Fig. 18.5 Clusters according to nationality

Fig. 18.6 Gender: female

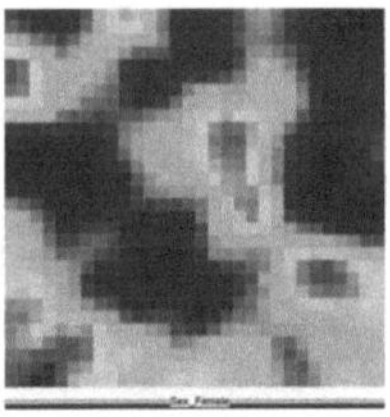

People from Middle East are concentrated in the MDMA group, while the East Europeans group has not a specific drug profile. The same is true for people from Asia and for the three groups of Africans.

European people are spread out on the map in many small clusters, while Jamaicans represent a strong group specialized in crack and heroin, two generic groups and one group more orientated toward cocaine.

Another novelty is represented by the distribution on the SOM map of the females arrested (Fig. 18.6):

Female are concentrated in four groups:

1. The first group in the northwest of the map: these females are Jamaican and specialize in crack (not necessary in heroin).
2. The second group is positioned in the southwest of the map and is fundamentally cocaine and cannabis oriented and composed of White and Dark Europeans.
3. The third group, in more or less the center of the map, is also specialized in cannabis but is completely different from the second one, because Arab and Asiatic women also belong to this group.
4. The fourth group in the southeast of the map seems to be composed only of dark European women who are completely dedicated to cocaine and MDMA.

Obviously, this analysis could go on in more detailed ways. For this reason, we prefer to show a series of tables in which each variable is shown with its most associated variables. The association index is the linear correlation of each variable with all the others in any specific codebook trained by the SOM. Naturally, all the SOM codebooks are generated in a nonlinear manner (see the SOM features above), and the linear correlation of the codebooks preserves the absolute nonlinearity of these findings. It is as to say as though we seek to establish a kind of the linear correlation that exists among a group of nonlinear dynamics.

Here are the SOM prototypes by means of some of the key variables. Note that the value in the right column is the linear correlation of the row with the table header, that is, the correlation between Tower_Hamlets and Male is 0.28 (Tables 18.1a and 18.1b):

UK citizens and people coming from Africa seem to segment the drug market of the prototypical male dealer. The prevalent male ethnic groups are Afro-Caribbean and Asiatic, and Tower Hamlet and Harrow are the boroughs where they work most of the time and in which they are arrested. Typically, they are young persons with an intensive criminal curriculum. At the moment of their arrest, they typically have a lot of cash, and this results in their making immediate bail. There is not a specific type of drug for the male prototype.

The female prototype is more complex. These females are preferentially Orientals or Dark Europeans. They are usually elderly (more than 51 years of age), very often are "clean," and cocaine is their favorite drug. Generally, they are non-UK people. They usually come from Europe, Vietnam, and South America (Tables 18.2a and 18.2b).

People coming from African countries are usually associated with one of the Afro-Caribbean group's expert for cannabis. People coming from Asia are too few in number to permit accurate inference, but in any case they seem to be slightly MDMA oriented (Tables 18.2c and 18.2d).

People coming from East of Europe are a nonspecific drug-oriented group, probably small and of recent immigration, whose unique common feature is to live in two boroughs: Ealing and Harrow.

What is named "Europeans" instead is a real group of Dark Europeans, mainly young females, strongly linked with persons from Turkey and Cyprus and with a strong inclination toward heroin trafficking (Tables 18.2e and 18.2f).

There are too few Irish people to allow some grounded inference. In any case, the few Irish individuals arrested present a very compact prototype: experienced delinquents, expert in different type of crimes, on bail at the time of the offence, and inclined to deal with MDMA (Tables 18.2g and 18.2h).

People coming from the Middle East, usually Arabs, are few and seem to have the inclination to hide their sex identity. Usually, they were arrested, because of cannabis possession, for the first time at a relatively older age (from 34 to 39 years old).

Table 18.1a

Male			
Places	Home borough	Tower_Hamlets	0.28
		Ealing	0.24
		Islington	0.22
	Borough of arrest	Arr_TowerHamlets	0.33
		Arr_Harrow	0.28
Anagraphic data	Gender		
	Nation group	Uk	0.37
		AFR	0.34
	Ethnic group	(EA3)_Afro-Caribbean	0.40
		(EA4)_Asia	0.28
	Age	Age(21–25)	0.47
Past criminal curriculum	Convictions	ConvictionsNumber(5–10)	0.44
		ConvictionsNumber(4)	0.34
	Offenses	OffensesNumber(6–10)	0.30
		OffensesNumber(11–20)	0.24
	Age at the first conviction	Off_FirstConvAge(19–21)	0.36
		Off_FirstConvAge(up-to-18)	0.30
	Offenses at the last conviction	Off_LastConvAge(1)	0.34
		Off_LastConvAge(2)	0.31
	Types and number of offenses	Off_Drug(3–5)	0.63
		Off_AgainstPerson(2)	0.52
		Off_OffensiveWeapons(1)	0.43
		Other_violent_offenses	0.41
		Offensive_Weapon_Offenses	0.40
		Off_Fraud(1)	0.39
		Off_RelatedToPolice(1)	0.38
		Off_OffensiveWeapons(2)	0.35
		Off_RelatedToPolice(2–5)	0.33
		Off_Sexual(1)	0.33
		Off_Total(6–10)	0.30
		Off_TheftKindred(1–5)	0.30
		NumOfArrests(3)	0.28
		Off_Drug(6–10)	0.27
		Off_Total(11–20)	0.24
		Off_AgainstProperty(1)	0.24
		Off_Fraud(2–5)	0.24
		Drug_trafficking_Offenses(1)	0.23
		Firearms_Offenses	0.23
		OffensesAgainstPerson	0.22

(continued)

Table 18.1a (continued)

Male			
Finding of the arrest	Types of seizure	Pounds(up5,000)	0.32
		NumOfCashSeizures(1)	0.32
		Pounds(up500)	0.30
		Pounds(over-5000)	0.23
		Pounds(up1,000)	0.23
	Drugs associated		
Tactics and arrest mode	Types of tactics	GenericTactic_Search_of_Person	0.53
		InOperation(0)	0.22
	Arrest mode	OnBailAtTimeOfOffence(1)	0.36
		OnBailAtTimeOfOffence(2)	0.30
		AR_OFF_Kidnapping_and_Abduction_offenses	0.28
		ArrMode_Other(1)	0.25

Table 18.1b

Female			
Places	Home borough	Bexley	0.37
		Lambeth	0.30
		Croydon	0.27
	Borough of arrest	Arr_Bexley	0.39
		Arr_Kingston_upon_Thames	0.35
Anagraphic data	Gender		
	Nation group	EU	0.51
		VTN	0.49
		NK	0.44
		SAME	0.30
	Ethnic group	(EA5)_Oriental	0.61
		(EA2)_Dark_European	0.46
		(EA1)_White _European	0.25
	Age	Age(UpTo18)	0.25
Past criminal curriculum	Convictions	ConvictionsNumber(0)	0.47
		ConvictionsNumber(1)	0.30
	Offenses	OffensesNumber(0)	0.48
		OffensesNumber(1)	0.28
		OffensesNumber(2)	0.27
	Age at the first conviction	Off_FirstConvAge(over-51)	0.60
		Off_FirstConvAge(34–39)	0.47
		Off_FirstConvAge(28–33)	0.44
		Off_FirstConvAge(40–45)	0.41
		Off_FirstConvAge(46–51)	0.39
		Off_FirstConvAge(22–27)	0.30

(continued)

Table 18.1b (continued)

Female			
	Offenses at the last conviction	Off_LastConvAge(0)	0.47
	Types and number of offenses	Off_OffensiveWeapons (0)	0.52
		Other_offenses	0.51
		Off_Drug (0)	0.50
		Off_Drug (1)	0.47
		Off_RelatedToPolice(0)	0.40
		Off_AgainstPerson (0)	0.34
		Off_TheftKindred (0)	0.32
		Off_Fraud(0)	0.31
		Off_Total (1)	0.28
		Off_Total (2)	0.26
		Off_Drug(1)	0.22
Finding of the arrest	Types of seizure	NumOfCashSeizures(0)	0.25
		NumOfDrugSeizures(0)	0.23
	Drugs associated	Cocaine	0.40
Tactics and arrest mode	Types of tactics	GenericTactic_Controlled_Delivery	0.28
		Non-Law_Enforcement_Agent	0.26
		InOperation(1)	0.25
		NumOfTactics(1)	0.21
	Arrest mode	OnBailAtTimeOfOffence(0)	0.47
		ViolentOnArrest(0)	0.33
		ArrMode_Given_into_custody	0.26
		ArrMode_Result_of_Enquiries(2)	0.23
		ArrMode_Direct(0)	0.22

Table 18.2a

Africa nation group			
Places	Home borough	Newham	0.39
		Redbridge	0.29
		Hammersmith_and_Fulham	0.28
		Walrham_Forest	0.26
	Borough of arrest		
Anagraphic data	Gender	Sex_Male	0.34
	Nation group	NK	0.32
	Ethic group	(EA3)_Afro-Caribbean	0.42
	Age		
Past criminal curriculum	Convictions		
	Offenses		
	Age at the first conviction	Off_LastConvAge(2)	
	Offenses at the last conviction	Off_LastConvAge(0)	0.28
	Types and number of offenses	NumOfDrugSeizures(2)	0.36
Findings of the arrest	Types of seizure		
	Drugs associated	Cannabis	0.25
Tactics and arrest mode	Type of tactics		
	Arrest mode	ArrMode_Other(1)	0.34

Table 18.2b

Asia nation group			
Places	Home borough	Hammeesmith_and_Fulham	0.45
		Hounslow	0.37
	Borough of arrest	Arr_Hammersmith_and_Fulham	0.56
		Arr_Hounslow	0.40
Anagraphic data	Gender		
	Nation group		
	Ethic group		
	Age	Age(25–35)	0.31
Past criminal curriculum	Convictions	ConvictionsNumber(1)	0.36
	Offenses	OffensesNumber(1)	0.45
	Age at the first conviction		
	Offenses at the last conviction		
	Types and number of offence	Off_Total(1)	0.45
		Off_Drug(1)	0.39
Findings of the arrest	Types of seizure	NumOf CashSeizures(0)	0.25
	Drugs associated	MDMA	0.14
Tactics and arrest mode	Type of tactics		
	Arrest mode		

Table 18.2c

East Europe nation group			
Places	Home borough	Ealing	0.48
		Harrow	0.43
Anagraphic data	Borough of arrest		
	Gender		
	Nation group		
	Ethic group		
	Age		
Past criminal curriculum	Convictions	ConvictionsNumber(3)	0.33
	Offenses	OffensesNumber(3–5)	0.52
	Age at the first conviction	Off_FirstConvAge(19–21)	0.42
	Offenses at the last conviction		
	Types and number of offence	NumOfArrests(1)	0.38
		Off_RelatedToPolice(0)	0.34
		Off_Total(3–5)	0.52
Findings of the arrest	Types of seizure		
	Drugs associated	Heroin	0.01
Tactics and Arrest Mode	Type of tactics	InOperation(1)	0.37
	Arrest mode	ArrMode_Direct(1)	0.43

Table 18.2d

Europe nation group			
Places	Home borough		
Anagraphic data	Gender	Sex_Female	0.51
	Nation group	TU-CY	0.39
	Ethic group	(EA2)_Dark_European	0.64
	Age	Age(UpTo18)	0.44
Past criminal curriculum	Convictions		
	Offenses	OffensesNumber(1)	0.43
	Age at the first conviction	Off_FirstConvAge(28–33)	0.43
		Off_FirstConvAge(34–39)	0.34
	Offenses at the last conviction		
	Types and number of offence	Off_Drug(1)	0.32
		Off_Total(1)	0.43
		Off_OffensiveWeapons(0)	0.41
Findings of the arrest	Types of seizure		
	Drugs associated	Heroin	0.26
Tactics and arrest mode	Type of tactics		
	Arrest mode		

Table 18.2e

Ireland nation group			
Places	Home borough		
	Borough of arrest		
Anagraphic Data	Gender		
	Nation group		
	Ethic group	(EA1)_White_European	0.42
	Age		
Past criminal curriculum	Convictions		
	Offenses	OffensesNumber(20–50)	0.46
	Age at the first conviction		
	Offenses at the last conviction	ConvictionNumber(11–20)	0.49
	Types and number of offence	Off_TheftKindred(6–10)	0.59
		Off_Total(20–50)	0.47
		Off_RelatedToPolice(2–5)	0.42
		Off_TheftKindred(11–20)	0.40
		Off_AgainstProperty(1)	0.42
		Off_Fraud(1)	0.47
Findings of the arrest	Types of seizure		
	Drugs associated	MDMA	0.24
Tactics and arrest mode	Type of tactics		
	Arrest mode	OnBailAtTimeOfOffence(1)	0.43

Table 18.2f

Jamaica nation group			
Places	Home borough	Harigey	0.65
	Borough of arrest	Arr_Hackney	0.68
Anagraphic data	Gender		
	Nation group		
	Ethic group	(EA3)_Afro_Caribbean	0.58
	Age		
Past criminal curriculum	Convictions		
	Offenses		
	Age at the first conviction	Off_FirstConvAge(28–33)	0.63
	Offenses at the last conviction		
	Types and number of offence	NumOfArrests(over-3)	0.58
Findings of the arrest	Types of seizure		
	Drugs associated	Crack	0.60
		Heroin	0.32
Tactics and arrest mode	Type of tactics	ArrMode_Result_of_Enquiries(0ver-2)	0.60
		NumOfTactics(0ver-3)	0.60
		NumOfTacticSequences(3–5)	0.57
	Arrest mode		

Table 18.2g

Middle East nation group			
Places	Home borough	Kingston_upon_Thames	0.45
		Sutton	0.31
		Brent	0.32
	Borough of arrest		
Anagraphic data	Gender	Sex_notknown	0.33
	Nation group		
	Ethic group	(EA6)_Arab	0.59
	Age		
Past criminal curriculum	Convictions		
	offenses		
	Age at the first conviction	Off_FirstConvAge(34–39)	0.37
	Offenses at the last conviction		
	Types and number of offence	Drug_Possession_Offenses	0.39
		NumOfArrests(2)	0.28
Findings of the arrest	Types of seizure		
	Drugs associated	Cannabis	0.29
		MDMA	0.09
Tactics and arrest mode	Type of tactics		
	Arrest mode	ArrMode_Direct(2)	0.31

Table 18.2h

South America nation group			
Places	Home borough	Kensington_and_Chelses	0.33
	Borough of arrest	Arr_Greenwich	0.44
		Arr_Bexley	0.43
		Arr_Kensington_and_Chelsea	0.39
		Arr_PL_NA	0.33
Anagraphic data	Gender		
	Nation group	VTN	0.43
	Ethic group		
	Age		
Past criminal curriculum	Convictions		
	Offenses		
	Age at the first conviction		
	Offenses at the last conviction		
	Types and number of offence	Off_FirstConvAge(40–45)	0.30
Findings of the arrest	Types of seizure	NumOfDrugSeizures(0ver-5)	0.31
	Drugs associated	Cannabis	0.53
		Heroin	0.27
Tactics and arrest mode	Type of tactics	InOperation(1)	0.37
	Arrest mode		

Table 18.2i

Turkey and Cyprus nation group (TU-CY)			
Places	Home borough		
	Borough of arrest		
Anagraphic data	Gender		
	Nation group	EU	0.39
		NK	0.38
	Ethic group	(EA2)_Dark_European	0.55
	Age		
Past criminal curriculum	Convictions	ConvictionsNumber(0)	0.43
	Offenses	OffensesNumber(0)	0.43
	Age at the first conviction	Off_FirstConvAge(over-51)	0.48
		Off_FirstConvAge(34–39)	0.43
	Offenses at the last conviction	Off_LastConvAge(0)	0.40
	Types and number of offenses	Off_Drug(0)	0.44
		Off_Total(0)	0.43
Findings of the arrest	Type of seizure		
	Drugs associated	Heroin	0.29
Tactics and arrest mode	Type of tactics		
	Arrest mode		

Table 18.2j

Vietnam nation group (VTN)			
Places	Home borough	Bexley	0.50
	Borough of arrest	Arr_Bexley	0.64
Anagraphic data	Gender	Sex_Female	0.49
	Nation group		
	Ethic group	(EA5)_Oriental	0.75
	Age		
Past criminal curriculum	Convictions		
	Offenses		
	Age at the first conviction	Off_FirstConvAge(over-51)	0.57
		Off_FirstConvAge(40–45)	0.53
		Off_FirstConvAge(46–51)	0.45
		Off_FirstConvAge(34–39)	0.44
	Offenses at the last conviction		
	Types and number of offenses	Other_offenses	0.62
		Off_Drug(0)	0.44
Findings of the arrest	Type of seizure		
	Drugs associated	Cocaine	0.29
Tactics and arrest mode	Type of tactics		
	Arrest mode		

South Americans are more defined: usually, they live in Kensington but very often are arrested outside of Kensington. They are very associated with a Vietnamese group and are specialized in large quantities of cocaine. Also, their first conviction is usually at an older age (around 40–45 years old). Sometimes, they are also involved in heroin trafficking (Tables 18.2i and 18.2j).

Aged Turkish and Cypriots seem to be the head of the young European female group trafficking in heroin. Vietnamese people are a group of aged women working in cocaine trafficking associated with a group of South American persons (see above) (Tables 18.3a and 18.3b).

Most of White Europeans arrested live in Sutton and in Barking and Dagenham, and they are mainly UK citizens with a high number of convictions. Typically, they have committed one or more property offenses. Their favorite drug is MDMA, and cocaine may be their second choice.

Dark Europeans are clustered as European females, typically coming from Turkey and Cyprus, without a criminal background and dedicated to cocaine and sometimes to heroin and cannabis (Tables 18.3c and 18.3d).

Afro-Caribbean are described as Jamaican and African males, often arrested in Camden and Hackney. Most are persons with over three arrests, mainly because of crack but also because of heroin. They are reluctant to deal with cannabis, cocaine, and MDMA. Police appear to need more than three tactics to catch them, and each arrest typically happens in direct and violent ways.

Table 18.3a

White Europeans ethnic group ((EA1)_White_European)			
Places	Home borough	Sutton	0.46
		Barking_and_Dagenham	0.44
	Borough of arrest		
Anagraphic data	Gender		
	Nation group	UK	0.53
	Ethnic group		
	Age		
Past criminal curriculum	Convictions	Convictions Number (11–20)	0.43
	Offenses		
	Age at the first conviction		
	Offenses at the last conviction		
	Types and number of offenses	Off_Against Property (1)	0.54
		Off_Theft Kindred (6–10)	0.46
		Off_Against Property (2)	0.45
		Other_offenses	0.45
Finding of the arrest	Type of seizure		
	Drugs associated	MDMA	0.27
		Cocaine	0.13
		Cannabis	0.10
		Heroin	−0.19
		Crack	−0.24
Tactics and arrest mode	Type of tactics	Other_Law_Enforcement_Agent	0.55
	Arrest mode		

Asian persons are arrested generally in two boroughs, Tower Hamlets and Ealing, where they also live. The age of their first conviction is when they are between 19 and 21 years old. They usually deal with cannabis. A couple of police tactics (typically "search a person") are needed to catch them (Tables 18.3e and 18.3f).

The ethnic group named "Oriental" is composed mainly of old Vietnamese females, and there is a suggestion of their involvement in cocaine traffic with some inclination toward cannabis and MDMA. The Arabs arrested have the same profile of the people coming from Middle East, described above.

It is useful to organize the SOM results from a perspective of the types of drugs. This view will illustrate the main profile of any drug in terms of:

1. Places where the persons are arrested and/or where they live
2. Basic anagraphic data of the arrested persons
3. Their fundamental criminal record
4. Findings of their arrest
5. Organization and modalities of their arrest

Table 18.3b

Dark Europeans ethnic group ((EA2)_Dark_European)			
Places	Home borough		
	Borough of arrest		
Anagraphic data	Gender	Sex_Female	0.46
	Nation group	EU	0.64
		TU-CY	0.55
	Ethnic group		
	Age		
Past criminal curriculum	Convictions		
	Offenses		
	Age at the first conviction		
	Offenses at the last conviction		
	Types and number of offenses	Off_Fraud(0)	0.52
		Off_Against Person(0)	0.51
		Off_Related To Police (0)	0.46
		Off_Sexual (0)	0.46
		Off_Offensive Weapons (0)	0.46
		OFF_AGAINST PROPERTY(0)	0.46
Finding of the arrest	Type of seizure		
	Drugs associated	Cocaine	0.27
		Heroin	0.18
		Cannabis	0.18
		MDMA	0.04
		Crack	−0.01
Tactics and arrest mode	Type of tactics		
	Arrest mode		

The following tables, consequently, should make explicit the *prototype* of the persons arrested in London from the *type of drug* viewpoint.

Persons Arrested Because of Cannabis

Tables 18.4a,18.4b, 18.4c, 18.4d, and 18.4e synthesizes the profile of persons arrested because of *cannabis* (but not limited only to cannabis). Many of these people generally live in *Sutton* and in *Richmond upon Thames* and in these boroughs are usually arrested. Gender is not meaningful, but many of them are people coming from the *Middle East and Africa*. From an ethnic point of view, they are often *Asiatic and Arabs*. Most of them are young (*from 21 to 25 years old*) with a *robust curriculum in offenses and convictions*, not only linked to drug problems. During the arrest, more seizures very often are executed, and sometimes, cannabis is associated *with MDMA*. The more effective tactics for this arrest is a *"generic search of*

Table 18.3c

Afro-Caribbean ethnic group ((EA3)_Afro-Caribbean)			
Places	Home borough		
	Borough of arrest	Arr_Camden	0.39
		Arr_Hackney	0.38
Anagraphic data	Gender	Sex_Male	0.40
	Nation group	JAM	0.58
		AFR	0.42
	Ethnic group		
	Age		
Past criminal curriculum	Convictions		
	Offenses		
	Age at the first conviction		
	Offenses at the last conviction		
	Types and number of offenses	Num Of Arrests (over-3)	0.40
Finding of the arrest	Type of seizure		
	Drugs associated	Crack	0.37
		Heroin	0.20
		Cannabis	−0.19
		Cocaine	−0.21
		MDMA	−0.28
Tactics and arrest mode	Type of tactics	Num Of Tactics (over-3)	0.43
	Arrest mode	Violent On Arrest (1)	0.45
		ArrMode_Direcct (3–5)	0.38

premises," a *"generic search of person,"* and a *"generic controlled delivery."* These tactics often have to be repeatedly activated as a result of many i*nquires*.

Persons Arrested Because of Cocaine

Tables 18.5a, 18.5b, 18.5c, 18.5d, and 18.5e synthesizes the profile of people arrested because of, but not limited to, *cocaine*. Mainly, these persons live in *Bexley, Bromley, Haringey*, and *Croydon*. Most of them are arrested in their home borough, but there are some exceptions: many people are arrested in *Kingston upon Thames*, but many also come from outside the area. There is some specific link between *women* and cocaine. The same specific link is present with people coming from *South America, Vietnam,* and *Jamaica*. The main ethnic groups of the people arrested because of cocaine seem to be constituted of *Dark Europeans, Orientals,* and *White Europeans*. Their age is between *25 and over 45*, and their criminal *curriculum tends to be clean*. Their first arrest is usually *one offence in elderly age* (beyond 40s). Their arrest is correlated to *many drug seizures and cash seizures. Heroin, crack, and MDMA* are associated with cocaine seizures. *Many tactics and*

Table 18.3d

Asiatic ethnic group ((EA4)_Asia)			
Places	Home borough	Ealing	0.52
		Tower_Hamlets	0.47
		Hounslow	0.46
	Borough of arrest	Arr_TowerHamlets	0.46
		Arr_Ealing	0.40
Anagraphic data	Gender		
	Nation group		
	Ethnic group		
	Age		
Past criminal curriculum	Convictions		
	Offenses		
	Age at the first conviction	Off_First Conv Age (19–21)	0.41
	Offenses at the last conviction		
	Types and number of offenses		
Finding of the arrest	Type of seizure		
	Drugs associated	Cannabis	0.18
		MDMA	0.03
		Cocaine	−0.15
		Heroin	−0.23
		Crack	−0.28
Tactics and arrest mode	Type of tactics	Generic Tactic_Search_of_Person	0.41
		Num Of Tactics (2)	0.39
	Arrest mode	ArrMode_Result_of_Enquiries (0)	0.39

tactics sequences are needed to find cocaine, and very often, they are the result of many *inquires*. "*Controlled delivery*" and "*covert purchase*" are shown to be the more effective tactics to discovering cocaine.

Persons Arrested Because of Heroin

Table 18.4c synthesizes the profile of people arrested because of, but not limited to, *heroin*. There is not a specific borough where they are arrested. Typically are groups of *Jamaican, Turkish,* and *Cypriots*.

These people seem to cluster into two subgroups: the first one is composed of *young people*, without relevant past offenses, while the second group is composed of *adult and very expert delinquents* with an impressive record in the field that addresses every kind of crime in robust quantity. In fact, their arrest is always associated with rich seizures of drugs.

Table 18.3e

Oriental ethnic group ((EA5)_Oriental)			
Places	Home borough		
	Borough of arrest		
Anagraphic data	Gender	Sex_Female	0.61
	Nation group	VTN	0.75
	Ethnic group		
	Age		
Past criminal curriculum	Convictions	Convictions Number (0)	0.55
	Offenses	Offenses Number (0)	0.55
	Age at the first conviction	Off_First Conv Age (over-51)	0.67
	Offenses at the last conviction	Off_Last Conv Age (0)	0.52
	Types and number of offenses	Other_offenses	0.59
		Off_Drug(0)	0.56
		Off_Total(0)	0.54
Finding of the arrest	Type of seizure		
	Drugs associated	Cocaine	0.24
		Cannabis	0.13
		MDMA	0.12
		Heroin	−0.04
		Crack	−0.21
Tactics and arrest mode	Type of tactics		
	Arrest mode		

Table 18.3f

Arab ethnic group ((EA6)_Arab)			
Places	Home borough	Kingston_upon_Thames	0.36
		Harrow	0.34
	Borough of arrest	Arr_Richmond_upon_Thames	0.33
		Arr_Kingston_upon_Thames	0.33
Anagraphic data	Gender		
	Nation group	ME	0.59
	Ethnic group		
	Age	Age (18–21)	0.39
Past criminal curriculum	Convictions	Convictions Number (3)	0.35
	Offenses		
	Age at the first conviction	Off_First Conv Age (34–39)	0.33
	Offenses at the last conviction		
	Types and number of offenses		
Finding of the arrest	Type of seizure		
	Drugs associated	Cannabis	0.17
		MDMA	0.10
		Cocaine	0.04
		Heroin	−0.03
		Crack	−0.07
Tactics and arrest mode	Type of tactics	Non-Law_Enforcement_Agent	0.32
	Arrest mode		

Table 18.4a

Cannabis			
Places	Home borough	Sutton	0.43
		Richmond_upon_Thames	0.41
		Hounslow	0.26
		Bexley	0.25
		Waltham_Forest	0.24
		Hammersmith_and_Fulham	0.22
		Barking_and_Dagenham	0.21
		Newham	0.19
		Bamet	0.18
		Brent	0.18
		Havering	0.17
		Ealing	0.16
		Lewisham	0.14
		Harrow	0.11
	Borough of arrest	Arr_Richmond_upon_Thames	0.43
		Arr_Sutton	0.33
		Arr_Place_Not Available	0.31
		Arr_Harrow	0.23
		Arr_Newham	0.22
		Arr_Redbridge	0.21
		Arr_Hammersmith_and_Fulham	0.18
		Arr_Bexley	0.15
		Arr_Kingston_upon_Thames	0.12
		Arr_Waltham Forest	0.12
		Arr_Merton	0.12
Anagraphic data	Gender	Sex_Male	0.01
		Sex_Female	−0.03
	Nation group	ME	0.29
		AFR	0.25
		VTN	0.13
		NK	0.10
	Ethnic group	(EA4)_Asia	0.18
		(EA6)_Arab	0.17
		(EA5)_Oriental	0.13
		(EA1)_White_European	0.10
	Age	Age (21–25)	0.29
		Age (18–21)	0.13
Past criminal curriculum	Convictions	Convictions Number (3)	0.39
		Convictions Number (4)	0.36
		Convictions Number (2)	0.13
	Offenses	Offenses Number (6–10)	0.30
		Offenses Number (3–5)	0.25
	Age at the first conviction	Off_First Conv Age (19–21)	0.12
		Off_First Conv Age (over-51)	0.10

(continued)

Table 18.4a (continued)

Cannabis			
	Years from the last conviction	Off_Last Conv Age(6–10)	0.27
		Off_Last Conv Age(2)	0.21
	Types and number of offenses	AR_OFF_Other_Drug_Offenses	0.55
		Other_offenses	0.48
		Drug_Possession_Offenses	0.36
		Offenses Against Person	0.26
		Off_Total (3–5)	0.25
		Drug_trafficking_Offenses(1)	0.25
		Theft_and_Kindred_Offenses	0.24
		Off_Related To Police(0)	0.24
		Firearms_Offenses	0.22
		Off_Drug(2)	0.22
		Off_Theft Kindred (1–5)	0.20
		Offensive_Weapon_Offenses	0.16
		Off_Drug(1)	0.14
Findings of the arrest	Arrests	Num Of Arrests (2)	0.34
		Number Of Arrests(3)	0.15
	Type of seizure	Num Of Drug Seizures (2)	0.47
		Num Of Cash Seizures (over-1)	0.17
		Num Of Drug Seizures (3–5)	0.12
	Drugs associated	MDMA	0.18
Tactics and arrest mode	Type of tactics	Generic Tactic_Search_of_Premises	0.52
		Num Of Tactic Sequences(1)	0.47
		Generic Tactic_Search_of_Person	0.35
		ArrMode_Result_of_Enquiries(2)	0.29
		Generic Tactic_Controlled_Delivery	0.21
		Num Of Tactics (2)	0.21
		Num Of Tactics (3)	0.16
		ArrMode_Result_of_Enquiries (1)	0.15
		Num Of Tactics (1)	0.14
	Arrest mode	ArrMode_Other (1)	0.18
		ArrMode_Other (2)	0.18
		On Bail At Time Of Offence (2)	0.17
		AR_OFF_Kidnapping_and_Abduction_offenses	0.16
		Violent On Arrest (more-1)	0.14
		ArrMode_Direct(2)	0.37

These people associate *heroin* traffic with *crack* and also with *cocaine*. Many tactics and tactics sequences, as *covert purchase*, are needed to arrest them, often in violent mode and at the end of many inquires.

Persons Arrested Because of Crack

Table 18.4d synthesizes the profile of people mainly arrested because of *crack*. The crack prototype seems to be much defined: it is an *Afro-Caribbean, Jamaican, and*

Table 18.4b

Cocaine			
Places	Home borough	Bexley	0.35
		Bromley	0.23
		Haringey	0.22
		Croydon	0.21
		Lambeth	0.18
		Kensington_and_Chelsea	0.16
		Borough not available	0.15
		Sutton	0.10
	Borough of arrest	Arr_Bexley	0.46
		Arr_Kingston_upon_Thames	0.39
		Arr_Bromley	0.35
		Arr_Haringey	0.33
		Arr_Kensington_and_Chelsea	0.32
		Arrest borough not available	0.30
		Arr_Hammersmith_and_Fulham	0.27
		Arr_Islington	0.25
		Arr_Westminster	0.25
		Arr_Southwark	0.21
		Arr_Lambeth	0.19
		Arr_Sutton	0.18
		Arr_Greenwich	0.17
		Arr_Brent	0.17
		Arr_Barnet	0.14
		Arr_Waltham Forest	0.13
		Arr_Hackney	0.12
		Arr_Lewisham	0.11
Anagraphic data	Gender	Sex_Female	0.40
		Sex_Male	−0.39
	Nation group	SAME	0.53
		VTN	0.29
		JAM	0.21
		EU	0.19
	Ethnic group	(EA2)_Dark_European	0.27
		(EA5)_Oriental	0.24
		(EA1)_White_European	0.13
	Age	Age (25–35)	0.24
		Age (35–45)	0.20
		Age(Over 45)	0.17
Past criminal curriculum	Convictions	Convictions Number (1)	0.28
		No_Convictions Number	0.18
	Offenses	Offenses Number (1)	0.28
		No_Offenses Number	0.19
		Offenses Number (20–50)	0.15

(continued)

Table 18.4b (continued)

Cocaine			
	Age at the first conviction	Off_First Conv Age (40–45)	0.39
		Off_First Conv Age (46–51)	0.34
		Off_First Conv Age (over-51)	0.30
		Off_First Conv Age (34–39)	0.25
		Off_First Conv Age (28–33)	0.19
		Off_First Conv Age (22–27)	0.17
	Years from the last conviction	No_Years From_Last Conv Age	0.16
	Types and number of offenses	Drug_trafficking_Offenses(over-5)	0.33
		Num Of Arrests (2)	0.28
		Other_offenses	0.28
		Off_Total (1)	0.28
		Num Of Arrests (over-3)	0.25
		Drug_Possession_Offenses	0.25
		No_Offensive Weapons	0.23
		Off_Drug(over-10)	0.21
		Theft_and_Kindred_Offenses	0.19
		Off_Total(20–50)	0.14
		Off_Drug(1)	0.13
		Off_Theft Kindred (11–20)	0.12
		Off_Fraud(over-5)	0.12
		Off_Against Property (more)	0.11
		Drug_trafficking_Offenses(2–5)	0.10
Findings of the arrest	Type of seizure	Num Of Drug Seizures (over-5)	0.44
		Num Of Cash Seizures (over-1)	0.14
		Pounds (up 1,000)	0.12
	Drugs associated	Heroin	0.32
		Crack	0.25
		MDMA	0.20
Tactics and arrest mode	Type of tactics	Generic Tactic_Controlled_Delivery	0.36
		Num Of Tactic Sequences(over-5)	0.34
		In Operation (over-5)	0.34
		Generic Tactic_Convert_Purchase	0.29
		In Operation (1)	0.28
		Num Of Tactics (over-3)	0.27
		ArrMode_Result_of_Enquiries (2)	0.22
		ArrMode_Result_of_Enquiries (over-2)	0.15
	Arrest mode	ArrMode_Direct(over-5)	0.33
		ArrMode_Direct (2)	0.29
		ArrMode_Other (over-2)	0.24
		ArrMode_Other (2)	0.19
		Violent On Arrest (more-1)	0.12

Table 18.4c

Heroin			
Places	Home borough	Kensington_and_Chelsea	0.64
		Haringey	0.51
		Bromley	0.35
		Southwark	0.17
		Camden	0.12
	Borough of arrest	Arr_Kensington_and_Chelsea	0.79
		Arr_Bromley	0.78
		Arr_Haringey	0.76
		Arr_Westminster	0.54
		Arr_Greenwich	0.52
		Arr_Lewisham	0.46
		Arr_Southwark	0.45
		Arr_Islington	0.40
		Arr_Barnet	0.38
		Arr_Hackney	0.33
		Arr_Camden	0.32
		Arr_Brent	0.28
Anagraphic data	Gender	Sex_notknown	0.16
		Sex_Female	0.08
		Sex_Male	−0.13
	Nation group	JAM	0.32
		TU-CY	0.29
		SAME	0.27
		EU	0.26
	Ethnic group	(EA3)_Afro-Caribbean	0.20
		(EA2)_Dark_European	0.18
	Age	Age (35–45)	0.27
		Age (25–35)	0.13
		Age (UpTo 18)	0.13
Past criminal curriculum	Convictions	Convictions Number (0)	0.17
		Convictions Number (over-20)	0.17
	Offenses	Offenses Number (over-50)	0.22
		Offenses Number (0)	0.17
	Age at the first conviction	Off_First Conv Age (34–39)	0.29
		Off_First Conv Age (28–33)	0.23
		Off_First Conv Age (over-51)	0.15
	Offenses at the last conviction	No_Years From_Last Conv Age	0.17
	Types and number of offenses	Drug_trafficking_Offenses(over-5)	0.79
		Off_Drug (over-10)	0.57
		Off_Drug (6–10)	0.27
		Off_Offensive Weapons (over-2)	0.24
		Off_Total (over-50)	0.22
		Off_Related To Police (over-5)	0.22
		Off_Against Property (more)	0.21
		Drug_trafficking_Offenses (2–5)	0.21

(continued)

Table 18.4c (continued)

Heroin			
		Off_Theft Kindred (over-20)	0.19
		Off_Against Person (1)	0.19
		Off_Against Person (3–5)	0.16
		Off_Sexual (over-2)	0.10
Findings of the arrest	Type of seizure	Num Of Drug Seizures (over-5)	0.81
		Num Of Drug Seizures (3–5)	0.14
	Drugs associated	Crack	0.73
		Cocaine	0.32
Tactics and arrest mode	Type of tactics	In Operation (over-5)	0.79
		Num Of Tactic Sequences(over-5)	0.79
		Generic Tactic_Convert_Purchase	0.79
		Num Of Arrests (over-3)	0.73
		Num Of Tactics (over-3)	0.66
		ArrMode_Result_of_Enquiries (over-2)	0.55
		In Operation (2–5)	0.32
		Num Of Tactic Sequences (3–5)	0.28
		Num Of Tactic Sequences (2)	0.24
		Generic Tactic_Other	0.23
	Arrest mode	ArrMode_Direct(over-5)	0.78
		ArrMode_Other (over-2)	0.71
		On Bail At Time Of Offence (over-2)	0.25
		Violent On Arrest (1)	0.18
		ArrMode_Direct (3–5)	0.13

between 25 and 35 years old (sometimes between 35 and 45 years old). These people seem divided into two subgroups: the first one whose individuals generally have collected *more than 20 convictions* and the second one with persons who have typically two convictions. The first group should also have *committed more than 50 offenses, in every kind of known crime*. The *average age* of their first crime is often late, *between 25 and 33 years old*. Most of them were found with more *than three and/or five doses* of crack, often associated with *heroin* and only sometimes with cocaine. *Covert purchase* is the most effective tactic used by the police with this kind of drug crime, and the *direct arrest* is one of the common ways of capturing these people, very often in *violent mode*.

Persons Arrested Because of MDMA

Table 18.4e synthesizes the profile of people mainly arrested because of *MDMA*. The majority of these people were arrested in the *South of London*. They are *White Europeans*, composed of a small group of Irish and more generally by *UK citizens with 1 or 2 offenses* in their background. These people sometimes are *young*

Table 18.4d

Crack			
Places	Home borough	Kensington_and_Chelsea	0.64
		Haringey	0.59
		Bromley	0.34
		Barnet	0.29
		Southwark	0.13
		Kingston_upon_Thames	0.10
		Borough not available	0.10
	Borough of arrest	Arr_Bromley	0.90
		Arr_Haringey	0.87
		Arr_Kensington_and_Chelsea	0.82
		Arr_Lewisham	0.77
		Arr_Hackney	0.73
		Arr_Barnet	0.64
		Arr_Islington	0.64
		Arr_Westminster	0.59
		Arr_Southwark	0.54
		Arr_Camden	0.46
		Arr_Greenwich	0.41
		Arr_Brent	0.23
		Arr_Croydon	0.17
		Arrest borough not available	0.16
		Arr_Lambeth	0.11
Anagraphic data	Gender	Sex_notknown	0.16
		Sex_Female	0.00
		Sex_Male	−0.05
	Nation group	JAM	0.60
		SAME	0.14
	Ethnic group	(EA3)_Afro-Caribbean	0.37
	Age	Age (25–35)	0.17
		Age (35–45)	0.13
Past criminal curriculum	Convictions	Convictions Number (2)	0.13
		Convictions Number (over-20)	0.13
	Offenses	Offenses Number (over-50)	0.17
	Age at the first conviction	Off_First Conv Age (28–33)	0.25
		Off_First Conv Age (34–39)	0.13
	Offenses at the last conviction	Off_Last Conv Age (1)	0.26
	Types and number of offenses	Drug_trafficking_Offenses(over-5)	0.91
		Off_Drug (over-10)	0.51
		Off_Drug (6–10)	0.44
		Drug_trafficking_Offenses (2–5)	0.40
		Off_Against Person (1)	0.22
		Off_Offensive Weapons (over-2)	0.20
		Off_Total (over-50)	0.17
		Off_Related To Police (over-5)	0.17
		Off_Against Property (more)	0.16
		Off_Against Person (3–5)	0.14

(continued)

Table 18.4d (continued)

Crack			
		Off_Theft Kindred (over-20)	0.14
		Off_Sexual (over-2)	0.12
		Off_Total (11–20)	0.11
Findings of the arrest	Type of seizure	Num Of Drug Seizures (over-5)	0.90
		Num Of Drug Seizures (3–5)	0.26
	Drugs associated	Heroin	0.73
		Cocaine	0.25
Tactics and arrest mode	Type of tactics	Num Of Arrests (over-3)	0.96
		Generic Tactic_Convert_Purchase	0.96
		Num Of Tactics (over-3)	0.91
		In Operation (over-5)	0.90
		Num Of Tactic Sequences (over-5)	0.90
		ArrMode_Result_of_Enquiries (over-2)	0.90
		Num Of Tactic Sequences (3–5)	0.57
		In Operation (2–5)	0.54
		Generic Tactic_Other	0.38
		Num Of Tactic Sequences (2)	0.28
	Arrest mode	ArrMode_Direct(over-5)	0.91
		ArrMode_Other (over-2)	0.85
		On Bail At Time Of Offence (over-2)	0.39
		Violent On Arrest (1)	0.32
		ArrMode_Direct (3–5)	0.28

(22–27 years old) and sometimes old (46–51 years old) and typically are expert in *drug offenses and theft and kindred offenses*. During seizure, a moderate quantity of *cash money is found, generally pounds*. Other drugs (cocaine and cannabis) are only occasionally associated. They are arrested usually *in violent and in direct ways by non-law enforcement agents*.

18.3 Explorative Analysis Using Auto-Contractive Maps

The SOM is a very well-known and effective ANN suitable for data mining. But it is not the only one. Auto-Contractive Map (Auto-CM), for instance, was recently shown to be a very robust and powerful ANN to discover hidden links within large dataset (see the Chap. 15 dedicated to Auto-CM theory and its equations in this book and Buscema 2007a, b; Buscema et al. 2008a, b; Buscema and Grossi 2008; Licastro et al. 2008).

Table 18.4e

MDMA			
Places	Home borough	Sutton	0.45
		Bexley	0.38
		Brent	0.33
		Borough not available	0.32
		Hounslow	0.29
		Merton	0.24
		Croydon	0.16
	Borough of arrest	Arr_Sutton	0.47
		Arr_Merton	0.44
		Arrest borough not available	0.37
		Arr_Wandsworth	0.33
		Arr_Bexley	0.29
		Arr_Kingston_upon_Thames	0.29
		Arr_Ealing	0.28
		Arr_Brent	0.26
		Arr_Enfield	0.25
		Arr_Lambeth	0.23
		Arr_Hounslow	0.18
		Arr_Waltham Forest	0.17
		Arr_Hammersmith_and_Fulham	0.15
		Arr_Croydon	0.14
		Arr_Newham	0.11
Anagraphic data	Gender	Sex_notknown	0.30
		Sex_Female	0.06
		Sex_Male	−0.15
	Nation group	IRE	0.24
		UK	0.15
		ASIA	0.14
		VTN	0.10
	Ethnic group	(EA1)_White_European	0.27
		(EA5)_Oriental	0.12
		(EA6)_Arab	0.10
	Age	Age (25–35)	0.22
Past criminal curriculum	Convictions	Convictions Number (1)	0.34
	Offenses	Offenses Number (1)	0.27
		Offenses Number (2)	0.12
	Age at the first conviction	Off_First Conv Age (22–27)	0.29
		Off_First Conv Age (46–51)	0.28
	Offenses at the last conviction	Off_Last Conv Age (1)	0.14
	Types and number of offenses	Drug_Possession_Offenses	0.58
		Theft_and_Kindred_Offenses	0.45
		Off_Total (1)	0.26
		Off_Drug (1)	0.23
		Offensive_Weapon_Offenses	0.22
		Drug_trafficking_Offenses(2–5)	0.19
		No_Offensive Weapons	0.19

(continued)

Table 18.4e (continued)

MDMA			
		AR_OFF_Kidnapping_and_ Abduction_offenses	0.16
		AR_OFF_Other_Drug_Offenses	0.15
		No_Theft Kindred	0.15
		Off_Total (2)	0.12
		Firearms_Offenses	0.12
		Other_offenses	0.12
		No_Related To Police	0.12
		No_Against Person	0.10
Findings of the arrest	Type of seizure	Num Of Drug Seizures (3–5)	0.45
		Pounds (up 1000)	0.33
		Pounds (up500)	0.29
		Num Of Cash Seizures (over-1)	0.24
		Num Of Cash Seizures (1)	0.24
		Pounds (over-5,000)	0.21
		Pounds (up5000)	0.13
	Drugs associated	Cocaine	0.20
		Cannabis	0.18
Tactics and arrest mode	Type of tactics	Num Of Arrest (2)	0.62
		Generic Tactic_Search_of_Premises	0.50
		Non-Law_Enforcement_Agent	0.46
		Generic Tactic_Search_of_Person	0.36
		Num Of Tactics (3)	0.31
		Num Of Arrests (3)	0.21
		Num Of Tactics (2)	0.18
		Generic Tactic_Search_of_Object	0.16
		Num Of Tactic Sequences (2)	0.15
		No_in_Operation	0.13
	Arrest mode	ArrMode_Direct (2)	0.63
		ArrMode_Other(2)	0.36
		Violent On Arrest (more-1)	0.27
		ArrMode_Result_of_Enquiries (1)	0.17
		ArrMode_Result_of_Enquiries (2)	0.15
		On Bail At Time Of Offence (2)	0.12
		ArrMode_Direct (3–5)	0.11
		ArrMode_Given_into_custody	0.10

Auto-CM provides a typology of drugs in a format that is slightly different from SOM. We present the tables with the Auto-CM prototypes of the five drugs with the same number used above for the SOM analysis. The only difference is in the number adjacent to each variable: in the SOM table, this number is the value of linear correlation between the variable with variables in the heading of the table; in Auto-CM, this adjacent number points out the strength of membership of the variables over the mean (zero representing average membership). We outline with boldface the shared associations between Auto-CM and SOM.

Table 18.5a

Cannabis			
Places	Home borough	NA_Borough	0.193
		Lewisham	0.176
		Havering	0.171
		Hounslow	0.147
		Lambeth	0.145
	Borough of arrest	Arr_PL_NA	0.188
		Arr_Hounslow	0.055
		Arr_Havering	−0.031
Anagraphic data	Gender	Sex_Male	0.726
		Sex_Female	0.212
	Nation group	UK	0.725
		NK	0.327
		VTN	0.297
		AFR	0.044
		JAM	0.033
	Ethnic group	**(ea1)_White_European**	0.618
		(EA3)_Afro-Caribbean	0.525
		(EA4)_Asia	0.483
	Age	**Age (21–25)**	0.622
		Age (25–35)	0.475
		Age (Over45)	0.456
Past criminal curriculum	Convictions	Convictions Number (5–10)	0.428
		Convictions Number (1)	0.407
		Convictions Number (4)	0.404
		Convictions Number (11–20)	0.388
	Offenses	**Offenses Number (6–10)**	0.537
		Offenses Number (3–5)	0.513
		Offenses Number (11–20)	0.367
		Offenses Number (0)	0.366
	Age at the first conviction	Off_First Conv Age (up-to-18)	0.633
		Off_First Conv Age (19–21)	0.552
		Off_First Conv Age (22–27)	0.478
		Off_First Conv Age (34–39)	0.140
	Offenses at the last conviction	Off_Last Conv Age (1)	0.613
		Off_Last Conv Age (2)	0.551
		Off_Last Conv Age (0)	0.474
	Types and number of offenses	**Other_offenses**	0.770
		Off_Against Person (0)	0.727
		Off_Fraud (0)	0.727
		Off_Sexual (0)	0.726
		Off_Against Property (0)	0.711

(continued)

Table 18.5a (continued)

Cannabis			
Findings of the arrest	Type of seizure	**Num Of Drug Seizures (3–5)**	0.791
		Num Of Drug Seizures (over-5)	0.779
		Pounds (0)	0.663
		Num Of Cash Seizures (0)	0.657
		Num Of Drug Seizures (2)	0.621
	Drugs associated	**MDMA**	−0.105
Tactics and arrest mode	Type of tactics	**Generic Tactic_Search_of_Premises**	1.030
		Generic Tactic_Search_of_Person	0.825
		In Operation (0)	0.779
		Num Of Tactic Sequences (1)	0.747
		Num Of Tactics (3)	0.632
	Arrest mode	Violent On Arrest (0)	0.727
		On Bail At Time Of Offence (0)	0.723
		ArrMode_Result_of_Enquiries(0)	0.715
		ArrMode_Other (0)	0.710
		ArrMode_Direct (1)	0.576

Table 18.5b

Cocaine			
Places	Home borough	Sutton	0.077
		Lewisham	0.042
		Islington	−0.078
		Bromley	−0.106
	Borough of arrest	**Arr_Bromley**	0.023
		Arr_Sutton	−0.110
		Arr_Lambeth	−0.132
Anagraphic data	Gender	**Sex_Female**	0.127
		Sex_Male	−0.078
	Nation group	**SAME**	0.225
		UK	−0.068
		EU	−0.164
	Ethnic group	**(EA2)_Dark_European**	0.272
		(EA1)_White_European	0.012
	Age	**Age (25–35)**	0.079
		Age (35–45)	−0.139
Past criminal curriculum	Convictions	**Convictions Number (0)**	−0.021
		Convictions Number (1)	−0.031
	Offenses	**Offenses Number (0)**	−0.018
		Offenses Number (1)	−0.092
	Age at the first conviction	**Off_First Conv Age (22–27)**	−0.002
		Off_First Conv Age (28–33)	−0.108
	Offenses at the last conviction	Off_Last Conv Age (1)	0.033
		Off_Last Conv Age (0)	−0.052

(continued)

Table 18.5b (continued)

Cocaine			
	Types and number of offenses	**Drug_trafficking_ Offenses (2–5)**	0.098
		Off_Related To Police (0)	0.048
		Off_Theft Kindred (0)	0.032
		Off_Sexual (0)	0.013
		Drug_trafficking_Offenses (over-5)	0.012
Findings of the arrest	Type of seizure	**Num Of Drug Seizures (over-5)**	0.615
		Num Of Cash seizures (over-1)	0.003
		Pounds (0)	−0.020
		Num Of Cash Seizures (0)	−0.037
		Num Of Drug Seizures (3–5)	−0.074
	Drugs associated		
Tactics and arrest mode	Type of tactics	In Operation (1)	0.352
		Generic Tactic_Search_of_Premises	0.173
		Generic Tactic_Other	0.070
		Num Of Tactics (2)	0.015
		Num Of Tactics (over 3)	0.012
	Arrest mode	**Arr Mode_Direct (over-5)**	0.079
		On Bail At Time Of Offence (0)	0.019
		Violent On Arrest (0)	0.012
		ArrMode_Direct (2)	0.012
		Arr Mode_Result_of_Enquiries (0)	0.006

Table 18.5c

Crack			
Places	Home borough	**Haringey**	0.261
		Kensington_and_Chelsea	0.191
		Southwark	0.189
		NA_Borough	0.087
		Camden	0.032
	Borough of arrest	**Arr_Hackney**	0.307
		Arr_Kensington_and_Chelsea	0.168
		Arr_Southwark	0.157
		Arr_Camden	0.046
		Arr_Lewisham	−0.015
Anagraphic data	Gender	**Sex_Male**	0.271
		Sex_Female	0.154
	Nation group	**JAM**	0.836
		UK	0.058
	Ethnic group	**(EA3)_Afro-Caribbean**	0.510
		(EA4)_Asia	−0.091
	Age	**Age(25–35)**	0.297
		Age(35–45)	0.191
		Age(Over45)	0.022

(continued)

Table 18.5c (continued)

Crack			
Past criminal curriculum	Convictions	ConvictionsNumber(1)	0.263
		ConvictionsNumber(2)	0.162
		ConvictionsNumber(5–10)	0.033
		ConvictionsNumber(3)	0.011
	Offenses	OffensesNumber(6–10)	0.241
		OffensesNumber(20–50)	0.204
		OffensesNumber(11–20)	0.164
		OffensesNumber(3–5)	0.108
	Age at the first conviction	**Off_FirstConvAge(28–33)**	0.206
		Off_FirstConvAge(up-to-18)	0.157
		Off_FirstConvAge(34–39)	0.125
		Off_FirstConvAge(22–27)	0.115
	Offenses at the last conviction	**Off_LastConvAge(1)**	0.285
		Off_LastConvAge(2)	0.276
	Types and number of offenses	Drug_trafficking_Offenses(over-5)	0.980
		NumOfArrests(over-3)	0.956
		Off_Drug(over-10)	0.601
		Off_Drug(6–10)	0.449
		Drug_trafficking_Offenses(2–5)	0.389
Findings of the arrest	Type of seizure	**NumOfDrugSeizures(over-5)**	0.927
		NumOfDrugSeizures(3–5)	0.306
		Pounds(0)	0.275
		NumOfCashSeizures(0)	0.263
		NumOfCashSeizures(1)	0.196
	Drugs associated	**Heroin**	0.460
Tactics and arrest mode	Type of tactics	**InOperation(over-5)**	0.979
		NumOfTacticSequences(over-5)	0.979
		GenericTactic_Covert_Purchase	0.927
		NumOfTactics(over-3)	0.835
		InOperation(2–5)	0.586
	Arrest mode	**ArrMode_Direct(over-5)**	0.683
		ArrMode_Other(over-2)	0.504
		OnBailAtTimeOfOffence(0)	0.293
		ViolentOnArrest(0)	0.287
		ArrMode_Other(0)	0.265

18.4 Data Mining Techniques Comparison

It is hard to make a comparison between two or more Autopoietic (non-supervised) ANN because in this case, a set of dependent variables that we can use as a "gold standard" are not present. Only the experimental findings can suggest something to us about their associative power, and even this is not true under every condition.

In any case, to compare the capability of SOM and Auto-CM on the same dataset, we will use a new validation strategy composed of the following steps:

Table 18.5d

Heroin			
Places	Home borough	Enfield	0.100
		Kensington_and_Chelsea	0.087
		Haringey	0.025
		Camden	0.019
	Borough of arrest	**Arr_Haringey**	0.013
		Arr_Kensington_and_ Chelsea	0.010
		Arr_Camden	−0.022
Anagraphic data	Gender	Sex_Male	0.122
	Nation group	**TU-CY**	0.611
		JAM	0.073
		UK	−0.047
	Ethnic group	**(EA2)_Dark_European**	0.257
		(EA3)_Afro-Caribbean	0.052
		(EA4)_Asia	−0.014
	Age	Age(Over45)	0.100
		Age(25–35)	0.062
		Age(35–45)	0.031
Past criminal curriculum	Convictions	ConvictionsNumber(1)	0.019
		ConvictionsNumber(0)	−0.004
	Offenses	OffensesNumber(20–50)	0.147
		OffensesNumber(0)	−0.005
		OffensesNumber(6–10)	−0.054
		OffensesNumber(11–20)	−0.089
	Age at the first conviction	**Off_FirstConvAge(over-51)**	0.323
		Off_FirstConvAge(40–45)	0.032
		Off_FirstConvAge(22–27)	−0.026
		Off_FirstConvAge(up-to-18)	−0.042
	Offenses at the last conviction	**Off_LastConvAge(0)**	0.044
		Off_LastConvAge(1)	0.014
	Types and number of offenses	**Drug_trafficking_ Offenses(over-5)**	0.578
		Off_Drug(over-10)	0.476
		NumOfArrests(over-3)	0.450
		Off_Total(20–50)	0.122
		Off_Sexual(0)	0.115
Findings of the arrest	Type of seizure	**NumOfDrugSeizures (over-5)**	0.817
		Pounds(0)	0.142
		NumOfCashSeizures(0)	0.132
		NumOfDrugSeizures(3–5)	−0.008
	Drugs associated	**Crack**	0.460

(continued)

Table 18.5d (continued)

Heroin			
Tactics and arrest mode	Type of tactics	**InOperation (over-5)**	0.591
		NumOfTacticSequences (over-5)	0.591
		GenericTactic_Covert_ Purchase	0.537
		NumOfTactics(over-3)	0.354
		GenericTactic_Search_of_ Premises	0.117
	Arrest mode	**ArrMode_Direct(over-5)**	0.251
		ArrMode_Other(over-2)	0.128
		ViolentOnArrest(0)	0.113
		OnBailAtTimeOfOffence(0)	0.108
		ArrMode_Result_of_ Enquiries(0)	0.097

Table 18.5e

MDMA			
Places	Home borough	**NA_Borough**	0.293
		Southwark	0.045
		Lambeth	0.028
	Borough of arrest	**Arr_Wandsworth**	0.050
		Arr_Westminster	−0.018
Anagraphic data	Gender	Sex_Male	0.125
		Sex_notknown	0.075
		Sex_Female	−0.002
	Nation group	EU	0.312
		UK	0.175
		ASIA	−0.052
	Ethnic group	**(EA1)_White_European**	0.368
		(EA2)_Dark_European	0.180
		(EA6)_Arab	−0.074
	Age	**Age(25–35)**	0.182
		Age(21–25)	0.121
		Age(35–45)	−0.015
Past criminal curriculum	Convictions	**ConvictionsNumber(1)**	0.236
		ConvictionsNumber(11–20)	0.012
		ConvictionsNumber(2)	−0.007
	Offenses	**OffensesNumber(1)**	0.072
		OffensesNumber(3–5)	−0.036
		OffensesNumber(11–20)	−0.051
		OffensesNumber(6–10)	−0.069
	Age at the first conviction	**Off_FirstConvAge(22–27)**	0.193
		Off_FirstConvAge(up-to-18)	0.010
	Offenses at the last conviction	Off_LastConvAge(2)	0.192
		Off_LastConvAge(1)	0.089

(continued)

Table 18.5e (continued)

MDMA			
	Types and number of offenses	**Drug_Possession_Offenses**	0.472
		NumOfArrests(3)	0.213
		Off_RelatedToPolice(0)	0.213
		AR_OFF_Other_Drug_Offenses	0.197
		Off_Fraud(0)	0.186
Findings of the arrest	Type of seizure	**NumOfDrugSeizures(3–5)**	0.355
		NumOfDrugSeizures(over-5)	0.338
		NumOfCashSeizures(1)	0.299
		Pounds(up100)	0.186
	Drugs associated	**Cannabis**	−0.105
Tactics and arrest mode	Type of tactics	**Non-Law_Enforcement_Agent**	0.799
		GenericTactic_Search_of_Person	0.548
		GenericTactic_Search_of_Premises	0.326
		NumOfTactics(over-3)	0.230
		InOperation(0)	0.229
	Arrest mode	**ArrMode_Given_into_custody**	0.281
		ViolentOnArrest(0)	0.195
		OnBailAtTimeOfOffence(0)	0.187
		ArrMode_Result_of_Enquiries(0)	0.179
		ArrMode_Other(0)	0.171

1. We analyze the dataset with three different, independent, limited, but grounded, techniques.

   (a) The linear correlation is calculated over all the couples of variables of the dataset (LC algorithm):

$$R_{i,j} = \frac{\sum_{k=1}^{N}(x_{i,k}-\bar{x}_i)\cdot(x_{j,k}-\bar{x}_j)}{\sqrt{\sum_{k=1}^{N}(x_{i,k}-\bar{x}_i)^2\cdot\sum_{k=1}^{N}(x_{j,k}-\bar{x}_j)^2}},$$

$$-1 \le R_{i,j} \le 1; \quad i,j \in [1,2,\ldots,M]. \tag{18.1}$$

   (b) Co-occurrence probability among all the couplet of the dataset variables (PP algorithm):

$$A_{i,j} = -\ln\frac{(1/N^2)\cdot\sum_{k=1}^{N}x_{i,k}\cdot(1-x_{j,k})\cdot\sum_{k=1}^{N}(1-x_{i,k})\cdot x_{j,k}}{(1/N^2)\cdot\sum_{k=1}^{N}x_{i,k}\cdot x_{j,k}\cdot\sum_{k=1}^{N}(1-x_{i,k})\cdot(1-x_{j,k})},$$

$$-\infty \le A_{i,j} \le +\infty; \quad x\in[0,1]; \quad i,j \in [1,2,\ldots,M]. \tag{18.2}$$

(c) Euclidean distance among all the dataset variables (ED algorithm):

$$d_{i,j}^{[E]} = \sqrt{\sum_{k=1}^{M} \left(x_{i,k} - x_{j,k}\right)^2}; \quad i,j \in [1,2,..N]; \quad x \in [0,1]. \qquad (18.3)$$

All three of these techniques are very robust: LC algorithm finds out the proportionality among the variables; PP algorithm defines their probability of co-occurrence; the ED algorithm measures their distances in a flat space. At the same time, all these techniques are very limited: the LC and the PP algorithms consider only the first-order effects among variables, and the ED algorithm assumes the Euclidean space as the only metric able to explain the closeness among variables. So, these three algorithms are three separate tools used to analyze what in a dataset should be three manifestations of evidence: *the evidence of linearity, the evidence of probability, and the evidence of distance*. But there is not a linear correlation among these three techniques: each technique can find what is evident to itself but is hidden from the others. Consequently, LC, PP, and ED algorithms are three robust, independent, and limited techniques. When a highly nonlinear and multivariate algorithm (like SOM or Auto-CM) identifies some association supported by at least one of these three techniques, we can say that this complex algorithm has discovered something that is trivial, but grounded.

2. We calculate for each linear algorithm (LC, PP, and ED) and for SOM and Auto-CM, the minimum spanning tree (Kruskal 1956; Cormen et al. 2001; Karger et al. 1995; Fredman and Willard 1990; Gabow et al. 1986); obviously, to do that requires some intelligent preprocessing.
3. We compare the agreement of the MST of each algorithm with the MST of the others; in this way, we can define for each algorithm three different basic indices and one composed index:

(a) *The Intersection Index*: how much does the association of any pair of algorithms agree:

$$Iindex_{i,j} = \sum_{k}^{M} \frac{(link_i = true \cap link_j = true)}{(link_i = true \cup link_j = true)}, \qquad (18.4)$$

where
$i, j \in N$
$k \in M$
$N =$ Number of different MST coming from different algorithms
$M =$ (Num Variables)^2.

(b) *The Evidence Index*: how much of the association associated with each algorithm is supported by the associations with the others:

$$Eindex_i = \frac{1}{2(N-1)(M-1)} \sum_{z=1}^{M} \sum_{k=1}^{M} \sum_{j=1;j \neq i}^{N} \times (link_{i,z,k} = true \cap link_{j,z,k} = true), \qquad (18.5)$$

where
$i, j \in N$
$k \in M$
N = Number of different MST coming from different algorithms
M = Number of Variables.

(c) *The Singularity Index*: how many times the association of each algorithm is only self-supported:

$$Sindex_i = \frac{1}{2(N-1)(M-1)} \sum_{z=1}^{M} \sum_{k=1}^{M} link_{i,z,k} = true \cap \left(\sum_{j=1;j \neq i}^{N} (link_{j,z,k} = false) \right) = 0; \qquad (18.6)$$

where
$i, j \in N$
$k \in M$
N = Number of different MST coming from different algorithms
M = Number of Variables.

(d) *The E-S Ratio Index*: for each algorithm, this determines the balance between associations supported by other algorithms and associations self-supported:

$$Ratio_i = -\ln\left(\frac{Sindex_i}{Eindex_i}\right); \quad -\infty \leq Ratio_i \leq +\infty. \qquad (18.7)$$

18.4.1 The Intersection Index

The Intersection Index shows the agreement and the disagreement for any pair of algorithms with respect to their fundamental associations among the dataset variables.

From a global point of view, the behavior of the five algorithms under consideration is shown in these Tables 18.6 and 18.7:

Table 18.6 points out that more than half of the Auto-CM connections are supported by linear correlation (LC) and more than one-third of SOM connections are supported by the Euclidean distance (ED). LC seems to behave as a bridge between Auto-CM and SOM, while the PP algorithm shows a completely different

Table 18.6

Algorithms	Agreement (%)
CM-LC	52.22
ED-SOM	37.01
ED-LC	34.73
LC-SOM	30.36
ED-CM	27.74
CM-SOM	23.37
LC-PP	14.10
CM-PP	14.06
ED-PP	11.36
PP-SOM	8.62

Table 18.7

Algorithms	Mean Agreement (%)
LC	32.85
CM	29.35
ED	27.71
SOM	24.84
PP	12.03

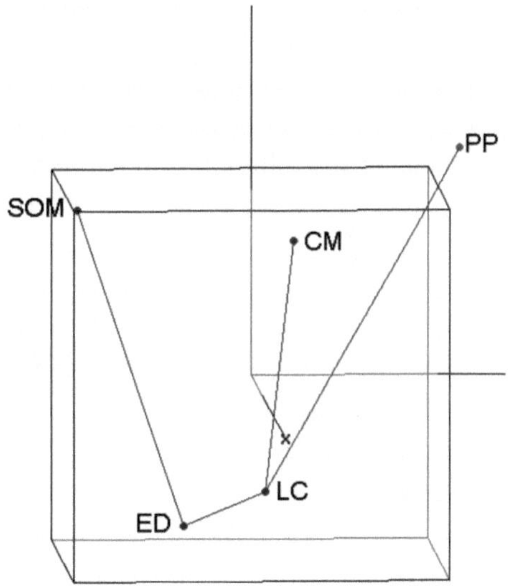

Fig. 18.7 Map of the five algorithms, according to their similarity of results in the "Persons" dataset. The error of projection is under 4%. The PP algorithm is located in an isolated position; the LC and ED algorithms seem to work as middle points among the others, while Auto-CM and SOM appear to be specialized in opposite areas. But Auto-CM collects a bigger number of supports than SOM

logic. If we translate the algorithm intersection values into a matrix distance and project this matrix distance into a three-dimensional space, we obtain the following map (Fig. 18.7):

We can conclude the analysis of this index with this observation:

Auto-CM is supported by LC as SOM is supported by LC and ED, while PP seems to be an outlier.

Table 18.8

Algorithms	Support	Singularities	Variety	Ratio
LC	0.7469	0.2531	0.4166	1.0824
CM	0.6816	0.3184	0.4142	0.7613
ED	0.5224	0.4776	0.4188	0.0899
SOM	0.4735	0.5265	0.4139	−0.1062
PP	0.1837	0.8163	0.4281	−1.4917

18.4.2 The Evidence and the Singularity Indexes

The Evidence Index measures the degree of association of each algorithm with the others. The Singularity Index measures the opposite: for each algorithm, how many associations are completely not supported by the others.

The Evidence Index is important in order to establish the degree of support for the connectivity of each algorithm that is shared by the others. The greater the support and sharing of connections, the greater the reliability.

The Singularity Index is fundamental to understanding the specificity of each algorithm and is hidden from the view of the others. Obviously, it is not possible to know a priori if the specificity index is grounded. For this reason, we have proposed a new index, the *ratio between evidence and singularity*. According to the *Ratio Index,* we can distinguish three classes of algorithms:

1. Conservative algorithms, where *Ratio* $\gg 0$.
2. Creative algorithms, where *Ratio* $\ll 0$.
3. Moderate algorithms, where *Ratio* $\cong 0$.

Furthermore, if, for each algorithm, we calculate the *variance* of the algorithms that support or lack support for all of its connections, then we can also define the *specificity* of the research area of each algorithm.

Table 18.8 shows the behavior of the five algorithms after the analysis of "Persons" dataset from the point of view of:

1. Evidence Index
2. Singularity Index
3. Ratio Index
4. Variance of Association Support

This table describes the basic profile of each algorithm:

1. The fundamental associations of the LC algorithm agree with most of the other algorithms, but its capability to discover new associations is limited.
2. The associations of Auto-CM algorithm are also well supported, but the number of its specific connections is also high.
3. Around 1/2 of the associations of ED algorithm are supported by the other algorithms, but most of these associations are singular (47.76%).

4. The basic associations of SOM algorithm are partially supported by the other algorithms, while its creativity is fairly high (52.65%).
5. The associations of PP algorithm are very creative (81.63%), while the number of its associations supported by the other algorithms is very low.

18.4.3 The Models Fusion Methodology (MFM)

When we use Autopoietic systems, such as non-supervised ANNs, it is not easy to establish which algorithm is more consistent than the others. In this kind of situation, the only "gold standard" is to explore every algorithm hypothesis on the field. But these algorithms should undergo an organization using some sort of detective strategy. Therefore, we need to define "a priori" criteria by which we can decide which of the proposed links are more believable.

Because each one of these presented algorithms follows a different mathematical foundation, the best way to decide on a ranking is to test them:

1. Each of these algorithms proposes a specific tree of dependencies (the MST) among the variables contained in the same dataset.
2. The goal is to select one graph, from all the trees, whose links to variables are the more robust and believable.
3. Thus, we overlap all the individual trees and retain only the connections selected at least by two different algorithms; in other words, if two different algorithms, using a different a mathematical basis, outlines the same link between two variables, then it is more probable that the link between these two variables is "real."

Consequently, the generate graph can contain some fascinating features:

1. There may be cycles in the resulting graph.
2. The resulting graph could be disconnected (see Fig. 18.8).

Working in this way, we have discovered a new scenario: the graph generated by the fusion of the different methods is a disconnected graph divided into at least four frames:

1. The variables without links: 16 of the analyzed variables belong to this group:

   Sexual Offenses (typically one)
   Sexual Offenses (typically two)
   Offenses against Property (typically one)
   Violence on the Arrest (typycally one)
   One year from the last Conviction

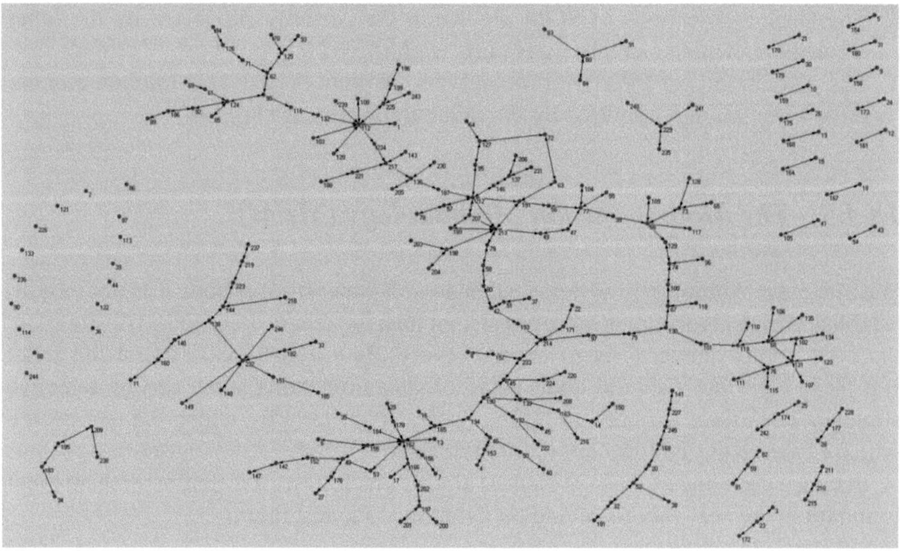

Fig. 18.8 The final graph of the associations among variables

Two years from the last Conviction
3–5 years from the last Conviction
11–20 years from the last Conviction
East Europe Nationality
Sex: Female
Fraud Offenses (typycally one)
Fraud Offenses (typycally between 2 and 5)
Residence borough not available
Non UK Nationality
Result of Enquiries (typically two)
On Bail at the time of Offence (typically one)

For this particular set of variables, it is not possible to make any specific inferences. Perhaps, in some cases, the number of records is too small, or too different, to be properly compared with each other. In any case, these variables lack even the minimum of convergence to be considered and are thus dismissed from further inquiry.

2. The apparent set of twin variables: 28 variables are clustered into 14 couples, but each couple is completely disconnected from all the other variables of the dataset; 12 of these couples represent the link between the "borough of the arrest" and "the borough where the arrested person lives":

Arrest: Lambeth	Home: Lambeth
Arrest: Islington	Home: Islington
Arrest: Hammesmith	Home: Hammesmith
Arrest: Hillingdon	Home: Hillingdon
Arrest: Croydon	Home: Croydon
Arrest: Sutton	Home: Sutton
Arrest: Havering	Home: Havering
Arrest: Merton	Home: Merton
Arrest: Southwark	Home: Southwark
Arrest: Barnet	Home: Barnet
Arrest: Enfield	Home: Enfield
Arrest: Camden	Home: Camden

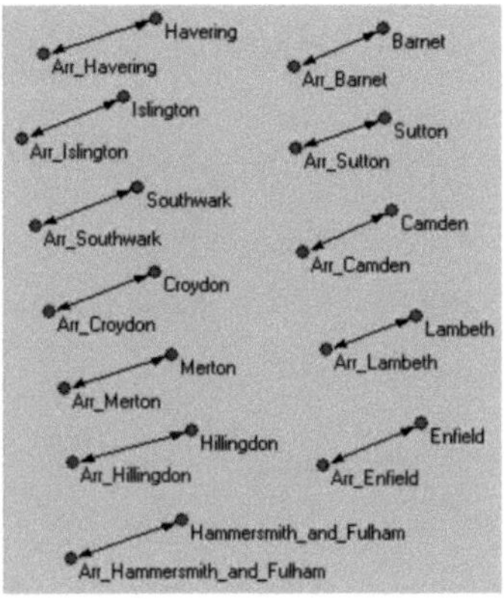

The remaining two couples point out a specific relationship between nationality and behavior, and nationality and ethnic group:

Irish	Offenses against persons (typically 2)
Middle East	Arab

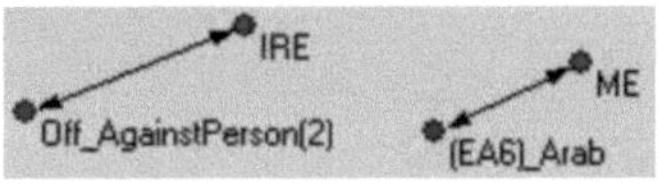

This clustering signifies that more than one algorithm has found a strong relationship between the two variables, but this link is the only one.

3. The next frame is formed by groups of three and four or more variables, with specific links to each other, clustered in small isolated worlds (see Fig. 18.9a–e):
   (a) "Other law enforcement agent" is the common point linking "controlled delivery" and "search of object."
   (b) Persons whose age is between 35 and 45 have their first conviction in that same interval of age.
   (c) Persons living in Kingston upon Thames are arrested in the same borough (and this seems to be typical), but with these persons, the agents typically declare they are not able to define their gender.
   (d) Persons arrested in Lewisham often live in the same borough but typically are arrested many times in a non-direct mode.
   (e) Persons arrested in Redbridge often live in the same borough, but typically their arrest mode is not declared by the police.
   (f) Persons over 45 received their last conviction 20 years ago and typically return to commit further offenses.
   (g) Persons not arrested in direct mode are obviously arrested in other ways, often as a result of inquiries.
   (h) MDMA is typically found by non-law enforcement agents. They arrest the people involved taking them in custody. This is typical in the borough of Wandsworth, the same place in which the persons arrested live.

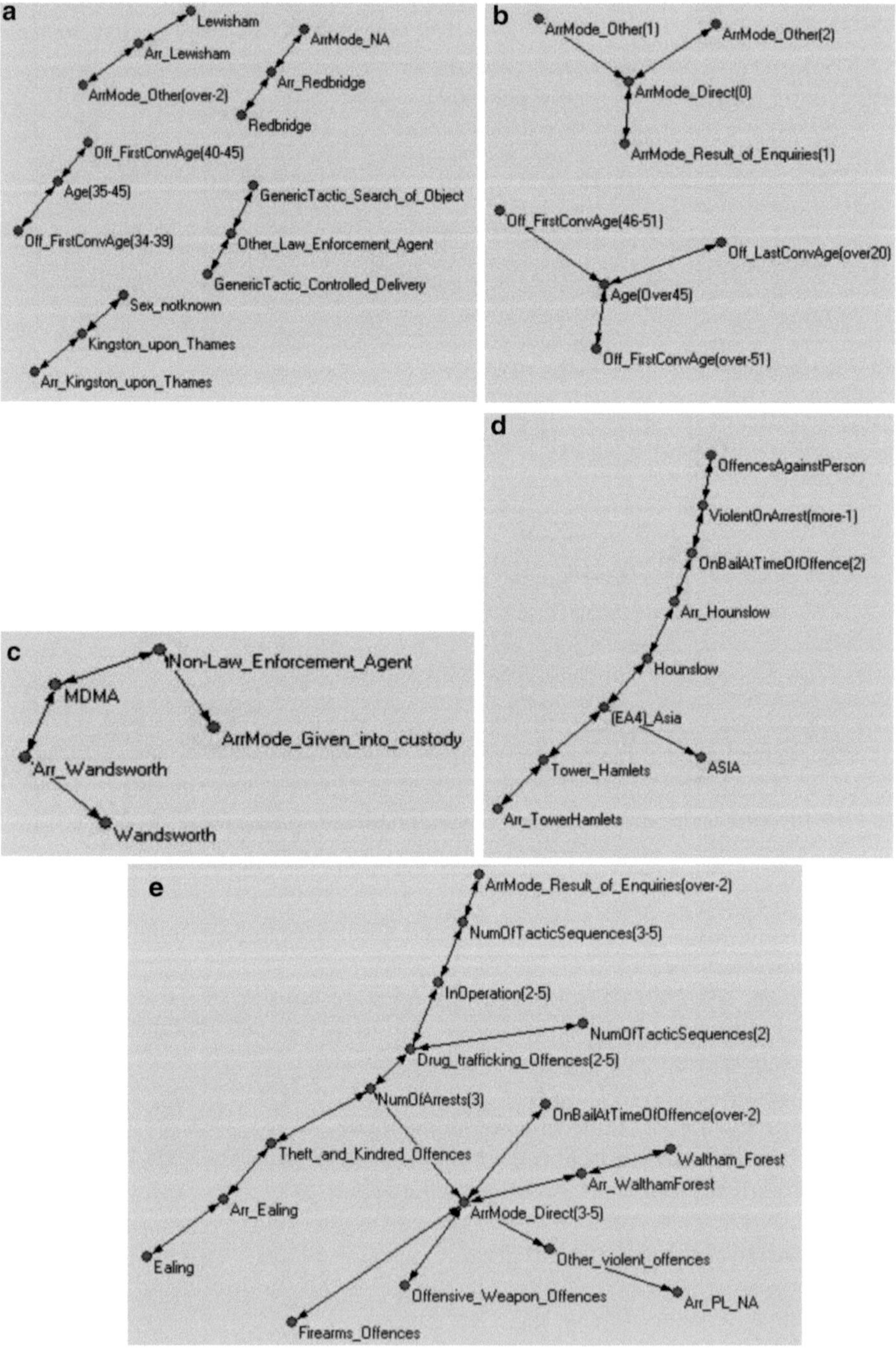

Fig. 18.9 a, b, c, d, e

(i) There is one graph about Asian world: Asian people, with Asian nationality, are arrested fundamentally in two boroughs: Tower Hamlets and Hounslow, where typically these people also live. But Asian people of Hounslow have interesting features: they have already been arrested more than one time in a violent mode because the offenses are against other people; this quite often resulted in them being on bail at the time of the offence.

(j) The last small world is about the direct arrest of violent people, with firearms and other weapons; the most of these people are older criminals with an impressive record in drug trafficking and other types of offenses. Typical places where these persons live and where they are arrested are Waltham Forest and Ealing.

Of course, the other 146 variables are connected in the one main graph. Each one of these connections has been established by at least two independent algorithms, so they should be quite strong and robust. This main graph (see Fig. 18.10) shows many cycles, and its deep analysis should take many pages. In any case, if we look only at the positions of the four drugs and their neighbors, we can define a robust and synthetic prototype of each drug (see Fig. 18.10a, b).

Table 18.9 should be a road map for the antidrug strategy on London: this tells us into which boroughs police should look and for which type of drug, which kind of person a particular criminal record, which tactic to use, etc.

18.5 Conclusion

This chapter provides a thorough example of how one should apply nonlinear auto-associative systems to data analysis. For this reason, we have shown both the data and the equations in sufficient detail such that a thorough understanding of the method and its operation could be understood.

Nonlinear auto-associative systems are often known with the generic name of non-supervised artificial neural networks (ANNs). These systems, indeed, represent a set of powerful techniques for data mining, and they do not deserve a generic name. We propose to name this set of ANNs "autopoietic ANNs" (i.e., systems that self-organize their behaviors).

Autopoietic ANNs constitute a complex domain of differing topologies, learning rules, signal dynamics, and cost functions. Consequently, their mathematics can be very different from one another, and their capability to discover hidden connections from the same dataset can also be very different. This aspect is both the strong point and the weak point of these algorithms.

All the autopoietic ANNs, in fact, tend toward one goal, that of taking a dataset and determining how each (independent) variable is associated with each other, taking into account the nonlinear associations involved in parallel, many-to-many relationships. But, because of the mathematics specific to each one of these algorithms, the final outcomes on any particular application taken on the same

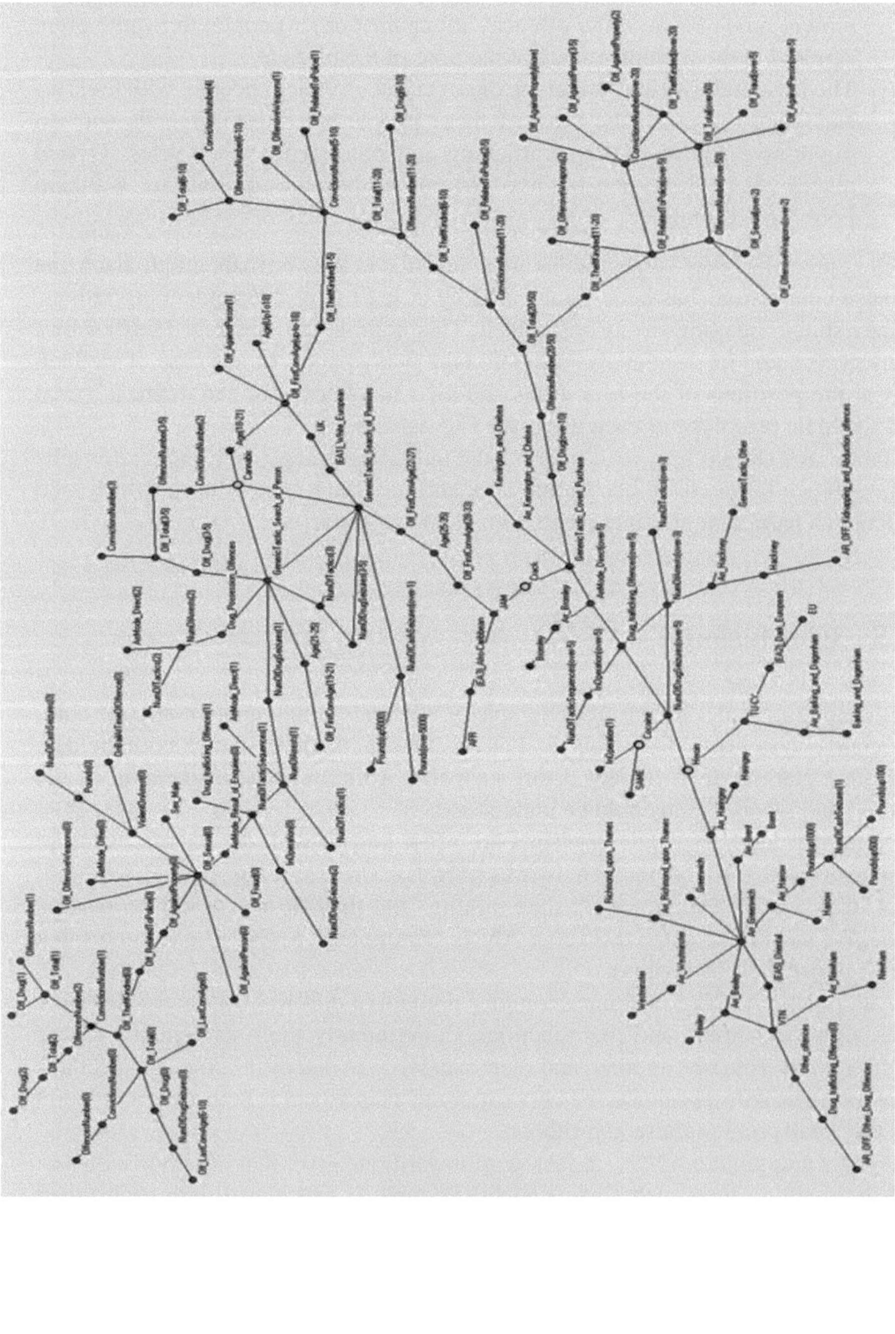

Fig. 18.10 The big graph

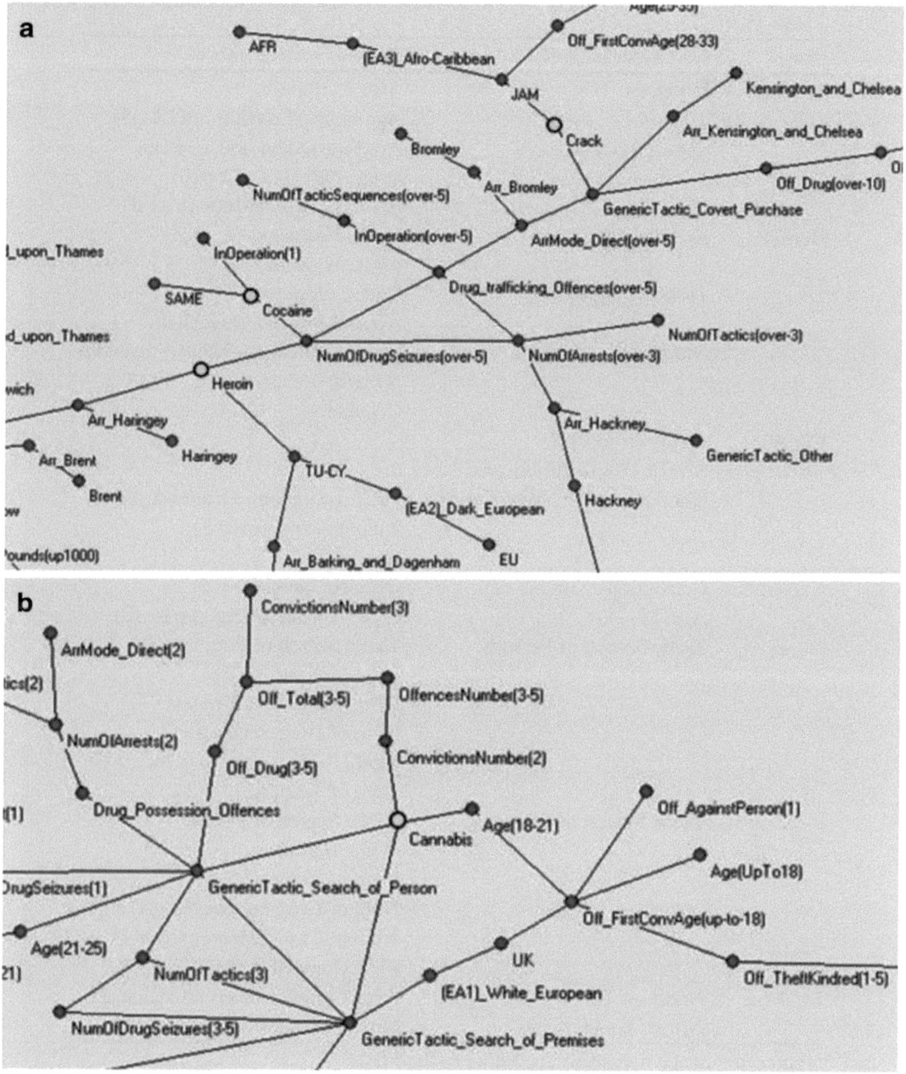

Fig. 18.10 **a** The three drugs zoomed. **b** Cannabis zoomed

dataset can be quite different. Consequently, when we apply different autopoietic ANNs to the same sample of data, we can find from the learning process different frames of associations existing among the same set of variables. The problem, at this point, is to determine which of these frames is more grounded? If the dataset represents a real situation, which one of these frames organizes a productive strategy of manipulation into the real world?

A weak, but politically correct, answer could simply be that every algorithm shows some different feature of the same world. It is up to the researcher to know

Table 18.9

Drugs[a]	First order neighbors	Second order neighbors
Crack	Jamaican	Afro-Caribbean
		Age of first Convictions(28–33)
	Covert Purchase	Arr_Kensignton_and_Chelsea
		Drugs Offenses (over 10)
		Arrest in Direct Mode (over 5)
Heroin	Arr_Haringay	Home Haringay
		Arr_Greenwhich
	Turkish-Cypriots	Dark-European
		Arr_Baking_and_Dagenham
	NumOfDrugSeizures(over 5)	Drug_Trafficking_Offenses(over 5)
		NumOfArrests(over 3)
		Cocaine
Cocaine	InOperation(1)	
	SAME (South Americans)	
	NumOfDrugSeizures(over 5)	Drug_Trafficking_Offenses(over 5)
		NumOfArrests(over 3)
		Heroin
Cannabis	Convictions Number (2)	Offenses Number(3–5)
	Age(18–21)	First_Conviction_Age(up to 18)
	Tactic Search of Persons	Tactic Search of Premises
		Drug Offenses (3–5)
		Drug Possesion Offenses
		Num of Drug Seizures (1)
		Age(21–25)
		Num of Tactics(3)
	Tactic Search of Premises	Tactic Search of Person
		White European
		Num of Tactics(3)
		Num of Drug Seizures (3–5)
		Num of Cash Seazures (over 1)
		First_Conviction_Age(22–27)
MDMA	Non-Law Enforcement Agent	Arrest Mode : Given into Custody
	Arrest: Wandsworth	Home: Wandsworth

[a]Prototypes of the five drugs according to the MFM

which of these features are the more robust and fundamental than the others. In fact, strategies and actions in the real world are expensive, and one should spend one's energy aiming directly at the critical points of a real situation. The main target of data analysis is exactly this point.

When we use supervised ANNs or other types of supervised classifiers, this kind of problem dissolves away because our dataset presents dependent variables. We need to believe that these dependent variables represent the "gold standard," and consequently, we can size the effectiveness of our supervised algorithms following different and robust validation protocols such as K-fold cross validation, 5×2 cross validation, training–testing-prediction protocol, etc.

Using autopoietic ANNs, we do not have dependent variables, so we lack a "gold standard."

Quite frankly, many validation protocols were proposed to analyze the performance of autopoietic systems (see Buscema 1998), but all these validation protocols seek to control the self-consistency and the flexibility (capability to react in similar ways to unseen inputs, when the last ones are similar to known inputs) of the autopoietic ANNs.

It is not enough.

The only effective validation protocol for these systems should consist of a field analysis in which the researcher is able to control the association schemes. This is the best of the possible worlds, following the right methodology to execute a validation on the field; but this way has some counter-indications: first, it is a very expensive method (the number of possible connections among N variables is 2^N, considering only the limit case where each connection can be present or absent), and the target of data mining should be known a priori.

For these reasons, we have proposed a new probabilistic way to validate the performance of autopoietic ANNs when used for data mining. The methodology is organized in the following steps:

1. Select a representative dataset from a real situation: Here, we have selected a database of those individuals arrested in London for drug trafficking during a sequence of 4 months.
2. Select the ANNs: We have chosen two very powerful and very different autopoietic ANNs, the Auto-Contractive Maps (Auto-CM) and the Self-Organizing Maps (SOM) networks. These two algorithms represent completely different mathematical foundations (topology, learning rule, signal dynamics, and cost function), but they have the same general goal: to analyze the global similarities of the records of a dataset according to their variables, including the nonlinear associations among the variables themselves.
3. Select a filter by which the results can be compared and visualized: We have chosen the minimum spanning tree (MST) as the filter to synthesize the main associations among variables that the two autopoietic ANNs have found at the end of their learning process. The MST has many suitable properties, especially its capability to put in evidence the fundamental backbone of a structure. In this case, the "structure" is the matrix of associations among variables generated by each one of our autopoietic ANNs.
4. Define a criteria by which the ANNs can be assessed as to their success in finding proper associations between variables: We have selected three simple linear algorithms, very different from each other and very robust and mathematically well grounded:
   (a) The linear correlations algorithm (LC) is based on the covariance of any pair of variables.

(b) The prior probability algorithm (PP) is based on the probabilistic co-occurrence of any pair of variables.
(c) The Euclidean distance (ED) is based on the assumption that the distance from any one variable to another is on a flat space of N dimensions.

These three algorithms are orthogonal to one another, but at the same time, each one of them is able to act very robustly, but they can also sometimes find trivial relationships among variables. We have used these three algorithms as "sapiens sauvage," that is to say, they have not been prepped to discover either nonlinear relations among variables or complex many-to-many associations, but they are very expert at discovering evidence in data. Most importantly, each of these algorithms is oriented to seek out evidence from the data but to do it in different areas of a data space consisting of all possible associations.

Using these three algorithms as "basic analysts," we are then able to understand when the associations found by the two complex autopoietic ANNs are:

(a) Evident for one or more of the "basic analysts"
(b) Original and supported by the two ANNs together
(c) Original and supported by only one of the two ANNs
(d) Evident for at least one of the "basic analysts," but unseen by one or both the ANNs

5. Compare and conclude: At this point, we have generated the MST of the five algorithms (the three basic analysts and the two ANNs), and we have made a point to point comparison among them. The goal of this match is to create a new graph in which only the associations among variables, supported at least by two algorithms, are present. Obviously, each connection will have a different membership of plausibility according to the number of algorithms supporting it. In the same way, each connection will also have a different grade of originality, if supported only by the ANNs (if the number of ANNs in our experiment are more than two, the cutoff of two algorithms to accept the connections will work all the same).

Using this methodology, which is named the models fusion methodology (MFM), we can produce a sparse graph (some nodes and/or groups of nodes disconnected from the others), with many varied and complex cycles. This is actually quite good. It is now possible to know which variables we can say nothing (nodes isolated), which ones we can say "something" (group of nodes but still isolated), and which simple or complex circuits (cliques) are grounded into the dataset.

Obviously, we might need to utilize more than two ANNs in order to deeply analyze a dataset. In the experiments presented here, if only *one* ANN finds a link between two variables, then there is some risk in removing the link for we have only used two ANNs. If, however, the artificial biodiversity of autopoietic ANNs is increased to three or more, then the possibility of arriving close to the "soul" of any dataset is possible.

References

Buscema, M. (1998). Special issue on artificial neural networks and complex social systems – Theory. *Substance Use & Misuse, 33*(1), 1–212.

Buscema, M. (2007a). *A novel adapting mapping method for emergent properties discovery in data bases: Experience in medical field.* In 2007 IEEE international conference on systems, man and cybernetics (SMC 2007), Montreal, Canada.

Buscema, M. (2007b). *Squashing theory and contractive map network* (Semeion Technical Paper #32), Rome.

Buscema, M., & Grossi, E. (2008). The semantic connectivity map: An adapting self-organizing knowledge discovery method in data bases. Experience in gastro-oesophageal reflux disease. *International Journal of Data Mining and Bioinformatics, 2*(4), 362–404.

Buscema, M., Grossi, E., Snowdon, D., & Antuono, P. (2008a). Auto-contractive maps: An artificial adaptive system for data mining. An application to Alzheimer disease. *Current Alzheimer Research, 5*(5), 481–498.

Buscema, M., Helgason, C., & Grossi, E. (2008b). *Auto contractive maps, H function and maximally regular graph: Theory and applications.* Special session on "Artificial adaptive systems in medicine: Applications in the real world", NAFIPS 2008 (IEEE), New York.

Cormen, T. H., Leiserson, C. E., Rivest, R. L., & Stein, C. (2001). *Introduction to algorithms* (2nd ed.). Cambridge, MA/New York: MIT Press/McGraw-Hill. Section 23.2: The algorithms of Kruskal and Prim, 567–574. ISBN 0-262-03293-7.

Fredman, M. L., & Willard, D. E. (1990). Trans-dichotomous algorithms for minimum spanning trees and shortest paths. In *31st IEEE symposium on foundations of computer science* (pp. 719–725). Washington, DC: IEE Computer Society Press.

Gabow, H. N., Galil, Z., Spencer, T., & Tarjan, R. E. (1986). Efficient algorithms for finding minimum spanning trees in undirected and directed graphs. *Combinatorica, 6*, 109–122.

Karger, D. R., Klein, P. N., & Tarjan, R. E. (1995). A randomized linear-time algorithm to find minimum spanning trees. *Journal of the ACM, 42*, 321–328.

Kohonen, T. (1990). The self-organizing map. *Proceedings of the IEEE, 78*, 1464–1480.

Kohonen, T. (1995a). Learning vector quantization. In M. A. Arbib (Ed.), *The handbook of brain theory and neural networks* (A Bradford book). Cambridge, MA/London: The MIT Press.

Kohonen, T. (1995b). *Self-organizing maps.* Berlin/Heidelberg: Springer.

Kruskal, J. B. (1956). On the shortest spanning subtree of a graph and the traveling salesman problem. *Proceedings of the American Mathematical Society, 7*(1), 48–50.

Licastro, F., Porcellini, E., Chiappelli, M., Forti, P., & Buscema, M. (2008). Multivariable network associated with cognitive decline and dementia. *International Journal of Neurobiology of Aging, 93*(2), 257.

Research Software

Buscema, M. (2002). *M Buscema, contractive maps* (Ver 1.0, Semeion Software #15), Rome, 2000–2002.

Buscema, M. (2007). *M Buscema, constraints satisfaction networks* (Ver 12.0, Semeion Software #14), Rome, 2001–2007.

Buscema, M. (2008). *M Buscema, MST* (Ver 5.1, Semeion Software #38), Rome, 2006–2008.

Massini, G. (2007a). *G Massini, trees visualizer* (Ver 3.0, Semeion Software #40), Rome, 2007.

Massini, G. (2007b). *G Massini, semantic connection map* (Ver 1.0, Semeion Software #45), Rome, 2007.

Chapter 19
Artificial Adaptive System for Parallel Querying of Multiple Databases

Massimo Buscema

19.1 Introduction

Artificial neural networks (ANN) have become a scientific asset acquired for drawing up complex databases (see Rumelhart et al. (1986) for a general presentation of the field, and Hebb (1961) and Hopfield (1982) for the first nonlinear auto-associative ANNs). Generally speaking, ANNs are designed to process bi-dimensional data matrices: a vector of rows, or records, and a vector of columns, or attributes (variables). Huge difficulties emerge if it is intended to process several databases through a single ANN consisting of variables and records that are conceptually different and of varying cardinality.

Such a requirement may seem arbitrary: a dataset that describes the aromatic characteristics of a sample of various wines does not seem to have anything to do with a dataset reporting the values of the variables of environmental pollution in various parts of a city. And yet, if the wines in question are produced by vineyards located near the city in question, it becomes relevant to understanding the complex interaction between the two datasets. Therefore, in order to process different databases through the same artificial system, there must be *semantic links* and *syntactic links* to which the different databases comply.

The main semantic link is that all databases that that are to be processed together involve data whose origin is in the same space-time region. In other words, they must be databases that sample the same reality from different points of view.

It is a conceptual link: if the different types of data of the different datasets are conceptually different samplings of the same reality, then it makes sense to reproduce in the analysis phase the interaction that these different types of data have in the same reality. The semantic link provides the necessary conditions which

M. Buscema (✉)
Semeion Research Center of Sciences of Communication, Rome, Italy
e-mail: m.buscema@semeion.it

M. Buscema and W.J. Tastle (eds.), *Intelligent Data Mining in Law Enforcement Analytics: New Neural Networks Applied to Real Problems*, DOI 10.1007/978-94-007-4914-6_19,

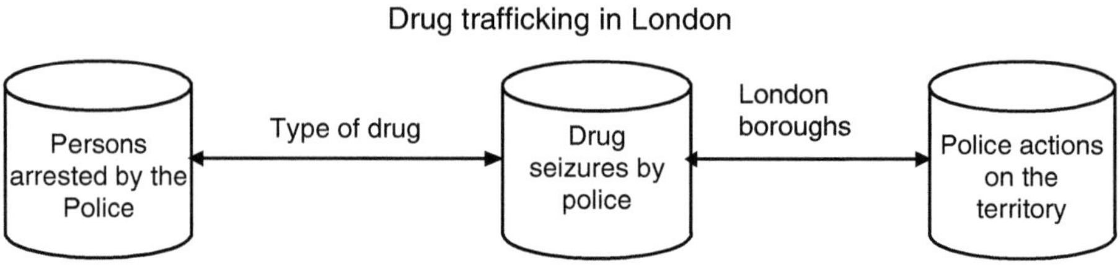

Fig. 19.1 The three datasets (an example)

permit several databases to be processed through a single artificial system, but this condition is not sufficient in itself. In order to make it possible, a syntactic link of this type must be satisfied:

> the different datasets must have some variables in common so that, assuming the dataset as nodes of a graph and the variables in common as arcs of the same graph, the graph that is obtained is at least a tree graph.

A variable can be considered common to two or more datasets if its format and its statistical nature are the same in the different dataset. The following example illustrates this point (Fig. 19.1).

Each of these three datasets has variables and records that are different in quality and quantity: the first contains the characteristics of people arrested for drugs; the second contains the characteristics of the different drug seizures in London; the third characterizes the different actions carried out by the police when combating drug trafficking in London.

However, all three datasets concern drug trafficking in the same space and time interval. Therefore, it can be hypothesized that the data contained in the three datasets come from the same reality, sampled from different perspectives (semantic link). Furthermore, it is possible to move from any one dataset to any other through the existence of common variables (syntactic link).

19.2 The ALOC System

We define an ALOC (At Least One Connection) system as an artificial adaptive system able to process several databases in parallel, when the databases in question satisfy the semantic and syntactic links defined above.

Having verified the two conditions that define the applicability of the ALOC system, the following steps must be completed:

(a) Choice of the equations establishing the relationships between each variable and each other variable in each dataset
(b) Choice of the equations allowing development of all the variables of each dataset in parallel
(c) Choice of the equations allowing development of each record of each dataset in parallel

19.2.1 The Prior Probability Algorithm

The prior probability algorithm (PPA) defines the strength of the link between all pairs of variables within a dataset, depending on the frequency of their concordances and discordances (for a first definition of this rule, see Rumelhart et al. (1986); for a more precise formalization in terms of equation, see Buscema (1995a, b)).

The more two variables are concordant, the more their link will be represented by a positive number proportional to that concordance. The more two variables are discordant, the more their link will be represented by a negative number proportional to their discordance. The greater the similarity between the number of concordances and discordances in a pair of variables, the more that particular link will be represented by a number close to zero.

PPA Equation:

$$w_{i,j} = -\ln \frac{(1/N^2) \cdot \sum_{k=1}^{N} x_{i,k} \cdot \left(1 - x_{j,k}\right) \cdot \sum_{k=1}^{N} (1 - x_{i,k}) \cdot x_{j,k}}{(1/N^2) \cdot \sum_{k=1}^{N} x_{i,k} \cdot x_{j,k} \cdot \sum_{k=1}^{N} (1 - x_{i,k}) \cdot \left(1 - x_{j,k}\right)};$$

$$w_{i,i} = -\ln \frac{(1/N) \cdot \sum_{k}^{N} (1 - x_{i,k})}{(1/N) \cdot \sum_{k}^{N} x_{i,k}}, \tag{19.1}$$

$-\infty \leqslant w_{i,j} \leqslant +\infty; \quad x \in [0, 1]; \quad i, j \in [1, 2, ..., M],$

where:

$W_{i,j}$ = association strength between any couple of variables x_i and x_j of the assigned dataset
x_i = value of the ith variable in the kth record, scaled between 0 and1
N = number of records of the assigned dataset
M = number of variables of the assigned dataset

The first summation, normalized with $1/N$ in the numerator, expresses with a fuzzy truth value (from zero to one) the presence (in the dataset) of discordant values in mode 1-0 for the variables ith and jth, respectively. The second summation, on the other hand, expresses a similar truth value for the opposite discordance of type 0–1. The numerator, obtained as a product of the normalized summations, then expresses a fuzzy truth value because of the simultaneous presence in the dataset of discordant variables in both modalities.

Similarly, the denominator expresses a fuzzy truth value because of the simultaneous presence in the dataset of concordant variables in the mode 0-0 and 1-1. The relationship between denominator and numerator therefore expresses the predominance of discordance over concordance with a value greater than one. The negative of its logarithm transforms nonlinearly the predominance of discordance into negative values and the predominance of concordance into positive values. Having assigned a dataset, the strength of the link between all the pairs of variables that can be formed by fixing a variable x_i with all other x_j, can be determined in terms of quantity of information together with the level of the pairs' concordance (x_i, x_j) in the mode 0-0 and 1-1, and discordance in the mode 1-0 and 0-1.

In practice, after having normalized all the values of the variables of all the records of each dataset between *0* and *1*, we apply the label *similarity* that deals with the values of each variable x_i with any other x_j the concordance of the values close to one another.

In each dataset, assuming variable x_i as a reference, for each pair (x_i, x_j) (with i and $j = 1, 2, \ldots, M$), there is a strength associated with each pair represented by the link $w_{i,j}$ between them, with all their values assumed in all the M records contributing to its definition. This strength is quantitatively given by the value $w_{i,j}$. It must be determined by the quantity of concordant and discordant information carried by all the pairs of values of the variables that it is possible to generate with each reference variable x_i and placed in a pair with the x_j's, for $j = 1, 2, \ldots, M$.

This is allowed because the argument of the logarithm shown in Eq. (19.1) can be interpreted as the ratio between the probability that the variables x_i and x_j of the entire dataset are concordant in the mode 0-0 and 0-1 and the probability that these variables are discordant in the mode 1-0 and 0-1.

In actual fact, the logarithm of each probability can be interpreted as a quantity of information, and so, these are added up algebraically, determining in this way the force of their link whose value is assigned to the weight $w_{i,j}$.

$$\begin{aligned} w_{i,j} &= -\ln\frac{P_{1-0}\cdot P_{0-1}}{P_{0-0}\cdot P_{1-1}} = \ln\frac{1}{P_{1-0}} + \ln\frac{1}{P_{0-1}} - \ln\frac{1}{P_{0-0}} - \ln\frac{1}{P_{1-1}} \\ &= I_{1-0} + I_{0-1} - I_{0-0} - I_{1-1}, \end{aligned}$$

where

$P_{1\text{-}0}$ *is* the probability that the pair of variables (x_i, x_j) are discordant in the mode 1-0

$P_{0\text{-}1}$ *is* the probability that the pair of variables (x_i, x_j) are discordant in the mode 0-1

$P_{0\text{-}0}$ *is* the probability that the pair of variables (x_i, x_j) are concordant in the mode 0-0

$P_{1\text{-}1}$ *is* the probability that the pair of variables (x_i, x_j) are concordant in the mode 1-1

Similar significance for the quantity of information $I_{1\text{-}0}$, $I_{0\text{-}1}$, $I_{0\text{-}0}$, $I_{1\text{-}1}$.

The values of the links between variables as defined by Eq. (19.1) represent the weights that regulate the activation of each variable in relation to each other in a specific dataset.

The weights matrix generated by Eq. (19.1) is a square symmetric matrix whose main diagonal presents the oriented threshold values with which each variable tends to be excited or inhibited.

Table 19.1 The first datasets

Dataset Name	Sex		Status	
Addicted	MALE	FEMALE	Single	Married
ART	1	0	0	1
AL	1	0	0	1
LARA	0	1	1	0
CLYDE	1	0	0	1
MIKE	1	0	0	1
JIM	1	0	0	1
GREG	1	0	0	1
JOHN	1	0	1	0
DOUG	1	0	0	1
NED	1	0	1	0
KARL	1	0	0	1
KEN	1	0	1	0
EARL	1	0	1	0
MARY	0	1	0	0
OL	1	0	0	1
ANNIE	0	1	1	0
DAVE	1	0	1	0
SARAH	0	1	0	1
SAM	1	0	0	1

Table 19.2 The second dataset

Dataset name	Gang		Education		Status	
Persons arrested	JET	SHARKS	JH	HS	Single	Married
LANCE	1	0	1	0	0	1
GEORGE	1	0	1	0	0	1
PETE	1	0	0	1	1	0
FRED	1	0	0	1	1	0
GENE	1	0	0	1	1	0
RALPH	1	0	1	0	1	0
PHIL	0	1	0	1	0	1
IKE	0	1	1	0	1	0
NICK	0	1	0	1	1	0
DON	0	1	0	1	0	1

For example, given three datasets (Tables 19.1, 19.2, and 19.3):

Let us suppose that these three datasets describe the same slice of reality in some specific city.

Furthermore, between the three datasets, there are common variables that allow the graph being formed by the three datasets to be coherent.

In the model that we are considering, it is not at all necessary that the commonality of two variables between two datasets has a semantic value of the father-son type between the records. In fact, the only requirement is that the two variables in the two datasets assume modalities (values) belonging to the same set.

Table 19.3 The third dataset

Dataset name	Sex		Family	
Police team	Male	Female	No children	With children
Adam	1	0	1	0
Mandeep	0	1	0	1
Zoe	0	1	0	1
Paul	1	0	1	0
Caroline	0	1	0	1
Geoff	1	0	1	0
Andy	1	0	0	1
Charles	1	0	1	0
Tommy	1	0	1	0

Table 19.4 The first dataset weights

Addicted	MALE	FEMALE	Single	Married
MALE	1.3218	−21.2313	−0.6931	1.7918
FEMALE	−21.2313	−1.3218	0.6931	−1.7918
Single	−0.6931	0.6931	−0.539	−12.9123
Married	1.7918	−1.7918	−12.9123	0.3185

Table 19.5 The second dataset weights

Persons arrested	JET	SHARKS	JH	HS	Single	Married
JET	0.4055	−21.5987	1.0986	−1.0986	0.6931	−0.6931
SHARKS	−21.5987	−0.4055	−1.0986	1.0986	−0.6931	0.6931
Junior_School	1.0986	−1.0986	−0.4055	−21.5987	−0.6931	0.6931
High_School	−1.0986	1.0986	−21.5987	0.405	0.6931	−0.6931
Single	0.6931	−0.6931	−0.6931	0.6931	0.4055	−21.5987
Married	−0.6931	0.6931	0.6931	−0.6931	−21.5987	−0.4055

Table 19.6 The third dataset weights

Police team	MALE	FEMALE	No_Children	With_Children
Male	0.6931	−21.5218	12.0238	−12.0238
Female	−21.5218	−0.6931	−12.0238	12.0238
No_Children	12.0238	−12.0238	0.2231	−21.6271
With_Children	−12.0238	12.0238	−21.6271	−0.2231

Applying Eq. (19.1) to each of the three datasets, the three weight matrices regulating in each dataset, the relationship between each variable, and the others are obtained (Tables 19.4, 19.5, and 19.6):

At this point, it is possible to represent the relationships between all the variables of each dataset through a single a-directional graph, connected and weighted, where the weights calculated through Eq. (19.1) ensure passage from one node to any other (Fig. 19.2):

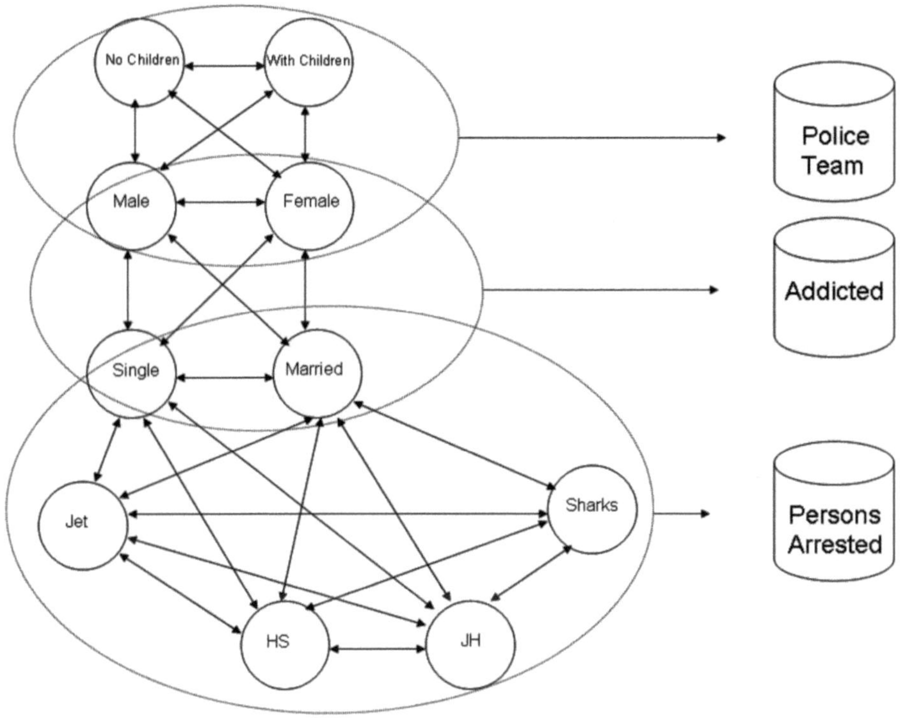

Fig. 19.2 The variables of all three datasets form a single-weighted graph connected through the common variables

At this point, the three datasets chosen as an example form, through their variables, a single network of ten nodes, weighted by bidirectional connections.

This network can be considered a monolayer neural network with a complete mesh if a *zero value* is assigned to all the possible connections but not defined through the algorithm PPA. This assignment is perfectly correct, as a connection with zero value indicates a relationship of indifference between the two connected nodes.

Therefore, the ALOC system makes it possible, under precise conditions, to pass from N datasets, differing in their type of data, to a monolayer network defined by the non-repetitive aggregation of the variables of each dataset and by symmetric connections with a complete mesh, weighted by a known interval of values.

19.2.2 The Constraint Satisfaction Networks

A monolayer network with a complete mesh with symmetric connections can be easily queried with a known maximization algorithm using the following cost function (Rumelhart et al. 1986; Buscema 1995a):

$$G_{(n)} = \sum_{i=1}^{N-1} \sum_{j=i+1}^{N} w_{i,j} \cdot u_{i(n)} \cdot u_{j(n)} + \sum_{i}^{N} w_{i,i} - \sum_{i}^{N} I_i \cdot u_{i(n)}, \tag{19.2}$$

where:

n = activation cycle of the system
w = weights and threshold for each node, defined by the PPA algorithm
u = internal and output state of each node
I = value of the external input to each node
G = function to be maximized
N = number of all the variables of all the datasets

The constraints satisfaction network (CS) is a neural network dedicated to the maximizing of Eq. (19.2).

The kind of activation equations of the CS nodes permits the maximization of the value of each node in compliance with the values of the weights that are fixed with the link function.

$$\begin{aligned}
Net_i^{(n)} &= \sum_{j}^{N} u_j^{(n)} \cdot w_{i,j} + w_{i,i} + I_i; \\
A_i^{(n)} &= \frac{e^{Net_i^{(n)}} - e^{-Net_i^{(n)}}}{e^{Net_i^{(n)}} + e^{-Net_i^{(n)}}}; \\
\delta_i^{(n)} &= A_i^{(n)} \cdot (1 - u_i^{(n)}) \quad A_i^{(n)} > 0; \\
\delta_i^{(n)} &= A_i^{(n)} \cdot u_i^{(n)} \quad A_i^{(n)} \leqslant 0; \\
u_i^{(n+1)} &= u_i^{(n)} + \delta_i^{(n)}.
\end{aligned} \tag{19.3}$$

The ALOC system uses the CS networks algorithm to query the network of all the datasets analyzed. The CS algorithm transforms the network of connections created through the PPA algorithm into a dynamic system capable of receiving external input (user requests) and reacting to these inputs by appropriately activating its own nodes (the variables).

When the system is activated from an external input, all the variables of the system begin to interact and modify their internal state in accordance with the value of the previously defined connections (weights). The variables connected by positive weights tend to have their state values increased by supporting one another, while receiving inhibitory thrusts from the variables connected to them by negative weights.

This complex dynamic of interrelationships occurs in parallel, and the entire system proceeds nonlinearly toward a global state of equilibrium in which the greatest number of variables is highly active in compliance with the excitatory and inhibitory values of the weights that link them.

Technically, the system is considered stable, and therefore closed, when the following equation is true:

$$G^{(n+1)} - G^{(n)} = \varepsilon \tag{19.4}$$

where

$\varepsilon =$ a positive value close to zero
$n =$ the system's parallel activation system
$G =$ the system's cost function defined in Eq. (19.2)

Therefore, the ALOC system acts as a content-addressable memory: if the user enters an external input of any variable at the maximum value (i.e., $Var = 1$), the entire system begins to evolve trying to activate all the variables significantly connected to the variable in question. The result will be a vector of active variables that constitute the "prototype" or the "frame" of the variable externally activated. In other words, it is as if the ALOC system defined the typical frame within which that variable becomes active.

From this point of view, the ALOC system acts as a structured query language (SQL) capable of interpreting the user's requests *semantically*. A traditional SQL searches in a database for the words requested and for the logical connections between those entered. The ALOC system, faced with a similar request, generates the frame of all the variables connected to those requested through semantic similarity.

19.2.3 Records Resonance Networks

Each managed variable of ALOC is distributed among the records of the dataset to which that variable belongs.

The various records may differ from one another in their conceptual nature, according to the dataset that they depend on: the tactical actions of the police in pursuing criminal activity are a different conceptual object from "person arrested for drugs." However, all these records, even in their diversity, are defined by specific variables that are connected with one another directly or indirectly. Therefore, each record, in theory, can be activated by the variables that define it and inhibited by the variables that do not define it. Furthermore, all the records, irrespective of their conceptual diversity, can enter into competition with one another, as different words compete with one another to enter and form part of the same phrase. If in the case of a phrase the criterion that selects the words in competition to express the sense of what is intended for expression, then in the case of the records, the selection criterion is the agreement of the records activated by the active vector in each cycle of the variables.

Therefore, in the ALOC system, the dynamic of the variables determines the dynamic of the records, but not vice versa. The algorithm that defines the development of the records is a competitive algorithm determining the value of each record on the basis of the variables value impacting on that record and according to the value of all the other records of its dataset.

In short, during the evolution of the system, the records coming into greater resonance with the set of all the variables tend to be activated.

The algorithm chosen to develop this evolution through resonance is already known in the literature as an IAC (interactive activation and competition) network (Buscema 1995b; Rumelhart and McClelland 1982; McClelland and Rumelhart 1988a, b; Massini 1998). However, the equations that describe IAC have undergone a specific reworking carried out on behalf of the ALOC system.

$$Ecc_{k,i}^{(n)} = \sum_{j=1}^{N_k} \frac{1}{N_k} \cdot u_{k,j}^{(n)} \cdot (2 \cdot R_{k,i,j} - 1) \quad R_{k,i,j} > \frac{1}{2}. \tag{19.5}$$

$$Inh_{k,i}^{(n)} = \sum_{j=1}^{N_k} \frac{1}{N_k} \cdot u_{k,j}^{(n)} \cdot (2 \cdot R_{k,i,j} - 1) \quad R_{k,i,j} < \frac{1}{2}. \tag{19.6}$$

$$Int_{k,i}^{(n)} = \sum_{s=1;s\neq i}^{M_k} \frac{1}{M_k} \cdot v_{k,s}^{(n)}; \tag{19.7}$$

$$Net_{k,i}^{(n)} = Ecc_{k,i}^{(n)} - Inh_{k,i}^{(n)} - Int_{k,i}^{(n)}. \tag{19.8}$$

$$\delta_{k,i}^{(n)} = \left(MaxAct - v_{k,i}^{(n)}\right) \cdot Net_{k,i}^{(n)} - Decay \cdot \left(v_{k,i}^{(n)} - Rest\right) \quad Net_{k,i}^{(n)} > 0. \tag{19.9a}$$

$$\delta_{k,i}^{(n)} = \left(v_{k,i}^{(n)} - MinAct\right) \cdot Net_{k,i}^{(n)} - Decay \cdot \left(v_{k,i}^{(n)} - Rest\right) \quad Net_{k,i}^{(n)} \leq 0. \tag{19.9b}$$

$$v_{k,i}^{(n+1)} = \frac{e^{\left(v_{k,i}^{(n)}+\delta_{k,i}^{(n)}\right)} - e^{-\left(v_{k,i}^{(n)}+\delta_{k,i}^{(n)}\right)}}{e^{\left(v_{k,i}^{(n)}+\delta_{k,i}^{(n)}\right)} + e^{-\left(v_{k,i}^{(n)}+\delta_{k,i}^{(n)}\right)}}. \tag{19.10}$$

$$Out_{k,i}^{(n)} = Scale^{(n)} \cdot v_{k,i}^{(n)} + Offset^{(n)}. \tag{19.11}$$

where:

N_k = number of the variables of the kth dataset
M_k = number of the records of the kth dataset
$R_{k,i,j}$ = original value of the ith record of the kth dataset – $R_{k,i,j} \in [0,1]$ – in the jth variable

$u_{k,j}^{(n)}$ = value of the *j*th variable of the *k*th dataset in cycle *n*
$Ecc_{k,i}^{(n)}$ = value of the excitations received by *i*th record from the variables of the *k*th dataset in cycle *n*
$Inh_{k,i}^{(n)}$ = value of the inhibitions received by *i*th record from the variables of the *k*th dataset in cycle *n*
MaxAct = maximum activation value of the records (*MaxAct* = 1)
MinAct = minimum activation value of the records (*MinAct* = 0)
Rest = rest value of the activation of the records (*Rest* = 0.1)
Decay = decay value of the activation force in each (*Decay* = 0.05)
$v_{k,i}^{(n)}$ = activation value of the *i*th record of the *k*th dataset

$$Scale^{(n)} = \frac{1}{H^{(n)} - L^{(n)}}; Offset^{(n)} = -\frac{L}{H^{(n)} - L^{(n)}}; H^{(n)} = Max\left\{v_{k,i}^{(n)}\right\}; L^{(n)} = Min\left\{v_{k,i}^{(n)}\right\}.$$

Equation (19.11) operates a dynamic ranking proportionate to each cycle to show the order with which the different records respond to the stimuli of their internal evaluative dynamic and of the variables defining them.

The power of these equations lies in many factors:

1. The records of each dataset interact as a pull of reciprocally inhibitory units (Eq. (19.7)). The value of reciprocal inhibition is equal to the mean of their activations in each cycle. Therefore, in the initial phase (cycle #1), when the activation of each record is nil, the inhibitory value between records of the same dataset is also correspondingly nil.
2. The records of different datasets do not interact with one another. In fact, they represent units that are compatible with one another, as each dataset samples a different aspect of the same reality.
3. The records of each dataset do not have feedback on the value of the variables. In fact, they represent *meta-units*, each of which is composed of other units that are the variables identifying them. Saying that the records are meta-units means that these nodes are recognizable only as vectors of variables in which the value of autonomous entities is recognized at a higher level (meta). Therefore, the records of each dataset interact with one another through the vectors of variables representing them.
4. The equations that govern the interaction between variables are different from the equations that govern the interactions between records. The variables, in fact, interact *globally* (all with all), through a *maximization algorithm* (CS algorithm); the records, on the other hand, interact *locally* (they receive input only from the variables and from the records of their datasets), through a *competitive reinforcement algorithm* of similarity and dissimilarity (IAC algorithm).

19.3 The ALOC System: Observations

The ALOC system represents a new and *semantically* relevant mode for finding hidden relationships between many datasets, different in terms of variables and records, but linked to the same reality.

The ALOC system operates in accordance with the principle common to all the semantic memories belonging to the biological type: given a detail, the system reconstructs its *context* (other variables) and is able to list the typical experiences (records) that support the validity of that reconstruction.

The ALOC system dynamically creates links between variables from different datasets according to the stimuli that it receives. These links are modified during the answering process according to the fuzzy similarity between variables and records, and to the competition and cooperation between the variables themselves. This process also allows the ALOC system to propose combinations of variables that are not present in any record of any datasets processed in that mode. However, these original combinations describe the best prototype that can be generated from the initial links ("external input"). Therefore, the ALOC system is able to make basic *abstractions* from the data.

The cost function that characterizes the ALOC system allows us to interpret its answers as a sort of optimum strategy aimed at *maximizing* the initial conditions from which the elaborative process started.

With this in mind, let us imagine having three datasets that are concerned with drug trafficking in London over a specific period of time. The first dataset concerns the *persons* arrested for drug trafficking, the second describes the different police *officers* involved against drug trafficking and the third addresses the drug *seizures* carried out. Let us also imagine that some common variables exist between these three datasets: for example, the type of drugs seized, what motivated the arrest, how it was targeted by the police action and the place (borough) in which the seizure took place, where the person was arrested and from what point or perspective did the police initiated the action. Lastly, let us imagine that among the data present in the dataset of the police actions, no action is present that targeted action against any type of drug, such as cocaine. At this point, we can force the ALOC system to maximize the value of the variable "cocaine" (external input), in order to find out what prototype context it will generate.

At the end of processing, we should have the following information:

1. Which variables characterize the prototypical cocaine pusher?
2. Which variables describe the seizures in which huge quantities of cocaine are typically found?
3. How must the police actions be explicitly aimed to combat cocaine effectively?
4. Which arrests, seizures, and police officers support an optimal strategy against cocaine trafficking?

The ALOC system can therefore be defined as a semantic memory capable of creating several cognitive maps from different data and coordinating these maps with one another in order to generate interpretative and abstract hypotheses based on the original data.

19.4 The Three Datasets About Drug Trafficking in London

From 2004 to 2006, the London Metropolitan Police in partnership with Semeion Research Centre of the Sciences of Communication (Rome, Italy) activated the Central Drugs Trafficking Database (CDTD). The main target of this project was to organize all the data about drug trafficking in London into a relational database to secure a new powerful method for intelligence gathering using a new set of artificial intelligence algorithms patented by Semeion over the last several years.

The results of this project were included in a special report dated March 2006. The report was evaluated enthusiastically in May 2006 by independent British academics. Actually, the CDTD project is awaiting management and continuance from the new MET Intelligence Bureau (MIB) (Source: MPS Drugs Strategy2007–2010 and Delivery Plan, Chapters 6.1–6.13).

In this chapter, we will use ALOC system to investigate three datasets:

(a) The first one is composed of 1,117 records of persons arrested in London due to drug trafficking. Each person is identified by 28 macro-variables taken from information gotten from five different sociological fields (see Table 19.7). These

Table 19.7 Sociological areas and macro-variables of the persons dataset

Persons	
Macro-variables	Number of atomic variables
Gender	3
Home residence	33
Arrest place	33
Nation group	12
Ethnicity	6
Age class	6
Convictions number	8
Offenses number	8
Age at the first conviction	8
Age from the last conviction	7
Type and number of offenses	57
Number of arrests	4
Number of drug seizures	5
Types of drug	5
Quantity of money	9
Tactic and modality of the arrest	42
Total variables	**246**

Table 19.8 Macro-variables and atomic variables of the seizures dataset

Seizures	
Macro-variables	Number of atomic variables
Type of tactics	16
Place of the operation	27
Type of drugs	7
Modality of the arrest	5
Ethnic data of the arrested	19
Number and gender of the arrested	3
Age of the arrested	4
Trafficking level of the arrested	3
Total variables	**84**

Table 19.9 Macro-variables and atomic variables of the officers dataset

Officers	
Macro-variables	Number of atomic variables
Agent age	4
Agent gender	2
Officer rank	4
Years of service	3
Agent ethnicity	14
Arrested ethnicity	6
Arrested gender	2
Agent tactic	5
Total variables	**40**

macro-variables have been expanded into 246 atomic variables. We have binned the numeric variables and have considered each option contained in categorical variables as independent variables (called the persons dataset for short).

(b) The second dataset is composed of 848 records. Each record organizes the same data of the first dataset but from the point of view of seizure (see Table 19.8): we have expanded each seizure into 84 atomic variables, with the same methodology that we have adopted for the dataset of the arrested persons (called the seizures dataset).

(c) The third dataset is composed of 467 police tactics describing the police officers (see Table 19.9) that the police used to arrest the persons of the first dataset and to secure the seizures of the second dataset. Each police officer involved in one tactic is characterized by 40 atomic variables (called the officers dataset).

These three datasets have many shared variables, connecting each dataset to the others:

(a) 39 variables shared between persons dataset and seizures dataset.
(b) 13 variables shared between persons dataset and officers dataset.
(c) 7 variables shared between officers dataset and seizures dataset.

This structure permits application of the ALOC system.

ALOC works globally with 370 variables and 2,432 records. In other words, the matrix of the connections of ALOC will be composed of 3,924,201 weights: 68,265 among variables, 2,956,096 among records, and 899,840 between variables and records.

19.4.1 The ALOC System and the Prototyping Questions

As we have previously stated, an auto-associative ANN, once trained, is able to dynamically answer many different types of questions that are content oriented (see previous applications in Diappi et al. 2004a, b; Buscema et al. 2006):

1. *Prototypical questions*: What is the prototype of a crack dealer or a cocaine dealer and similar kinds of inquiries?
2. *Virtual questions*: Give me the prototype of a cocaine dealer without any precedent convictions; these questions are possible even if this combination of features is not present at all.
3. *Impossible questions*: Give me the prototype of an arrested person, young and old at the same time, even if this combination does not exist in the assigned dataset. But this question could make sense: we seek the common links, if they are present, among young and old dealers.

In the ALOC system, these questions take on a special interest because we can take information coming from a variety of datasets and connect them to form one prototype. Consider this fact: in the person dataset, the arrests that occurred due only to cocaine are really few. The reason is that a dealer usually has small quantity of cocaine combined with a much larger quantity of other drugs. At this point, it might be useful to ask ALOC to maximize only the cocaine seizures (in the seizures dataset), disregarding the seizures of the other drugs.

ALOC will design the profile of a dealer specializing in cocaine, including the places in which he/she can be arrested, where it is possibly that he/she lives, which police tactics have a greater chance of success, and also the optimal composition of the police team to find this profiled dealer.

19.4.2 The Cocaine Prototype

The first question to pose to ALOC is to arrange the prototype of cocaine dealer. The cocaine profile is quite complex for many reasons:

(a) The MPS itself declares that the cocaine network is very hard to understand.
(b) We did not find specific strategies against cocaine trafficking in New Scotland Yard.

Table 19.10 The cocaine prototype (in **bold** the differences with cannabis prototype)

Cocaine dealer prototype	
Persons dataset	
Gender	Female
Home	Redbridge, Greenwich
Nation group	**Europe, NonUK**,Turkey-Cyprious,Veitnam
Etnicity	Dark European or Oriental
Age of the persons	Over 45
Convitcions number	Zero or one conviction
Offenses number	Zero or one offence
Age at the first conviction	**From 28 to over 51**
Drug offenses number	Zero or one drug offence
Other offenses	**None**
Place of arrest	Barking and Dagenham,Bexley, Newham,Redbridge
Other drugs	**Cannabis**
Cash at the arrest	No cash
Arrest mode	Given into custody
Type of police	**No-Law Enforcement Agent**
Type of tactic	Search of premises, controlled delivery
Type of Arrest	Direct arrest, no violent
Seizures sataset	
Type of tactic	Search of premises with warrant
Place of the arrest	Newham
Other drugs	**Cannabis**
Nunber of persons arrested	**one male british**
Etnicity	**Asian,** Oriental, **Pakistan, India, Bangladesh**
Age of arrested person	**less than 25**
Level of trafficking	**Third Level**
Officers dataset	
Agent age	From 25 to 35
Agent gender	**Female**
Agent service years range	Between 5 and 15 years
Agent official rank	DC
Agent etnicity	White British
Arrested etnicity	Dark European or Oriental
Arrested gender	Female
Type of tactic	Search of premises

(c) The majority of seizures found only small quantities of cocaine, and then it was always mixed with other drugs.

When we asked ALOC to define the prototype of the cocaine dealer, the algorithm attained stability after a number of cycles and designed this profile (Table 19.10):

Many considerations are possible at this point:

1. The typical cocaine dealer is an aged female (over 45), completely away from the usual criminal activity (no other offenses and convictions, in general). She is an unsuspected person.
2. She comes from Europe (not UK), typically a Turkish or Cypriot citizen (so a Dark European); she can be also an Oriental, from Vietnam, Pakistan India, or Bangladesh.
3. Typically, these persons live in Redbridge or Greenwich, but usually, they are arrested in Barking and Dagenham, Bexley, Newham, and Redbridge.
4. When there is a seizure, particularly in Newham, they are arrested also for cannabis, but in this case, usually they "become" males under 25 years old. This strange transformation, obviously, has to be explained.
5. In any case, these persons are labeled as criminals belonging to the "third level" of drug trafficking. That is the most dangerous level of drug trafficking activity for it represents the international one.
6. Regardless of their relevance for drug trafficking investigations, these persons are arrested, by chance, first by private agents (no MPS Agents), and then they are given into the custody of young female agents, with medium experience and low officer rank.
7. No specific tactics or operations are planned for these persons, other than a generic search of premises or a controlled delivery.

This is the final frame of the prototypical cocaine dealer. But ALOC system can be more informative. In fact, the frame shown in Fig. 19.6 is the final attractor of the algorithm: ALOC reaches this state after 370 cycles of dynamic negotiation among all the variables. During this many-to-many interaction, two simple types of dynamic can occur:

1. A variable starts to monotonically increase its value up to its maximum activation; in this case, it will be active until the algorithm becomes stable. Therefore, these types of variables will be present in the final prototypical frame. By watching their dynamics, ALOC allows us to distinguish three fuzzy sets of variables:

   (a) Those more quickly to be activated, that is, those that are more associated with the prototype (strong associations).
   (b) Those more slowly to be activated and thus are weakly associated with the prototype (weak association).
   (c) Those that are activated more lately in the process. They represent side effects of the dynamics (indirect associations).

2. A variable begins to increase and, after a while, begins to decrease: in this case, we have discovered a hidden signal within the process but not of sufficient strength to maintain its activation up to the end. For this kind of variable, we have either of the two choices:

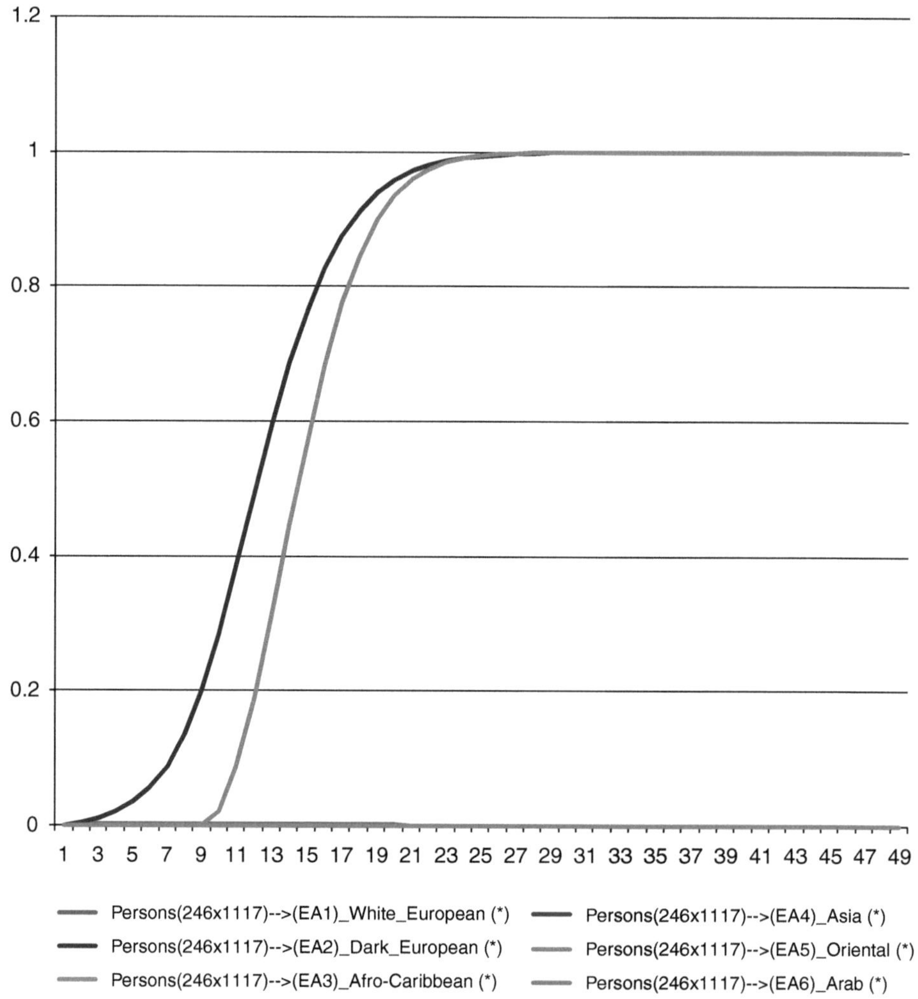

Fig. 19.3 Cocaine dealer prototype: ethnic group dynamics

(a) This hidden signal is a flickering noise designated to be absorbed by the global dynamics.
(b) It is key information about the process, but not strong enough in terms of frequency, to make itself fully evident.

All these cases are present in the dynamical process leading to the definition of the cocaine dealer prototype.

In Fig. 19.3, the feature "Dark European" seems to be strongly associated with the cocaine dealer, while the feature "Oriental" starts later as a side effect of the activation of other variables.

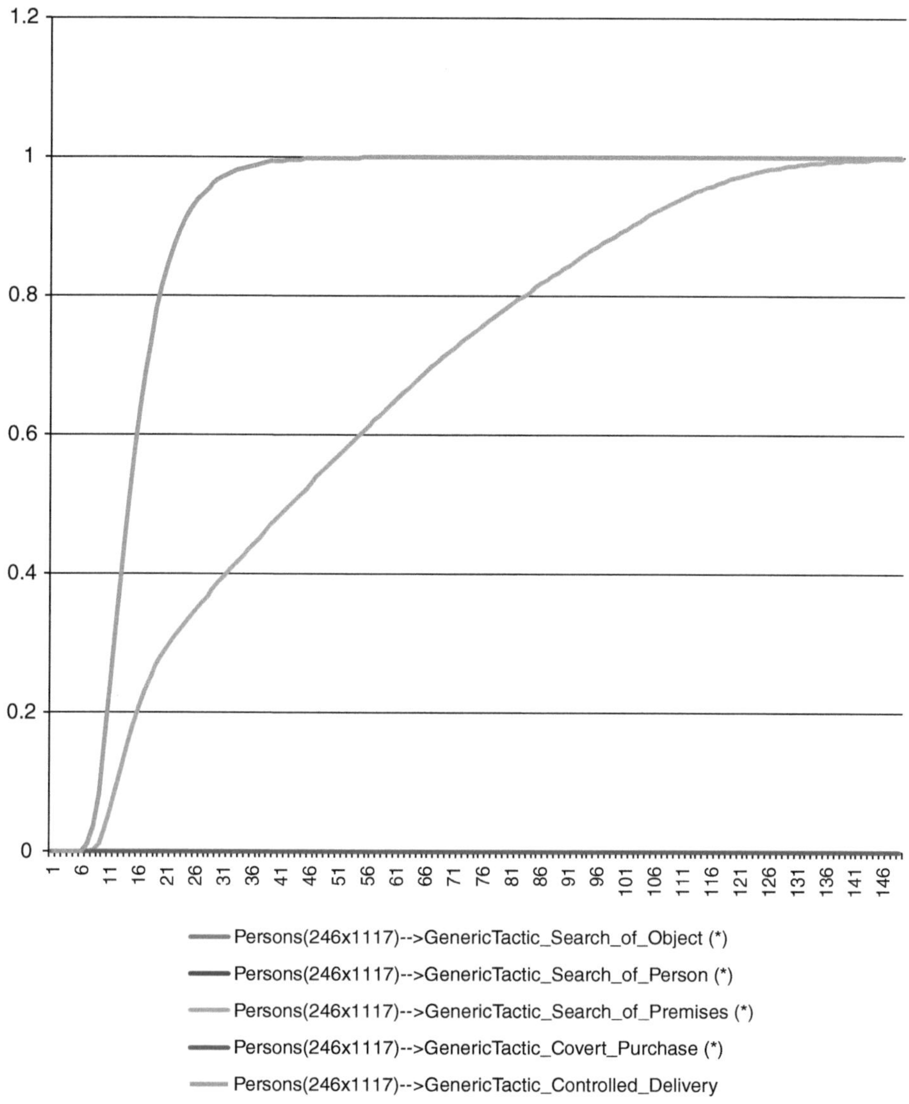

Fig. 19.4 Cocaine dealer prototype: types of tactic dynamics

In Fig. 19.4, the tactic "delivery control" has an immediate and strong association with the cocaine dealer, while the tactic "search of premises" shows a weaker association with the prototype.

Figure 19.5 shows that the cannabis activation is a classic late side effect of the global process: because the number of cocaine arrests is small, while the number of cannabis arrests is huge, and because, many times, cocaine and cannabis are found together, it is an expected consequence that the cannabis dealer profile tends to overlap the cocaine dealer profile.

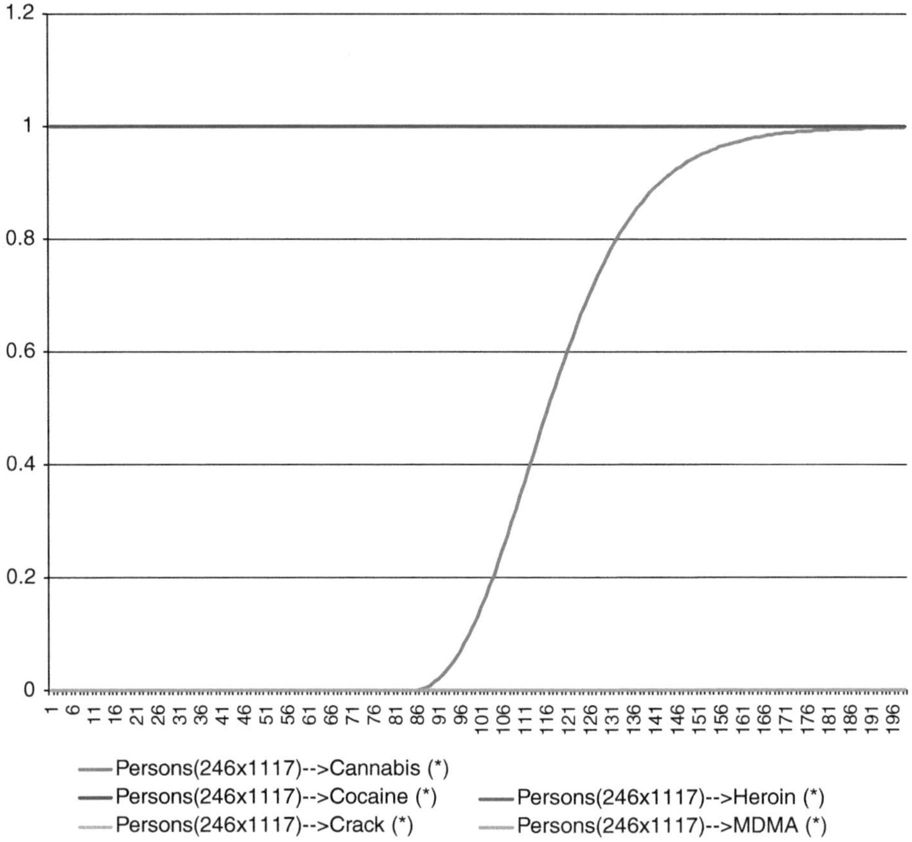

Fig. 19.5 Cocaine dealer prototype: types of drugs dynamics

Figure 19.6 shows an interesting "hidden signal": the feature "SAME" (South American Nation Group), at the beginning, increases its activation very speedily only to maintain its activation state in a second step, and then at the end it decreases rapidly, as if the process had changed its initial dynamic. The variable "EASTEU" (East Europe Nation Group), at a smaller scale, presents the same dynamic.

It is known that South American people are the main actors in the cocaine trafficking in London. What ALOC allows us to discover is a new fact: South American people stay in the background of the cocaine traffic and probably use Dark European females and young Oriental persons as the visible actors of their hidden activities.

The representation of cocaine seems to be clear: cannabis covers cocaine and Oriental and Dark European cover the South American group, as well as white-collar females apparently cover male criminals. The result of the game is the following: random seizures of cocaine, young and inexperienced female agents with low officer rank, to contrast one of the more aggressive drugs in the criminal market.

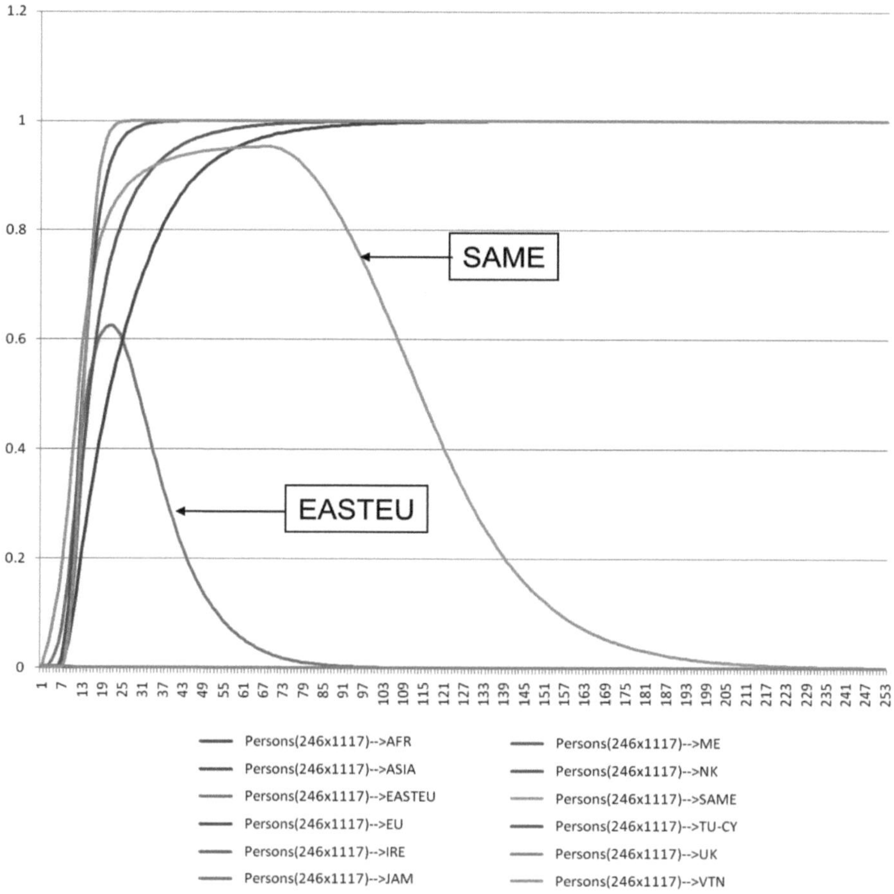

Fig. 19.6 Cocaine dealer prototype: types of drugs dynamics

19.4.3 The Crack Prototype

The prototypical profile of a crack dealer is shown in Table 19.11.

Crack dealer is the perfect picture of a very experienced delinquent Afro-Caribbean male having problems with justice for a long time, a UK citizen, with many offenses and convictions of different types possessing violent and aggressive behavior.

The police use complex tactics to arrest him utilizing the more experienced and high-ranking agent. The seizures and the arrests show a strong link between crack and heroin, and very often the places of the arrest are the same places where this dealer lives.

In any case, they are classified as local or regional dealers (levels 1 and 2) and do not work at international trafficking level.

Table 19.11 The crack prototype

Crack dealer prototype	
Persons dataset	
Gender	Male
Home	Camden, Kensington and Chelsea, Westminster
Nation group	UK
Ethnicity	Afro-Caribbean
Age of the persons	35–45 and over 45
Convictions number	From 5 to10 to more than 20
Offenses number	From 11 to 20 to more than 50
Age at the first conviction	Before 18
Time from the last conviction	Less than 1 year
Drug offenses number	From 3 to 5 to more than 10
Theft-kindred offenses	From 6 to 10 to more than 20
Offenses against person	From 1 to more than 5
Offenses with offensive weapons	From 1 to more than 2
Sexual offenses	More than 2
Offenses against police	From 1 to more than 5
Fraud offenses	From 1 to more than 5
Offenses against property	From 1 to more than 2
Drug trafficking offenses	From 2 to more than 5
Other violent offenses	Yes
Total offenses	From 11 to more than 50
Number of arrests	From 3 to more than 3
Number of drug seizures	From 3 to more than 5
Place of arrest	Camden, Haringey, Kensington and Chelsea, Southwark, Westminster
Other drugs	Heroin
Cash at the arrest	No cash
Number of tactics	Over 3
Type of tactic	Covert purchase
Number of tactics sequences	From 2 to more than 5
Arrest in operation	From 2 to more than 5
Violent on arrest	Yes
Direct arrest	From 3 to more than 5
Arrest on result of inquiries	More than 2
On bail at the time of the offence	From 1 to more than 2
Seizures dataset	
Type of tactic	Covert purchase test in operation, detailed tactics
Place of the arrest	Camden, Haringey, Kensington and Chelsea, Southwark, Westminster
Other drugs	Heroin
Number of persons arrested	1 Male
Ethnicity	Not British, Afro-Caribbean, Black-Caribbean

(continued)

Table 19.11 (continued)

Crack dealer prototype	
Age of arrested person	Between 25 and 35
Level of trafficking	Level 1, level 2
Officers dataset	
Agent age	35–45, over 45
Agent gender	Male
Agent service years range	More than 15 years
Agent official rank	DC, PS
Agent ethnicity	White British, White and Black African
Arrested ethnicity	Afro-Caribbean
Arrested gender	Male
Type of tactic	Covert purchase

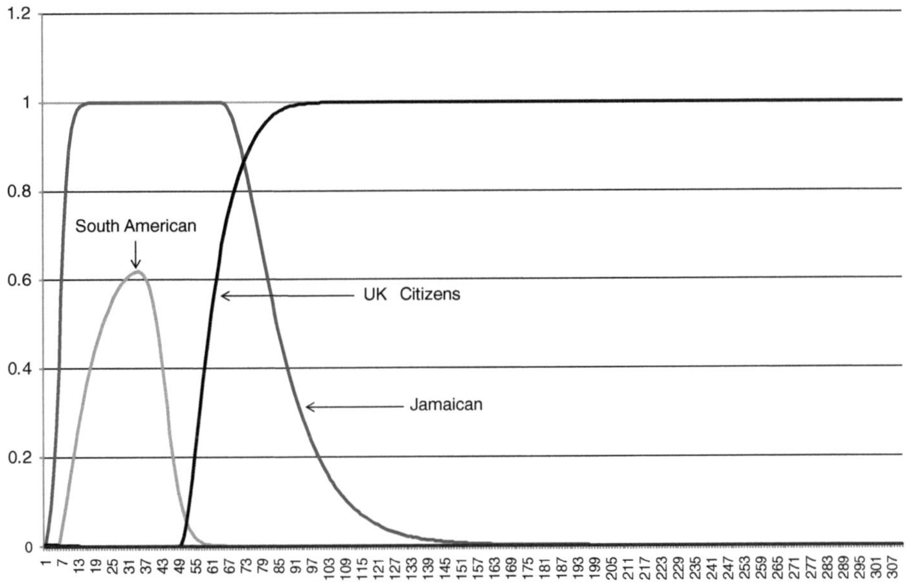

Fig. 19.7 Crack prototype: nation group dynamics

A more detailed analysis of ALOC dynamics shows another possible profile of the crack dealer, visible only when watching the hidden signals of the algorithm evolution:

1. In Fig. 19.7, we can see that the variable "Jamaican" grows rapidly at the beginning, reaches up the top of its activation, and maintains this state for a long time. In second position, the variable "UK citizens" is activated, but at this point the "Jamaican" begins to decrease and suddenly disappears as if these two variables were linked by a nonlinear inverse association.

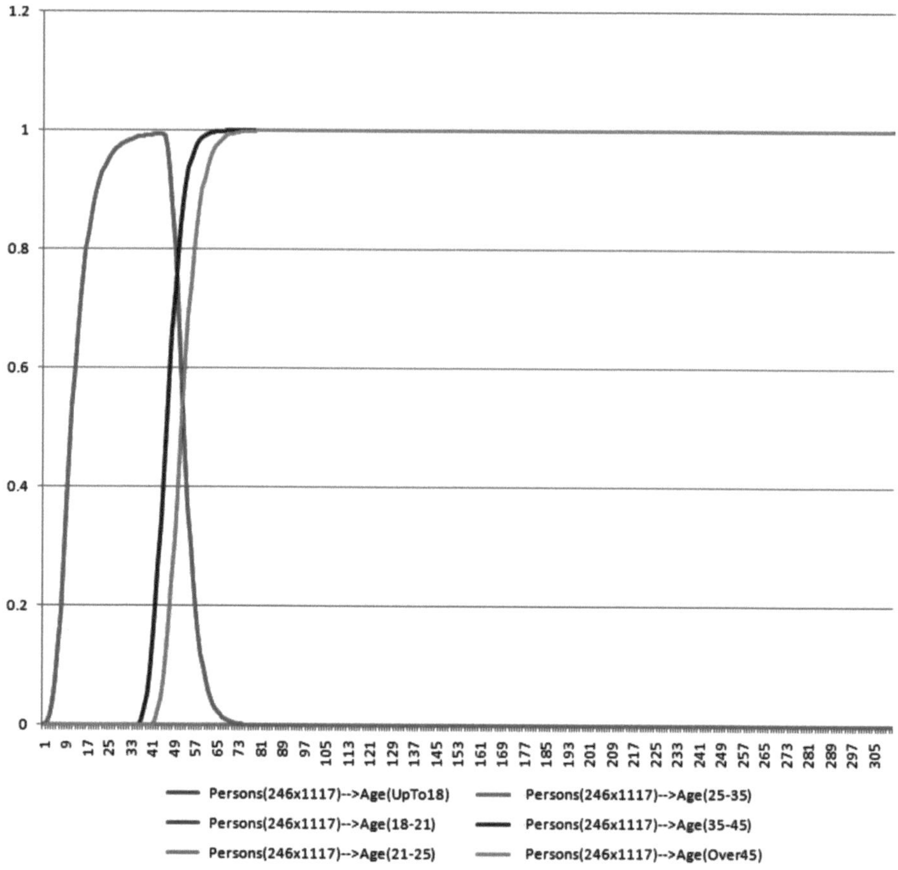

Fig. 19.8 Crack prototype: class of age dynamics

2. Figure 19.8 shows the same typical dynamics of the hidden signals about the class of age: young persons (25–35) are suddenly substituted by more adult persons (35–45 and over 45).
3. Figure 19.9 presents the same process: this time, the persons with one or no convictions are substituted by professional delinquents with a record full of convictions.

This situation opens other scenarios with two prototypes of crack dealer: the prototype described in Table 19.11 of a professional UK delinquent and a younger crack pusher coming from Jamaica and without special problems within the UK justice system. Both are Black-Caribbean, but the Jamaicans are probably the working class of the more aged and experienced group of criminals.

A confirmation of this interpretation is provided by Fig. 19.10: the agents involved in crack trafficking are at the beginning the officers with the lowest rank (DC), but later, they are supported by agents with the highest rank (PS).

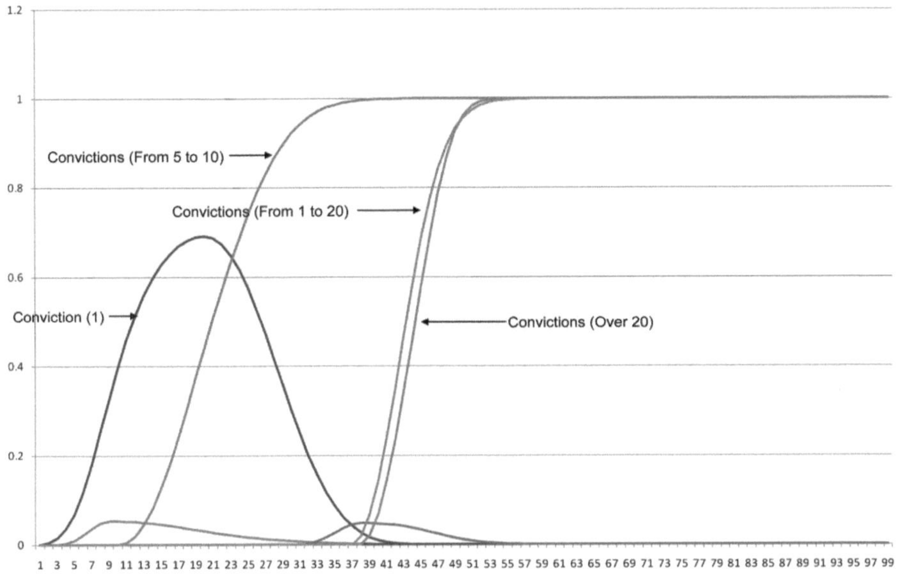

Fig. 19.9 Crack prototype: class of age dynamics

This double crack dealer prototype makes sense especially if we again consider Table 19.11; the trafficking level of these persons is also doubled: local (level 1) and regional (level 2). It is easy to infer that the Jamaicans are involved in the streets at the local level, while the more aged and experienced Black-Caribbean manage the regional networks.

This job distribution is interesting particularly if we reinspect Fig. 19.7: from the ALOC point of view, the South Americans are activated in the crack trafficking area but for a short period of time. The hidden signal of their presence is small but clear. If this link is true, we must consider differently the relationships of South Americans with cocaine and crack trafficking: they could be the hidden meta-levels of the whole drug trafficking network in UK.

19.4.4 The Cannabis Prototype

The Cannabis prototype generated by ALOC is similar to the cocaine prototype. It seems that cannabis trafficking is an effective way to cover cocaine trafficking. The differences between the two are few, but they are important:

1. Cocaine dealers usually are not UK citizens (European) while cannabis dealers are in general persons with a UK passport.
2. Cannabis dealers are very young (18–21) and live sometime in Richmond upon Thames.

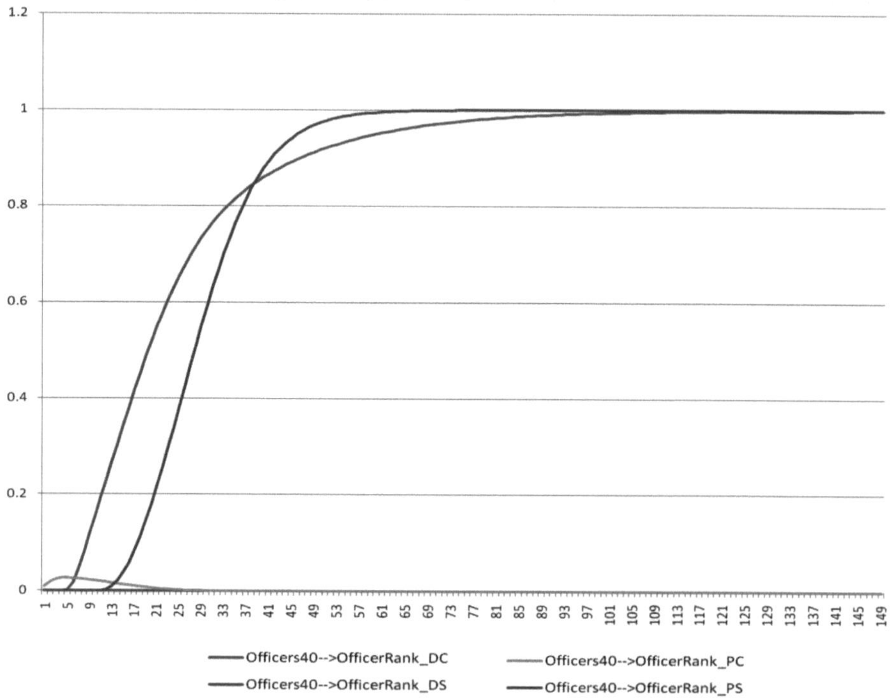

Fig. 19.10 Crack prototype: police agent rank

3. Cannabis is not associated with cocaine, but cocaine is often associated with cannabis.
4. Cannabis dealers are usually arrested by MPS agents, while with cocaine dealers, this is not typical.
5. Groups of males and females are arrested in cannabis seizures, while in cocaine seizures, only one person at the time is generally arrested.
6. In seizures, the typical age of the arrested for cannabis is between 35 and 45, while in seizures because of cocaine, the person arrested is very young (under 25) and that seems to be a masking strategy.
7. The ethnic group of people arrested in seizures is completely different: Afro-Caribbean, Black-Caribbean, White British, and Oriental in cannabis seizures and people from India, Pakistan, and Bangladesh in the case of cocaine seizures.
8. The persons arrested because of cocaine belong to the international level of trafficking (third level), while for the people arrested for cannabis, the level of trafficking is not defined.

However important these differences, the typology of agents used for these tactics is essentially the same: MPS seems not to see the differences between these two dealer prototypes (Table 19.12).

Table 19.12 The cannabis prototype (in **bold** the differences with cocaine prototype)

Cannabis dealer prototype	
Persons dataset	
Gender	Female
Home	Redbridge, Greenwich,Bexley, **Richmond upon Thames**
Nation group	**UK**,Turkey-Cyprious,Veitnam
Etnicity	Dark European or Oriental
Age of the persons	**18–21,** Over 45
Convitcions number	Zero or One Conviction
Offenses number	Zero or One Offence
Age at the first conviction	**From 34 to over 51**
Drug offenses number	Zero or one drug offence
Other offenses	**One**
Place of arrest	Barking and Dagenham,Bexley, Newham,Redbridge,Richmond upon Thames
Other drugs	**None**
Cash at the arrest	No cash
Arrest mode	Given into custody
Type of tactic	Search of premises, controlled delivery
Type of arrest	Direct arrest, no violent
Seizures dataset	
Type of tactic	Search of premises with warrant and without warrant
Place of the arrest	Newham
Other drugs	**None**
Nunber of persons arrested	**2 or more than 2 male and female british**
Etnicity	**White European**,Oriental, **Afro-Caribbean, Black-Caribbean, Persons decline to define**
Age of arrested person	**From 35 to over 45**
Level of trafficking	**No define**
Officers dataset	
Agent age	From 25 to 35
Agent gender	**No define**
Agent service years range	Between 5 and 15 years
Agent official rank	DC
Agent etnicity	White British
Arrested etnicity	Dark European or Oriental
Arrested gender	Female
Type of tactic	Search of premises

19.4.5 The Heroin Prototype

Table 19.13 shows the prototype of the heroin dealer. The similarities with the crack dealer are many, but there are also a number of differences. The heroin dealer is not as common and an experienced a criminal as the crack dealer. The heroin dealer is very often a female, from Jamaica and sometimes South America. She/he is a

Table 19.13 The heroin prototype

Heroin dealer prototype	
Persons dataset	
Gender	Female
Home	Haringey, Kensington and Chelsea, Southwark
Nation group	Jamaica, South American
Etnicity	Afro-Caribbean
Age of the persons	25–35
Convitcions number	One or zero
Offenses number	None
Age at the first conviction	From 22 up to 51
Time from the last conviction	Less than 1 year
Drug offenses number	None or more than 10
Theft kindred offenses	None
Offenses against person	None
Offenses with offensive wepons	None
Sexual offenses	None
Offenses against police	None
Fraud offenses	None
Offenses against property	None
Drug trafficking offenses	From 2 to more than 5
Other violent offenses	No
Total offenses	None
Number of arrests	Over 3
Number of drug seizures	From 3 to more than 5
Place of arrest	Hackey, Haringey, Southwark, Westminster
Other drugs	Crack
Cash at the arrest	No cash
Number of tactics	Over 3
Type of tactic	Covert purchase, controlled delivery
Number of tactics sequences	From 3 to more than 5
Arrest in operation	From 2 to more than 5
Violent on arrest	No
Direct arrest	None or more than 5
Arrest on result of enquiries	More than 2
On bail at the time of the offence	None
Seizures dataset	
Type of tactic	Covert purchase test in operation, detailed tactics
Place of the arrest	Hackney, Haringey, Southwark, Westminster
Other drugs	Crack
Nunber of persons arrested	1 Male
Etnicity	No British, Afro-Caribbean, Black-Caribbean
Age of arrested person	Between 25 and 35
Level of trafficking	Level 1, level 2

(continued)

Table 19.13 (continued)

Heroin dealer	
Officers dataset	
Agent age	35–45, over 45
Agent gender	Male
Agent service years range	15 years
Agent official rank	DC, DS
Agent etnicity	White British
Arrested etnicity	Afro-Caribbean
Arrested gender	Female
Type of tactic	Covert purchase

different kind of Afro-Caribbean. In fact, MPS use a different team of agents to arrest this type of drug trafficker. There is a contiguity and a similarity of places for crack and heroin, but ALOC suggests to us that these two drugs are managed by two different networks: Hackney and Southwark, as example, seem to be two boroughs specific for heroin trafficking while Kensington and Chelsea is a typical place to arrest crack dealers.

Also, in the case of heroin and crack, two different populations of dealers work side by side, the less dangerous covering the more dangerous.

19.5 Conclusions

The ALOC system is a new adaptive system which is able to connect the contents of different datasets presenting different views of the same reality. This is fundamental when a problem is represented with different statistical observations and different variables for any (data) set of observations.

In order for ALOC to work, some variables in the datasets must be shared in a way such that it is possible to create a tree structure from among the assigned datasets.

When this prerequisite is satisfied, then the ALOC system can transform the datasets into a multifocal dynamic memory able to connect each variable and each record of the assigned datasets to any others using statistical contents.

In short, ALOC transforms all the assigned datasets in a content-addressable memory, CAM (see Hopfield 1982, 1984).

This transformation is particularly useful in discovering hidden connections and side effects among the datasets.

The ALOC system is able to perform this action using three components and one-stop criterion:

1. A group of equations able to approximate the implicit function of each one of the assigned datasets

2. A group of equations able to maximize the activation values of the variables of all the datasets in constraint environments (the constraint environments are the different weights matrices representing the implicit functions of each dataset and an external input by which the system is activated by control)
3. A group of equations able to create a resonance among the dynamic activations of all the variables and the records of the datasets
4. A simple equation to decide when the process has reached a new stable state

The way to exploit this embedded knowledge, in practice, is by means "questions": a question in this context means to activate from the outside one or more variables of one dataset and to give the necessary freedom for ALOC to work dynamically over all the datasets to reach a stable attractor.

At the end of this process, ALOC will present the best prototype that satisfies the initial question.

From one perspective, we can define ALOC as a complex device that is content oriented with an ability to generate prototypes.

But the dynamics of this prototyping process is also meaningful. In fact, during the ALOC evolution, all the variables and records of the assigned datasets dynamically negotiate its reciprocal activation values, through a game of competition and cooperation: the activation of some variables will activate other variables which support and/or inhibit yet other variables, until this complex dynamic machine reaches a stabilization point.

The analysis of this process provides new key information about all the datasets:

1. Which variables and records are strongly or weakly associated?
2. Which variables and records are activated as a side effect of the process itself?
3. Which variables and records represent hidden signals of a transient prototype? This can happen when some variables in a first step increase, and after a while, they decrease according to a parabolic shape.

Future research about the ALOC system will address the understanding of the meaning of complex many-to-many dynamics. In other words, how are we able to automatically capture the complex cause-effect relationship existent among variables during the ALOC evolutionary process?

We think that a new type of intelligent data mining technique may emerge in response to this question.

References

Buscema, M. (1995a). *Constraint Satisfaction and Recirculation Neural Networks* (Technical Paper n. 18). Semeion, Rome.

Buscema, M. (1995b). Self-reflexive networks. Theory, topology, applications. *Quality and Quantity, 29*(4), 339–403. Dordrecht: Kluwer Academic Publishers.

Buscema, M., Terzi, S., Maurelli, G., Capriotti, M., & Carlei, M. (2006). The smart library architecture of an orientation portal. *Quality and Quantity, 40*, 911–933, Springer.

Diappi, L. P., Bolchim, P., & Buscema, M. (2004a). Improved understanding of urban sprawl using neural networks. In J. P. Van Leeuwen & H. J. P. Timmermans (Eds.), *Recent advances in design and decision support systems in architecture and urban planning*. Dordrecht: Kluwer Academic Publishers.

Diappi, L., Buscema, M., & Ottana, M. (2004b). Complexity in sustainability: An investigation of the Italian urban system through self-reflexive neural networks. In L. Diappi (Ed.), *Evolving cities*. England: Ashgate Publishing.

Hebb, D. O. (1961). *The organization of behavior*. New York: Wiley.

Hopfield, J. J. (1982). Neural networks and physical systems with emergent collective computational abilities. *Proceedings of the National Academy of Sciences USA, 79*, 2554–2558.

Hopfield, J. J. (1984). Neurons with graded response have collective computational properties like those of two-state neurons. *Proceedings of the National Academy of Sciences USA, 81*, 3088–3092.

Massini, G. (1998). Interactive activation and competition neural networks. *Substance Use & Misuse, 33*(2), 463–479.

McClelland, J. L., & Rumelhart, D. E. (1988a). Interactive activation and competition, Chapter 2. In *Explorations in PDP. A handbook for models, programs and exercises* (pp. 11–47). Cambridge, MA: The MIT Press.

McClelland, J. L., & Rumelhart, D. E. (1988b). *Explorations in parallel distributed processing*. Cambridge, MA: The MIT Press.

Rumelhart, D., & McClelland, J. L. (1982). An interactive activation model for context effects in letter perception: Part 2. The contextual enhancement effect and some tests and extensions of the model. *Psychological Review, 89*, 60–64.

Rumelhart, D. E., Smolensky, P., McClelland, J. L., & Hinton, G. E. (1986). Schemata and sequential thought processes in PDP models. In J. L. McClelland & D. E. Rumelhart (Eds.), *PDP, exploration in the microstructure of cognition* (Vol. II). Cambridge, MA: The MIT Press.

Index

M. Buscema and W.J. Tastle (eds.), *Intelligent Data Mining in Law Enforcement Analytics: New Neural Networks Applied to Real Problems*, DOI 10.1007/978-94-007-4914-6,

MIX
Papier aus verantwortungsvollen Quellen
Paper from responsible sources
FSC® C105338

If you have any concerns about our products,
you can contact us on
ProductSafety@springernature.com

In case Publisher is established outside the EU,
the EU authorized representative is:
Springer Nature Customer Service Center GmbH
Europaplatz 3, 69115 Heidelberg, Germany

Printed by Libri Plureos GmbH
in Hamburg, Germany